Frontiers of Virology   1

Y. Becker · G. Darai (Eds.)

Diagnosis of Human Viruses by Polymerase Chain Reaction Technology

With 72 Figures and 48 Tables

Springer-Verlag
Berlin Heidelberg New York
London Paris Tokyo
Hong Kong Barcelona
Budapest

Editors:

Professor Yechiel Becker
Department of Molecular Virology
Faculty of Medicine
Hebrew University of Jerusalem
POB 1172
Jerusalem, Israel

Professor Gholamreza Darai
Institut für Medizinische Virologie
der Universität Heidelberg
Im Neuenheimer Feld 324
W-6900 Heidelberg, FRG

ISBN-13: 978-3-642-84768-4   e-ISBN-13: 978-3-642-84766-0
DOI: 10.1007/978-3-642-84766-0

Typesetting: Triltsch, Würzburg
27/3145-5 4 3 2 1 0 – Printed on acid-free paper

Frontiers of Virology: Aims and Plans

Virology at the end of the twentieth century is still a frontier of the biological sciences with new viruses emerging as disease-causing agents of man, animal, and plant. Ever since 1796 when Edward Jenner named the causative agent of smallpox "virus" (toxin), knowledge on virus diseases has slowly accumulated. The vast influenza epidemic (the "Spanish flu") at the end of World War I revealed that virologists lacked the scientific tools to identify the epidemic and protect against it. During the middle of the twentieth century, the methods for the diagnosis of viruses and for research were markedly improved. New technologies were developed to identify the structure and organization of viruses, ranging from bacterial to plant viruses and from human to animal viruses. In the past 50 years, virology has developed as a part of immunology and molecular biology, providing these biological sciences with tools and at the same time utilizing their biochemical, molecular, and immunological advancements to further knowledge on the mechanisms by which viruses cause a variety of diseases. Thus, virology remains in the forefront of science.

With this idea that virology is constantly developing, the present series, *Frontiers of Virology*, was conceived. We intend to select topics on which knowledge from a number of biological fields of research, including conceptual and technical breakthroughs, can merge in the field of virology and move it further ahead. We plan to put the emphasis on discoveries which will help to curb virus diseases. We hope that *Frontiers of Virology* will be of interest not only to virologists, molecular biologists, and immunologists, but also to physicians with expertise in infectious diseases.

We wish to express our thanks to Dr. Jürgen Wieczoreck, Springer-Verlag, for his general support and encouragement to develop the series *Frontiers of Virology*.

Y. Becker, Jerusalem
G. Darai, Heidelberg

Preface

The basis for the effective treatment and cure of a patient is the rapid diagnosis of the disease and its causative agent, which is based on the analysis of the clinical symptoms coupled with laboratory tests. Although rapid advancements have been made in the laboratory diagnosis of virus diseases, the necessary isolation of the causative virus from the clinical specimens is a relatively long procedure. Viruses which integrate into the cellular DNA (such as human immunodeficiency virus, HIV-1, or hepatitis B virus) are difficult to identify by molecular techniques, while viruses which exist in the clinical material in low concentrations are even more formidable to identify. Recently, the application of the polymerase chain reaction (PCR) technique developed by K. D. Mullis and detailed in the study by Saiki et al. (1985) led to a revolution in virus diagnosis. The PCR technique was rapidly applied to the diagnosis of viruses in clinical material.

Volume 1 of *Frontiers of Virology* provides new information on the advantages of the use of the PCR for the diagnosis of many human disease-causing viruses, as well as on some problems with its use. The volume is divided into sections which deal with the diagnosis of human viruses: (1) human lentivirus (HIV-1, acquired immunodeficiency syndrome virus, and retroviruses human T-cell leukemia virus I and II); (2) hepatitis viruses A, B, C, and delta; (3) herpesviruses (herpes simplex virus, Varicella-Zoster virus, human cytomegalovirus and Epstein-Barr virus, and human herpesvirus HHV-6); (4) papillomaviruses and papova viruses BK and J1; (5) airborne and respiratory viruses, rubella, measles, influenza, rhinoviruses, parvovirus B19, adenoviruses, and coronaviruses; and (6) additional disease-causing RNA viruses (flaviviruses, hantaviruses, and intestinal picorna viruses, rota virus, and rabies). In these chapters, the authors deal not only with the advantages of PCR technology but also with its current limitations and provide basic information on the virus and the disease caused by it as well as the accurate PCR diagnosis.

The present book provides the excitement of exploring a new frontier and is the first attempt to accumulate the knowledge of this new diagnostic technology into one volume. No doubt, as this book goes to press, new data on the diagnosis of additional, clinically important viruses are being reported. We hope to include this additional information in one of the future volumes of *Frontiers of Virology*.

We hope that this volume will be of interest to virologists, molecular biologists, clinical virologists, and physicians treating patients affected with infectious diseases.

We wish to thank all the contributors. The editorial assistance of Waltraud Janssen is much appreciated.

Y. Becker, Jerusalem
G. Darai, Heidelberg

Contents

List of Autors

Albert, J., Department of Virology, National Bacteriological Laboratory, 10521 Stockholm, Sweden

Allard, A., Department of Virology, University of Umea, 90185 Umea, Sweden

Arnoldus, E. P. J., Department of Neurology, University Hospital Leiden and Department of Cytochemistry and Cytometry, State University Leiden, Postbus 9600, 2300 RC Leiden, The Netherlands

Arthur, R. R., Department of Immunology and Infectious Diseases, School of Hygiene and Public Health, John Hopkins University, 615 N. Wolfe Street, Baltimore, MD 21205, USA

Auer, H., Institute of Molecular Pathology, Dr. Bohr-Gasse 7, 1030 Wien, Austria

Bachmann, E., Division of Haematology, Department of Medicine, University Hospital CHUV, 1011 Lausanne, Switzerland

Bachmann, F., Division of Haematology, Department of Medicine, University Hospital CHUV, 1011 Lausanne, Switzerland

Bautz, E. K. F., Zentrum für Molekulare Biologie Heidelberg und Institut für Molekulare Genetik, Universität Heidelberg, Im Neuenheimer Feld 230, W-6900 Heidelberg, FRG

Becker, Y., Department of Molecular Virology, Faculty of Medicine, Hebrew University of Jerusalem, Jerusalem, Israel

Blaas, D., Institut für Biochemie, Universität Wien, Währinger Straße 17, 1090 Wien, Austria

Bloem, B. R., Department of Neurology, University Hospital Leiden, Postbus 9600, 2300 RC Leiden, The Netherlands

Boerman, R. H., Department of Neurology, University Hospital Leiden, Postbus 9600, 2300 RC Leiden, The Netherlands

Bourhy, H., Unité de la Rage, Institut Pasteur, 25, rue du Docteur Roux, 75724 Paris Cedex 15, France

Brechot, C., Hybridotest, Institut Pasteur, rue du Dr. Roux 28 and INSERM U75 CHU Necker, Service d'Hepatologie, Hôpital Laennec, 75015 Paris, France

Bryan, R. L., Department of Histopathology, East Birmingham Hospital, Bordesley Green East, Birmingham B9 5ST, UK

Chang, G.-J., Molecular Biology Branch, Division of Vector-Borne Infectious Diseases, Centers for Disease Control, P.O. Box 2087, Foothills Campus, Fort Collins, CO 80522, USA

Chapman, N. M., Department of Pathology and Microbiology, University of Nebraska Medical Center, 600 S. 42nd Street, Omaha, NE 68198, USA

Clewley, J. P., PHLS Virus Reference Laboratory, 61 Colindale Avenue, London NW9 5HT, UK

Cohrs, R., Department of Neurology, University of Colorado School of Medicine, Campus Box B-183, 4200 East 9th Avenue, Denver, CO 80262, USA

Crocker, J., Department of Histopathology, East Birmingham Hospital, Bordesley Green East, Birmingham B9 5ST, UK

Darai, G., Institut für Medizinische Virologie der Universität Heidelberg, Im Neuenheimer Feld 324, W-6900 Heidelberg, FRG

Deny, P., Hybridotest, Institut Pasteur, rue du Dr. Roux 28, 75015 Paris, France

Dueland, A. N., Department of Neurology, University of Colorado School of Medicine, Campus Box B-183, 4200 East 9th Avenue, Denver, CO 80262, USA

Dusheiko, G. M., University Department of Medicine, Royal Free Hospital School of Medicine, Rowland Hill Street, London NW3 2PF, UK

Fenyö, E. M., Department of Virology, Karolinska Institute, c/o National Bacteriological Laboratory, 105 21 Stockholm, Sweden

Flügel, R. M., Human Retrovirus Group, Deutsches Krebsforschungszentrum, Im Neuenheimer Feld 280, W-6900 Heidelberg, FRG

Gentilini, P., Istituto di Clinica e Terapia Medica II, Viale Morgagni 85, 50134 Firenze, Italy

Giebel, L. B., ARRIS Pharmaceutical Corporation, 385 Oyster Point Blvd., Suite 4, South San Francisco, CA 94080, USA

Gilden, D. H., Department of Neurology, and Microbiology and Immunology, University of Colorado School of Medicine, Campus Box B-183, 4200 East 9th Avenue, Denver, CO 80262, USA

Glass, R. I., Viral Gastroenteritis Unit, Division of Viral and Rickettsial Diseases, Center for Infectious Diseases, Centers for Disease Control, 1600 Clifton RD., NE Atlanta, GA 30333, USA

Godec, M. S., Laboratory of Central Nervous System Studies, National Institute of Neurological Disorders and Stroke, National Institutes of Health, Rm. 5B21, Bldg. 36, Bethesda, MD 20892, USA

Gouvea, V., Molecular Biology Branch, Division of Microbiology, Center for Food Safety and Applied Nutrition, Food and Drug Administration, 200 C St. SW, Washington, DC 20204, USA

Hall, J. E., Chiron Corporation, 4560 Horton St., Emeryville, CA 94608, USA

Han, J., Chiron Corporation, 4560 Horton St., Emeryville, CA 94608, USA

Hartmuth, K., Institut für Biochemie, Universität Wien, Währinger Straße 17, 1090 Wien, Austria

Hayoz, D., Division of Haematology, Department of Medicine, University Hospital CHUV, 1011 Lausanne, Switzerland

Hjelle, B., Department of Pathology, University of New Mexico School of Medicine, 915 Camino de Salud NE, Albuquerque, NM 87131, and United Blood Services, 1515 University Blvd., Albuquerque, NM 87125, USA

Ho, L., Department of Medical Microbiology and Department of Chemical Pathology, Faculty of Clinical Sciences, University College London, 46 Cleveland St., London W1P 6DB, UK

Hodinka, R. L., Division of Infectious Diseases and Immunology Departments of Pathology and Pediatrics, Children's Hospital of Philadelphia and the University of Pennsylvania, Philadelphia, PA 19104, USA

Houghton, M., Chiron Corporation, 4560 Horton St., Emeryville, CA 94608, USA

Inkster, M., Department of Microbiology, St. Louis University Medical Center, 1402 South Grand Blvd., St. Louis, MO 63104, USA

Jansen, R. W., Department of Medicine, The University of North Carolina at Chapel Hill, 547 Burnett-Womack CBS, CB 7030 Chapel Hill, NC 27599-7030, USA

Joske, D. J. L., Division of Haematology, Department of Medicine, University Hospital CHUV, 1011 Lausanne, Switzerland

Jug, G., Children's Cancer Research Institute, St. Anna Kinderspital, Kinderspitalgasse 6, 1090 Wien, Austria

Knecht, H., Division of Haematology, Department of Medicine, University Hospital CHUV, 1011 Lausanne, Switzerland

Kondo, K., Department of Virology, Research Institute for Microbial Diseases, Osaka University, 3-1 Yamada-oka, Suita, Osaka 565, Japan

Kondo, T., Department of Virology, Research Institute for Microbial Diseases, Osaka University, 3-1 Yamada-oka, Suita, Osaka 565, Japan

Kovar, H., Children's Cancer Research Institute, St. Anna Kinderspital, Kinderspitalgasse 6, 1090 Wien, Austria

Krivine, A., Laboratoire de Virologie, Hôpital Saint-Vincent-de-Paul, 82 avenue Denfert-Rochereau, 75674 Paris Cedex 14, France

Lemon, S. M., Department of Medicine, The University of North Carolina at Chapel Hill, 547 Burnett-Womack CBS, CB 7030 Chapel Hill, NC 27599-7030, USA

van Loon, A. M., Department of Virology, Rijksinstituut voor de Volksgezondheit en Milieuhygiëne, Bilthoven, The Netherlands

Mahalingam, R., Department of Neurology, University of Colorado School of Medicine, Campus Box B-183, 4200 East 9th Avenue, Denver, CO 80262, USA

Mangano, M. F., The Wistar Institute, 3601 Spruce Street and Division of Infectious Diseases and Immunology, Children's Hospital of Philadelphia, Philadephia, PA 19104, USA

Manger, I. D., Department of Microbiology, St. Louis University Medical Center, 1402 South Grand Blvd., St. Louis, MO 63104, USA

Matsumoto, C., The Japanese Red Cross Central Blood Center, Hiroo 4-1-31, Shibuya-ku, Tokyo 150, Japan

Mounir, S., Virology Research Center, Institut Armand-Frappier, Université du Québec, 531 boulevard des Prairies, Laval, Québec, Canada H7N 4Z3

Mukai, T., Department of Virology, Research Institute for Microbial Diseases, Osaka University, 3-1 Yamada-oka, Suita, Osaka 565, Japan

Muranyi, W., Human Retrovirus Group, Deutsches Krebsforschungszentrum, Im Neuenheimer Feld 280, W-6900 Heidelberg, FRG

Nishioka, K., The Japanese Red Cross Central Blood Center, Hiroo 4-1-31, Shibuya-ku, Tokyo 150, Japan

Odermatt, B. F., Department of Pathology, University Hospital, Zürich, Switzerland

Peters, A. C. B., Department of Neurology, University Hospital Leiden, Postbus 9600, 2300 RC Leiden, The Netherlands

Pistillo, J. M., Department of Pathology and Microbiology, University of Nebraska Medical Center, 600 S. 42nd Street, Omaha, NE 68198, USA

van der Ploeg, M., Department of Cytochemistry and Cytometry, State University Hospital Leiden, Postbus 9600, 2300 RC Leiden, The Netherlands

Raab, K., Institut für Medizinische Virologie der Universität Heidelberg, Im Neuenheimer Feld 324, W-6900 Heidelberg, FRG

Raap, A. K., Department of Cytochemistry and Cytometry, State University Leiden, Postbus 9600, 2300 RC Leiden, The Netherlands

Rajakumar, A., Department of Microbiology, St. Louis University Medical Center, 1402 South Grand Blvd., St. Louis, MO 63104, USA

Sacramento, D., Unité de la Rage, Institut Pasteur, 25, rue du Docteur Roux, 75724 Paris Cedex 15, France

Sahli, R., Institute of Microbiology, University Hospital CHUV, 1011 Lausanne, Switzerland

Schulz, I. T., Department of Microbiology, St. Louis University Medical Center, 1402 South Grand Blvd., St. Louis, MO 63104, USA

Shaw, P., Institute of Pathology, University Hospital CHUV, 1011 Lausanne, Switzerland

Shyamala, V., Chiron Corporation, 4560 Horton St., Emeryville, CA 94608, USA

Skern, T., Institut für Biochemie, Universität Wien, Währinger Straße 17, 1090 Wien, Austria

Spivack, J. G., The Wistar Institute, 3601 Spruce Street and Division of Infectious Diseases and Immunology, Children's Hospital of Philadelphia, Philadelphia, PA 19104, USA

Stewart, J. N., Virology Research Center, Institut Armand-Frappier, Université du Québec, 531 boulevard des Prairies, Laval, Québec, Canada H7N 4Z3

Stohwasser, R., Zentrum für Molekulare Biologie Heidelberg und Institut für Molekulare Genetik, Universität Heidelberg, Im Neuenheimer Feld 230, W-6900 Heidelberg, FRG

Swierkosz, E. M., Diagnostic Virology Laboratory, Cardinal Glennon Children's Hospital and Department of Pediatrics/Adolescent Medicine and Pathology, St. Louis University Medical Center, 1465 South Grand Blvd., St. Louis, MO 63104, USA

Syrjänen, K., Department of Pathology and Kuopio Cancer Research Centre, University of Kuopio, POB 1627, 70211 Kuopio, Finland

Syrjänen, S., Department of Pathology and Kuopio Cancer Research Centre, University of Kuopio, POB 1627, 70211 Kuopio, Finland

Talbot, P. J., Virology Research Center, Institut Armand-Frappier, Université du Québec, 531 boulevard des Prairies, Laval, Québec, Canada H7N 4Z3

Terry, G., Department of Medical Microbiology and Department of Chemical Pathology, Faculty of Clinical Sciences, University College London, 46 Cleveland St., London W1P 6DB, UK

Tomomori, C., Department of Virology, Research Institute for Microbial Diseases, Osaka University, 3-1 Yamada-oka, Suita, Osaka 565, Japan

Tordo, N., Unité de la Rage, Institut Pasteur, 25, rue du Docteur Roux, 75724 Paris Cedex 15, France

Torgersen, H., Institut für Biochemie, Universität Wien, Währinger Straße 17, 1090 Wien, Austria

Tracy, S., Department of Pathology and Microbiology, University of Nebraska Medical Center, 600 S. 42nd Street, Omaha, NE 68198, USA

Trent, D. W., Molecular Biology Branch, Division of Vector-Borne Infectious Diseases, Centers for Disease Control, P.O. Box 2087, Foothills Campus, Fort Collins, CO 80522, USA

Wadell, G., Department of Virology, University of Umea, 90185 Umea, Sweden

Weiner, A. J., Chiron Corporation, 4560 Horton St., Emeryville, CA 94608, USA

Xu, J., University Department of Medicine, Royal Free Hospital School of Medicine, Rowland Hill Street, London NW3 2PF, UK

Yamamoto, T., Department of Virology, Research Institute for Microbial Diseases, Osaka University, 3-1 Yamada-oka, Suita, Osaka 565, Japan

Yamanishi, K., Department of Virology, Research Institute for Microbial Diseases, Osaka University, 3-1 Yamada-oka, Suita, Osaka 565, Japan

Zignego, A. L., Istituto di Clinica e Terapia Medica II, Viale Morgagni 85, 50134 Firenze, Italy and Hybridotest, Institut Pasteur, 25, rue du Dr. Roux, 75015 Paris, France

Zuckerman, A. J., University Department of Medicine, Royal Free Hospital School of Medicine, Rowland Hill Street, London NW3 2PF and WHO Collaborating Center for Reference and Research of Viral Diseases, London, UK

Section I Human Retroviruses

Chapter 1 Diagnosis of Human Immunodeficiency Virus Infection by Polymerase Chain Reaction*

Jan Albert[1] and Eva Maria Fenyö[2]

Summary

Infection by the human immunodeficiency virus (HIV) is routinely diagnosed by the identification of specific antibodies in serum. In some cases, as in primary infection, HIV can be detected prior to seroconversion by the polymerase chain reaction (PCR), among other methods. Furthermore, there is a great need for accurate and early detection of perinatal HIV infection, as early diagnosis is complicated by the persistence of maternal antibodies. Another important question is posed by PCR positivity in seronegative individuals at risk of HIV infection, as available data are conflicting. We have developed a simple PCR with nested primers for the detection of HIV-1 and HIV-2. By this PCR protocol, 97% of unselected HIV-1 seropositive individuals were positive. HIV-1 infected persons who had an uninfected sexual partner or child were somewhat less often PCR positive (79%), indicating that contagiousness may be related to virus load. Seronegative persons at risk of HIV-1 infection ($n = 55$) were PCR negative. Also, in children born to HIV-1 infected mothers, there was excellent agreement between the results of PCR and other signs and symptoms of HIV-1 infection. However, PCR was the most sensitive method for the early diagnosis of perinatal transmission of HIV-1. In conclusion PCR is of great diagnostic value in selected cases. PCR positivity in persons who remain HIV seronegative for prolonged periods of time appears to be very rare.

Introduction

Infection by the human immunodeficiency virus (HIV) is routinely diagnosed by the identification of specific antibodies in serum (Centers for Disease Control 1986). However, some cases can be earlier or more accurately identified by

[1] Department of Virology, National Bacteriological Laboratory, 105 21 Stockholm, Sweden
[2] Department of Virology, Karolinska Institute, c/o National Bacteriological Laboratory, 105 21 Stockholm, Sweden
* This work was supported by the Swedish Medical Research Council (grant no. 16H-7737) and by the Labour Market Insurance Company (AFA).

direct detection of HIV in clinical specimens using virus isolation (Ho et al. 1985; Albert et al. 1987) and antigen assay (Goudsmit et al. 1986). More recently, the polymerase chain reaction (PCR) has been used for the detection of HIV-1-specific DNA sequences in clinical specimens (Ou et al. 1988). Examples of cases in which the direct detection of HIV-1 may be of value are those with suspected primary HIV-1 infection who have not yet seroconverted and children born to HIV-1-infected mothers, who passively acquire maternal antibodies even if they are uninfected. Furthermore, some HIV-1-infected persons have been reported to remain seronegative for prolonged periods (Salahuddin et al. 1984; Ranki et al. 1987; Loche and Mach 1988; Imagawa et al. 1989; Wolinsky et al. 1989). Obviously, there is an urgent need to establish whether or not HIV-1 infection is common among HIV-1 seronegative individuals at risk of infection. Available data are conflicting; several studies report that PCR positivity is not infrequent among seronegative individuals (Loche and Mach 1988; Imagawa et al. 1989; Wolinsky et al. 1989), but others have failed to document such cases (Horsburgh et al. 1989; Jackson et al. 1989; Lifson et al. 1990; Lefrère et al. 1991; Busch et al. 1991).

PCR has several advantages over other methods for the direct detection of HIV-1. It is highly sensitive and allows the identification of a single molecule against a background of 10^5 cells (Saiki et al. 1988). This high sensitivity is an advantage, as some infected persons have very low frequencies ($<1:10000$) of infected peripheral blood mononuclear cells (PBMC) (Psallidopoulos et al. 1989; Simmonds et al. 1990; Brinchmann et al. 1991). The only other method with a comparable, but somewhat lower, sensitivity is virus isolation. Virus isolation is, however, slower and more cumbersome to perform than PCR.

Two potential problems in using PCR for the diagnosis of HIV-1 infection need to be considered. The first derives from the enormous genetic diversity of HIV-1 (Alizon et al. 1986; Coffin 1986). Since every HIV-1 genome is unique, it is necessary to describe HIV-1 isolates as populations of closely related genomes, referred to as quasispecies (Meyerhans et al. 1989). This means that false-negative PCR results may result from divergent nucleotide sequences in some isolates. However, by careful selection of PCR primers in regions of the HIV-1 genome with limited sequence variation, it is possible to amplify HIV-1 sequences from virtually all HIV-1 seropositive individuals (Ou et al. 1988; Jackson et al. 1990; Albert and Fenyö 1990; Lifson et al. 1990). Secondly, a well-known problem with PCR is the high risk of false-positive results due to contamination of sample or reagents (Kwok and Higuchi 1989). This problem is especially troublesome in HIV-1 diagnosis since the low frequency of infected cells makes it necessary to use a highly sensitive PCR. On the other hand, a false-positive HIV-1 diagnosis is unacceptable because of the nature of the disease. False-positive PCR results may be checked by using strict laboratory routines and always including relevant controls in each PCR run (Kwok and Higuchi 1989).

Early PCR protocols could not easily be applied to large-scale clinical diagnostics, as they relied on DNA extraction as well as Southern blotting and hybridization with a radiolabelled probe for the detection of the PCR product

(Ou et al. 1988; Loche and Mach 1988; Jackson et al. 1989; Imagawa et al. 1989). Also, simpler PCR protocols, in which the handling of samples and PCR products is reduced, minimize the risk of false-positive results.

We have developed a simple, sensitive, and specific PCR for the detection of HIV-1 in clinical specimens (Albert and Fenyö 1990). The protocol has been used in clinical practice for more than 2 years. Furthermore, we have developed a PCR for the detection of HIV-2 and simian immunodeficiency virus (SIV) based on the same general protocol (Putkonen et al. 1991).

Materials and Methods

Patients

Unselected samples were collected from HIV-1 seropositive individuals who were participating in different longitudinal studies; some of them were receiving antiviral therapy. Selected samples were collected from seropositive individuals who had an uninfected sexual partner (Albert et al. 1988) and from mothers who gave birth to an uninfected child (Scarlatti et al. 1991). Samples were also collected from seronegative individuals who were considered to be at risk of HIV-1 infection because they either had had unprotected sexual intercourse or had shared needles with an infected individual. Additional samples were collected from children born to HIV-1 seropositive mothers. HIV-1 antibodies were detected by enzyme-linked immunosorbent assay (ELISA) (Organon Teknika) and confirmed by Western blotting (Biotech-Dupont).

Primers specific for regions of *gag* and *pol* genes, which are conserved between HIV-2 and SIV were developed. The HIV-2/SIV primers were tested on blood donor PBMC which were infected in vitro by 17 different HIV-2 isolates as well as SIV of sooty mangabey (SIV_{sm}) and macaque (SIV_{mac}) origin. The characteristics of the HIV-2 isolates have been described elsewhere (Albert et al. 1990). The HIV-2/SIV primers were then used to analyze PBMC samples from monkeys infected by $HIV-2_{SBL-6669}$ and SIV_{sm} in ongoing vaccine research (Putkonen et al. 1991).

Sample Preparation

PBMC were isolated from ethylenediaminetetra-acetic acid (EDTA)-treated blood by Ficoll-Paque (Pharmacia) centrifugation according to the recommendations of the manufacturers. The PBMC were either processed immediately or were stored in liquid nitrogen in fetal calf serum with 10% dimethyl sulfoxide and thawed before use. The cells were washed once in phosphate-buffered saline (PBS) and counted; 4×10^6 cells were then pelleted, the PBS was carefully removed, and the cells were resuspended and lysed in 400 µl of lysis buffer (10 mM TRIS-HCl pH 8.3, 1 mM EDTA, 0.5% NP-40, 0.5% Tween 20, and 300 µg/ml proteinase K). The lysed cells were digested with proteinase K for 1 h at 55 °C or overnight at 37 °C, at which time the enzyme was inactivated by 15 min at 94 °C. The lysates were stored at -20 °C until they were used. Sample preparation was done in a biosafety level 3 facility used for HIV isolation, but infected cultures were not handled in the biosafety cabinet used for preparation of the samples for PCR. Plasmid DNA or PCR products have never been used in this laboratory.

Primers specific for the β-globin gene of human DNA (PC03/04) (Saiki et al. 1985) were used to verify that the cell lysates, which were negative for HIV-1 DNA, were suitable for PCR amplification. Cell lysates which were negative in PCR with these primers were excluded from the study.

Primers

DNA oligonucleotide primers were synthesized using phosphoramidite chemistry (Albert et al. 1990). The nucleotide sequences of the primers and their locations in the HIV-1 and HIV-2 genomes are shown in Table 1.

Amplification

The PCR was performed in a two-step reaction, first with a pair of outer primers, then with a pair of inner (i.e., nested) primers. The PCR reactions were performed in 0.5 ml microfuge tubes (Perkin Elmer) in a total volume of 50 µl. The reaction mixtures contained (final concentrations) 10 mM Tris-HCl pH 8.3, 40 mM KCl, MgCl$_2$ according to titered optimum (7 mM in the first PCR and 3 mM in the second nested PCR), 0.1 µM of each primer, 50 µM of each nucleotide (ATP, CTP, GTP, and TTP) (Pharmacia), 1 unit Taq polymerase (Perkin-Elmer) per 50 µl. We prepared master mixes (usually for 20 samples) for each of the two steps of the PCR; the mixes were divided into samples and overlaid with mineral oil (Sigma no. M3516). The reaction mixtures for the second PCR were stored at $+4$ °C. Then, 10 µl cell lysate (corresponding to 10^5 PBMC) were added to the reaction mixtures for the first PCR. Lysis buffer was added to every third sample, instead of cell lysate, to check the reagents for contamination. The samples were first denatured for 5 min at 94 °C in a thermal cycler (Perkin-Elmer), then cycled 24 times at 95 °C for 30 s, annealing (at temperatures described in Table 1) 30 s, 75 °C for 30 s, and finally incubated at 72 °C for 5 min. After the first PCR one-tenth (5 µl) of the product was amplified for 30 cycles with the corresponding inner primers. The product from the second PCR (10 µl) was analyzed by electrophoresis on a 1.5% agarose gel stained with ethidium bromide. The reaction mixtures were set up in a sterile laboratory into which neither PCR product nor plasmid DNA was brought. All reagents were aliquoted into volumes sufficient for 20 samples and stored at -20 °C. The cell lysates were first amplified once with

Table 1. Primer sequences, location in the HIV-1$_{BRU}$ or HIV-2$_{ISY(SBL-6669)}$ genomes, and annealing temperatures

Primer	Sequence (5′–3′)	Gene and location	Annealing temperature
JA4 (outer sense)	GAA GGC TTT CAG CCC AGA AG	HIV-1 *gag* (818– 837)	56 °C
JA5 (inner sense)	ACC ATC AAT GAG GAA GCT GC	HIV-1 *gag* (945– 964)	50 °C
JA6 (inner antisense)	TAT TTG TTC CTG AAG GGT AC	HIV-1 *gag* (1076–1057)	50 °C
JA7 (outer antisense)	TCT CCT ACT GGG ATA GGT GG	HIV-1 *gag* (1114–1095)	56 °C
JA13 (outer sense)	TTC CTT GGG TTC TTG GGA GC	HIV-1 *env* (7372–7391)	54 °C
JA14 (inner sense)	GCA GCA GGA AGC ACT ATG GG	HIV-1 *env* (7390–7409)	58 °C
JA15 (inner antisense)	CCA GGA CTC TTG CCT GGA GC	HIV-1 *env* (7561–7542)	58 °C
JA16 (outer antisense)	AGG TAT CTT TCC ACA GCC AG	HIV-1 *env* (7577–7558)	54 °C
JA17 (outer sense)	TAC AGG AGC AGA TGA TAC AG	HIV-1 *pol* (1909–1928)	50 °C
JA18 (inner sense)	GGA AAC CAA AAA TGA TAG GG	HIV-1 *pol* (1959–1978)	50 °C
JA19 (inner antisense)	ATT ATG TTG ACA GGT GTA GG	HIV-1 *pol* (2088–2069)	50 °C
JA20 (outer antisense)	CCT GGC TTT AAT TTT ACT GG	HIV-1 *pol* (2175–2156)	50 °C
JA43 (outer sense)	AAG CCA AGG AGT AGT GGA AGC AAT G	HIV-2/SIV *pol* (4490–4514)	60 °C
JA44 (inner sense)	GAA GGG GAG GAA TAG GGG ATA TGA C	HIV-2/SIV *pol* (4612–4636)	60 °C
JA45 (inner antisense)	CCA CAG CTG ATC TCT GCC TTC TCT G	HIV-2/SIV *pol* (4757–4733)	60 °C
JA46 (outer antisense)	TCT GTC CCT ACC TTT ACG ATG ACT G	HIV-2/SIV *pol* (4819–4795)	60 °C
JA47 (outer sense)	TGG TGC ATG CAC GCA GAA GAG AAA G	HIV-2/SIV *gag* (802– 826)	60 °C
JA48 (inner sense)	CAA GTA GAC CAA CAG CAC CAC CTA G	HIV-2/SIV *gag* (926– 950)	60 °C
JA49 (inner antisense)	CCT GAA ATC CTG GCA CTA CTT CTG C	HIV-2/SIV *gag* (1072–1048)	60 °C
JA52 (outer antisense)	ATA TCA TAG GGC GTG CAG CCT TCT G	HIV-2/SIV *gag* (1103–1079)	60 °C

each of the primer sets JA4–JA7, JA13–JA16, and JA17–JA20 for the HIV-1 samples and JA43–JA46 and JA47–JA52 for the HIV-2/SIV samples. The tests were repeated in positive samples with some but not all primer sets. If the repeated tests yielding opposite results, a third test was done, and the PCR was regarded as positive if two of three tests with that primer set gave positive reactions.

Human Immunodeficiency Virus Isolation

Virus isolation (Albert et al. 1990) was attempted in all samples from HIV-1 seropositive individuals and blood donors. In short, we cocultivated 2×10^6 patient PBMC with 4×10^6 phytohemagglutinin-P-stimulated PBMC from each of two normal blood donors. Blood donor PBMC (3×10^6) was added to the cultures every week. The cultures were maintained in RPMI-1640 (Gibco) supplemented by 10% fetal calf serum (Flow), 5 units/ml recombinant inter-leukin-2 (Amersham), 2 µg/ml polybrene (Sigma), and antibiotics. Every 3–4 days the cell culture supernatants were harvested and assayed by in house HIV-1 (Sundqvist et al. 1989) or HIV-2 (Thorstensson et al. 1990) antigen ELISAs.

Results

Diagnosis in Adults

The PCR protocol described was successfully used for amplification of HIV-1 DNA sequences in 231 of 237 samples (97%) obtained from unselected HIV-1 seropositive individuals and individuals selected on the basis of their African origin (Table 2). The primer sets used in this study were well suited for routine diagnostic applications, as samples which could be expected to contain HIV-1 variants with divergent nucleotide sequences (i.e., African samples) were read-ily detected. Negative samples and reagent controls were consistently negative for HIV-1 DNA (data not shown). Furthermore, we were unable to amplify HIV-1 DNA in any of the 55 persons at risk of HIV-1 infection (Table 2).

The frequency of positive PCR results was somewhat lower among the patients who were selected because they had an uninfected sexual partner or uninfected child, 72% and 85%, respectively. Conceivably, they have fewer infected PBMC than other patients. Interestingly, most of the negative PCR results (4 of 6) among the unselected seropositive patients occurred in the asymptomatic group. It should be stressed that only 5 of the 18 seropositive patients, who were not classified as PCR positive, were completely nonreactive with PCR. The remaining 13 patients showed PCR reactivity with one or the other of the primer sets, but they could not be classified as PCR positive because they did not show consistent positive amplification with at least two

Table 2. Results of polymerase chain reaction (PCR) and virus isolation in HIV-1-infected individuals and individuals at risk of HIV-1 infection

Classification of subjects	PCR (no. positive/ total no.)	Virus isolation (no. positive/ total no.)
HIV-1 seropositive unselected patients		
Asymptomatic (CDC group II)	59/63 (94%)	49/63 (78%)
Lymphadenopathy (CDC group III)	62/62	47/62 (76%)
AIDS (CDC group IVC-1)	18/18	18/18
Other symptoms (CDC group IVA, IVC-2, IVE)	41/41	35/41 (85%)
Unknown symptoms	22/23 (96%)	18/23 (78%)
HIV-1 seropositive selected patients		
Of African origin [a]	39/40 (98%)	
With an uninfected sexual partner	13/18 (72%)	
With an uninfected child	33/40 (85%)	23/40 (58%)
HIV-1 seronegative individuals at risk [b]	0/55	0/55

The samples were collected and analyzed between the fall of 1989 and the spring of 1991. Patients were classified according to the Centers for Disease Control (CDC) classification system for HIV infection (CDC 1986)

[a] A majority of the samples were collected in Uganda

[b] Individuals who either were sexual partners of seropositive persons or shared needles with seropositive persons

AIDS, acquired immunodeficiency syndrome

of three primer sets. Such PCR indeterminate results were not seen in samples from blood donors or in samples from persons at risk of HIV infection.

Diagnosis of Mother-to-Child Transmission

Detection of HIV-1 by PCR, virus isolation, and HIV-Ag assay was attempted in 143 blood samples from 110 children born to HIV-1-infected mothers. HIV-1-specific DNA sequences were amplified by PCR in samples from 13 children (Table 3). Seven of these children had already been diagnosed as HIV-1 infected (CDC P1 and CDC P2), and 6 were not yet diagnosed as infected or uninfected (CDC P0). HIV-1 was also detected by virus isolation in 12 of the 13 PCR-positive children and by HIV-Ag assay in 9.

The 97 PCR negative children were all clinically healthy. Virus isolation and HIV-Ag assay results were negative in these children. Some 58 PCR-negative children were antibody-positive, but these antibodies were likely to be of maternal origin since the children were all younger than 15 months. In fact, 18 of these children had become antibody negative by ELISA but were still positive by the more sensitive Western Blotting (WB) test, indicating that clearing of maternal antibodies was in progress. Repeated tests, when available, gave concordant results.

Table 3. CDC classification, results of HIV-1 isolation, and HIV-antigen test in PCR-negative and PCR-positive children

	PCR negative	PCR positive
No. of samples	128	15
No. of children	97	13
CDC stage		
P0	58[a]	6
P1	0	5
P2	0	2
seronegative	39	0
Positive HIV isolation	0	12
Positive HIV antigen assay	0[b]	9

CDC P0, P1, and P2 correspond to the Centers for Disease Control (CDC) classification of pediatric HIV infection (P0 indeterminate status, P1 asymptomatic infection, and P2 symptomatic infection)
[a] Eighteen children had become HIV-ELISA negative but were still WB positive
[b] $n = 83$ children tested
HIV, human immunodeficiency virus; PCR, polymerase chain reaction; ELISA, enzyme-linked immunosorbent assay

Diagnosis of Human Immunodeficiency Virus Type 2 (HIV-2)

Two sets of primers were selected in regions of the *gag* and *pol* genes, which are conserved between HIV-2 and SIV. These primers were successfully shown to amplify 17 different HIV-2 isolates from blood donor PBMC infected in vitro (data not shown). These primers also amplify SIV_{sm} and SIV_{mac} (data not shown). These primers have not yet been clinically evaluated on patient specimens. However, they have proven very efficient for the detection of HIV-2 and SIV in PBMC samples from experimentally infected cynomolgus macaques (Putkonen et al. 1991).

Discussion

Technical Considerations

We present a simple, sensitive, and specific PCR protocol for the amplification of HIV-1 DNA sequences. Patient PBMC were directly lysed, and this crude cell lysate was amplified in a two-step PCR, first with outer primers and then with inner primers nested within the first ones. The PCR product was visualized by agarose gel electrophoresis and ethidium bromide staining. By this PCR protocol we were able to amplify HIV-1 sequences from almost all HIV-1 seropositive individuals.

This protocol has several advantages over conventional ones. Sample preparation by direct lysis of PBMC has at least two obvious advantages over conventional DNA extraction. First, it minimizes sample handling and thus the risk of contamination; second, it saves time. PCR with nested primers (Simmonds et al. 1990; Albert and Fenyö 1990; Brinchmann et al. 1991) has increased sensitivity and specificity, so that the amplified product can be visualized directly by agarose gel electrophoresis. The sensitivity of our PCR protocol is demonstrated by the amplification of HIV-1 sequences from almost all seropositive individuals. In most previous publications, the detection of the amplified product has involved hybridization with a radioactively labelled probe (Ou et al. 1988; Loche and Mach 1988; Imagawa et al. 1989; Wolinsky et al. 1989; Horsburg et al. 1989; Jackson et al. 1989). Our protocol eliminates the need for radioactive material and thus has advantages in terms of safety, cost, and simplicity.

We have shown that our PCR protocol detects one HIV-1 proviral copy against a background of 10^5 PBMC, and this has enabled us to quantify HIV-1 in clinical samples by limiting dilution (Brinchmann et al. 1991). When we adopted our previously published (Albert and Fenyö 1990) PCR protocol to quantification, we had to make some adjustments to detect reproducibly one single molecule. This was done by increasing the primer annealing temperatures (to those shown in Fig. 1) and by increasing the number of amplification cycles in the second PCR from 24 to 30. We also noticed that PCR with longer primers (25-meres instead of 20-meres) can be optimized more easily to give maximum sensitivity (detection of one molecule). Furthermore, it should be mentioned that our initial PBMC lysis buffer contained 0.001% sodium dodecyl sulfate, which may potentially inhibit the Taq polymerase. This potential problem is eliminated by using the lysis buffer described in Materials and Methods.

The crude cell lysates were well suited for amplification of HIV-1 DNA sequences. In fact, the quality of the samples is high enough to allow direct DNA sequencing of the PCR amplified material without further purification (data not shown). A small proportion of the samples could not be amplified by β-globin primers (Saiki et al. 1985). Some of these samples were visibly contaminated by erythrocytes; such contamination has been reported to inhibit the PCR (R. Higuchi, Amplifications, issue 2, Perkin Elmer Cetus). This problem can be solved by selective lysis of the erythrocytes (for instance by ammonium chloride) prior to lysis of the PBMC. We now routinely lyse erythrocytes if the Ficoll separation does not yield sufficiently pure PBMCs.

The elimination of the need for hybridization for the detection of the amplified material is a clear advantage, but even simple agarose gel electrophoresis may be cumbersome to perform on a large scale. Different methods for a colorimetric detection of the amplified product have been published (Kemp et al. 1989; Lundeberg et al. 1991), and they are likely to prove useful for screening purposes. When PCR is used for the diagnosis of HIV infection, the correct size of the amplified product should be confirmed by gel electrophoresis in positive samples.

Diagnosis of Human Immundeficiency Virus Infection

We recommend that multiple PCR analyses should be performed on each sample. The risk of false-negative results due to HIV-1 sequence variations will be diminished by using several primer sets rather than repeated tests with one primer set. In our opinion, PCR analysis with three different primer sets (JA4–7, JA13–16, and JA17–20) is sufficient. A positive PCR analysis should require successful amplification with at least two of three primer sets. By these criteria, 277 of 295 (94%) HIV-1 seropositive individuals included in this study were PCR positive (Table 2). Of the remaining 18 individuals, 13 were PCR indeterminate. Twelve of these patients were selected because they had an uninfected sexual partner or an uninfected child, and another 4 were asymptomatic seropositive individuals. This indicates that contagiousness may be related to virus load and that some selected patients may have fewer than 10^5 infected PBMC. The latter finding is not surprising as it has been shown that some asymptomatic patients have a very low frequency of infected cells in the blood (Psallidopoulos et al. 1989; Simmonds et al. 1990; Brinchmann et al. 1991). These data indicate that more than 10^5 PBMC may have to be analyzed for maximal PCR sensitivity. One way to achieve this is to add more PBMC to the first PCR reaction, but we have not yet established how many PBMC can be added to a 50 µl PCR reaction mixture without reducing sensitivity. Another possibility would be to increase the reaction volume of the first PCR so as to be able to analyze more PBMC or to enrich for $CD4^+$ lymphocytes before cell lysis. In one study, we used magnetic beads to prepare $CD4^+$ lymphocytes for PCR analysis (Brinchmann et al. 1991), but we do not routinely use this method at present.

There is an urgent need to establish whether or not HIV-1 infection is common among HIV-1 seronegative individuals at risk of infection. Available data are conflicting; several studies report that PCR positivity is not infrequent among seronegative individuals (Loche and Mach 1988; Imagawa et al. 1989; Wolinsky et al. 1989), but others have failed to document such cases (Horsburgh et al. 1989; Jackson et al. 1989; Lifson et al. 1990; Lefrère et al. 1991; Busch et al. 1991). We have analyzed samples from 55 seronegative persons at risk of HIV-1 infection without finding any cases of PCR positivity. Thus, our data support the notion that PCR positivity among seronegative persons at risk is very rare. Similarly, we found no evidence for HIV-1 infection among children who were born to seropositive mothers and who had become seronegative. In fact, the PCR results in children born to HIV-1 infected mothers correlated very well to other markers of HIV-1 infection, such as virus isolation, antigen assay, persistent antibody positivity, and presence of clinical symptoms. This confirms our previous findings in a smaller group of children from the same cohort (Scarlatti et al. 1991). Thus, while the presented PCR protocol is highly sensitive, the risk of false-positive results is very low in our hands. This is probably due to (a) strict laboratory routines (Kwok and Higuchi 1989) and (b) the simple PCR protocol. In addition, nested PCR only generates large amounts of product from the second, and not from the first,

PCR. Since the product of the inner primers cannot be amplified by the outer primers in later PCR reactions, the risk of false-positive results due to contamination by PCR products is reduced.

Taken together, our data suggest that PCR positivity among seronegative persons at risk, adults or children, is very rare. Conventional antibody assays are therefore reliable in most diagnostic situations. However, PCR is a valuable complement to virus isolation and antigen detection in certain selected cases. With these assays, an earlier diagnosis can be reached in cases of primary HIV-1 infection and most importantly in children born to HIV-1-infected mothers. PCR or other methods for the direct detection of HIV will probably not completely replace the antibody assay in the near future. It should be stressed that a HIV-1 seronegative individual should not be regarded as PCR-positive unless the positive result is confirmed on a separate sample from the same individual.

PCR may prove even more useful in other clinical situations. One is the quantification of HIV-1 DNA or RNA to follow the progression of the disease and the effect of antiviral therapy (Schnittman et al. 1990). Other methods are either too insensitive (antigen detection) or too cumbersome to be used on a large scale (virus isolation). Another important application of PCR is to monitor the emergence of mutations in the HIV-1 genome, for instance, the selection of resistant variants during antiviral therapy. This can be achieved by identifying specific mutations directly by PCR (Boucher et al. 1990) or by determining the nucleotide sequence of a DNA fragment generated by PCR (Larder and Kemp 1989).

In conclusion, we have shown that HIV-1 can be detected in a majority (97%) of crude cell lysates from infected individuals by PCR with nested primers and simple visualization of the product in agarose gels. Seronegative individuals at risk of HIV-1 infection were PCR negative, indicating that HIV-1 infection among persons who remain antibody negative for a prolonged period is very rare. Our PCR protocol is generally applicable for almost all types of PCR diagnostics. Here we have presented the use of PCR nested primers for the detection of HIV-1, HIV-2, and SIV. In our laboratory the same basic concept has been used for the detection of human lymphotropic virus types I and II, cytomegalovirus, herpes simplex virus types 1 and 2, and parvovirus B19.

Acknowledgements. We thank Gabriella Scarlatti, Kajsa Aperia, Ulla Grandell, Lena Benthin, Bengt Abrahamsson, and Per Putkonen for valuable contributions.

References

Albert J, Fenjö EM (1990) Simple, sensitive and specific detection of HIV-1 in clinical specimens by polymerase chain reaction with nested primers. J Clin Microbiol 28:1560–1564

Albert J, Gaines H, Sönnerborg A et al. (1987) Isolation of human immunodeficiency virus (HIV) from plasma during primary HIV infection. J Med Virol 23:67–73

Albert J, Pehrson PO, Schulman S et al. (1988) HIV isolation and antigen detection in infected individuals and their seronegative sexual partners. AIDS 2:107–111

Albert J, Nauclér A, Böttiger B et al. (1990) Replicative capacity of HIV-2, like HIV-1, correlates with severity of immunodeficiency. AIDS 4:291–295

Alizon M, Wain-Hobson S, Montagnier L et al. (1986) Genetic variability of the AIDS virus: nucleotide sequence analysis of two isolates from African patients. Cell 46:63–74

Boucher CAB, Tersmette M, Lange JMA et al. (1990) Zidovudine sensitivity of human immunodeficiency virus from high-risk, symptom-free individuals during therapy. Lancet ii:585–590

Brinchmann JE, Albert J, Vartdal F (1991) Few infected $CD4^+$ T cells but a high proportion of replication-competent provirus copies in asymptomatic human immunodeficiency virus type 1 infection. J Virol 65:2019–2023

Busch MP, Eble BE, Khayam-Bashi H et al. (1991) Evaluation of screened blood donations for human immunodeficiency virus type 1 infection by culture and DNA amplification of pooled blood. N Engl J Med 325:1–5

Centers for Disease Control (1986) Classification system for human T-lymphotropic virus type III/lymphadenopathy-associated virus infections. Morbid Mortal Weekly Rep 35:334–339

Coffin JM (1986) Genetic variation in AIDS viruses. Cell 46:1–4

Goudsmit J, de Wolf F, Paul DA et al. (1986) Expression of human immunodeficiency virus antigen (HIV-Ag) in serum and cerebrospinal fluid during acute and chronic infection. Lancet ii:177–180

Ho DD, Sarngadharan MG, Resnick L et al. (1985) Primary human T-lymphotropic type III infection. Ann Intern Med 103:880–883

Horsburgh Jr CR, Ou CY, Jason J et al. (1989) Duration of human immunodeficiency virus infection before detection of antibody. Lancet 2:637–640

Imagawa DT, Lee MH, Wolinsky SM et al. (1989) Human immunodeficiency virus type 1 infection in homosexual men who remain seronegative for prolonged periods. N Engl J Med 320:1458–1462

Jackson JB, Kwok SY, Hopsicker JS et al. (1989) Absence of HIV-1 infection in antibody-negative sexual partners of HIV-1 infected hemophiliacs. Transfusion 29:265–267

Jackson JB, Kwok SY, Sninski et al. (1990) Human immunodeficiency virus type 1 detected in all seropositive symptomatic and asymptomatic individuals. J Clin Microbiol 28:16–19

Kemp DJ, Smit DB, Foote SJ et al. (1989) Colorimetric detection of specific DNA segments amplified by polymerase chain reactions. Proc Natl Acad Sci USA 86:2423–2427

Kwok S, Higuchi R (1989) Avoiding false positives with PCR. Nature 339:237–238

Larder BA, Kemp SD (1989) Multiple mutations in HIV-1 reverse transcriptase confer high-level resistance to Zidovudine (AZT). Science 246:1155–1158

Lefrère JJ, Mariotti M, Vittecoq D et al. (1991) No evidence of frequent human immunodeficiency virus type 1 infection in seronegative at-risk individuals. Transfusion 31:205–211

Lifson AR, Stanley M, Pane J et al. (1990) Detection of human immunodeficiency virus DNA using the polymerase chain reaction in a well-characterized group of homosexual and bisexual men. J Infect Dis 161:436–439

Loche M, Mach B (1988) Identification of HIV-infected seronegative individuals by a direct diagnostic test based on hybridisation to amplified viral DNA. Lancet 2:418–421

Lundeberg J, Wahlberg J, Uhlén M (1991) Rapid colorimetric quantification of PCR-amplified DNA. Biotechniques 10:68–75

Meyerhans A, Cheynier R, Albert J et al. (1989) Temporal fluctuations in HIV quasispecies in vivo are not reflected by sequential HIV isolations. Cell 58:901–910

Ou CY, Kwok S, Mitchell SW et al. (1988) DNA amplification for direct detection of HIV-1 in DNA of peripheral blood mononuclear cells. Science 239:295–297

Psallidopoulos MC, Schnittman SM, Thompson LM et al. (1989) Integrated proviral human immunodeficiency virus type 1 is present in $CD4^+$ peripheral blood lymphocytes in healthy seropositive individuals. J Virol 63:4626–4631

Putkonen P, Thorstensson R, Ghavamzadeh L et al. (1991) Prevention of HIV-2 and SIV_{sm} infection by passive immunization in cynomolgus monkeys. Nature 352:436–438

Ranki A, Valle SL, Krohn M et al. (1987) Long latency precedes overt seroconversion in sexually transmitted human-immunodeficiency-virus infection. Lancet ii:589–593

Saiki RK, Scharf S, Faloona F et al. (1985) Enzymatic amplification of β-globin genomic sequences and restriction site analysis for diagnosis of sickle cell anemia. Science 230:1350–1354

Saiki RK, Gelfand DH, Stoffel S et al. (1988) Primer-directed enzymatic amplification of DNA with a thermostable DNA polymerase. Science 239:487–491

Salahuddin SZ, Groopman JE, Markham PD et al. (1984) HTLV-III in symptom-free seronegative persons. Lancet i:1418–1420

Scarlatti G, Lombardi V, Plebani A et al. (1991) Polymerase chain reaction, virus isolation and antigen assay in HIV-1 antibody positive mothers and their children. AIDS 5:1173–1178

Schnittman SM, Greenhouse JJ, Psallidopoulos MC et al. (1990) Increasing viral burden in $CD4^+$ T cells from patients with human immunodeficiency virus (HIV) infection reflects rapidly progressive immunosuppression and clinical disease. Ann Intern Med 113:438–443

Simmonds P, Balfe P, Beutherer JF et al. (1990) Human immunodeficiency virus-infected individuals contain provirus in small numbers of peripheral mononuclear cells and at low copy numbers. J Virol 64:864–872

Sundqvist V-A, Albert J, Ohlsson E et al. (1989) HIV-1 p24 antigenic variation in tissue culture of isolates with defined growths characteristics. J Med Virol 29:170–175

Thorstensson R, Wahlter L, Putkonen P et al. (1990) A capture enzyme immunoassay for detection of HIV-2/SIV antigen. J AIDS 4:374–379

Wolinsky SM, Rinaldo CR, Kwok S et al. (1989) Human immunodeficiency virus type 1 (HIV-1) infection a median of 18 months before a diagnostic western blot. Evidence from a cohort of homosexual men. Ann Intern Med 111:961–972

Chapter 2 Human Immunodeficiency Virus Type 1 (HIV-1) Detection by Polymerase Chain Reaction in Children of Infected Mothers

Anne Krivine

Summary

The results of two studies involving 54 children born to HIV-1 seropositive women are reported. In the first one, I tested 24 children by PCR and viral isolation from peripheral blood mononuclear cells: Results were concordant in 96% of the cases, showing that PCR is a valid alternative to viral culture for the diagnosis of perinatal HIV-1 infection. In the second study, I tested prospectively 30 newborns at birth, 4–9 weeks, and 6–9 months: diagnosis of HIV infection by PCR was possible at birth for 3 infants compared with 12 one month later. No changes in the results were observed between the second and third testing. I conclude that, in most cases, PCR will not detect the presence of HIV-1 proviral DNA at birth, but it is a powerful technique that can be used for the diagnosis of perinatal HIV-1 infection from the second month of life.

Introduction

Since the beginning of the acquired immunodeficiency syndrome (AIDS) epidemic, the prevalence of human immunodeficiency virus (HIV)-specific antibodies among pregnant women has been increasing, especially in urban areas, e.g., 0.2% in Massachusetts (Hoff et al. 1988) and 5%–7% in Kinshasa, Zaire (Ryder et al. 1989). Like some other viruses, HIV-1 may be transmitted from a mother to her offspring, but the timing and the factors involved are still unknown. The rate of perinatal HIV transmission is estimated at between 13% and 40% (European Collaborative Study 1991; Blanche et al. 1989; Ryder et al. 1989, Hutto et al. 1991). Pediatric AIDS is a devastating disease: Scott et al. (1989) report a median survival of 38 months from the time of diagnosis in 172 infected children, with a 17% mortality within the first year of life. The main clinical manifestations of the disease may be divided into two categories: nonspecific conditions such as failure to thrive, lymphadenopathy, hepatosplenomegaly, oral candidiasis, or recurrent infections; and specific mani-

Laboratoire de Virologie, Hôpital Saint-Vincent-de-Paul, 82, Avenue Denfert-Rochereau, F-75674 Paris, France

festations including opportunistic infections (*Pneumocystis carinii* pneumonia, *Candida* esophagitis), specific encephalopathy, lymphoid interstitial pneumonitis (Scott et al. 1989; Blanche et al. 1989). Since zidovudine therapy has been shown to reduce the level of p24 antigenemia and the neurological impairment of infected children (McKinney et al. 1991), it is crucial to be able to make the diagnosis of perinatal HIV-1 infection as early as possible.

Diagnosis of HIV-1 infection in children aged 15 months or older is based, like in adults, on the detection of HIV-specific antibodies in the serum. Since all children born to HIV seropositive mothers carry passively transmitted maternal antibodies at birth, alternative methods of diagnosis have been developed that directly address the presence of virus or proviral DNA in the sample; these methods include virus isolation from peripheral blood mononuclear cells (PBMC) (Jackson et al. 1988), which is still the reference technique, antigen detection in serum or plasma by immunocapture assay (Borkowsky et al. 1989), "in vitro" production of HIV antibodies (Amadori et al. 1990), and detection of proviral HIV-1 DNA by polymerase chain reaction (PCR). This last technique, like antigen detection, has the great advantage of being partially automated and not requiring cell culture. The use of PCR for the detection of HIV-1 proviral sequences in seropositive adults has been described by Ou et al. in 1988 and applied to the diagnosis of HIV-1 infection in newborns by Laure et al. (1988).

To try to validate the use of this new technology for the early diagnosis of HIV infection in children, my approach was first to compare its sensitivity and specificity with the other methods of HIV-1 detection and then to evaluate its sensitivity during the first weeks of life. I report on a retrospective study conducted in Boston in 1988–1989 and a prospective study conducted in Paris in 1990.

Comparative Study of Virus Isolation, PCR, and Antigen Detection

Material and Methods

In this study (Krivine et al. 1990), 24 children (from 4 days to 4 years old) born to seropositive women were evaluated by PCR, viral isolation, and antigen detection on the same blood sample.

The isolation of HIV was performed as described by Jackson et al. (1988) and the presence of p24 antigen in the plasma was tested by commercial immunocapture assay (DuPont).

For PCR, 1 ml of heparinized blood was fractionated on Ficoll-Hypaque solution; DNA was obtained from PBMC after a single phenol-chloroform extraction followed by ethanol precipitation. The test was performed with the three sets of primers described by Ou et al. (1988): SK 38/39 (*gag*), SK 68/69 (*env*), and SK 29/30 (LTR). Specimens were always tested in duplicate, each reaction tube containing 1 µg of DNA in 10 mM TRIS pH 8.3, 50 mM KCl, 1.5 mM MgCl$_2$, 1 nM of each primer, 200 µM of all dNTP, and 2.5 U of Taq polymerase (Perkin-Elmer Cetus). Forty cycles of amplification were performed in an automatic DNA thermal cycler (Perkin-Elmer), 45 s at 94 °C, 60 s at 55 °C, and 45 s at 72 °C, followed by 10 min at 72 °C. Revelation of the PCR product was done by Southern blot and hybridization with an [32]P-5′-end-labeled oligoprobe complementary to each amplified fragment. A sample was considered positive for one set of primers if an amplified fragment of expected length was detected by specific hybridization in duplicate tubes in two separate experiments.

Results

Among the 24 children tested, 10 were positive by PCR and HIV culture, 1 had a positive result by PCR but negative culture result, and 13 were negative on both tests (Table 1). The concordance between PCR and viral isolation was thus 96%. The child who had a positive PCR but negative culture result exhibited the clinical criteria of AIDS. In the 6 children over 15 months of age, PCR and culture results correlated with serological and clinical status (4 positive, 2 negative), and all the children who became seronegative had negative PCR and culture results. In contrast, measurement of p24 antigenemia was less sensitive: among 11 PCR-positive children, only 5 had detectable antigenemia. From this work, I concluded that HIV-1 detection by PCR was a valid alternative to viral culture for the diagnosis of pediatric HIV-1 infection.

Early Diagnosis of HIV Infection by PCR During the First Weeks of Life

I conducted a prospective longitudinal study of infants born to seropositive mothers in two hospitals in Paris, beginning January 1990 (Krivine et al. 1992). Blood samples were obtained soon after birth, during the second month of life, and then between 6 and 9 months; this showed that among 30 children a diagnosis of HIV-1 infection by PCR was possible in the neonatal period in only three children, compared with 12 one month later. No changes were observed between 4–9 weeks and 6–9 months. I conclude that in most of the cases PCR is unable to detect the presence of HIV-1 proviral DNA at birth, but it is a powerful diagnostic tool that can be applied from the second month of life.

Material and Methods

Thirty children were included in the study; they were born between January and December 1990 in two hospitals in Paris. Three blood samples (2–3 ml on heparin) were scheduled during the first year of life: between 4 and 10 days of life for PCR and p24 antigenemia; between 4 and 9 weeks for PCR, p24

Table 1. Sensitivity and specificity of PCR, HIV culture, and p24 antigenemia in 24 children born to seropositive women

Comparison	Reference test	Positive	Negative	Sensitivity/ specificity (%)
PCR vs. culture	Culture	PCR	PCR	100/93
	Positive	10	0	
	Negative	1	13	
p24 vs. PCR	PCR	p24 antigen	p24 antigen	45/100
	Positive	5	6	
	Negative	0	13	

PCR, polymerase chain reaction; HIV, human immunodeficiency virus

antigenemia, and HIV culture; between 6 and 9 months for the same parameters. A last sample was scheduled between 12 and 18 months for serological testing.

For PCR, washed PBMC were lyzed at a concentration of 10 millions/ml at 56 °C for 2 h in 10 mM TRIS (pH 8.3), 50 mM KCl, 2.5 mM MgCl$_2$, 0.45% Nonidet P40, 0.45% Tween 20, 8 µg/ml proteinase K (Higuchi 1989). Proteinase K was then inactivated by heat for 10 min at 95 °C, and lysates were stored at −20 °C. A total of 40 cycles of amplification was performed in a Perkin-Elmer Cetus thermal cycler, 1 min at 94 °C, 1 min at 60 °C, 1 min at 72 °C, followed by 10 min at 72 °C. Each sample was tested in duplicate, with at least 2 different HIV-1 primer pairs (SK 38/39 and SK 68/69 or pol 3/4; Laure et al. 1988). An internal β-globin control was done in a separate tube. Some 25 µl of lysate were used in each reaction, equivalent to 200 000 cells. After amplification, 15 µl of the PCR product were loaded on a 2% agarose gel stained with ethidium bromide. The gel was transferred to a nylon membrane (Hybond N$^+$, Amersham) and then hybridized to a specific synthetic oligoprobe labeled with digoxigenin dUTP (Boehringer) at the 3′-end by means of terminal transferase. After two washes, the membrane was incubated with a digoxigenin-specific antibody bound to alkaline phosphatase, and a colored reaction was performed in the dark for 3 h after addition of an enzyme substrate.

For HIV coculture, 2−6 million patients PBMC were used, as described by Jackson et al. (1988). The presence of p24 antigen in the supernatant was monitored at least once a week with a commercial immunocapture assay (Abbott).

Results

All the children ($n = 30$) underwent the first two tests, and 24 of them were tested at 6−9 months (five were lost to follow-up, one died when 4 months old).

Table 2. Evolution of three markers of HIV-1 during infancy in 12 infected children

Case	Age at testing								
	4–10 days		4–9 weeks			6–9 months			
	PCR	p24	PCR	p24	culture	PCR	p24	culture	
1	+	+	+	+	+	+	+	+	
2	+	−	+	−	+	+	+	+	
3	+	+	+	+	+	na	+	na	
4	−	−	+	−	nd	+	+	+	
5	−	−	+	+	nd	na	na	na	
6	−	−	+	+	+	na	na	na	
7	−	−	+	−	+	na	na	na[a]	
8	−	−	+	−	+	+	+	+	
9	−	−	+	−	+	+	+	+	
10	−	−	+	−	+	+	−	+	
11	−	−	+	−	+	+	−	+	
12	−	−	+	+	+	+	+	+	

nd, not done; na, sample not available; PCR, polymerase chain reaction; HIV, human immunodeficiency virus
[a] Died when 4 months old with positive p24 antigenemia

Out of 30 12 (35%) children were found positive by PCR during the second month of life; 10 of them had a positive HIV culture, and five had detectable p24 antigenemia (Table 2). In contrast, only three were positive by PCR on the first blood sample, two of them having p24 antigenemia. Thus, 9 children turned from negative to positive during the first weeks of life. Among the 24 children tested at 6–9 months of age, no changes were observed in the results (eight remained positive, 16 negative, and the baby who died had p24 antigenemia). In our series, PCR was able to detect one-fourth of the infected infants within the first week of life, and presumably all of them before the end of the second month. A similar trend was observed with p24 antigenemia, which was able to detect 2/12 infected babies at birth and 5/12 on subsequent sample.

Discussion

Although several published studies report an interest in PCR for the diagnosis of HIV-1 infection in children born to seropositive mothers (Laure et al. 1988; Rogers et al. 1989; Chadwick et al. 1989), only a few of them show comparative results between PCR and HIV culture (Edwards et al. 1989; Krivine et al. 1990). My first study shows that PCR is as sensitive and specific as HIV culture for the diagnosis of HIV-1 infection in 24 children less than 48 months old. This was in agreement with the findings of Edwards et al. (1989) showing that for 25 children at risk of HIV infection, simultaneously grown HIV-1 cultures concurred with PCR results in all cases (7 positive and 18 negative).

In my second study, technical modifications were made:

– The use of crude cell lysate instead of fully extracted DNA for PCR
– The use of a different third primer pair (*pol* 3/4)
– The use of nonisotopic detection of amplified product

Direct amplification from crude cell lysates has been described by Higuchi (1989) and used by several authors (Rogers et al. 1989; Chadwick et al. 1989; Edwards et al. 1989). For this study, I performed sensitivity experiments by diluting a known number of 8E5 cells in uninfected fresh PBMC. The 8E5 cells contain a single copy of the HIV-1 genome and therefore one can prepare cell lysates containing 1, 5, 10, or 100 copies of proviral DNA per 25 µl. I was able to detect reproducibly 5 copies of HIV-1 genome with the *gag* (SK38/39) primers and 5–10 copies with the other primer pairs. This degree of sensitivity was similar to that obtained in the first study with extracted DNA and isotopic detection of the amplified product. The use of digoxigenin dUTP-labeled oligoprobe for the detection of the PCR product avoids handling of hazardous material and does not cause any significant decrease of sensitivity or specificity. However, the sensitivity of PCR on crude cell lysates may be decreased if the PBMC pellet is "contaminated" with a great amount of red blood cells: In this case, it is necessary first to lyse the red cells in 0.32 M sucrose, 1% Triton X-100, 5 mM MgCl$_2$, and 10 mM TRIS pH 7.5 (Higuchi 1989). To ensure the good quality of the sample to be tested by PCR, all the lysates must be amplified with an internal control primer set; I have successfully used the β-globin pair PC03/PCO4 described by Saiki et al. (1988).

Because of the high degree of variability of the HIV-1 genome, each sample must be tested with at least 2 primer pairs to reduce the risk of a false-negative result. In the first study, 10/11 samples were positive with three primer pairs, 1 was negative with SK 68/69 (*env*). In the second study, among 12 positive samples, 11 were positive with SK 38/39 (*gag*), 11 with SK 68/69 (*env*), and 11 with *pol* 3/4. Overall, 9 were positive with all 3 primer sets, but 3 were positive with only 2. Constant care must be taken to avoid false-positive results due to carry-over of PCR product from a previously amplified sample to a tube to be tested. To minimize the risk, I conducted the experiments in 4 separate rooms (one for separation and lysis of PBMC, one for preparation of the PCR mixture, one for amplification, and one for analysis of the PCR product); no piece of equipment or supplies went back and forth between the different rooms.

The sensitivity of PCR for the diagnosis of perinatal HIV-1 infection within the first weeks of life has not been fully studied by other authors. Laure et al. (1988) found 6 PCR-positive newborns among 14 infants born to seropositive women, but 1 of them was positive with only 1 primer pair, and the results were not confirmed by a positive HIV culture or by clinical and virological follow-up. Rogers et al. (1989) has retrospectively tested 24 newborns, and among 12 who later developed criteria of HIV infection, 6 were detected within the first 2 weeks of life (5/6 developed AIDS between 5 and 16 months). Although the results of PCR within the neonatal period were encouraging, no data were shown regarding sequential samples within the first 2 months of life.

In our prospective study in which all the infants were tested twice within this period of time, we were able to detect 3 infected children soon after birth and 9 others between 4 and 9 weeks of life. The high rate of perinatal HIV transmission in this small sample may reflect local characteristics of the seropositive women followed in this hospital.

The lack of sensitivity of PCR within the first week might have been related to a possible lesser quality of the sample at birth. Therefore, I performed serial dilutions of the lysates and amplified them with the β-globin primer pair: No difference was observed between the samples obtained at birth and at 4–9 weeks of age, suggesting that there was no lack of sensitivity of the technique. These results suggest that the number of copies of proviral DNA avalaible for PCR in PBMCs increases during the first weeks of life, indicative of an active viral replication; they could also be consistent with a late transmission of the virus, at birth and not during gestation. Since no changes in the PCR/culture results were observed between 4–9 weeks and 6–9 months, I conclude that in most of the cases the diagnosis of HIV-1 infection in children born to seropositive mothers can be made by PCR before the end of the second month of life.

References

Amadori A, De Rossi A, Chieco-Blanchi L et al. (1990) Diagnosis of human immunodeficiency virus 1 infection in infants: in vitro production of virus-specific antibody in lymphocytes. Pediatr Infect Dis J 9:26–30

Blanche S, Rouzioux C, Guihard-Moscato ML et al. (1989) A prospective study of infants born to women seropositive for human immunodeficiency virus type 1. N Engl J Med 320:1643–1648

Borkowsky W, Krasinski K, Paul D et al. (1989) Human immunodeficiency virus type 1 antigenemia in children. J Pediatr 114:942–945

Chadwick EG, Yogev R, Kwok S et al. (1989) Enzymatic amplification of the human immunodeficiency virus in peripheral blood mononuclear cells from pediatric patients. J Infect Dis 160:954–959

Edwards JR, Ulrich PP, Weintrub PS et al. (1989) Polymerase chain reaction compared with concurrent viral cultures for rapid identification of human immunodeficiency virus infection among high-risk infants and children. J Pediatr 115:200–203

European Collaborative Study (1991) Children born to women with HIV-1 infection: natural history and risk of transmission. Lancet 1:253–260

Higuchi R (1989) Rapid, efficient DNA extraction for PCR from cells or blood. Amplifications 2:1–3

Hoff R, Berardi VP, Weiblen BJ et al. (1988) Seroprevalence of human immunodeficiency virus among childbearing women. N Engl J Med 318:325–330

Hutto C, Parks WP, Lai S et al. (1991) A hospital-based prospective study of perinatal infection with human immunodeficiency virus type 1. J Pediatr 118:347–353

Jackson JB, Coombs RW, Sannerud K et al. (1988) Rapid and sensitive viral culture method for human immunodeficiency virus type 1. J Clin Microbiol 26:1416–1418

Krivine A, Yakudima A, LeMay M et al. (1990) A comparative study of virus isolation, polymerase chain reaction, and antigen detection in children of mothers infected with human immunodeficiency virus. J Pediatr 116:372–376

Krivine A, Firtion G, Cao L et al. (1992) HIV replication during the first weeks of life. Lancet 339:1187–1189

Laure F, Courgnaud V, Rouzioux C et al. (1988) Detection of HIV-1 DNA in infants and children by means of the polymerase chain reaction. Lancet 2:538–540

McKinney RE, Maha M, Connor EM et al. (1991) A multicenter trial of oral zidovudine in children with advanced human immunodeficiency virus disease. N Engl J Med 324:1018–1025

Ou CY, Kwok S, Mitchell SW et al. (1988) DNA amplification for direct detection of HIV-1 in DNA of peripheral blood mononuclear cells. Science 239:295–297

Rogers MF, Ou CY, Rayfield M et al. (1989) Use of the polymerase chain reaction for early detection of the proviral sequences of human immunodeficiency virus in infants born to seropositive mothers. N Engl J Med 320:1649–1654

Ryder RW, Nsa W, Hassig SE et al. (1989) Perinatal transmission of the human immunodeficiency virus type 1 to infants of seropositive women in Zaire. N Engl J Med 320:1637–1642

Saiki RK, Gelfand DH, Stoffel S et al. (1988) Primer directed enzymatic amplification of DNA with a thermostable DNA polymerase. Science 239:287–291

Scott GB, Hutto C, Makuch RW et al. (1989) Survival in children with perinatally acquired immunodeficiency virus infection type 1. N Engl J Med 321:1791–1796

Chapter 3 Detection of Human T-cell Leukemia Virus Type 1 Provirus: Semiquantitative, Nested, Double Polymerase Chain Reaction

Chieko Matsumoto and Kusuya Nishioka

Summary

The polymerase chain reaction (PCR) is a powerful tool for diagnosing human T-cell leukemia virus type 1 (HTLV-I) infection, particularly for samples difficult to assess by immunological methods. We used the nested double PCR method to detect HTLV-I provirus in DNA obtained from an infected cell line and peripheral blood mononuclear cells from donated blood. This method yielded clear results even if only one template DNA was amplified. Furthermore, the amplified products showed one, two, or three bands after electrophoresis, depending on the amounts of template DNA present. The semiquantitative, nested, double PCR and its applications are described in this chapter.

Introduction

Following the initial description of adult T-cell leukemia (ATL) (Uchiyama et al. 1977) and the discovery of human T-cell leukemia virus type 1 (HTLV-I) as an etiological agent of ATL (Hinuma et al. 1981), virus infection was shown to be associated with tropical spastic paraparesis (TSP) or HTLV-I-associated myelopathy (HAM) (Gessain et al. 1985; Osame et al. 1986). Recently, the association of HTLV-I infection with other types of clinical manifestations such as alveolitis (Sugimoto et al. 1989), Sjögren's syndrome (Vernant et al. 1988), polymyositis (Wiley et al. 1989), and chronic inflammatory arthropathy (Nishioka et al. 1989) has been suggested. The main routes of transmission of HTLV-I described to date are blood transfusion (Okochi et al. 1984), breast milk, and sexual contact from male to female via semen (Komuro et al. 1983; Tajima et al. 1982). In all three cases, lymphocytes infected with HTLV-I have been thought to play the crucial role. Moreover, the provirus form of HTLV-I is present in these lymphocytes. Therefore, a highly sensitive method which can detect the provirus in infected cells and provide a means of quantitative evaluation is essential.

The Japanese Red Cross Central Blood Center, Hiroo 4-1-31, Shibuya-ku, Tokyo, 150 Japan

Differentiation of HTLV-I from HTLV-II infection cannot be carried out by serological methods because of cross-reactivity (Lee et al. 1991).

Because of the necessity of preventing HTLV-I infection by blood transfusion, screening for HTLV-I infection in blood donors began in HTLV-I endemic areas, such as Japan, and in countries where recent spreading has occurred, such as the USA. Evaluation of routine serological screening methods for HTLV-I antibody such as particle agglutination (PA), enzyme immunoassay, and indirect immunofluorescence (IF) are urgently required in these situations for confirming the presence of HTLV-I in peripheral leukocytes (Matsumoto et al. 1990).

Based on these considerations, we evaluated the polymerase chain reaction (PCR) method as a means of detecting proviral DNA. We introduced a modified nested double PCR method to detect HTLV-I provirus. In this method, two amplification stages are used under nonradioactive conditions. This is accomplished first by amplifying a DNA sample with *Thermus aquaticus* polymerase and one pair of primers. Next, a portion of the products is amplified again with another pair of primers called nested primers that are located inside the first pair. After the second amplification stage, almost all of the nonspecific background observed at the first stage disappears, and the desired product is confirmed using the nested primers. Finally, the amplified products, originating from a very small number of templates, are visualized by polyacrylamide gel electrophoresis (PAGE) stained with ethidium bromide. We also will discuss the possibility that this PCR method can be used to estimate the quantity of template DNA since one, two, or three bands appear on the PAGE gel depending on the amount of template DNA.

HTLV-I provirus integrated into the host T-cell genome is consistent with *gag, pol, env, pX*, and long terminal repeat (LTR). The *gag, pol*, and *env* genes encode capsid proteins, reverse transcriptase, and envelope proteins, respectively. The *pX* gene encodes *rex* and *tax* proteins which regulate *trans*-provirus replication (Fujisawa et al. 1985). A portion of the HTLV-I provirus *pX* region was thought to be the most appropriate target for amplification, because it overlaps the coding regions of p40^{tax} and p27^{rex}, which regulate replication of HTLV-I at the transcriptional and posttranscriptional levels, respectively (Chen et al. 1985; Inoue et al. 1987). Therefore, this region is considered indispensable for HTLV-I provirus multiplication in vivo. We believe that the region we amplified is suitable for detecting HTLV-I provirus in peripheral mononuclear cells (PBMC) of HTLV-I-infected individuals.

The following is a recommended protocol.

Method

Synthesis of Oligonucleotides. Oligonucleotides were synthesized on an Applied Biosystems 381A DNA synthesizer by the phosphoramidate method and purified with oligonucleotide purification cartridges (Applied Biosystems).

Source and Isolation of DNA. DNA from two human T-cell lines or PBMC was isolated by the phenol-chloroform method. One of the cell lines, CEM, was uninfected with HTLV-I, and the other, HUT 102, was HTLV-I positive (Poiesz et al. 1980). PBMC were obtained from voluntary blood donors in our blood center.

Nested Double PCR. The *pX* region of the HTLV-I genome was amplified as shown in Fig. 1 using DNA sequence data from Seiki et al. (1983). The first amplification stage was carried out with primers ① (5′-AGGGTTTGGACA-GAGTCTT-3′, nucleotide position 7312–7330) and ② (5′-AAGGACCTT-GAGGGTCTTAG-3′, nucleotide position 7548–7567). The second ampli-

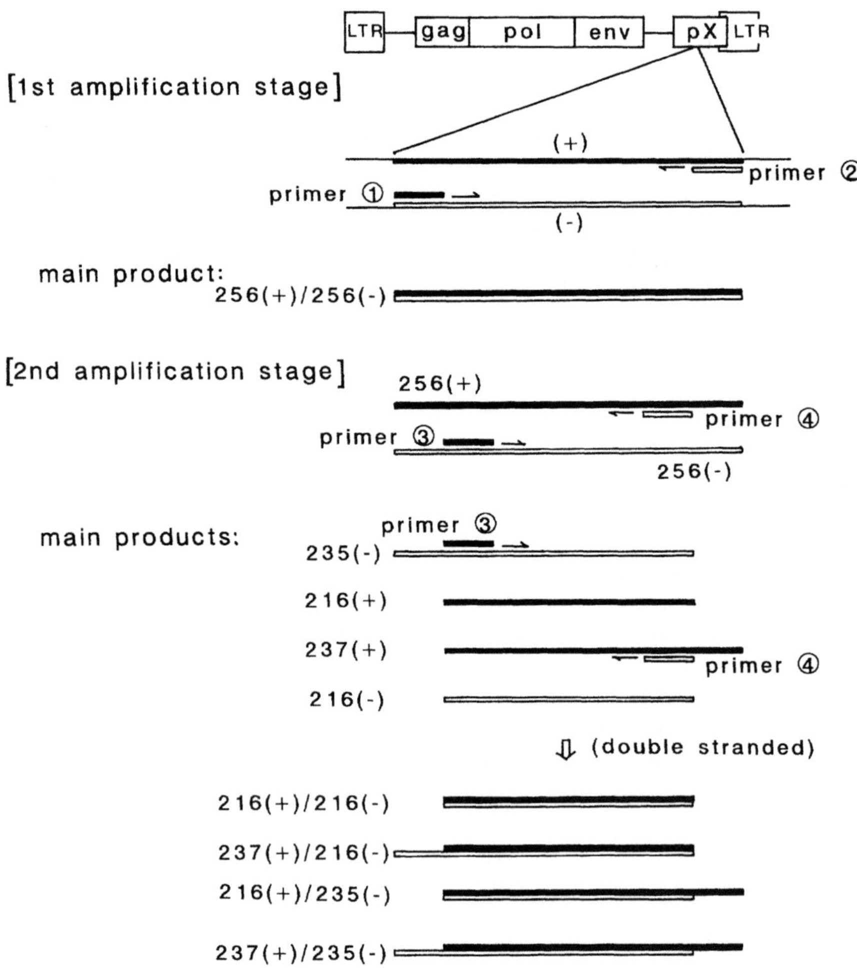

Fig. 1. Nested double polymerase chain reaction (PCR) of human T-cell leukemia virus (HTLV-I) in *pX* region

fication stage was carried out with nested primers ③ (5'-CTTTTCGGA-TACCCAGTCTAC-3', nucleotide position 7331–7351) and ④ (5'-GGTTCTCTGGGTGGGGAAGGAG-3', nucleotide position 7525–7546). Then, 10 µl of the first amplification stage products were used as template DNA. All reactions were performed in a volume of 100 µl containing 50 pmol of each primer, 50 mM TRIS hydrochloride (pH 8.8), 10 mM MgCl$_2$, 10 mM (NH$_4$)$_2$SO$_4$, template DNA, 2.5 U of *T. aquaticus* polymerase (Perkin-Elmer Cetus), and 1.5 mM each of dATP, dCTP, dGTP, and dTTP. Reactions were carried out for 30 cycles during each amplification stage at an annealing temperature of 60 °C for 1 min, a polymerization temperature of 72 °C for 2 min, and a heat denaturation temperature of 94 °C for 1 min using a Perkin-Elmer Cetus DNA thermal cycler.

Analysis of the PCR Products. A portion (4 µl) from each of the completed PCR reactions was mixed with 4 µl of loading buffer and subjected to PAGE on 5% polyacrylamide gels. PAGE was performed in TRIS-borate buffer (pH 8.0). After completion, the gel was stained with ethidium bromide and photographed under UV transillumination.

Comparison of Single PCR, Repeated PCR, and Nested Double PCR

The HTLV-I genome in HUT 102 cell DNA plus 1 µg of CEM cell DNA was detected by the single PCR, repeated PCR, and nested double PCR methods. With single PCR, cellular DNA was amplified for 30 cycles with primers ① and ②. With repeated PCR, portions of the single PCR products were amplified for a further 30 cycles with the same primers ① and ②. With the nested double PCR, portions of the products of the first single PCR were amplified for a further 30 cycles with the nested primers ③ and ④, which were located inside the first pair.

One HUT 102 cell contains about 10 copies of the *pX* region of HTLV-I (Watanabe et al. 1984), and 150 human diploid cells contain 1 ng of genomic DNA (assuming a haploid genome size of 3×10^9 bp). Calculation reveals that 1 ng of HUT 102 cell DNA contains 1500 molecules of the *pX* region DNA.

PAGE patterns of products of single PCR, expected to be 256 bp in length, are shown in Fig. 2a. A band corresponding to 256 bp was obtained from 130 pg of HUT 102 cell DNA calculated to contain 195 copies of the *pX* region. Nonspecific background can be seen in the upper region. The products of repeated PCR are shown in Fig. 2b. Only 256 bp bands with several nonspecific background bands were observed, with an endpoint of 15 pg of HUT 102 cell DNA, which was nine times more sensitive than that obtained with single PCR. Products of the nested double PCR are shown in Fig. 2c. A distinct 216 bp band, as expected by the positions of primers ③ and ④, was observed with 1.6 pg of HUT 102 cell DNA, which is calculated to contain

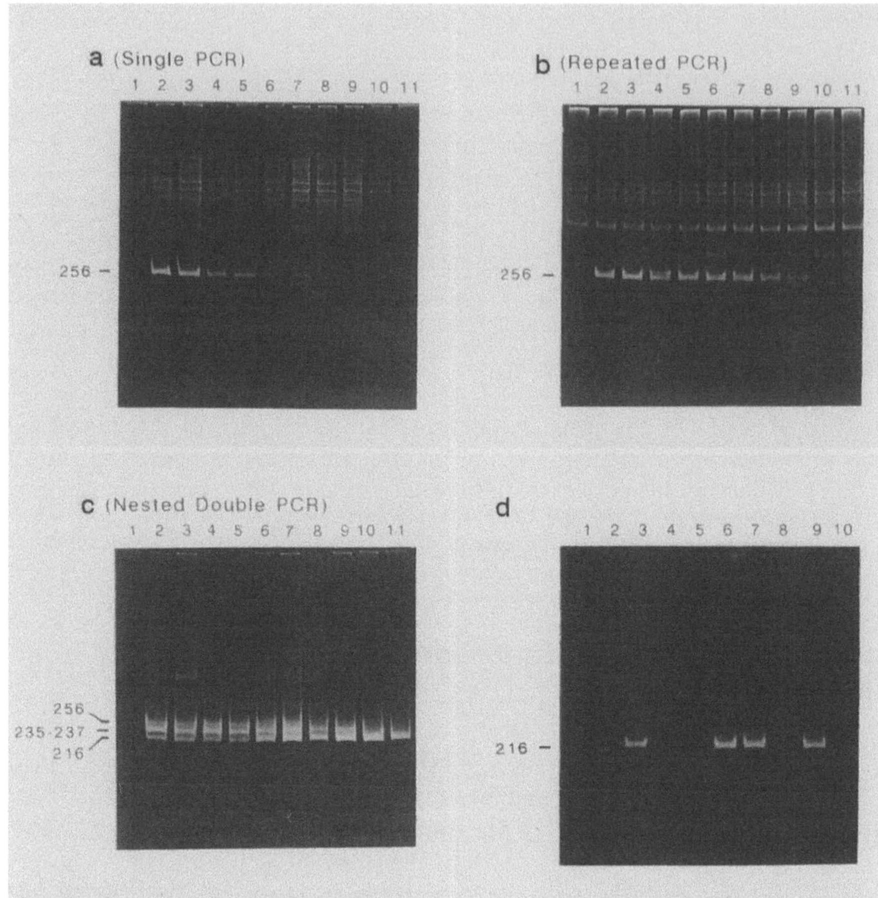

Fig. 2a–d. Polyacrylamide gel electrophoresis of polymerase chain reaction (PCR) products. **a** HUT 102 cell and CEM cell DNAs were amplified for 30 cycles with primers ① and ②. Portions of these were amplified for a further 30 cycles with the same primers ① and ② (**b**) or with the nested primers ③ and ④ (**c**). In each amplification in **a–c**, 1 µg of CEM cell DNA plus 0, 3.3×10^4 pg, 1.1×10^4 pg, 3.7×10^3 pg, 1.2×10^3 pg, 4×10^2 pg, 1.3×10^2 pg, 45 pg, 15 pg, 5 pg, or 1.6 pg of HUT 102 cell DNA (lanes 1–11, respectively) were used as template DNA. **d** HUT 102 cell and CEM cell DNA was amplified by the nested double PCR with primers ① and ② in the first amplification stage and with nested primers ③ and ④ in the second amplification stage. HUT 102 cell DNA (0.5 pg) and 1 µg CEM cell DNA were used as template DNA in each lane

about 2 copies of the *pX* region. Therefore, the sensitivity of the nested double PCR was more than 80 times higher than that of the single PCR.

In the double PCR method with nested primers, only a real target DNA region was reamplified from the products of the first amplification stage. Therefore, it is possible to avoid reamplifing extra products from the first amplification stage. The nested primers would be consumed mainly to amplify a real target DNA, and much larger amounts of desired products could be produced than in the repeated PCR. Thus, the amplified products are made detectable by PAGE without a hybridization step with a probe.

Detection of One Template DNA

We attempted to detect one molecule of *pX* region DNA by the nested double PCR. The initial DNA sample was 0.5 pg of HUT 102 cell DNA plus 1.0 µg of CEM cell DNA. It was calculated that 0.5 pg of HUT 102 cell DNA contains 0.75 molecules of *pX* region DNA, in other words, one or no molecule of the *pX* region DNA was present. According to Poisson's distribution, of the 10 DNA samples containing 0.5 pg of HUT 102 cell DNA, 52% of those tested should contain one or more *pX* regions.

The amplified products gave four positive results out of ten PCR tests and appeared as distinct single bands of 216 bp without nonspecific background as shown in Fig. 2 d. This result indicates that one template DNA can be detected by the nested double PCR without ambiguity.

Quantity of Template DNA and Number of Bands
Constructed by Nested Double PCR

In Fig. 2 c, one, two, and three bands can be seen. With less than 5 pg of HUT 102 cell DNA, one band (216 bp in length) appears, with 15–130 pg of HUT 102 cell DNA, two bands (216 and 235–237 bp in length) appear, and with more than 400 pg of cell DNA, three bands (216, 235–237, and 256 bp in length) appear.

The question arose as to whether primers ① and ② from the first stage were carried over, producing 235–237 bp or 256 bp bands after the nested double PCR. If so, the amounts of primers ① and ② carried over to the second stage should decrease as the initial template DNA number increases. Opposite results were obtained. As the amount of initial template DNA increased, 235–237 bp and 256 bp bands were observed.

Four different, single-stranded DNAs schematically shown in Fig. 1 would be produced during the second amplification stage. In the second 30-cycle amplification stage, products of the 237-base positive strand and 235-base negative strand might have been created in amounts 30-fold greater than the template 256-base products which had been constructed during the first stage.

Where the number of the template DNA, 256-base positive or negative strand, is a and the amplification cycle number is n, $a \times (2^n - n)$ 216-base DNAs and $a \times n$ 235- or 237-base DNA should be produced. The 216-base positive strand coupled with the 216-base negative strand product, shown as $216(+)/216(-)$, should be constructed and completely double-stranded. At the same time, $237(+)/216(-)$, $216(+)/235(-)$, and $237(+)/235(-)$ products would be constructed depending on the quantities of each four single-stranded DNA, and portions of these products should remain partially as single strands. If number a is large, $a \times n$ would be large. Value a depends on the quantities of initial template DNA, that is, HTLV-I provirus. The $237(+)/216(-)$ and $216(+)/235(-)$ products should appear as the amount of 256 bp DNA constructed in the first amplification stage increases. Finally, $237(+)/235(-)$ DNA would be observed. In practice, the efficiency of amplification is not 100%, and the amount of amplified products would be within the limit of the amount of added primers. In the second amplification stage, template DNA, 256-base positive or negative strand, should be annealed with the 216-base negative or positive strand DNA, respectively, and should extend to produce $256(+)/235(-)$ or $237(+)/256(-)$.

With a limited quantity of template provirus DNA, the amount of $237(+)/216(-)$ and $216(+)/235(-)$ was not measurable, and only one band, 216 bp in length, was observed. With moderate amounts of template provirus DNA, two bands, 216 and 235–237 bp in length, were noted. The second band of 235–237 bp consisted of $237(+)/216(-)$ and $216(+)/235(-)$. It was difficult to distinguish the two products, and they appeared as a single band on PAGE. With excess template provirus DNA, three bands of 216, 235–237, and 256 bp in length were found. Annealed products, $237(+)/235(-)$, and possibly small amounts of $237(+)/256(-)$ and $256(+)/235(-)$ likely make up the third band of 256 bp.

To prove that the assumption described above is true, the amplified products of the nested double PCR shown in Fig. 2c were treated with nuclease S1 which degrades single-stranded DNA. The amplified products, appearing in lane 7 as three bands and in lane 8 as two bands, became only one band, 216 bp in length, after treatment in which the single-stranded region was removed from the annealed products.

Quantity of Provirus in PBMC and Antibody Titer of Blood Donors

We carried out the nested double PCR on 1 µg of DNA samples obtained from PBMC of blood donors. A total of 256 blood donors were tested by the nested double PCR for PBMC DNA and by PA and IF for serum antibody against HTLV-I. A PA assay was used for mass screening for HTLV-I-specific antibody in sera (Ikeda et al. 1984; Kamihira et al. 1987). A final dilution of 1:16 or more capable of causing agglutination of antigen-coated particles was considered positive. IF is the method used to detect antibody bound to an HTLV-I-infected cell line using an indirect immunofluorescence technique (Hinuma et al. 1981; Chen et al. 1985). It has been widely employed in Japan to confirm HTLV-I infection. None of the 101 samples taken from individuals with PA titers lower than 8 and IF negative were PCR positive. Some 57 DNA samples were designated HTLV-I provirus-positive, and as expected, their amplified products by nested double PCR products showed one, two, or three bands on PAGE.

According to the nested double PCR results of HUT 102 cell DNA showed in Fig. 2c using a limited amount (1–7 genomes) of template provirus DNA, only one band, 216 bp in length, was observed. With moderate amounts (22–70 genomes) of template provirus, two bands, 216 and 235–237 bp in length, were observed. With excess amounts (200 provirus genomes or more), three bands of 216, 235–237, and 256 bp were detected.

The number of bands on a PAGE gel of the amplified product, antibody titer by PA, and the expected amount of provirus from the DNA of the 256 samples tested are summarized in Table 1. For each PCR, 1 µg of DNA from

Table 1. Number of bands on PAGE gels of nested double PCR products and antibody titer of HTLV-I in adult carriers

No. of bands	No. of samples with PA titer[a]			Estimated provirus number in 1.5×10^5 PBMC[b]
	<8	16–256	>512	
0	101 (100[c])	96 (84.2)	2 (4.9)	
1	0	2 (1.8)	0	2–7
2	0	5 (4.4)	2 (4.9)	22–70
3	0	11 (9.6)	37 (90.2)	>200
Total	101 (100)	114 (100)	41 (100)	

[a] The particle agglutination titer represents the final serum dilution
[b] Number of *pX* region DNA was estimated from data in Fig. 2c, and provirus number was assumed
[c] Percentage of sample number in each PA titer group

150000 cells was used. It is thought that the expected number of *pX* regions indicates the amount of provirus in 150000 PBMC of HTLV-I carriers. The levels of antibody titer 16–256 are classified as moderately positive and over 512, as strongly positive. As discussed above, among 101 samples of the negative group, no bands were observed by the nested double PCR. In the moderately positive group, 96 (84.2%) were negative by PCR, while in a few cases, one, two, and three bands were observed. In 37 (90.2%) of the strongly positive group, three bands were observed. Two were negative by PCR, and in both of them, two bands were observed. Therefore, there was a tendency for a correlation between the amount of provirus in PBMC of blood donors and antibody titer determined by PA.

Clinical Applications

Diagnostic Confirmation Testing

To screen or diagnose HTLV-I infection, antibody testing is generally used. However, there is sometimes a discrepancy in results obtained by different methods. In such cases, information obtained by PCR testing is very useful.

To prevent infection by HTLV-I, detection by PA assay of HTLV-I-specific antibody has been used since 1986 to screen donated blood in all Japanese Red Cross blood centers. However, some samples were judged to be antibody negative by IF in spite of a positive PA assay (Matsumoto et al. 1990; Yoshida et al. 1986), particularly in low-titer sera. It was therefore necessary to find some means of determining the presence of HTLV-I in such PBMC to prevent post transfusion HTLV-I infection. Accordingly, we attempted detection of HTLV-I provirus DNA by nested double PCR directly.

We carried out a comparative study by performing PCR on PBMC and PA and IF on sera of 256 blood donors. Some 57 of the 155 samples were antibody positive by IF, and provirus was detected in all 57 samples without exception.

The presence of provirus in PBMC was not found in 96 (84.2%) of the moderately PA-positive group and in 2 (4.9%) of the strongly PA-positive group. All of these samples were also negative by IF. The results of IF for HTLV-I antibody coincided completely with the results of the nested double PCR for detection of provirus DNA in PBMC. Therefore, the IF test correlates well with the presence of provirus in PBMC by PCR, and it is considered as the most appropriate confirmatory test routinely available for screening out donated blood infected with HTLV-I. As for PA, a correlation of the PA titer and the amount of proviral DNA in PBMC was observed. However, in most of the low titer group and even in a small number of the high titer group, the presence of proviral DNA was not detected. In these cases, three possibilities can be considered. First, provirus DNA might have been present in cells other than the 150 000 PBMC used for PCR. Secondly, antibody reactivity with antigenic epitopes capable of causing neutralization or protection of HTLV-I infection might have been detected by PA but not by IF. Thirdly, a nonspecific reaction may have occurred. These problems need to be clarified.

Distinction Between HTLV-I and -II

Another major type of HTLV, HTLV-II, has been isolated (Kalyanaraman et al. 1982; Rosenblatt et al. 1986). HTLV-II infection has been found to be more prevalent than previously thought (Lairmore et al. 1990; Lee et al. 1989), and a high correlation of coinfection of HTLV-I and HTLV-II was demonstrated in HAM (Kira et al. 1991). It is known that HTLV-I and -II are not distinguishable by the usual serological techniques. HTLV-II-positive sera from 19 American donors were tested by our PA and IF methods for HTLV-I-specific antibody and were judged positive. By nested double PCR, as expected for its high specificity, we could distinguish HTLV-I and -II completely. To learn more about HTLV infection, particularly HTLV-II, infected populations, and association with various diseases, it is necessary to distinguish HTLV-I from HTLV-II, and PCR is the most useful method for accomplishing this at present.

Study of HTLV-I-Associated Disease

The presence of HTLV-I has been associated with myelopathy or tropical spastic paraparesis. Other diseases have also been suspected of being associated with HTLV-I. The presence of chronic inflammatory arthropathy associated with HTLV-I has been reported (Iwakura et al. 1991). HTLV-I provirus was detected in multiple sclerosis patients by PCR (Reddy et al. 1989). In HTLV-I *tax* transgenic mice, development of exocrinopathy similar to Sjögren's syndrome was observed (Green et al. 1989). There is also a report on a case of HTLV-I-associated lymphoproliferative disease which was misdiagnosed as Hodgkin's disease (Duggan et al. 1988). Some investigation would be helpful to determine whether the HTLV-I genome has been integrated into nonlymphocytic cells (Bhagavati et al. 1988; Kitajima et al. 1991). If HTLV-I provirus were present, it would be very important to know both the nature and quantity of provirus present. Quantitative evaluation of the HTLV-I genome may be of value in treatment efficacy. The PCR technique provides a useful means of detecting HTLV-I. Furthermore, use of the nested double PCR method will greatly aid the estimation of amount of provirus present.

By carrying out the nested double PCR method on hepatitis B virus and hepatitis C virus, we have observed one, two, and three bands on PAGE gel depending on the number of virus genomes. Application of the nested double PCR method for the detection and quantification of the virus would provide more information on pathogenesis and may be a useful tool for the prevention, diagnosis, and treatment of various viral infections.

Acknowledgements. We thank S. Mitsunaga for invaluable discussions and E. Tokunaga, Y. Mitomi, and J. Watanabe for support in this study.

References

Bhagavati S, Ehrlich G, Kula RW, Kwok S, Sninsky J, Udani V, Poiesz BJ (1988) Detection of human T-cell lymphoma/leukemia virus type I DNA and antigen in spinal fluid and blood of patients with chronic progressive myelopathy. N Engl J Med 318:1141–1147

Chen ISY, Slamon DJ, Rosenblatt JD, Shah NP, Quan SG, Wachsman W (1985) The *x* gene is essential for HTLV replication. Science 229:54–58

Duggan DB, Ehrlich GD, Davey FP, Kwok S, Sninsky J, Goldberg J, Baltrucki L, Poiesz BJ (1988) HTLV-I-induced lymphoma mimicking Hodgkin's disease. Diagnosis by polymerase chain reaction amplification of specific HTLV-I sequences in tumor DNA. Blood 71:1027–1032

Fujisawa J, Seiki M, Kiyokawa T, Yoshida M (1985) Functional activation of the long terminal repeat of human T-cell leukemia virus type I by a *trans*-acting factor. Proc Natl Acad Sci USA 82:2277–2281

Gessain A, Barin F, Vernant JC, Gout O, Maurs L, Calender AG (1985) Antibodies to human T-lymphotropic virus type-I in patients with tropical spastic paraparesis. Lancet 2:407–409

Green JE, Hinrichs SH, Vogel J, Jay G (1989) Exocrinopathy resembling Sjögren's syndrome in HTLV-1 *tax* transgenic mice. Nature 341:72–74

Hinuma Y, Nagata K, Hanaoka M, Nakai M, Matsumoto T, Kinoshita K, Shirakawa S, Miyoshi I (1981) Adult T-cell leukemia: antigen in an ATL cell line and detection of antibodies to the antigen in human sera. Proc Natl Acad Sci USA 78:6476–6480

Ikeda M, Fujino R, Matsui T, Yoshida T, Komoda H, Imai J (1984) A new agglutination test for serum antibodies to adult T-cell leukemia virus. Gann 75:845–848

Inoue J, Yoshida M, Seiki M (1987) Transcriptional (p40^x) and post-transcriptional (p27^{x-III}) regulators are required for the expression and replication of human T-cell leukemia virus type I genes. Proc Natl Acad Sci USA 84:3653–3657

Iwakura Y, Tosu M, Yoshida E, Takiguchi M, Sato K, Kitajima I, Nishioka K, Yamamoto K, Takeda T, Hatanaka M, Yamamoto H, Sekiguchi T (1991) Induction of inflammatory arthropathy resembling rheumatoid arthritis in mice transgenic for HTLV-I. Science 253:1026–1028

Kalyanaraman VS, Sarngadharan MG, Robert-Guroff M, Miyoshi I, Blayney D, Golde D, Gallo RC (1982) A new subtype of human T-cell leukemia virus (HTLV-II) associated with a T-cell variant hairy cell leukemia. Science 218:571–573

Kamihira S, Nakashima S, Oyakawa Y, Moriuti Y, Ichimaru M, Okuda H, Kanamura M, Oota T (1987) Transmission of human T cell lymphotropic virus type I by blood transfusion before and after mass screening of sera from seropositive donors. Vox Sang 52:43–44

Kira J, Koyanagi Y, Hamakado T, Itoyama Y, Yamamoto N, Goto I (1991) HTLV-II in patients with HTLV-I-associated myelopathy. Lancet 338:64–65

Kitajima I, Yamamoto K, Sato K, Nakajima Y, Nakajima T, Maruyama I, Osame M, Nishioka K (1991) Detection of HTLV-I proviral DNA and its gene expression in synovial cells in chronic inflammatory arthropathy. J Clin Invest 88:1315–1322

Komuro A, Hayami M, Fujii H, Miyahara S, Hirayama M (1983) Vertical transmission of adult T-cell leukaemia virus. Lancet i:240

Lairmore MD, Jacobson S, Gracia F, De BK, Castillo L, Larreategui M, Roberts BD, Levin PH, Blattner WA, Kaplan JE (1990) Isolation of human T-cell lymphotropic virus type 2 from Guaymi Indians in Panama. Proc Natl Acad Sci USA 87:8840–8844

Lee H, Swanson P, Shorty VS, Zack JA, Rosenblatt JD, Chen ISY (1989) High rate of HTLV-II infection in seropositive IV drug abusers in New Orleans. Science 244:471–475

Lee HH, Swanson P, Rosenblatt JD, Chen ISY, Sherwood WC, Smith DE, Tegtmeier GE, Fernando LP, Fang CT, Osame M, Kleinman SH (1991) Relative prevalence and risk factors of HTLV-I and HTLV-II infection in US blood donors. Lancet 337:1435–1439

Matsumoto C, Mitsunaga S, Oguchi T, Mitomi Y, Shimada T, Ichikawa A, Watanabe J, Nishioka K (1990) Detection of human T-cell leukemia virus type I (HTLV-I) provirus in an infected cell line and in peripheral mononuclear cells of blood donors by the nested double polymerase chain reaction method: comparison with HTLV-I antibody tests. J Virol 64:5290–5294

Miyoshi I, Kubonishi I, Yoshimoto S, Akagi T, Ohtsuki Y, Shiraishi Y, Nagata K, Hinuma Y (1981) Type C virus particles in a cord T-cell line derived by co-cultivating normal human cord leukocytes and human leukaemic T cells. Nature 294:770–771

Nishioka K, Maruyama I, Sato K, Kitajima I, Nakajima Y, Osame M (1989) Chronic inflammatory arthropathy associated with HTLV-I. Lancet 1:441

Okochi K, Sato H, Hinuma Y (1984) A retrospective study on transmission of adult T cell leukemia virus by blood transfusion: seroconversion in recipients. Vox Sang 46:245–253

Osame M, Usuku K, Izumo S, Ijichi N, Amitani H, Igata A, Matsumoto M, Tara M (1986) HTLV-I associated myelopathy, a new clinical entity. Lancet 1:1031–1032

Poiesz BJ, Ruscetti FW, Gazdar AF, Bunn PA, Minna JD, Gallo RC (1980) Detection and isolation of type C retrovirus particles from fresh and cultured lymphocytes of a patient with cutaneous T-cell lymphoma. Proc Natl Acad Sci USA 77:7415–7419

Reddy EP, Sandberg-Wollheim M, Mettus RV, Ray PE, DeFreitas E, Koprowski H (1989) Amplification and molecular cloning of HTLV-I sequences from DNA of multiple sclerosis patients. Science 244:529–533

Rosenblatt JD, Golde DW, Wachsman W, Giorgi JV, Jacobs A, Schmidt GM, Quan S, Gasson JC, Chen ISY (1986) A second isolate of HTLV-II associated with atypical hairy-cell leukemia. N Engl J Med 315:372–377

Seiki M, Hattori S, Hirayama Y, Yoshida M (1983) Human adult T-cell leukemia virus: complete nucleotide sequence of the provirus genome integrated in leukemia cell DNA. Proc Natl Acad Sci USA 80:3618–3622

Sugimoto M, Nakashima H, Matsumoto M, Uyama E, Ando M, Araki S (1989) Pulmonary involvement in patients with HTLV-I-associated myelopathy: increased soluble IL-2 receptors in bronchoalveolar lavage fluid. Am Rev Respir Dis 139:1329–1335

Tajima K, Tominaga S, Suchi T, Kawagoe T, Komoda H, Hinuma Y, Oda T, Fujita K (1982) Epidemiological analysis of the distribution of antibody to adult T-cell leukemia-virus-associated antigen: possible horizontal transmission of adult T-cell leukemia virus. Gann 73:893–901

Uchiyama T, Yodoi J, Sagawa K, Takatsuki K, Uchino H (1977) Adult T cell leukemia: clinical and hematological features of 16 cases. Blood 50:481–492

Vernant JC, Buisson G, Magdeleine J, DeThore J, Jouannelle A, Vernant CN, Monplaisir N (1988) T-lymphocyte alveolitis tropical spastic paresis and Sjögren's syndrome. Lancet 1:177

Watanabe T, Seiki M, Yoshida M (1984) HTLV type I (US isolate) and ATLV (Japanese isolate) are the same species of human retrovirus. Virology 133:238–241

Wiley CA, Nerenberg M, Cross D, Soto-Aguilar MC (1989) HTLV-I polymyositis in patient also infected with the human immunodeficiency virus. N Engl J Med 320:992–995

Yoshida T, Kobayashi S, Yamamoto N (1986) Discrepancy between gelatin particle agglutination assay and indirect immunofluorescence assay for antibodies to adult T cell leukemia associated antigen (ATLA). J Clin Exp Med (Igaku No Ayumi) 136:387–388

Chapter 4 Detection of Human T-cell Leukemia Viruses

Brian Hjelle

Summary

Human T-cell leukemia/lymphoma virus type I (HTLV-I) is the cause of adult T-cell leukemia/lymphoma (ATL) and a chronic neurologic syndrome. It occurs worldwide but with clusters of high seroprevalence in Japan, Africa, and the Caribbean. A related virus known as HTLV-II occurs in endemic clusters in New World Indian populations and is very common among intravenous drug users in some regions. HTLV-II accounts for about half of all HTLV seropositivity in the USA. HTLV-II is not known to be a pathogen.

Serologic reagents for the detection of these viruses, made from lysates of HTLV-I-infected cells, are believed to be very sensitive to HTLV-I infection. Their sensitivity to HTLV-II is largely unknown. Highly sensitive polymerase chain reaction (PCR) detection systems are available and are particularly useful for the evaluation of serologic tests, for distinguishing HTLV-I from HTLV-II, for determining whether or not infection is present in patients with incomplete or absent seroreactivity or in the obligately seropositive children of seropositive mothers.

Introduction

HTLV-I

Since the discovery of the first oncogenic retrovirus in nature by Peyton Rous (1910), numerous examples of such agents have been uncovered in birds and animals, including primates. The discovery of human T-cell leukemia/ lymphoma virus type I (HTLV-I) was the result of many years of searching for evidence of RNA tumor viruses in man. HTLV-I was discovered by Poiesz and colleagues (1980), with the discovery of virus in the cells of a patient with cutaneous T-cell leukemia.

Department of Pathology, University of New Mexico School of Medicine, 915 Camino de Salud NE, Albuquerque, NM 87131, and United Blood Services, 1515 University Blvd., Albuquerque, NM 87125, USA

HTLV-I occurs in highly focal geographic clusters throughout the world. Foci of seroprevalence as high as 15%–30% have been described in Southwestern Japan (Blattner et al. 1983; Hinuma 1985). Other high-prevalence regions include the Caribbean and Africa. The virus is transmitted by cell-to-cell contact. Thus, sexual transmission occurs but strongly in favor of transmission from men to women. Perinatal transmission occurs predominantly via breast feeding (Hino et al. 1990). Transfusion of cellular blood products and needle-sharing have been important modes of spread. Blood banks have been testing donors for antibodies to HTLV-I since the mid-1980s in Japan and since late 1988 in the USA. Both "screening" and "confirmatory" serologic tests are in routine use by blood centers and hospitals (reviewed in Hjelle 1991).

HTLV-II

In vitro culture of cells from the spleen of a patient with hairy cell leukemia gave rise to a cell line, MO-T, that was found to bear T-cell markers. Kalyanramen later demonstrated that this cell line produced a virus, HTLV-II, with antigenic properties similar to HTLV-I but clearly distinguishable (Kalyanramen et al. 1982; Rosenblatt et al. 1986).

The transmission routes for HTLV-II are less well described but include blood transfusion and sexual transmission (Hjelle et al. 1990a; Donegan et al. 1991; B. Hjelle 1990, unpublished). It is endemic in geographically and linguistically diverse New World Indian tribes (Hjelle et al. 1991; Heneine et al. 1991). A high fraction of "HTLV-I" seropositive intravenous drug users and blood donors in the USA are actually infected with HTLV-II (Lee et al. 1989; Hjelle et al. 1990a).

Disease Associations

The association between HTLV-I infection and adult T-cell leukemia (ATL) is convincing. Monoclonally integrated provirus can be detected in tumor DNA, and virus has been isolated from ATL cells (Seiki et al. 1984; Miyoshi et al. 1981). The epidemiologic linkage between HTLV-I prevalence and ATL has also proved strong. ATL is a rapidly fatal malignancy in most cases, although less serious variant forms have been described.

HTLV-I is also linked to a chronic and progressive neurologic disease known as tropical spastic paraparesis/HTLV-associated myelopathy (TSP/ HAM) (Gessain et al. 1985; Osame et al. 1986). Other diseases (polymyositis, infectious dermatitis, and lymphadenopathy) may occur with increased frequency among HTLV-I-seropositive patients (Morgan et al. 1989; LeGrenade et al. 1990).

Although HTLV-II has been isolated twice from patients with hairy cell leukemia, in neither case were the hairy cells themselves demonstrated to be infected in vivo (Kalyanaramen et al. 1982; Rosenblatt et al. 1986). Follow-up studies have shown that most patients with hairy cell leukemia are not infected with HTLV-II and that a population with endemic HTLV-II infection does not experience an excess incidence of hairy cell leukemia (Rosenblatt et al. 1987; Hjelle et al. 1991). HTLV-II is one of many candidate etiologic agents for chronic fatigue syndrome (DeFreitas et al. 1991). However, in my experience in interviewing >50 HTLV-II-positive individuals, none have professed to have excessive fatigue, fevers, or weight loss.

Biologic Characteristics

The HTLV genome is composed of single-stranded (+) RNA. Upon entry into the cell by an unknown receptor, the viral RNA is converted to DNA through the action of a virion-associated reverse transcriptase. Viral reverse transcriptase removes hybridized viral RNA and replaces it with a plus-stranded DNA copy. The resulting double-stranded DNA then integrates into the host genome, producing the provirus. The provirus serves as the template for production of the viral mRNA and genomic RNA. Genomic RNA is packaged by virion proteins encoded by *gag* and *env* for further cycles of infection. Reverse transcriptase for the virion is synthesized by the *pol* gene (reviewed in Hjelle 1991).

The organization of the HTLV-I and -II genome is diagrammed in Fig. 1. Nucleotide sequence variability among different isolates of HTLV-I is indicated as a function of map location along the proviral genome.

Unlike many other retroviruses, HTLV replication in vitro is relatively sluggish. Reverse transcriptase activity is minimal in infected cultures. Primary isolates are propagated by coculture of peripheral blood mononuclear cells (PBMC) from infected patients with normal donor PBMC. Cell-free virus is produced in a majority of primary cultures (B. Hjelle et al. 1991, unpublished). Syncytia production is dramatic, and resulting syncytia stain strongly for HTLV antigens.

In vivo HTLV infection lacks a viremic phase. Exposure is believed to result in lifelong seropositivity and latent infection. Typically, very small amounts of virus are produced; a small fraction of cells contain proviral DNA (estimates of <1 in 500 to >1 in 750 have been published) (Rosenblatt et al. 1990; Matsumoto et al. 1990).

Serologic Diagnosis and Isolation

Numerous methods for serologic diagnosis have been described, and many early reports of HTLV-I seroprevalence undoubtedly erroneously overestimated the rate of infection with this virus. This occurred because a reliable "confirmatory" test for infection was not available until the late 1980s. In the USA enzyme immunoassay (EIA) with wells coated with lysate from the HTLV-I-infected cell line HUT-102 is used as a screening test. The Western blot,

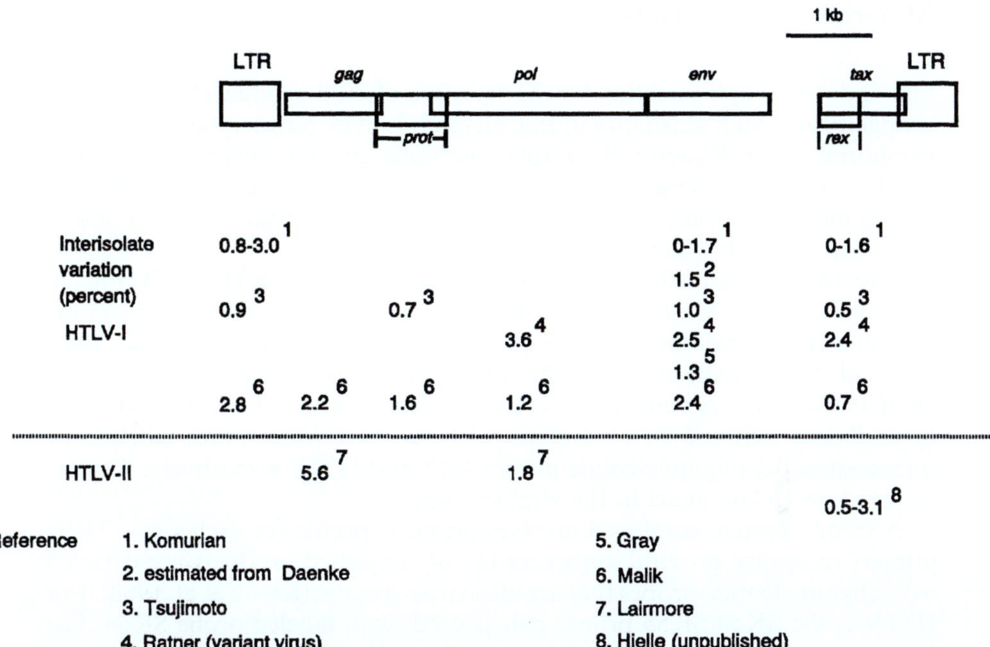

Fig. 1. Genetic organization of HTLV-I provirus. The HTLV-II provirus is similar, but the genes differ slightly in size. Nucleotide sequence variability from prototype clones is shown below, as determined in the studies indicated

augmented by supplementation with envelope protein gp21^e (produced via recombinant DNA technology), is the method of choice for confirmatory serologic testing (Lillehoj et al. 1990). Western blots are positive if antibodies are detected against p24^{gag} and gp21^e; reactivity to any antigen(s) other than both of these is an indeterminate pattern. In some cases, a radioimmunoprecipitation assay is necessary to diagnose infection in patients with indeterminate Western blots. Standard serologic techniques have not generally been capable of distinguishing HTLV-I from HTLV-II. However, it is now known that infection with HTLV-I usually produces a very different pattern of Western blot reactivity than does HTLV-II when using the HUT-102 Western blot (Lillehoj et al. 1990; Wiktor et al. 1990). Methods to distinguish HTLV-I and -II using short synthetic peptide epitopes specific for each virus are under development.

In the past, HTLV culture had been difficult because positive cultures produced very little reverse transcriptase; there were thus few convenient and reliable means to monitor in vitro replication of virus. The recent commercial availability of antigen-capture EIA kits has made culture a much more practical option.

Materials and Methods

I favor dot-blot hybridization for the detection of PCR products because of the simplicity and high sensitivity of this method. Unlike polyacrylamide gel electrophoresis (PAGE)/autoradiographic methods, one can increase PCR cycle number without generating a spurious signal or losing a specific product among numerous nonspecific products. Dot-blot hybridization is easily adapted to nonradioactive detection methods.

I use two different strategies to distinguish HTLV-I from HTLV-II (Fig. 2). The method used for *tax/rex* amplification involves the use of primers designed to recognize either HTLV-I or -II with equal facility (Kwok et al. 1988; Lee et al. 1989; Hjelle et al. 1990b). These primers produced PCR product in 44 of 46 donors or patients with HTLV-II infection and in 6 of 6 with HTLV-I (overall sensitivity of 96%). I distinguish the two viruses with labeled, sequence-specific, oligonucleotide probes HTI and HTII, respectively. The *tax* primers are 159 bp apart in the viral genome.

A second system employed involves primers specific for each virus. These primers recognize proviral sequences in *pol*; amplified products are detected with oligonucleotide probes that are also virus-specific (Kwok et al. 1988). For HTLV-I, the SK54/SK55 primer pair is used, with labeled probe SK56. For HTLV-II, the product produced by the SK58/SK59 primer pair is hybridized to probe SK60. The combined sensitivity of these primers is about 75%–80%

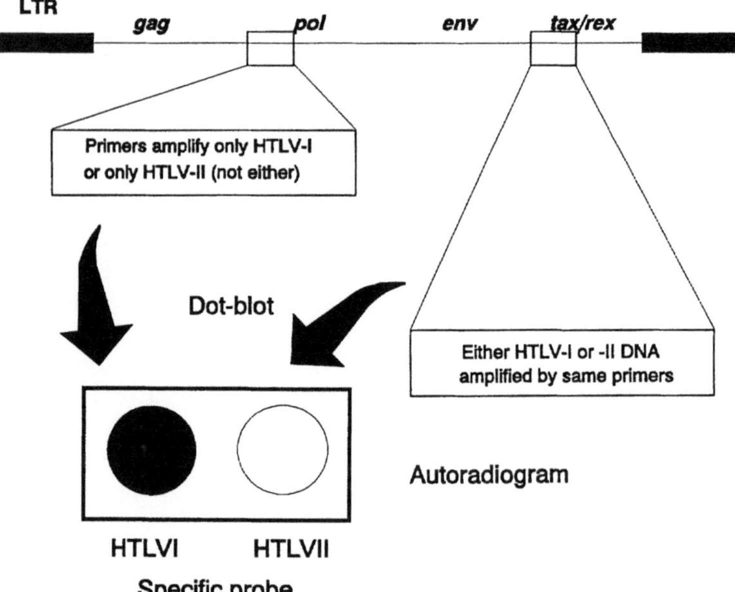

Fig. 2. Two methods used by this lab in the amplification and distinction of HTLV-I and HTLV-II infection. Both methods use dot-blots with virus-specific probes but differ in whether virus-specific primer pairs are used in amplification. *LTR*, long terminal repeat

with PBMC DNA target without "boosting" (see below). The sensitivity rises to 90–95% when boosting is performed. These primers reside 119 bp and 103 bp apart, respectively, in the HTLV-I and -II genomes.

In performing PCR for single-copy genes from 1 μg of genomic DNA of higher mammals, primers need only anneal to one or a few of the 3×10^5 copies of target DNA for amplification to occur. By contrast, there may be as few as 10^2–10^3 target DNA sequences in a preparation of PBMC DNA from an HTLV-infected person. It is common for primers to "miss" the target in the first few cycles. Thus, one may increase PCR sensitivity by increasing cycle number to the highest practical level. This is particularly true when using a hybridization-based detection method, because nonspecific products do not interfere with the interpretation of results. I routinely go to 45–50 cycles of PCR for diagnostic usage, since there is no disadvantage to doing so.

PCR Cycling Parameters

For the *tax* primers, I use a modification of the method recommended by Lee et al. (1989). A 25 μl reaction mixture is prepared containing 2 μg of target DNA (e.g., from PBMC); 50 pmol of each primer; 50 m*M* KCl; 25 m*M* TRIS-HCl, pH 7.5; 0.25 μ*M* each of the 4 dNTPs (Pharmacia); 3 m*M* MgCl$_2$; and 2.5 units of Taq polymerase. After covering the mixture with about 60 μl of mineral oil, 45 cycles of PCR are conducted in an Ericomp cycler (91 °C for 1 min followed by 65 °C for 2 min). A 6-min extension at 70° is the last step.

For the *pol* primers, I have found a slightly greater sensitivity when the SK54/SK55 (HTLV-I) primer pair is not used in the same tube as the SK58/SK59 (HTLV-II) primer pair. Reactions are carried out in 100 μl total volume containing 100 pmol of each primer; 2 μg of target DNA; 10 m*M* TRIS HCl, pH 8.3; 0.2 m*M* each of the 4 dNTPs; 4 m*M* MgCl$_2$; and 2.5 units of Taq polymerase. Then, 100 μl of mineral oil is used, and 50 cycles of PCR (95°, 55°, and 72 °C, each for 1 min), followed by a 6-min extension at 70 °C. "Booster" PCR improves sensitivity. In this modification, 15 cycles of amplifications are carried out using only 1 pmol of each primer per tube. Another 100 pmol of each primer is added to the tube before proceeding to the next 35 cycles (Ruano et al. 1989).

Dot-Blotting

Nylon membranes (Zetaprobe, Bio-Rad) are marked with pencil into a grid of 1 by 1 cm squares corresponding to the number of samples to be spotted. The membrane is briefly wetted in water, then immersed for 15 min in 2XSSC (20XSSC = 3 *M* NaCl and 0.3 *M* sodium citrate, pH 7.0), and blotted dry with filter paper. Then, 6 μl of PCR product is spotted onto each 1 by 1 cm section of the grid. After spotting, the membrane is layered, DNA side up, onto a sheet of filter paper that had been prewetted with 0.5 *N* NaOH/0.6 *M* NaCl. After 5 min, it is transferred to a new sheet of filter paper that was prewetted with 0.5 *M* TRIS-HCl, and allowed to neutralize for 5 min.

The membrane is then placed on a sheet of filter paper and allowed to dry at room temperature or 37 °C (usually 5 min). It is turned DNA side down onto a sheet of plastic wrap and irradiated at 302 nm with a ultraviolet light box for 5 min, then placed into a 1-pint (0.5-l) resealable freezer bag containing 5–10 ml of prehybridization mix (25% formamide, 3XSSC, 100 µg/ml denatured salmon testis DNA, 20 µg/ml yeast RNA, 0.1%, sodium dodecyl sulfate (SDS), 1 mM ethylene diamine tetra-acetic acid (Na$_2$EDTA), 0.2% Ficoll, 0.2% polyvinylpyrrolidone, 0.05% sodium pyrophosphate, 100 µg/ml bovine serum albumin, and 10 mM HEPES pH 7.5). The bag is incubated with gentle agitation at 37°–42 °C for 30 min to 3 h.

Oligonucleotide Probes

For dot-blot detection, I label oligonucleotide probes with gamma-[^{32}P] ATP as described. Probes are useful if specific activity is $>4 \times 10^5$ dpm/pmol and can be reused until activity decays below that level. The prehybridization mix is poured out of the resealable bag and replaced with hybridization mix (5 ml), which consists of prehybridization mix supplemented with $5–10 \times 10^6$ dpm of end-labeled oligonucleotide probe. Hybridization is from 1 h to overnight. After hybridization, stringent wash conditions are preferable. For most probes, 70 °C in 6XSSC buffer for 1 min reduces background sufficiently to produce clean hybridization signals. The dried membrane is exposed to X-OMat film (Kodak) for 2–100 h (depending upon the specific activity of the probe) at −70 °C with intensifying screens before the film is developed.

Discussion

PCR Diagnosis

Since PCR directly examines proviral DNA, it has been the "gold standard" technique for testing the validity of serologic techniques, for distinguishing HTLV-I from -II, and for measuring the in vivo distribution of provirus. For research purposes, PCR is the only practical method for detecting and quantitating proviral DNA in vivo.

Viral Subclassification. HTLV-I and -II are 60% similar in nucleotide sequence. While this makes serologic distinction of the two viruses challenging, this degree of divergence is easily exploited with sequence-specific oligonucleotide primers and probes to differentiate the two agents.

Serologically Ambiguous Infection. Several groups have reported apparent HTLV-I or -II infection (by PCR) of persons with serologically indeterminate patterns of Western blot reactivity, and even some with negative screening

EIAs. For HTLV-II in particular, there is at present no way to know how many persons in a population are infected but fail to make antibodies that cross-react with HTLV-I in high enough titer to allow diagnosis.

Special Situations. PCR can theoretically be used to diagnose infection in persons in whom serologic diagnosis is not possible. For example, the obligately seropositive newborn children of infected mothers might be diagnosed in this manner. Other situations in which PCR would be especially valuable include detection of infection during the "window" period between exposure and seroconversion and in fixed tumor tissue.

Problems with PCR Diagnosis

False-positive diagnoses of HTLV infection or incorrect subtyping of virus may occur if even minute amounts of previously amplified DNA or cloned target sequences get into the amplification mixture or any of the reagent stocks. Usually, no amount of controls can allow one to determine with absolute confidence that a particular positive signal was not produced by contamination. This is a problem of sufficient magnitude and frequency to merit constant attention. Recommendations to prevent contamination have been published (Kwok and Higuchi 1989).

Combination of PCR and Culture: Increased Sensitivity and Use in Confirmation of PCR Results

Although the *tax* primers discussed above are very sensitive in detecting proviral DNA in PBMC specimens, in many cases 2–4 weeks of coculture of PBMC with uninfected cells greatly increases (10^2–10^4 fold) the amount of *tax* DNA template and allows one to detect proviral DNA more readily. This frequently occurs in cocultures in which p24^{gag} antigen is not elaborated into the media. This occurs because HTLV can presumably produce incomplete reverse transcription products in in vitro infection, in a manner analogous to HIV-1 (B. Hjelle et al. 1991, unpublished; Zack et al. 1990). The block to HTLV replication occurs between *tax* and *pol*, so primers in *pol* do not detect a similar increase in target abundance in antigen-nonproducing cultures. In those rare cases in which HTLV *tax* DNA is not detectable in PBMC from a seropositive person, growth of cells in coculture followed by *tax* PCR will elicit a positive signal. PCR of cocultured PBMC offers an additional advantage to PCR of PBMC alone, since one may "confirm" a positive result obtained with PBMC by repeating the assay on an independently prepared DNA sample.

The diagnosis of HTLV infection in seronegative or serologically indeterminate persons by PCR alone is sufficiently problematic to require the use of coculture for confirmation. Antigen production is monitored by antigen-capture EIA and/or immunofluorescence. Positive and negative cultures are both subject to repeat PCR. We have been able to diagnose unequivocally several individuals with serologically indeterminate results as infected with HTLV-I or -II by these means (B. Hjelle et al. 1990, unpublished).

References

Blattner WA, Takatsuki K, Gallo RC (1983) Human T-cell leukemia-lymphoma virus and adult T-cell leukemia. JAMA 250:51–58

Daenke S, Nightingale S, Cruickshank JK, Bangham CRM (1990) Sequence variants of human T-cell lymphotropic virus type I from patients with tropical spastic paraparesis and adult T-cell leukemia do not distinguish neurological from leukemic isolates. J Virol 64:1278–1282

DeFreitas E, Hilliard B, Cheney PR, Bell DS, Kiggundu E, Sankey D, Wroblewska Z, Palladino M, Woodward JP, Koprowski H (1991) Retroviral sequences related to human T-lymphotropic virus type II in patients with chronic fatigue immune dysfunction syndrome. Proc Natl Acad Sci USA 88:2922–2926

Donegan E, Busch MP, Galleshaw JA et al. (1990) Transfusion of blood components from a donor with human T-lymphotropic virus type II (HTLV-II) infection. Ann Intern Med 113:555–556

Gessain A, Vernant JC, Maurs L, Barin F, Gout O, Calender A, de The G (1985) Antibodies to human T-lymphotropic virus type-I in patients with tropical spastic paraparesis. Lancet 2:407–409

Gray GS, White M, Bartman T, Mann D (1990) Envelope gene sequence of HTLV-1 isolate MT-2 and its comparison with other HTLV-1 isolates. Virology 177:391–395

Heneine W, Kaplan JE, Gracia F, Lal R, Roberts B, Levine PH, Reeves WC (1991) HTLV-II endemicity among Guaymi Indians in Panama. N Engl J Med 324:565

Hino S (1990) Maternal-infant transmission of HTLV-I: implications for disease. In: Blattner WA (ed) Human retrovirology: HTLV. Raven, New York, pp 363–375

Hinuma Y (1985) A retrovirus associated with a human leukemia, adult T-cell leukemia. Curr Top Microbiol Immunol 115:127–141

Hjelle B (1991) Human T-cell leukemia/lymphoma viruses: life cycle, pathogenicity, epidemiology and diagnosis. Arch Pathol Lab Med 115:440–450

Hjelle B, Chaney R (1992) Sequence variation of functional HTLV-II *tax* alleles among isolates from an endemic population: Lack of evidence for oncogenic determinant in *tax*. J Med Virol 36:136–141

Hjelle B, Mills R, Mertz G, Swenson S (1990a) Transmission of HTLV-II by transfusion. Vox Sang 59:119–122

Hjelle B, Scalf R, Swenson S (1990b) High frequency of human T-cell leukemia-lymphoma virus type II infection in New Mexico blood donors: determination by sequence-specific oligonucleotide hybridization. Blood 76:450–454

Hjelle B, Mills R, Swenson S, Mertz G, Key C, Allen S (1991) Incidence of hairy cell leukemia, mycosis fungoides, and chronic lymphocytic leukemia in first known HTLV-II-endemic population. J Infect Dis 163:435–440

Kalyanramen VS, Sarngadharan MG, Robert-Guroff M, Miyoshi I, Blayney D, Golde DW, Gallo RC (1982) A new subtype of human T-cell leukemia virus (HTLV-II) associated with a T-cell variant of hairy cell leukemia. Science 218:571–573

Komurian F, Pelloquin F, de The G (1991) In vivo genomic variability of human T-cell leukemia virus type I depends more upon geography than upon pathologies. J Virol 65:3770–3778

Kwok S, Higuchi R (1989) Avoiding false positives with PCR. Nature 339:237–238

Kwok S, Kellogg D, Ehrlich G, Poiesz B, Bhagavati S, Sninsky JJ (1988) Characterization of a sequence of human T cell leukemia virus type I from a patient with chronic progressive myelopathy. J Infect Dis 158:1193–1197

Lairmore MD, Jacobson S, Gracia F, De BK, Castillo L, Larreategui M, Roberts BD, Levine PH, Blattner WA, Kaplan JE (1990) Isolation of human T-cell lymphotropic virus type 2 from Guaymi Indians in Panama. Proc Natl Acad Sci USA 87:8840–8844

Lee H, Swanson P, Shorty VS, Zack JA, Rosenblatt JD, Chen ISY (1989) High rate of HTLV-II infection in seropositive IV drug abusers in New Orleans. Science 244:471–475

LeGrenade L, Hanchard B, Fletcher V, Cranston B, Blattner W (1990) Infective dermatitis of Jamaican children: a marker for HTLV-I infection. Lancet 336:1345–1347

Lillehoj EP, Alexander SS, Dubrule CJ, Wiktor S, Adams R, Tai C-C, Manns A, Blattner WA (1990) Development and evaluation of a human T-cell leukemia virus type I serologic confirmatory assay incorporating a recombinant envelope polypeptide. J Clin Microbiol 28:2653–2658

Malik KT, Even J, Karpas A (1988) Molecular cloning and complete nucleotide sequence of an adult T cell leukaemia virus/human T cell leukaemia virus type I (ATLV/HTLV-I isolate of Caribbean origin: relationship to other members of the ATLV/HTLV-I subgroup). J Gen Virol 69:1695–1710

Matsumoto C, Mitsunaga S, Oguchi T, Mitomi Y, Shimada T, Ichikawa A, Watanabe J, Nishioka K (1990) Detection of human T-cell leukemia virus type I (HTLV-I) provirus in an infected cell line and in peripheral mononuclear cell of blood donors by the nested polymerase chain reaction method: comparison with HTLV-I antibody tests. J Virol 64:5290–5294

Miyoshi I, Kubinishi I, Yoshimoto S, Akagi T, Ohtsuki Y, Shiraishi Y, Nagata K, Hinuma Y (1981) Type-C virus particles in a cord T-cell line derived by cocultivation of normal human cord leukocytes and human leukemic T-cells. Nature 294:770–771

Morgan O StC, Mora C, Rodgers-Johnson P, Char G (1989) HTLV-I and polymyositis in Jamaica. Lancet 2:1184–1187

Osame M, Usuku K, Izumo S, Ijichi N, Amitani H, Igata A, Matsumoto M, Tara M (1986) HTLV-I associated myelopathy, a new clinical entity. Lancet 1:1031–1032

Poiesz BJ, Ruscetti FW, Gadzar AF, Bunn PA, Minna JD, Gallo RC (1980) Detection and isolation of type-C retrovirus particles from fresh and cultured lymphocytes of a patient with cutaneous T-cell lymphoma. Proc Natl Acad Sci USA 77:7415–7419

Ratner L, Josephs SF, Starcich B, Hahn B, Shaw GM, Gallo RC, Wong-Staal F (1985) Nucleotide sequence analysis of a variant human T-cell leukemia virus (HTLV-I b) provirus with a deletion in pX-I. J Virol 54:781–790

Rosenblatt JD, Golde DW, Wachsman W, Jacobs A, Schmidt G, Quan S, Gasson JC, Chen ISY (1986) A second isolate of HTLV-II associated with atypical hairy-cell leukemia. N Engl J Med 313:372–377

Rosenblatt JD, Gasson JC, Glaspy J, Bhuta S, Abud M, Chen ISY, Golde DW (1987) Relationship between human T-cell leukemia virus-II and atypical hairy cell leukemia: a serologic study of hairy cell leukemia patients. Leukemia 1:397–401

Rosenblatt JD, Plaeger-Marshall S, Giorgi JV, Swanson P, Chen ISY, Chin E, Wang H, Canavaggio M, Black AC, Lee H (1990) A clinical, hematologic, and immunologic analysis of 21 HTLV-II-infected intravenous drug users. Blood 76:409–417

Rous P (1910) A transmissable avian neoplasm: sarcoma of the common fowl. J Exp Med 12:696–705

Ruano G, Fenton W, Kidd KK (1989) 13. phasic amplification of very dilute DNA samples via 'booster' PCR Nucl Acids Res 17:5407–5415

Seiki M, Eddy R, Shows TB, Yoshida M (1984) Nonspecific integration of the HTLV provirus into adult T-cell leukemia cells. Nature 309:640–642

Tsujimoto A, Teruuchi T, Imamura J, Shimotohno K, Miyoshi I, Miwa M (1988) Nucleotide sequence analysis of a provirus derived from HTLV-1-associated myelopathy (HAM). Mol Biol Med 5:29–42

Wiktor SZ, Alexander SS, Shaw GM, Weiss SH, Murphy L, Wilks RJ, Shortly VJ, Hanchard B, Blattner WA (1990) Distinguishing between HTLV-I and HTLV-II by Western blot. Lancet 335:1533

Zack JA, Arrigo SJ, Weitsman SR, Go AS, Haislip A, Chen ISY (1990) HIV-1 entry into quiescent primary lymphocytes: molecular analysis reveals a labile, latent viral structure. Cell 61:213–222

Chapter 5 Detection of Human Spumaviruses by Polymerase Chain Reaction *

Walter Muranyi and Rolf M. Flügel

Summary

The detection of human spumaretrovirus (HSRV)-specific RNA and DNA sequences by polymerase chain reactions (PCR) in human cells is described. Various oligodeoxynucleotides of different chain lengths were derived from the prototypic HSRV sequence, synthesized, and employed as sense and antisense primers in several modified versions of a standard PCR. PCR seems to be the method of choice for detecting HSRV nucleic acids and is superior to other techniques. The only serious problem encountered is the oversensitivity of this method that can lead to false-positive results due to laboratory contaminations. Controls and the complete physical separation of all reagents and equipment are required to avoid any contaminations.

Introduction

Spumaviruses are complex retroviruses of approximately 12–13 kbp in size (Flügel 1991). The regulatory genes of spumaviruses (called *bel* genes) contribute substantially to the greater genome size when compared with other retroviruses groups. The *bel* genes are characteristic and distinguish the spumaviruses from the other two groups of complex retroviruses (Cullen 1991). The lentiviruses and the human T-cell lymphotropic/bovine leukemia virus (HTLV/BLV) group also encode a number of additional and regulatory genes besides the common genes *gag, pol*, and *env* of the simple retroviruses. The *bel* genes, however, are clearly different from the lentiviral accessory genes, although there is a functional equivalence between the *trans*-activating *bel 1* and either human immunodeficiency virus (HIV-1) *tat* or *tax* of HTLV (Flügel 1991).

Spumaviruses share with other known retroviruses the ability to reverse transcribe a single-stranded RNA genome into a double-stranded DNA. The

Human Retrovirus Group, German Cancer Research Center, Applied Tumor Virology, Im Neuenheimer Feld 280, W-6900 Heidelberg, FRG
* This research project was financed by a grant KI 18901/6 from the Bundesministerium für Forschung und Technologie.

double-stranded DNA can exist in several forms in retrovirus-infected cells: It can be integrated into the host cell genome as a provirus, and it can occur in an unintegrated form (linear or circular).

There are only a few reports on the isolation of spumaviruses from human tissue (Flügel 1991). The first human spumaretrovirus (HSRV) was isolated from the lymphoblastoid cells of a nasopharyngeal carcinoma patient by Achong et al. (1971). HSRV is considered to be the prototype spumavirus, since it has been characterized more extensively by nucleotide sequencing and transcriptional mapping (Flügel et al. 1987; Maurer et al. 1988; Muranyi and Flügel 1991). The results of phylogenetic analyses by several groups unambiguously showed that spumaviruses are clearly distinct from all known retroviruses and distantly related to the complex groups of retroviruses (Maurer and Flügel 1988; Lewe and Flügel 1989; Xiong and Eickbush 1988; Doolittle et al. 1989; Myers and Pavlakis 1991). The analysis of the splicing pattern revealed that the HSRV genomic RNA is characteristically spliced into a unique arrangement of exons when compared with the transcriptional patterns of the more related lentiviruses, in particular that of HIV-1. This finding opens the way for designing and performing diagnostic PCRs that are gene-specific for human spumavirus sequences as presented in the following section.

Experimental Method

Reverse Transcription of Viral RNA and Amplification of cDNAs by PCR

HSRV DNA was synthesized with Moloney murine leukemia virus reverse transcriptase for 1 h at 37 °C or with avian myeloblastosis virus reverse transcriptase for 60 min at 42 °C from 5 µg total RNA from HSRV-infected cells using conditions given by the manufacturer (Gibco, Karlsruhe and Boehringer, Mannheim). Reverse transcriptions were started by adding an antisense oligodeoxynucleotide (150 pmol, 20-mer) derived from the 3′-region of the HSRV genome and random hexa oligodeoxynucleotides (0.5 µg/reaction; Pharmacia, Freiburg) according to Kinoshita et al. (1989). Alternatively, reverse transcriptions were started with random hexa primers only, and the sense and antisense primers were added before the PCR to avoid amplification of false priming events. Concentrations of dNTPs were 2.5 m*M* in a reaction volume of 10 µl. The reaction products were denatured at 94 °C for 5 min. Viral cDNAs were added to a solution that contained in 90 µl 0.25 m*M* of the four dNTPs, 150 pmol of a sense primer, and 10 × Taq buffer. The PCR was started by adding 2.5 units of Taq DNA polymerase (Cetus Perkin-Elmer or Stratagene). Paraffin oil, 50 µl (Merck), was layered on top of the reaction mixture. The reaction mixture was incubated at 94 °C, 55 °C, and 72 °C for 1 min, 2 min, and 3 min. The cycle was repeated 35 times in a DNA thermal cycler (Cetus Perkin-Elmer). Total RNA from uninfected HEL cells was run as a control under the same conditions but did not result in any HSRV-specific

DNA bands. PCR products were analyzed on agarose gels by visualizing with ethidium bromide or alternatively by hybridizing to an oligodeoxynucleotide located in both exons. PCR-amplified DNA bands in the range of 150 to about 1000 bp were isolated by electrophoresing onto DEAE membrane type NA45 (Schleicher & Schüll), purified, and cloned into the polylinker site of the Bluescript vector according to standard methods (Davis et al. 1986). Larger and distinct DNA fragments were directly isolated from agarose gels and purified. Some viral DNAs were prechecked by restriction enzyme analysis. Nucleotide sequencing was performed with the method of Maxam and Gilbert or the dideoxy-mediated chain termination method (Sambrook et al. 1989).

Direct Detection of Spumaviral DNA by PCR

Total DNA (0.1–1.0 µg) extracted from patients' biopsies or virus-infected cells was used in a reaction volume of 50 or 100 µl of PCR reaction buffer as given by the manufacturer (Stratagene), dNTPs were added at 250 μM each plus 2.5 units of Taq DNA polymerase. Primers SCP1 + DU5 (Table 1) at 150 pmol were used and overlaid with paraffin oil. The amplification was performed in 40 cycles of 1 min at 92°, 1 min at 50°–55 °C, and 2 min at 72 °C. PCR products were analysed as described above.

Results and Discussion

Guidelines

The design and the application of PCR suitable for the specific detection of spumavirus sequences in patients' biopsies and sera requires a few points that are useful as guidelines.

1. Primer pairs for PCR were derived from the published HSRV sequence (Flügel et al. 1987; Maurer et al. 1988) including those that are located in HSRV exon sequences (Muranyi and Flügel 1991), see below for specific examples.
2. Reverse transcription of spumaviral RNA into cDNA was carried out prior to amplification by PCR in the presence of hexa random oligodeoxynucleotides and an antisense primer (Muranyi and Flügel 1991; Kinoshita et al. 1989).
3. Direct detection of spumaviral DNA sequences by PCR amplification using primer pairs SCP 1 plus DU 5 from the 5′-LTR (long terminal repeat) or CBS plus ZBA from the *gag* region (Table 1).
4. The selection of primer pairs was done by restricting the size of PCR products to about 150–500 bp. In exceptional cases primer pairs with larger distances of up to 1500 bp between the genetic locations were used.

Table 1. Characteristic features of spumaretroviral (SRV) primer pair sequences

Name of primer	Sequences of primer pairs (5′–3′)	Genomic localization[a]	Gene assignment	Size of PCR products (bp)
SCP 1	GCTCTTCACTACTCGCTGCGT	3– 23	LTR (R)	268 (RU5)
DU 5	CACTAGATGTCTCCCTTAGCA[b]	251– 271	LTR (U5)	
CBS	CTCTATACCTGGGATTCATCC	1066–1086	*gag*	332 (CA)
ZBA	GGAGGCTCTCCAGTTACAGAT	1378–1398	*gag*	
SCP 1	GCTCTTCACTACTCGCTGCGT	3– 23	LTR (R)	572, 501, 264
E1A	CCAGGCCAATACTCTTGAGCT	5906–5929	*env*	
SCP 1	GCTCTTCACTACTCGCTGCGT	3– 23	LTR (R)	711, 640, 440
GAP	CAATCAGATACTGACCCTGACTG	9015–9038	*bel 1*	
SCP 1	GCTCTTCACTACTCGCTGCGT	3– 23	LTR (R)	810, 434
B1CA	ATCGATGGATCGTCTCCTGG	9312–9331	*bet*	
SCP 1	GCTCTTCACTACTCGCTGCGT	3– 23	LTR (R)	207, 162
B1CA	ATCGATGGATCGTCTCCTGG	9312–9331	*bel 2*	

[a]   In nucleotide positions of the HSRV genome (Muranyi and Flügel 1991)
[b]   Antisense primer is given as used for polymerase chain reaction (PCR)
LTR, long terminal repeat; R, repeat region; U5, unique region of 5′-LTR; cA, capsid antigen

5. The PCR products were routinely analyzed after electrophoresis on agarose gels by staining with ethidium bromide. Alternatively, and in order to increase the sensitivity of detection of PCR-amplified DNA bands, blot hybridization with a labeled spumavirus-specific oligodeoxynucleotide as probe was used. The sequence of the oligonucleotide was derived from a location between the two primers used.

6. PCR amplification of viral mRNAs expressed in human tissues was done by using total RNA as starting material that had been extracted from virus-infected human cells and subsequently subjected to the same procedure as described above.

7. The PCR technique to be used for the detection of proviral sequences also depends on the source of the cells or biopsies. The salient point is that if a naturally occurring virus that has never been passaged in in vitro cell cultures is to be detected, it cannot be expected that the sequence of a field virus is identical to that of the known prototype genome. Accordingly, degenerate PCR primer pairs should be designed and used by taking again the aforelisted rules into account. In addition to field isolates, this is also valid for variant and defective viruses. Degenerate primers should be invariant at the termini but can vary in the central part of their sequences (Frohman 1990).

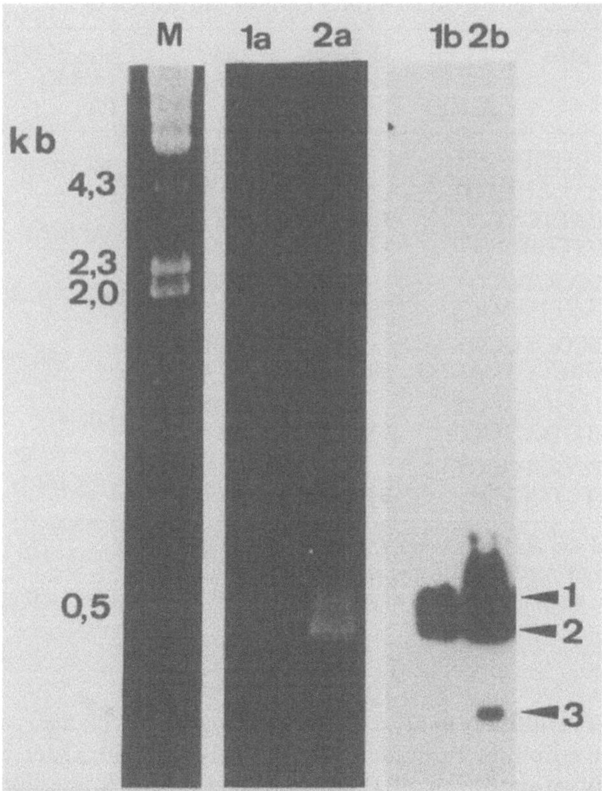

Fig. 1. Analysis of reverse-transcribed and PCR-amplified human spumaretrovirus (HSRV)-specific *env* mRNAs with primer pair SCP 1 + E1A. One tenth of the reaction mixture was electrophoretically separated on a 1% agarose gel and stained with ethidium bromide. Starting material was total RNA extracted from HSRV-infected human embryonic fibroblast cells 24 h (*lane 1a*) and 72 h (*lane 2a*) after infection. *Lanes 1b* and *2b* show the corresponding autoradiogram after blot hybridization under stringent conditions with an oligodeoxynucleotide (SCP 2) derived from the R region of the 5'-LTR of HSRV (Muranyi and Flügel 1991). The arrows mark three HSRV *env* DNA bands. *Lane M* contains *lambda* *Hin*dIII DNA fragments as markers

Selection of Defined Primer Pairs

The genetic structure of the HSRV DNA genome is in its shortest form: 5'-LTR-*gag-pol-env-bel 1-bet, bel 2-bel 3*-3'-LTR.

In general, genomic regions that are required for viral replication and transcription are suitable for PCR amplification. Some examples of sequences of spumaviral primer pairs that were used for PCR amplification of HSRV transcripts from virus-infected human cells are compiled in Table 1. Sense primers that are located at or just downstream of the start site of viral tran-

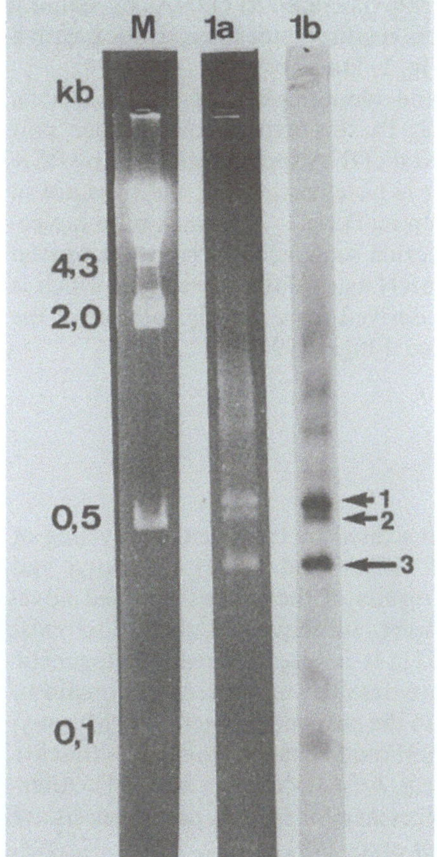

Fig. 2. Analysis of reverse-transcribed and PCR-amplified HSRV-specific *bel 1* transcripts with primer pair SCP 1 + GAP. Conditions were as described in legend to Fig. 1. The *arrows* indicate three *bel 1-*specific DNAs after staining with ethidium bromide (*lane 1 a*) and autoradiography (*lane 1 b*)

scription are those of choice, since they can be used for the detection of viral DNA, genomic RNA, and subgenomic transcripts. Thus, the SCP 1 sense primer is the one most frequently used (Table 1). The antisense primer should be selected by taking the expected value of the PCR product into account. Product sizes of 150–500 bp are optimal and relatively easy to detect on agarose gels. The design of primer pairs that result in PCR products that exceed 900 bp should be avoided, since more sophisticated methods are required for the synthesis and detection of large viral genomic sequences. The lengths of the primers that were used varied within 18–23 bp. The actual length is not of primary importance; however, to adjust the molar $(G + C)$ contents of a given primer to approximately 55%, primers of various lengths can be selected to approach that value (Frohman 1990). Figure 1 illustrates PCR amplification and the detection of rare and more abundant spumaretroviral transcripts in one set of experiments. The arrows mark three simultaneously detected HSRV cDNA bands that correspond to *env* mRNAs. Whereas the more abundant mRNAs were directly visualized as cDNAs by staining with ethidium bromide, the rare transcripts required blot hybridization with a radiolabeled oligonucleotide as probe (Fig. 1, lane 2b).

Another example for the detection and identification of HSRV-specific transcripts by PCR is shown in Fig. 2. In this experiment, primer pair SCP1+GAP was used (Table 1). Three viral cDNA species amplified by PCR were detectable as marked by arrows. It is noteworthy that the sizes are in agreement with the calculated values given in Table 1. All bands were molecularly cloned and characterized by nucleotide sequencing. The result proved that the DNAs represent *bel 1*-specific mRNAs of different lengths which is due to the presence of extra small exons derived from the central part of the HSRV genome (Flügel 1991; Muranyi and Flügel 1991).

Conclusions

The power of the PCR method is well illustrated by the recent unraveling of the complex transcriptional patterns of HSRV and HIV-1 (Muranyi and Flügel 1991; Schwartz et al. 1990). The results of the studies revealed novel gene products for both viruses that were subsequently shown to exist (Schwartz et al. 1990; Löchelt et al. 1991). It is also of great phylogenetic interest that several noncoding exons were detected in the central domain of both virus genomes, some of which occur in the *pol* gene (Flügel 1991; Muranyi and Flügel 1991). In general, PCR primers should be taken from genes that are absolutely required for viral replication, e.g, *bel 1* (Löchelt et al. 1991). Alternatively, genes like HIV-1 *nef* that are necessary for in vivo pathogenesis are also suitable for PCR diagnostics (Kestler et al. 1991).

The methods described above were particularly useful for the detection of HSRV DNA and RNA in virus-infected human cells. Doubtless, improved PCR techniques shaped according to the purpose and to a particular problem will be of great value for identifying spumaviral sequences in patients' biopsies and sera.

Acknowledgements. We thank Harald zur Hausen for support.

References

Achong G, Mansell PWA, Epstein MA, Clifford P (1971) An unusual virus in cultures from a human nasopharyngeal carcinoma. J Natl Cancer Inst 42:299–307

Cullen B (1991) Human immunodeficiency virus as a prototype complex retrovirus. J Virol 65:1053–1056

Davis LG, Dibner MD, Battey JF (1986) Basic methods in molecular biology. Elsevier, New York, pp 79–152

Doolittle RF, Feng D-F, Johnson MS, McClure MA (1989) Origins and evolutionary relationship of retroviruses. Q Rev Biol 64:1–30

Flügel RM (1991) Spumaviruses: a group of complex retroviruses. J AIDS 4:739–750

Flügel RM, Rethwilm A, Maurer B, Darai G (1987) Nucleotide sequence analysis of the *env* gene and its flanking regions of the human spumaretrovirus reveals two novel genes. EMBO J 6:2077–2084

Frohman MA (1990) Race: rapid amplification of cDNAs ends. In: Innis MA, Gelfand DH, Sninsky JJ, White TJ (eds) PCR protocols, a guide to methods and applications. Academic, San Diego, pp 28–53

Kestler HW, Ringler DJ, Mori K, Panicalli DL, Sehgal P, Daniel MD, Desrosiers RC (1991) Importance of the *nef* gene for maintenance of high virus loads and for development of AIDS. Cell 65:651–662

Kinoshita T, Shimoyana M, Tobinai K, Ito M, Ito S-I, Ikeda S, Tajima K, Shimotono K, Sugimura T (1989) Detection of mRNA for the *tax/rex* gene of human T-cell leukemia virus type 1 in fresh peripheral blood mononuclear cells of adult T-cell leukemia patients and viral carriers by using polymerase chain reactions. Proc Natl Acad Sci USA 86:5620–5624

Lewe G, Flügel RM (1989) Tracing genetic events in the evolutionary pathway of retroviruses: comparative analysis of the *pol* and *env* protein sequences. Virus Genes 2:195–204

Löchelt M, Zentgraf H, Flügel RM (1991) Construction of an infectious DNA clone of the full-length human spumaretrovirus genome and mutagenesis of the *bel 1* gene. Virology 184 (in press)

Maurer B, Bannert H, Darai G, Flügel RM (1988) Analysis of the primary structure of the LTR, the *gag* and the *pol* gene of the human spumaretrovirus. J Virol 62:1590–1597

Maurer B, Flügel RM (1988) Genomic organization of the human spumaretrovirus and its relatedness to AIDS and other retroviruses. AIDS Res Hum Retroviruses 4:467–473

Muranyi W, Flügel RM (1991) Analysis of splicing patterns by polymerase chain reaction of the human spumaretrovirus reveals complex RNA structures. J Virol 65:727–735

Myers G, Pavlakis GN (1991) Evolutionary potential of complex retroviruses. In: Wagner RR, Frenkel-Conrat H (eds) Viruses. Levy J (ed) The retroviridae: Plenum, New York, pp 1–37

Sambrook J, Fritsch EF, Maniatis T (1989) Molecular cloning. A laboratory manual, 2nd edn. Cold Spring Harbor Laboratory, Cold Spring Harbor

Schwartz S, Felber B, Benko DM, Fenyö E-M, Pavlakis G (1990) Cloning and functional analysis of multiply spliced mRNA species of human immunodeficiency virus type 1. J Virol 64:2519–2529

Xiong Y, Eickbush TH (1988) Origin and evolution of retroelements based upon their reverse transcriptase sequences. EMBO J 10:3353–3362

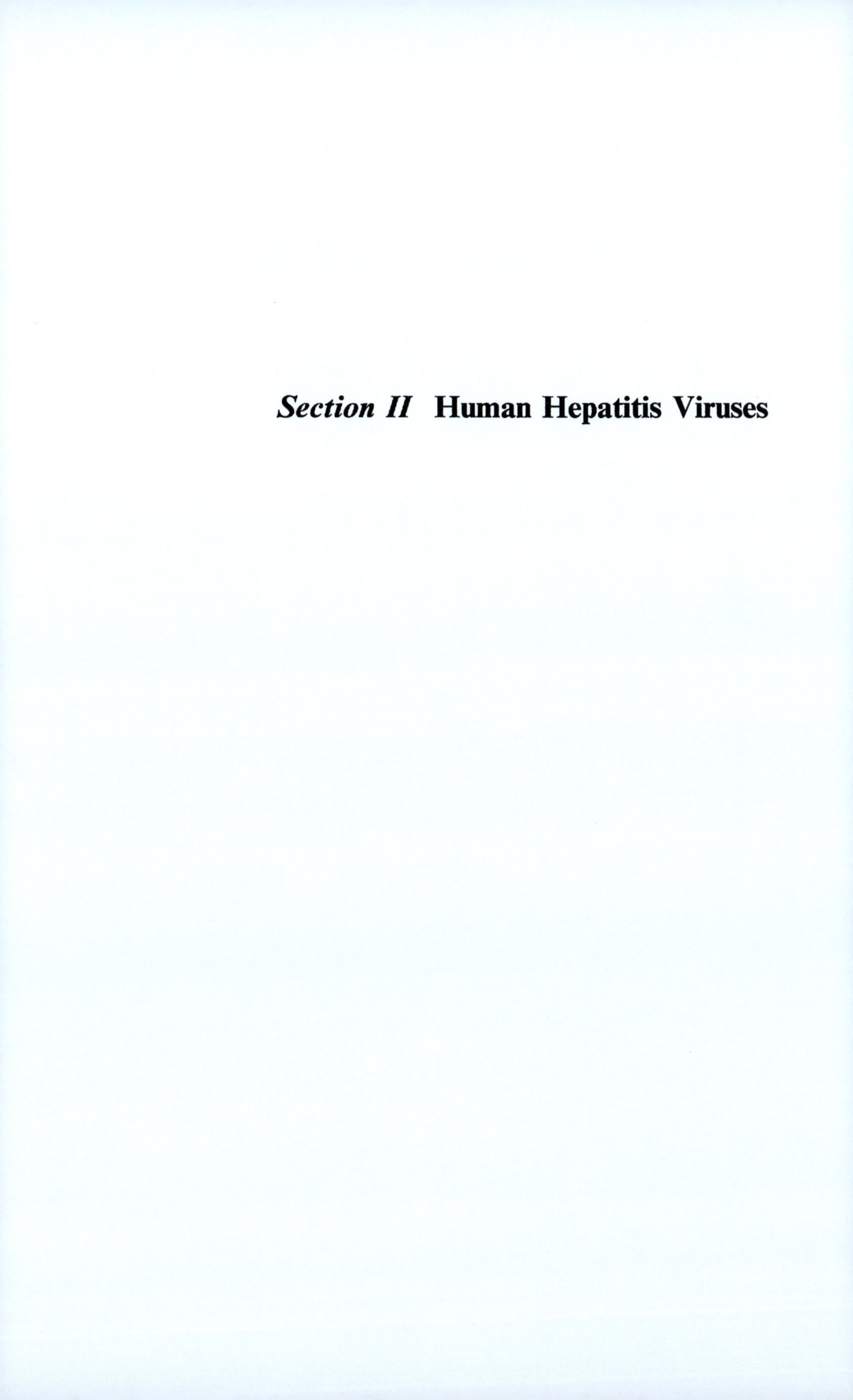# *Section II* Human Hepatitis Viruses

Chapter 6 Antigen Capture/Polymerase Chain Reaction for the Detection of Hepatitis A Virus in Human Clinical Materials

Robert W. Jansen [1,2] and Stanley M. Lemon [1]

Introduction

Human hepatitis A virus (HAV) is the single most common cause of acute viral hepatitis in many regions of the world. Recently classified within the genus Hepatovirus of the family Picornaviridae, this positive-strand RNA virus has several notable biological features. Unlike many picornaviruses, HAV has a protracted, most often noncytolytic replication cycle in cell culture which makes the isolation of this virus from clinical materials exceptionally difficult. To facilitate the detection of HAV in clinical samples and to allow us to determine the genetic relatedness of epidemiologically unrelated virus strains, we have developed an antigen capture/polymerase chain reaction (AC/PCR) method which utilizes a simple immunoaffinity-based virus purification scheme prior to isolation of viral RNA, reverse transcription, and PCR (Jansen et al. 1990). This method involves the "capture" of virus by a monoclonal antibody bound to a solid-phase support (the PCR reaction tube itself), heat denaturation of the virus which results in release of its RNA, and reverse transcription followed immediately by PCR. Only a single reaction tube is used throughout the entire process, and thus the method involves fewer manipulations than conventional PCR methods for detection of RNA viruses. In addition, the immunoaffinity virus purification step results in a greatly increased abundance of the target RNA relative to extraneous nucleic acids subjected to PCR. The specificity is thus enhanced, and AC/PCR usually generates a single PCR product which is readily visualized on agarose gels containing ethidium bromide. As little as 5 μl of a clinical sample (usually a crude fecal suspension) is required. The general method and its application to the study of the molecular epidemiology of HAV have been described in detail elsewhere (Jansen et al. 1990).

Throughout the AC/PCR procedure, it is important to conform to the normal precautions associated with PCR amplification of viral nucleic acid sequences present in clinical materials (Kwok and Higuchi 1989). In particular, one must pay attention to the avoidance of cross-contamination during the

[1] Department of Medicine, 547 Burnett-Womack CSB, CB # 7030, The University of North Carolina at Chapel Hill, Chapel Hill, NC 27599-7030, USA

[2] *Current address*: Division of Experimental Therapy, Burroughs-Wellcome Co., Research Triangle Park, NC 27709, USA

pre-PCR immunoaffinity capture steps. Unless the washing steps are carried out with care, excess virus in one high-titered sample may be inadvertently transferred to a tube containing little or no virus, and some of this contaminating virus may bind to the capture antibody. However, we have found a greater problem to be the presence of cloned cDNA in the PCR laboratory. Contamination can be partly controlled by performance of replicate AC/PCR assays, but we prefer the greater specificity afforded by sequence analysis of the amplified PCR product. Nonetheless, when two or more clinical samples produce PCR products with 100% sequence identity (or identify to a cloned cDNA maintained in the laboratory), the entire method must be repeated to assure validity.

Experimental Approach

Antigen Capture

Monoclonal HAV-specific IgG diluted in pH 9.6 carbonate buffer (100 μl) is allowed to coat the surface of PCR reaction tubes (sterile 1.5 ml or 0.5 ml microcentrifuge tubes) at 35 °C for 4 h, as described previously for the coating of microplates for immunoaffinity hybridization (Jansen et al. 1985). We use 3–10-fold higher concentrations of capture antibodies to coat polypropylene tubes than we normally use to coat polyvinylchloride or polystyrene microplates, however, because of the lower protein-binding abilities of polypropylene. After "blocking" unbound sites on the plastic surface with a solution of 1% bovine serum albumin, the tubes are washed three times with 300 μl of phosphate buffered saline (PBS) containing 0.05% Tween-80. Clinical samples (25–80 μl), suspended in PBS or TRIS-ethylene diamine tetra-acetic acid (EDTA), are added directly to the washed tubes and incubated at 4 °C overnight. Using care to avoid cross-contamination, the tubes are washed separately six times with 300 μl of 20 mM TRIS, pH 8.4, 75 mM KCl, and 2.5 mM MgCl$_2$. For washing, we utilized a simple apparatus consisting of a thick-walled plastic tube clamped to a ring stand at one end and attached to a vacuum trap at the other. A disposable micropipette tip is placed at the open end of the tube, where it is held in place by the vacuum and serves as a readily replaceable aspiration probe. After washing is completed, the tubes can be stored frozen or taken immediately into the subsequent reverse transcription/ PCR steps.

Reverse Transcription and PCR

The captured virus is denatured by placing the tubes at 95 °C for 5 min after the addition of 80 μl of reverse transcriptase (RT)/Taq universal buffer (25 mM TRIS, pH 8.4, 75 mM KCl, 2.5 mM MgCl$_2$, 0.25 mM each dNTP)

Table 1. Hepatitis A virus VP1/2A region primers for antigen capture/polymerase chain reaction

---

Positive-sense primers:
+2891 5′-GGTTTCTATTCAGATTGCAAATTA
+2933 5′-TTTGTCTTTTAGTTGTTATTTGTCTGT
+2948 5′-GTTATTTGTGTGTCACAGAACAATC
+2984 5′-TCCCAGAGCTCCATTGAA

Negative-sense primers:
−3192 5′-AGGAGGTGGAAGCACTTCATTTGA
−3265 5′-CATTATTTCATGCTCCTCAG
−3375 5′-AGTAAAAACTCCAGCATCCATTTC
−3285 5′-AGTCACACCTCTCCAGGAAAACTT

---

and 5 µl (50 pmols) of the negative-strand primer. The positive-strand primer is not added at this step so as to minimize possible nonspecific primer extension by the RT. As the thermal cycler temperature reaches 42 °C, 5 µl (2 units) of RT is added to each tube, and the temperature is held at 42 °C for an additional 30 min. The reaction tubes are placed on ice for the addition of 50 pmols plus-strand primer and 2 units heat-stable Taq DNA polymerase (10 µl total). The low temperature at this step may minimize nonspecific extension products ("cold deoxyribonuclear fusion"). The reaction mix is then overlaid with mineral oil and placed at 95 °C for 5 min, before the initiation of 30–40 PCR cycles involving denaturation at 95 °C for 15 s, annealing for 5 s at 42 °C, and extension for 1 min at 72 °C. The last cycle extension is for 5 min, to allow full completion of double-stranded PCR products. Oligonucleotide primers used for AC/PCR are shown in Table 1.

Analysis of AC/PCR Products

A 10%–100% aliquot of the AC/PCR product is subjected to electrophoresis in 3% NuSieve GTG (FMC) agarose gels [containing 1% Seakem ME agarose (FMC) if for Southern blotting] in TRIS-acetate/EDTA buffer. PCR products are visualized by ethidium bromide staining. For maximum sensitivity, the gel may be subjected to Southern transfer for hybridization analysis. Alternatively, the individual PCR product bands may be excised, purified, and sequenced by the dideoxynucleotide method, as described previously (Jansen et al. 1990).

Results and Discussion

Specificity and Sensitivity

It is not likely than any method for the detection of viral antigen or viral RNA will ever supplant serological methods for the diagnosis of hepatitis A. The

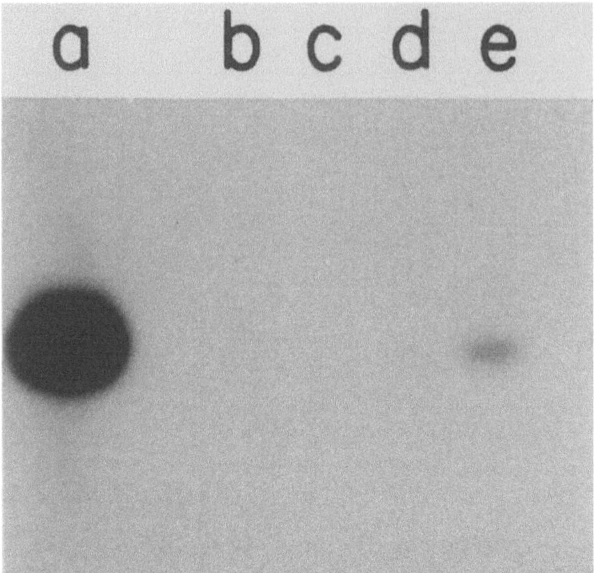

Fig. 1. Southern hybridization of agarose gel-fractionated antigen capture/polymerase chain reaction products obtained with different capture antibody specificities. Reaction tubes were coated with 1:100 dilutions of ascitic fluids containing monoclonal antibodies hepatitis A virus (HAV) (*lane a*, monoclonal K3-2F2), poliovirus type 1 (*lanes b and c*), or respiratory syncytial virus (*lane d*) or with carbonate buffer only (*lane e*), prior to loading with an HAV-positive 10% human fecal suspension (GR-1). (From Jansen et al. 1990)

IgM antibody response to HAV is brisk and substantial in patients who experience acute hepatitis A infection. Such antibody is present in >99% of patients by the time they present to their physicians with symptoms of infection (Lemon et al. 1980). HAV-specific IgM is readily detected by commercially available antibody-capture immunoassays and represents the diagnostic standard for HAV infection. Nonetheless, the detection of HAV in fecal materials of acutely infected patients by AC/PCR allows the particular strain of HAV causing the infection to be characterized in great detail and thus offers novel opportunities for advancing our understanding of its molecular epidemiology (Jansen et al. 1990).

We have shown that the AC step in AC/PCR enhances the overall specificity of PCR-based procedures for the detection of HAV in clinical specimens (Jansen et al., 1990). An HAV-containing fecal specimen was placed into tubes coated with either HAV-specific monoclonal antibody or other non-HAV-related monoclonal antibodies. AC/PCR was carried out as described above, and the products separated on an agarose gel. The PCR products were then transferred to nitrocellulose and probed with a ^{32}P-labelled oligonucleotide probe complementary to the predicted sequence of the amplification product. Hybridization signals obtained with the AC/PCR products from the reaction

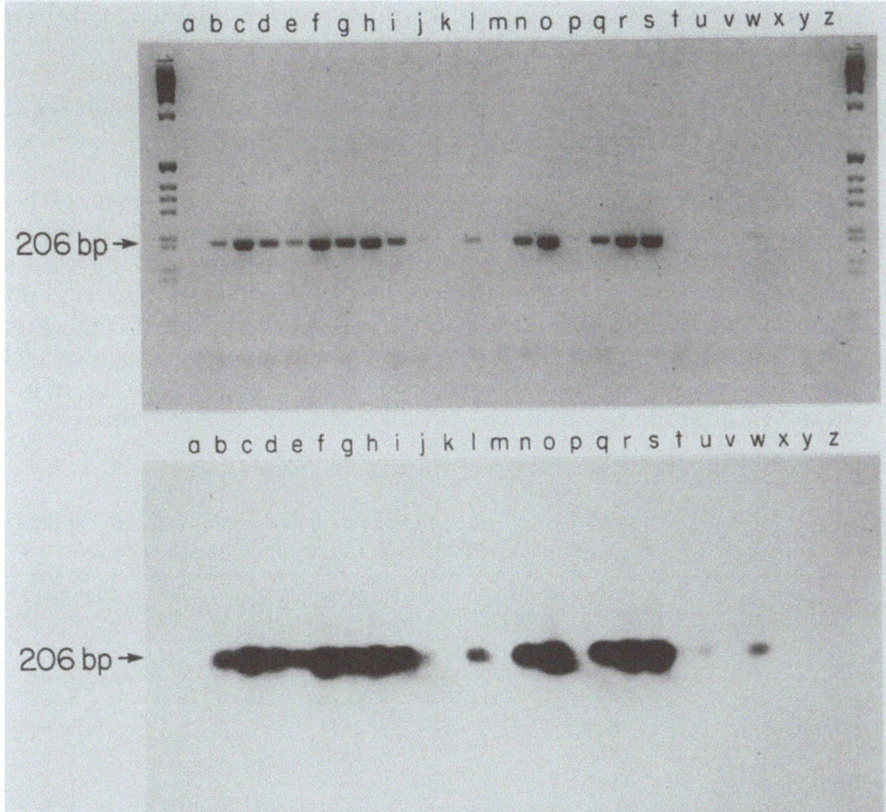

Fig. 2. Agarose gel analysis of AC/PCR products obtained with fecal samples collected from patients with hepatitis A. The primer set utilized was +2020/−2211, which results in amplification of a 206-bp segment encoding part of the capsid protein VP3. In the *top panel*, reaction products are visualized by ethidium bromide staining (negative image). *Lane a*, phosphate buffered saline (PBS) control and *b*, GR-7 fecal suspension immunoaffinity captured by monoclonal antibody to poliovirus (compare with Fig. 1, where washing was more extensive after antigen binding); *c–w*, acute hepatitis A fecal suspensions captured by HAV-specific monoclonal K3-2F2 (*lane h* is GR-7); *x*, HAV-negative fecal suspension from normal human volunteer, *y*, *z*, additional PBS PCR controls. In the *lower panel* are the results of Southern hybridization with an oligonucleotide probe (+2189, 5′-TATGGATGTTACTA-CACA). (Figure modified from Jansen et al. 1990)

tubes coated with non-HAV-related antibodies were weak or undetectable, while a very strong signal was obtained from the tube coated with the HAV-specific monoclonal antibody (Fig. 1). The level of random signals obtained from the non-HAV-coated tubes was not unreasonable given the amount of HAV present in the original sample, and the extreme sensitivity of PCR in general. The significant point is the differential signal strength that is evident upon examination of the blot shown in Fig. 1.

We compared the sensitivity of AC/PCR with that of older methods for the detection of HAV in clinical materials (Jansen et al. 1990). This analysis revealed a significant advantage in terms of the sensitivity obtained with AC/PCR (Table 2). Fecal suspensions collected from 20 patients with acute hepatitis A who were involved in a common source outbreak of this disease were studied for the presence of HAV antigen by solid-phase radioimmunoassay, RIA (Lemon et al. 1982) or HAV RNA by combined immunoaffinity cDNA-RNA slot blot hybridization (Jansen et al. 1985) and AC/PCR. AC/PCR products were transferred to nitrocellulose for Southern hybridization with an oligonucleotide probe which was complementary to an internal region in the expected amplicon. Seventeen (85%) of the specimens yielded AC/PCR products of the expected size which were visible on ethidium-stained agarose gels. Furthermore, products from 19 of the 20 (95%) fecal specimens produced positive hybridization signals when hybridized with the labelled probe (Fig. 2).

Table 2. Comparison of radioimmunoassay (RIA), cDNA-RNA immunoaffinity hybridization, and AC/PCR for detection of hepatitis A virus 1 (HAV) in human fecal specimens

Specimen	RIA (S/N)	cDNA-RNA hybridization blot intensity	AC/PCR
GR15	59.0	6419	+ + + +
GR7	42.4	2694	+ + +
GR19	30.0	2470	+ + + +
GR5	33.8	2428	+ + +
GR1	21.2	123	+ + +
GR17	5.9	17	+ +
GR14	3.5	64	+ +
GR4	3.1	78	+
GR6	2.7	131	+ +
GR9	2.2	1	+ +
GR2	1.9 (−)	45	+ +
GR21	2.0 (−)	+/−	(+)
GR20	1.0 (−)	+/−	(+)
GR16	1.7 (−)	−	+
GR12	1.1 (−)	−	+
GR24	1.1 (−)	−	+
GR10	1.1 (−)	−	+
GR11	1.1 (−)	−	+
GR13	0.8 (−)	−	+
GR23	0.8 (−)	nd	−

Note: Fecal samples, collected from 20 individuals involved in a common source outbreak of hepatitis A (Lednar et al., 1985), were tested for HAV by RIA (Lemon et al. 1982), immunoaffinity hybridization with a probe complementary to the P1 region of the HAV genome (Jansen et al. 1985), and AC/PCR (Jansen et al. 1990). S/N is the signal-to-noise ratio in the RIA (≥ 2.1 is considered positive). Hybridization intensity was determined by laser densitometry of the autoradiogram. AC/PCR results are shown by intensity of the amplified transcripts in an ethidium-stained agarose gel; (+) indicates that the transcripts were identified only by Southern hybridization; nd = not done

In contrast, only 10 of 20 (50%) and 13 of 19 (68%) specimens were positive by RIA and cDNA-RNA hybridization, respectively (Table 2). Subsequent sequencing studies demonstrated a single-base mismatch between the oligonucleotide detection probe used in this experiment and the complementary nucleotide sequence of the virus causing this outbreak of disease (Jansen et al. 1990). We would thus expect that even greater sensitivity might have been achieved with a perfectly matching probe, or perhaps a probe of greater length. The RIA signal and the cDNA-RNA hybridization intensities correlated roughly with the quantities of PCR products (Table 2). This correlation suggests that AC/PCR is at least semiquantitative under these conditions.

We compared the detection of HAV by AC/PCR with that by a conventional PCR approach. This latter method involved the isolation of viral nucleic acid by treating clinical samples with proteinase K and SDS, followed by extraction with phenol-chloroform. While this method was much more labor intensive than AC/PCR, the sensitivities of the two methods were found to be comparable (Jansen et al. 1990). In addition to facilitating the testing of a larger number of samples, the fewer manipulations involved in AC/PCR may lessen risks of cross-contamination of reaction mixes or even carry-over between serial assays.

We have estimated that the detection limit of AC/PCR, when coupled with the detection of amplified products in Southern blots, approaches 3–30 virus particles per 80-µl sample (Jansen et al. 1990). However, we emphasize that this is only an approximation and that it is based in part on less-than-perfect information concerning the particle/infectivity ratio of the virus. Previous data suggest that there are approximately 100 RNA-containing particles of HAV for every infectious unit of virus present in clinical samples (Jansen et al. 1988). Thus, it is not surprising to see that AC/PCR may be more sensitive than virus isolation in cell culture and may be capable of detecting less than one infectious unit of virus. Such sensitivity is particularly important when dealing with a virus like HAV which is especially difficult to isolate in cell cultures. The AC/PCR method thus provides a high degree of specificity without loss of sensitivity while conveniently simplifying technical manipulations prior to PCR.

Selection of Sample Type

All of the work described above has been carried out using fecal suspensions prepared from infected patients or laboratory animals. However, a significant viremia accompanies acute hepatitis A (Lemon et al. 1990), and it would be simpler in many ways to have a method capable of detecting HAV in serum samples. While this has been accomplished by conventional PCR strategies involving isolation of the viral RNA prior to reverse transcription (Robertson et al. 1991), we have not been successful in detecting HAV in serum specimens by AC/PCR. Indeed, we have found that the addition of a small amount of serum to a fecal suspension containing HAV results in significant inhibition of the AC/PCR signal. The reasons for this inhibition remain undefined, but it is possible that it reflects reduced capture of virus due to the blocking effect of certain serum proteins which have high affinity for the HAV particle (Zajac et al. 1991). Alternatively, it may reflect carry-over of serum components that are inhibitory to RT or Taq polymerase.

Application of AC/PCR to Molecular Epidemiology of HAV

We have utilized AC/PCR followed by sequencing of the amplified product to study the genetic relatedness of HAV strains recovered from patients in different epidemiologic settings (Jansen et al. 1990, 1991). We compared the relatedness between strains that was present in two regions of the genome, one which encodes the carboxyl end of VP3 and the other which encodes the putative cleavage region at the VP1/2A junction. This latter region is analogous to the segment of the poliovirus genome which was studied by Rico-Hesse et al. (1987) in determining the genetic relatedness of poliovirus strains recovered in different geographic regions, and it contains greater sequence diversity than the VP3 region (Jansen et al. 1990). A sample dendrogram depicting a maximum parsimony analysis of the relatedness among various strains of HAV within the 172-base region encoding the putative VP1/2A domain (nucleotide positions 3020–3191, wild-type HM175 numbering) is shown in Fig. 3.

Worth noting is the choice of oligonucleotide primers employed for detection of HAV in epidemiologically diverse specimens. Of the several regions of the genome which have been examined, better resolution of strain differences may be obtained from sequence analysis of PCR products derived from the VP1/2A junction (Jansen et al. 1990). VP1/2A primer sets include oligonucleotides +2948 (or +2984) and −3265 (or −3285) (Table 1). With genetically divergent HAV strains, the use of alternative primers (+2891 or +2933, and −3192 or −3375, see Table 1) may be necessitated by amplification failures observed with primer sets based on the HM175-strain. When maximum sensitivity is required, however, and the ability to distinguish between different strains is not important, it may be reasonable to consider selection of primers representing much more conserved segments within the 5′-nontranslated (5′-NTR) region of the genome (Brown et al. 1991). Thus, primer selection is largely driven by the specific aims of the AC/PCR procedure: detection of the presence of virus or determination of strain relatedness.

Application of AC/PCR to Other Viruses

Because AC/PCR combines the sensitivity of a PCR-based procedure with the potential specificity of an antigen–antibody interaction, it may prove especially useful for the rapid detection and serotypic differentiation of viruses which are closely related genetically but widely divergent antigenically. Among the picornaviruses, HAV is relatively unique in having a highly conserved anti-

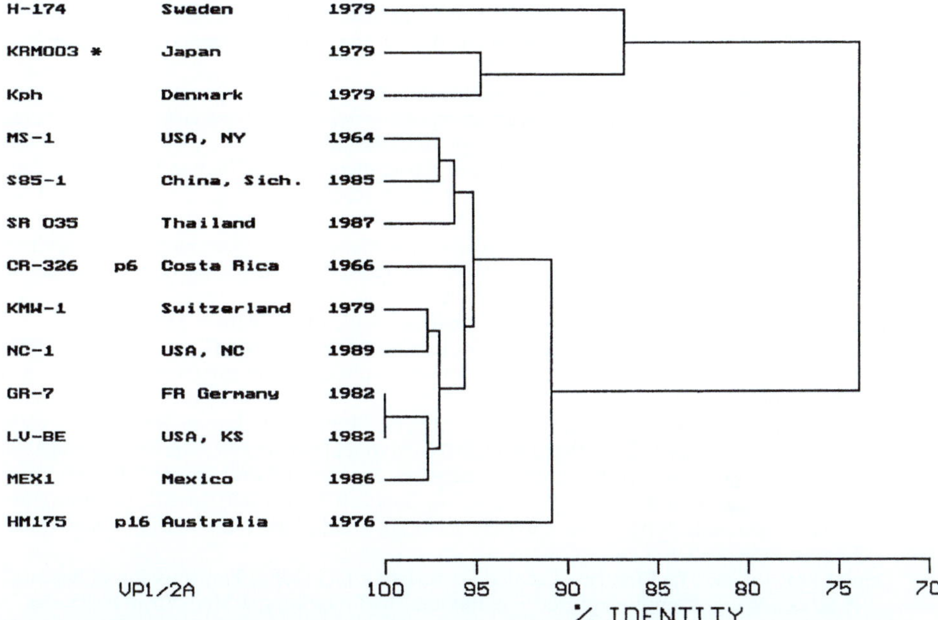

Fig. 3. Dendrogram showing approximate genetic relatedness among 13 strains of HAV recovered from epidemiologically and geographically different sources. The virus strain and geographic site and year of recovery are indicated at the *left*. CR326 and HM175 were studied from cell culture materials (passage level also noted). The approximate degree of relatedness between any two strains (percentage nucleotide identity in a 172-base region at the putative VP1/2A junction) is represented by the distance from the left of the chart to the first common node. GR-7 and LV-BE have identical sequences in this area, although they were recovered from widely separated geographic regions. Their close relatedness suggests a common epidemiologic source. Both virus strains were recovered during outbreaks of hepatitis A in American soldiers (Jansen et al. 1990). The sequence of KRM003 was provided by Yasuo Moritsugu of the National Institute of Health, Japan

genic structure (Lemon and Ping 1989). In contrast, the enteroviruses and rhinoviruses comprise a large group of medically important viruses with diverse antigenic phenotypes but closely related nucleotide sequences, particularly in the 5′-NTR of their genomes. It should prove possible to develop rapid and practical methods for the detection and serotypic characterization of these viruses when they are present in human clinical materials by combining AC (perhaps with pools of monospecific antibodies) and PCR with generic 5′-NTR primers.

References

Brown EA, Day SP, Jansen RW, Lemon SM (1991) The 5′ nontranslated region of hepatitis A virus: secondary structure and elements required for translation in vitro. J Virol 65:5828–5838

Jansen RW, Newbold JE, Lemon SM (1985) Combined immunoaffinity cDNA-RNA hybridization assay for detection of hepatitis A virus in clinical specimens. J Clin Microbiol 22:984–989

Jansen RW, Newbold JE, Lemon SM (1988) Complete nucleotide sequence of a cell culture-adapted variant of hepatitis A virus: comparison with wild-type virus with restricted capacity for in vitro replication. Virology 163:299–307

Jansen RW, Siegl G, Lemon SM (1990) Molecular epidemiology of human hepatitis A virus defined by an antigen-capture polymerase chain reaction method. Proc Natl Acad Sci USA 87:2867–2871

Jansen RW, Siegl G, Lemon SM (1991) Molecular epidemiology of human hepatitis A virus. In: Hollinger FB, Lemon SM, Margolis HS (eds) Viral hepatitis and liver disease. Williams and Wilkins, Baltimore, pp 58–62

Kwok S, Higuchi R (1989) Avoiding false positives with PCR. Nature 339:237–238

Lednar WM, Lemon SM, Kirkpatrick JW, Redfield RR, Fields ML, Kelley PW (1985) Frequency of illness associated with epidemic hepatitis A virus infections in adults. Am J Epidemiol 122:226–233

Lemon SM, Binn LN, Marchwicki R, Murphy PC, Ping L-H, Jansen RW, Asher LVS, Stapleton JT, Taylor DG, LeDuc JW (1990) In vivo replication and reversion to wild-type of a neutralization-resistant variant of hepatitis A virus. J Infect Dis 161:7–13

Lemon SM, Brown CD, Brooks DS, Simms TE, Bancroft WH (1980) Specific immunoglobulin M response to hepatitis A virus determined by solid-phase radioimmunoassay. Infect Immun 28:927–936

Lemon SM, LeDuc JW, Binn LN, Escajadillo A, Ishak KG (1982) Transmission of hepatitis A virus among recently captured Panamanian owl monkeys. J Med Virol 10:25–36

Lemon SM, Ping L-H (1989) Antigenic structure of hepatitis A virus. In: Semler B, Ehrenfeld E (eds) Molecular aspects of picornavirus infection and detection. American Society for Microbiology, Washington, pp 193–208

Rico-Hesse R, Pallansch MA, Nottay BK, Kew OM (1987) Geographic distribution of wild poliovirus type 1 genotypes. Virology 160:311–322

Robertson BH, Khanna B, Nainan OV, Margolis HS (1991) Epidemiologic patterns of wild-type hepatitis A virus determined by genetic variation. J Infect Dis 163:286–292

Zajac AJ, Amphlett EM, Rowlands DJ, Sangar DV (1991) Parameters influencing the attachment of hepatitis A virus to a variety of continuous cell lines. J Gen Virol 72:1667–1675

Chapter 7 Clinical Diagnosis of Hepatitis B Infection: Applications of the Polymerase Chain Reaction

Geoffrey Dusheiko[1], Jianye Xu[1], and Arie J. Zuckermann[1,2]

Summary

Hepatitis B virus (HBV) infection serves as a good example of the use of the polymerase chain reaction (PCR) for the diagnosis of viral infection. The technique is at least 10^4 times more sensitive than current dot blot hybridization assays for HBV DNA. PCR overcomes the disadvantage of low concentrations of HBV DNA in patients with a low level of viraemia, and permits the detection, cloning, and sequencing of HBV genomes from such patients. Use of this assay indicates that loss of HBeAg is associated with a decrease rather than complete disappearance of HBV DNA.

A number of other applications of PCR have been informative. The most interesting application of PCR has been to study the genetic heterogeneity of HBV. These investigations have identified variants of HBV by generating fragments of DNA that could be sequenced either directly or by subcloning the amplified fragments into M13 and sequencing. Critical point mutations have been detected by this means. Variants of HBV have been described which are not recognized by the classical serological reagents for HBsAg. Vaccine escape mutants, characterized by a point mutation from guanosine to adenosine, have been recognized. These mutants are not neutralized in anti-HBs-positive vaccines.

Genetic mutants have recently been described in serum of anti-HBe and HBV DNA positive patients. PCR amplification and subsequent sequencing of genomic DNA in patients with this form of chronic hepatitis B has indicated that one or more nucleotide substitutions in the pre-C region of the genome account for the absent expression of HBeAg.

Thus PCR methodology has already altered our perceptions of the natural history of chronic hepatitis B infection, and is likely to shed further light on this complex disease. The technique is proving a useful utility in the diagnosis of hepatitis B and a valuable research tool for advancing our understanding of HBV variants, vaccine escape mutants, and hepatocellular carcinoma.

[1] Royal Free Hospital School of Medicine, Rowland Hill Street, London NW3 2PF
[2] WHO Collaborating Center for Reference and Research on Viral Diseases, London, UK

Introduction

A number of viral infections have been usefully studied using the polymerase chain reaction (PCR) to amplify viral genomes. Hepatitis B virus (HBV) infection serves as a good example of the use of this technique for the diagnosis of viral infection (Lampertico et al. 1990; Zeldis et al. 1989; Brechot 1990). HBV is a 42 nm long, enveloped, double-stranded DNA virus causing acute and chronic hepatitis infection. The virus contains an outer envelope, hepatitis B surface antigen (HBsAg), and an inner nucleocapsid (HBcAg). In addition to virions, 20-nm spheres and tubules, which consist of HBsAg without HBcAg or DNA, are found in excess in serum. The genome of HBV is a small, circular, double-stranded DNA of about 3200 bases. Four open reading frames (ORFs) encoding the *pol*, C, S and X genes have been characterised. The complete sequence of several HBV isolates are known (Tiollais et al. 1985).

Similar viruses are found in woodchucks, ground squirrels and ducks. This family of viruses, termed hepadnaviridae, uses a mode of replication peculiar to it, involving a cycle of RNA synthesis from covalently closed DNA, reverse transcription of pregenomic RNA to minus-strand DNA, and DNA plus-strand synthesis. HBV DNA may become integrated, but this process is not a necessary step in the life cycle. These viruses could therefore be grouped within a superfamily of retroid viruses or reversiviruses.

The surface protein consists of three envelope proteins. These originate from one gene by the alternative use of three start codons for protein synthesis (Neurath et al. 1985) (Fig. 1). These sequences, known as pre-S1, pre-S2, and S, all constitute part of the HBsAg. The terms LHBS, MHBS and SHBS have been suggested for these proteins. It is likely that attachment to hepatocytes and infectivity are related to pre-S1 sequences.

The two nucleocapsid proteins, HBcAg and HBeAg, are the products of a single gene region on the viral genome (Stahl et al. 1982). The C gene has two initiation codons and two gene regions (pre-C and C) encoding these two potential molecular forms (Figs. 1, 2). Initiation of translation at the first site (nucleotide 1814) produces a 312 amino acid polypeptide (P25) which has a signal peptide directing it to the endoplasmic reticulum where the signal piece is removed by signal peptidases, cleaving the N-terminal 19 amino acid residues. With subsequent processing, the C-terminal 34 amino acid residues are removed, and the resultant polypeptide is secreted as HBeAg (p15–18), a soluble protein that is the product of 10 residues coded by the pre-C region and 149 residues coded by the C gene (Ou et al. 1986). Part of the HBeAg protein may be expressed in the cell membrane, and secreted HBeAg may induce immune tolerance against HBeAg (Schlicht et al. 1991). HBeAg does not, however, appear necessary for HBV replication and infectivity.

Translation from the second initiation codon (nucleotide 1901) results in unprocessed core polypeptides which are assembled into HBcAg particles. RNA or DNA is packaged within the core particles, rendering the genome resistant to nucleases. An envelope is then acquired, and the virion is secreted.

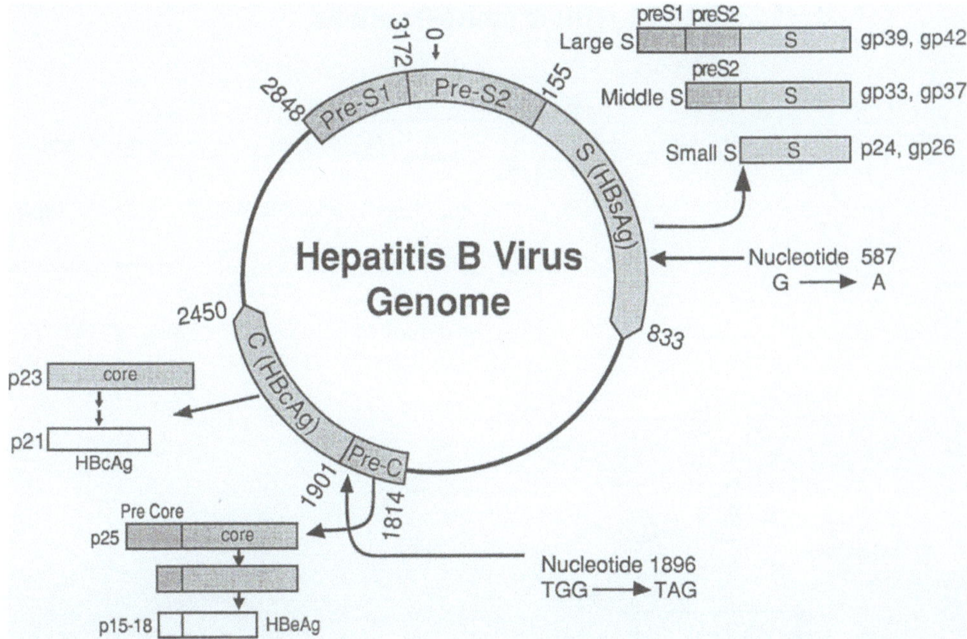

Fig. 1. Genomic map of hepatitis B virus (HBV) showing the sites of mutation of vaccine escape and precore mutants (modified from Peters et al. 1991)

HBeAg therefore has substantial homology to HBcAg, and the proteins are cross-reactive at the T-cell level. Core particles of HBV are potent immunogens, inducing T-cell-independent production of anti-HBc.

HBV may cause acute or chronic hepatitis B. Acute hepatitis is defined as a self-limiting disease of the liver marked by acute inflammation and hepatocellular necrosis in association with a transient HBV infection. Chronic hepatitis B is defined as persistent HBV infection accompanied by evidence of hepatocellular injury and inflammation. The disease may be life-long. The proportion of infected individuals who develop chronic hepatitis varies inversely according to the age of the patient. The disease may vary from silent or asymptomatic disease to severe symptomatic chronic active hepatitis, cirrhosis or hepatocellular carcinoma (HCC). Extrahepatic manifestations such as glomerulonephritis, vasculitis or polyarteritis may also occur.

HBsAg carriers vary in infectivity, but highly viraemic patients may circulate up to 10^8 particles per milliliter of serum. There is epidemiological variation in the prevalence of HBV infection, reflecting differences in the modes and ages of transmission in developed and developing regions. For example, in sub-Saharan Africa and China, transmission occurs predominantly in childhood, whereas in Western Europe and North America, the disease occurs predominantly in high-risk adults.

HBV C gene Products

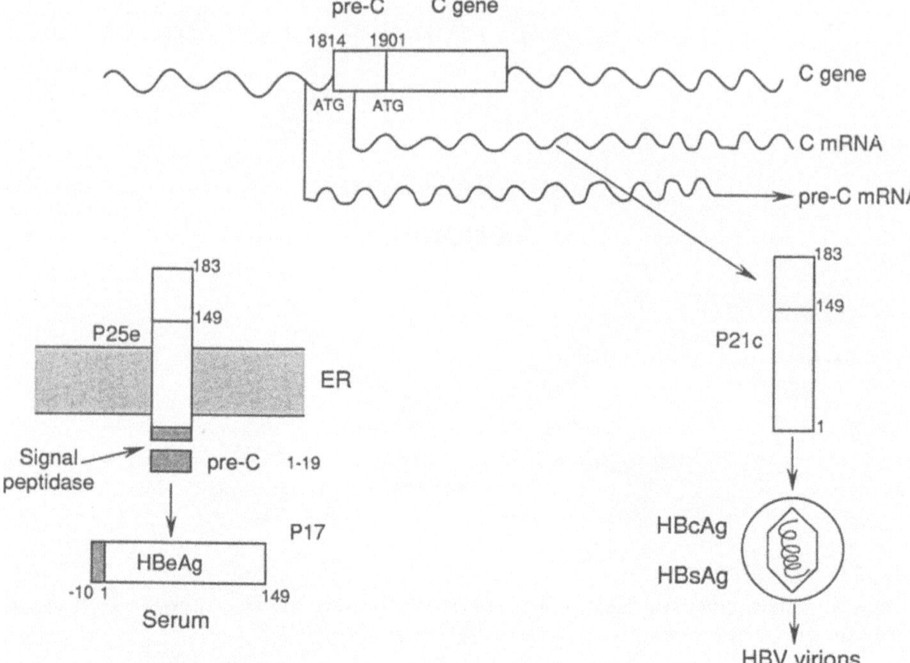

Fig. 2. Schematic representation of the two different synthetic pathways of the hepatitis B virus (HBV) core gene products (modified from Schlicht et al. 1989)

The pathogenesis of the disease is incompletely understood. HBV may perhaps cause damage by a direct cytopathic action. However, immunologic mechanisms appear more important, since most evidence suggests that HBV is not cytopathic and that the expression of disease involves a poorly understood interplay between viral and host factors. There is a propensity to chronic infection if infection occurs at a young age.

Clinical Diagnosis of Hepatitis B

Serological Markers

Serological markers for the clinical diagnosis of HBV have been well characterised. With the advances in molecular biology, newer techniques complementing the established immunoassays have been developed which facilitate making a diagnosis and have accelerated hepatitis research. The most widely used test for the diagnosis of hepatitis B is assay for HBsAg. Current immunoassays for HBsAg detect 100–200 pg HBsAg per ml of serum, corresponding to roughly 3×10^7 particles per ml. Immunoassays for HBeAg and anti-HBe are commercially available. Most HBeAg-positive carriers have more than 10^5 genomes per ml serum, whereas most anti-HBe-positive carriers have less than this number. Immunity to hepatitis B is characterised by the presence of anti-HBs and anti-HBc in serum.

Assay of Viral DNA

The dot blot hybridisation test for serum HBV DNA correlates with infectivity and is an important means of determining the presence of viral genomes. Typically, a serum sample is lysed by sodium hydroxide and filtered through a nitrocellulose membrane; specific binding of DNA occurs. The membrane is then incubated with cloned labelled HBV DNA, and if the test sample contains HBV DNA, annealing to the membrane-bound DNA permits detection by autoradiography. Usually probes are labelled by nick translation using deoxynucleotide [^{32}P] triphosphate. Non-radioactive probe labelling is also possible (Seibl et al. 1990). HBV DNA in serum is expressed in pg/ml or, alternatively, genome equivalents per ml. One picogram corresponds to 2.86×10^5 genome equivalents if a molecular weight of 2.1×10^6 is assumed for full-length, double-stranded HBV DNA. The usual range of sensitivity is $0.1-1$ pg or 10^5 genome equivalents per ml of serum (Bonino et al. 1981). The calculated weight of HBV DNA equivalent to one virion is approximately 3×10^{-6} pg. Thus, at least 10^{3-5} virions have to be present in a specimen to be detected by dot blot assay. One problem with the dot blot assay is their variable sensitivity when performed in different laboratories.

This problem is partly overcome by a liquid phase hybridization assay for HBV DNA, in which HBV DNA extracted by virion lysis is mixed with ^{125}I-labelled nucleic acid probe and allowed to hybridize, and then the free and hybridized probes are separated by column chromatography. Hybrid molecules are eluted by this system, whereas unhybridized [^{125}I]-HBV probe is retained by the column. The sample and eluate are counted in a gamma-counter, and samples with counts above the cut-off value determined by running positive and negative controls are considered positive. The linearity of the assay allows the positive control to be used as a standard for quantitation of HBV DNA levels in pg/ml. Although better standardised, the assay is costly and cannot be easily used to process a large number of samples.

Southern Blot Hybridisation

This assay is most useful in the study of liver DNA samples, allowing detection of integrated or free episomal HBV DNA. Integrated HBV DNA can be detected if a restriction enzyme is used which does not cleave HBV DNA (*Hind*III) (Shafritz et al. 1981). Integrated HBV DNA has been found in hepatocellular carcinoma and in HBsAg-negative samples (Brechot et al. 1981 a).

Polymerase Chain Reaction

PCR overcomes the disadvantage of low concentrations of HBV DNA in patients with a low level of viraemia and permits the detection, cloning and sequencing of HBV genomes from such patients. As in other applications, PCR detection of HBV DNA depends upon the in vitro production of large amounts of HBV DNA fragments of defined length and sequence from small amounts of DNA using short oligonucleotide sequences as primers. The reaction typically involves an extraction step, thermal denaturation of double-stranded target molecules and the addition of synthetic oligonucleotides as primer pairs. The primers anneal to complementary sequences, and these are extended by enzymatic synthesis with DNA polymerase. The procedure is performed in a programmable thermal cycler which automatically repeats the cycles of heat denaturation, primer annealing and Taq polymerase enzymatic extension of HBV DNA.

PCR is extremely sensitive, theoretically allowing the detection of one genome equivalent per sample. It is at least 10^4 times more sensitive than dot blot hybridisation assays for HBV DNA. It is a somewhat laborious procedure, depending upon the extraction method used, which usually involves proteinase K and chloroform-phenol extraction and ethanol precipitation. Amplified DNA can be detected by ethidium bromide staining after gel electrophoresis, but this is relatively insensitive. The size of the fragment is determined by the distance between the primer pairs. In practice, HBV DNA can be detected by ethidium bromide staining after 50 cycles of PCR of 3×10^{-6} pg cloned HBV (Yokosuka et al. 1991). The sensitivity can be improved to detection of less than 0.1 fg (100 ag) of original DNA if the amplified DNA is hybridised to labelled probe and detected by autoradiography. Indeed, the identity of the PCR amplification product should be confirmed by hybridisation. The probe should contain only the sequences between the primer sequences to ensure specificity. Contamination remains the major difficulty of the procedure.

Modifications of the method include capture of viral genome from serum using a high-affinity IgM monoclonal antibody directed against a common determinant of HBsAg and the subsequent amplification of viral DNA by PCR (Liang et al. 1989). By using a nested primer technique, the sensitivity and specificity of the test are improved (1 molecule per 10^6 cells), and the amplified product can be visualised in a same day test.

The quantitation of the test remains approximate, although methods have been developed to assess semi-quantitatively the amounts of target DNA detected by PCR.

Diagnosis of Acute Hepatitis B

The symptoms and signs of acute hepatitis are non-specific, but serological testing provides an accurate diagnosis. The typical serological course of acute

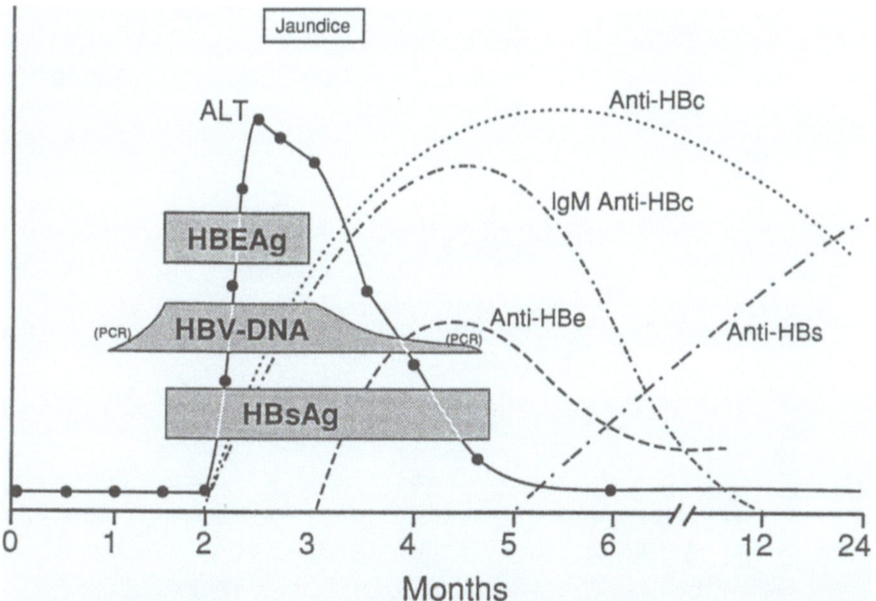

Fig. 3. Serologic pattern of acute hepatitis B infection indicating the period of detection of hepatitis B virus (HBV) DNA by polymerase chain reaction (PCR)

hepatitis using available immunoassays is shown in Fig. 3. HBsAg is the first marker to appear in serum. Shortly thereafter, HBeAg and HBV DNA become detectable (Krogsgaard et al. 1985). Levels of HBV DNA usually reach 10^5 to 10^8 genome equivalents per ml with the onset of symptoms, after which the levels decrease, becoming undetectable within a few weeks. In contrast in patients who develop chronic hepatitis B, levels of HBV DNA remain high. Measurement of HBV DNA by PCR in acute hepatitis has demonstrated the greater sensitivity of this technique. In the majority of patients, HBV DNA measured by PCR persists as long as HBsAg is present. In experimentally infected chimpanzees, HBV DNA could be detected by PCR before HBsAg and for an average period of 3 weeks after detection of anit-HBs (Kaneko et al. 1990).

Diagnosis of Chronic Hepatitis B

The typical serological course of a patient with chronic hepatitis B is shown in Fig. 4. The disease is complex, and the course in individual patients is variable over time. Some patients may develop cirrhosis over 5–20 years or even HCC. A spontaneous remission in disease activity may occur in approximately 10% – 15% HBeAg-positive carriers per year, characterised by the disappearance of

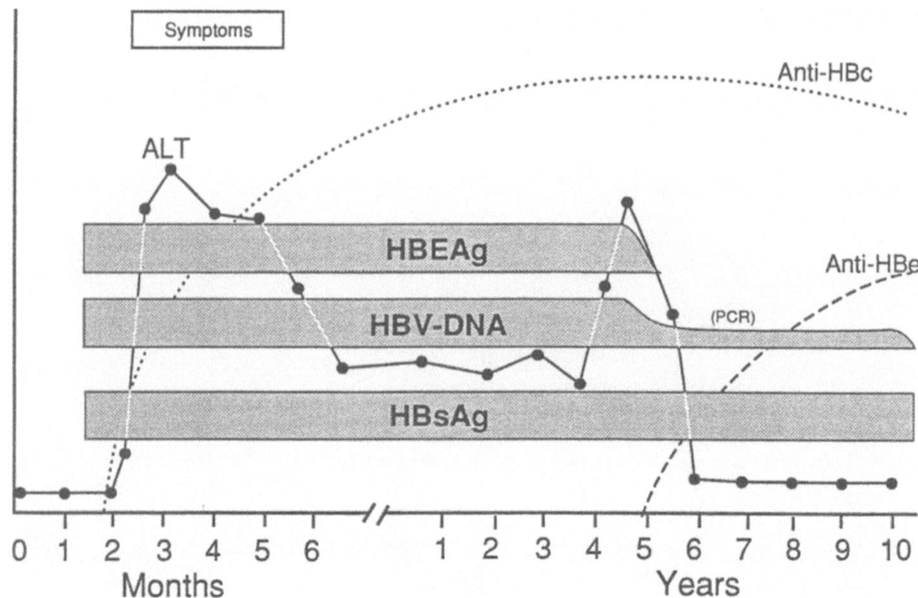

Fig. 4. Serologic pattern of chronic hepatitis B infection indicating the period of detection of hepatitis B virus (HBV) DNA by polymerase chain reaction (PCR)

HBV DNA from the serum, followed by loss of HBeAg. This may occur following a sudden exacerbation in serum aminotransferases. Once HBeAg is cleared, the disease remits, and serum aminotransferases become normal.

Currently, PCR is the most sensitive technique for the measurement of viraemia in hepatitis B. Use of this assay indicates that loss of HBeAg is associated with a decrease rather than a complete disappearance of HBV DNA. With loss of HBeAg, HBV DNA becomes undetectable by dot blot hybridisation but is still detectable by PCR (Kaneko et al. 1990). This finding apparently represents low levels of DNA replication but is frequently not associated with active liver disease; the continued persistence of HBV infection without disease is unexplained and may be due to infection with defective genomes or containment by the immune system.

Reactivation of active viral replication can occur in 10%–15% of patients after clearance of HBeAg; the residual persistence of HBV DNA by PCR measurement may explain this phenomenon (Marcellin et al. 1991). Vaccine failure may also be explained by the presence of HBV DNA in the breast milk of mothers and cord blood of infants (Mitsuda et al. 1989). When HBV DNA has been measured by PCR in carriers, viral genomes have generally been found in the majority of HBsAg-positive carriers. Virtually all HBeAg-positive patients and most anti-HBe-positive carriers are positive (Kaneko et al. 1989). HBV DNA has also been found for a period in up to one-third of prospectively followed chronic carriers who lost HBsAg but has not usually been found in patients who lost HBsAg after acute hepatitis B (Kaneko et al. 1990). These results indicate the potential infectivity of anti-HBe-positive carriers (Krogsgaard et al. 1986; Liaw et al. 1991).

Further Applications

A number of other applications of PCR have been informative. After successful anti-viral therapy, HBV DNA measured by dot blot hybridisation disappears, but in our experience and that of others, HBV DNA can be detected for more than a year after loss of HBeAg (Gerken et al. 1991). Using PCR, HBV DNA has been detected in white blood cells. Recurrence of HBV infection is a major disadvantage of liver transplantation undertaken for cirrhosis due to hepatitis B (Muller et al. 1991; Samuel et al. 1991). HBV DNA has been demonstrated in the liver tissue of patients in whom HBsAg reappeared and in the peripheral blood mononuclear cells of patients whose serum and liver were negative by PCR for HBV DNA, identifying a potential mechanism of re-infection from extrahepatic sites (Feray et al. 1990). The successful detection of HBV DNA in formalin-fixed, paraffin-embedded liver tissue by PCR has enabled the detection of low levels of viral sequences and viral deletions (Lampertico et al. 1990). The general availability of embedded specimens provides a vast resource for future studies.

Typing of specimens can provide evidence for intrafamilial transmission, and this technique will undoubtedly be further exploited (Lin et al. 1990).

Genetic Variability

The most interesting application of PCR has been to study genetic heterogeneity of HBV. These investigations have identified variants of HBV either by generating fragments of DNA that could be sequenced directly or by subcloning the amplified fragments into M13 and sequencing. Critical point mutations have been detected by this means.

Envelope Protein Variability

Generally, no more than 10% nucleotide sequence variation is found within human isolates of HBV. HBs epitopes present in all previously described HBsAg subtypes have been grouped together as the *a* determinant. Hypervariable regions have been identified in HBs proteins, particularly the pre-S1 domain. The resulting differences in the amino acid sequence determine the known HBsAg subtype alleles *d/y* or *w/r*. These exchanges have been mapped. For example, codon 122 determines *d/y*, i.e. a lysine is found in subtype *d* compared with an arginine for subtype *y*. Codon 160 determines *w/r*, namely *w*/lysine and *r*/arginine (Okamoto et al. 1987). These amino acid exchanges have recently been confirmed by PCR and direct sequencing (Yokosuka et al. 1991).

HBV replication has been detected by PCR in anti-HBc- and anti-HBs-positive carriers. In analyses of blood donors and outpatients in Taiwan, between 4% and 11% of HBsAg-negative individuals were positive for HBV DNA by PCR; based on the incidence of posttransfusion hepatitis in recipients, approximately 0.04% of HBsAg-negative donors were considered infectious. Although this has obvious implications for blood screening, the practicality of testing donors in this way remains doubtful (Wang et al. 1991; Jackson 1991; Shih et al. 1990).

Variants of HBV have been described which are not recognised by the classic serological reagents for HBsAg. The finding of HBV DNA in persons negative for all serological markers of HBV has been taken to suggest that there are genetic variants of HBV accounting for some cases of hepatitis not attributable to hepatitis B or non-A, non-B hepatitis (NANB). A particularly high prevalence of HBV DNA has been found in HBsAg-negative, alcoholic patients with HCC (Brechot et al. 1981 b, 1982, 1985). In addition, HBsAg determinants have been identified in HBsAg-negative sera with HBs-specific monoclonal antibodies and HBV DNA by dot blot hybridisation (Liang et al. 1990; Wands et al. 1986). Inoculation of such HBsAg-negative human sera to chimpanzees has induced hepatitis in the infected animals, and subsequent cloning of the isolates, after PCR amplification and comparison of the nucleotide sequence with that of known HBV subtypes, revealed point mutations in the 3'-end of the S gene (Thiers et al. 1988). PCR was able to identify variation in HBV genomes, but the reason for the different serological patterns remains unexplained.

The finding of HBV DNA in HBsAg-negative persons has several possible explanations: this may perhaps reflect HBsAg hidden in circulating immune complexes, absent expression of HBV genes, absence of humoral responses, infection caused by different strains of HBV (see below) or, most likely, the greater sensitivity of PCR than immunological assays. Mutations in the genome may affect viral replication or secretion of the virus.

Escape Mutants

HBsAg has been identified in vaccinated children who developed HBV infection despite the presence of protective titres of anti-HBs in serum. In a survey in Italy, HBsAg was found in 44 (2.8%) of 1590 vaccinated children, all of whom were anti-HBs-positive. In these children, anti-HBs was uncomplexed to the *a* determinant. Analysis of HBsAg with monoclonal antibodies showed that the circulating antigen did not carry the *a* determinant or that this determinant was masked. HBsAg sequences were determined by PCR in a carrier child by Carman et al. (1990). A point mutation from guanosine to adenosine was found at nucleotide 587, with the result that a glycine was substituted for an arginine at position 145 in the *a* determinant, thus masking the *a* determinant and preventing virus neutralisation. Recently, McMahon et al. (1990) described an identical mutation in a liver transplant recipient who re-developed HBV infection after treatment with monoclonal anti-HBs. Such escape mutants have also been described from vaccine recipients from Singapore, and clearly further surveillance of vaccinated individuals is indicated. A modification of future HBV vaccines to prevent the emergence of this escape mutant may be necessary.

Recently, a second type of HBV (HBV2) was described in West Africa (Coursaget et al. 1987). This variant appeared unable to induce HBc-specific antibody. However, a more recent appraisal suggests that this is accounted for by lack of anti-HBc response rather than by a serotype of HBV.

C Gene Mutations

Genetic mutants have recently been described in serum of anti-HBe- and HBV DNA-positive patients who have HBcAg in hepatocytes and histological evidence of chronic active hepatitis but who lack HBeAg in serum (Hadziyannis et al. 1983; Carman et al. 1989; Brunetto et al. 1990). Typically, such patients are more common in Southern Mediterranean countries, but these mutant HBVs have been diagnosed from several areas worldwide. The carriers tend to have severe disease, with raised alanine aminotransferase (ALT) concentrations and acute exarcerbations of hepatitis (Raimondo et al. 1990), associated with intermittent serum HBV DNA positivity and focal HBcAg histologically, which is predominantly cytoplasmic. They are frequently IgM anti-HBc positive. PCR amplification and subsequent sequencing of genomic DNA in patients with this form of chronic hepatitis B have indicated that one or more nucleotide substitutions in the pre-C region of the genome account for the absent expression of HBeAg. The most common mutation is a point mutation from guanine to adenine creating an in-frame stop codon, i.e. converting codon 28 for tryptophan (TGG) to a termination or nonsense codon (TAG) (Fig. 1). An example comparing a patient with a mixture of virions, including virions with a TAG sequence, with a patient with wild-type sequence at position 1896 is shown in Fig. 5. Additional point mutations in succeeding codons may or may not be present. A mutation in the translation initiation codon preventing expression of the HBeAg two stop codons and frameshift mutations have also been described in these patients (Raimondo et al. 1990a, b; Fiordalisi et al. 1990; Tong et al. 1990; Akahane et al. 1990). Most samples tested contain a mixture of wild-type and mutated viruses.

Distinct viraemic phases can be detected in these chronically infected patients. It is not clear whether these defective viruses represent de novo infections with a mutant-type virus or, as is perhaps more likely, whether these mutations have arisen during infection as a result of immune selection pressure. Clearly, however, the results suggest that the HBV genome contains segments of high variability that could have been selected during evolution to

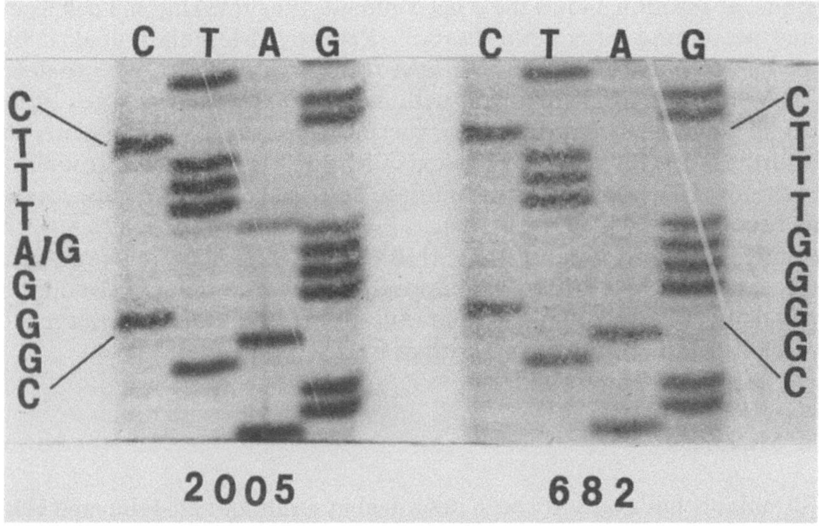

Fig. 5. Direct sequencing of two isolates of hepatitis B virus (HBV) DNA from serum. The results show sequence analysis of the pre-C region. Patient 2005 has a mixture of virions in serum including virions with a TAG (termination sequence) and TGG (wild type) at position 1896, while patient 682 shows wild-type TGG sequence

favour the segregation of mutants with a functional advantage capable of avoiding host immunity. HBV may be evolving continuously during persistent infection. The process of HBV replication is prone to mutations, as HBV replicates via transcription of an RNA intermediate using reverse transcriptase, an enzyme lacking in proof-reading capacity.

In order for phenotypes different from predecessor HBV to become dominant, a selective force would have to operate in their favour. It has been suggested that hepatocytes infected with wild type produce HBeAg molecules which would be exposed on the cell surface; most evidence suggests that the nucleocapsid antigens expressed on the cell membrane are the important target of the immune response and cytolytic T cells. HBeAg expressed at the cell membrane via the secretory pathway represents a potential target for antibody-mediated elimination of virus-infected cells (Schlicht et al. 1991). However, hepatocytes infected with pre-C mutants would be unable to produce HBeAg, and such liver cells without expression of HBeAg might escape immune elimination. Immune selection may thus assist pre-C defective mutants with an HBeAg-negative phenotype to exceed HBeAg-positive phenotypes.

In HBeAg-positive patients, the ratio of pre-C-minus variants to wild-type HBV can increase during the course of the infection (Okamoto et al. 1990); pre-C region defects have not generally been predominant from sera studied in the HBeAg phase, but a population of pre-C mutants emerges in patients seroconverting to anti-HBe. Mixed infections can occur in both HBeAg- and anti-HBe-positive patients. Usually, seroconversion to anti-HBe is accompanied by a remission in disease activity: why some patients develop severe disease and not others has yet to be explained.

Pre-C mutant HBV genomes have been also detected in patients seroconverting after interferon treatment. The defective virions were found in patients who remained positive for anti-HBe and in whom hepatitis resolved. In contrast, the defect was not found in patients who seroconverted to anti-HBe but who then re-developed HBeAg with reactivation of the hepatitis (Takeda et al. 1990). It has been suggested that the presence of circulating pre-C mutants could predict which patients would respond favourably to interferon; using this approach, we analysed whether the presence of pre-C mutants is a determinant of responsiveness to α-interferon therapy. Fifteen carriers (9 responders and 6 non-responders) who were treated with interferon were examined. Serum samples were collected before and after therapy. After extraction of DNA, the pre-C region was amplified by PCR and the product identified by gel electrophoresis, ethidium bromide staining and then Southern blotting and molecular hybridisation (Fig. 6). The products were amplified by a modified PCR and the pre-C region directly sequenced. Circulating HBeAg(−) mutants were not identified prior to treatment in either responders or non-responders. These results suggest that a predominance of HBeAg(−) virions can be considered the overriding factor in determining response to treatment.

The ability to detect the most prevalent form of pre-C-defective mutants in a larger number of patients has been facilitated by using selective oligonucleotide hybridisation. In this technique, the pre-C region is amplified and detect-

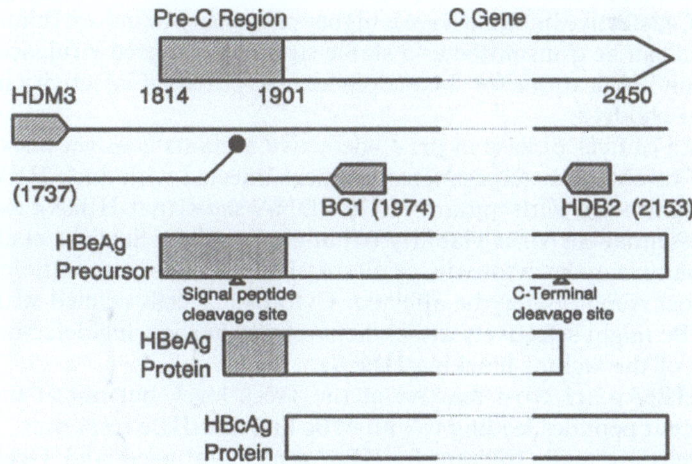

Fig. 6. Primers used to amplify and sequence for pre-C mutants. HDM3:5′GCGCTGC-AGGAGTTGGGGGGAGGAGATTA, at map position 1737–1755 of the HBV genome. HDB2:5′GCGAAGCTTAGATCTCTGGATGCTGGA at map position 2153–2133. For asymmetric amplification, 10 μl of the amplified product was used for nested assymetrical PCR with the primers HDM3 and BC1 (sequence position 1974–1954). The concentration of primer HDM3 was 50 pmol and of primer BC1, as limiting, primer, was 0.5 pmol. BC1 was also used as sequencing primer

ed by PCR, and stringent hybridisation and washing conditions (63 °C) are used to anneal the product to 21 or 22 mer non-mutated or point-mutated oligonucleotide probes (Li et al. 1990). Using this approach to study a large number of patients, a mixed but predominantly wild-type viral population was found in persistently HBeAg-positive patients. The predominant viral population in anti-HBe-positive patients with exacerbated disease consisted of HBeAg pre-C-defective HBV (Brunetto et al. 1991).

Fulminant Hepatitis

Fatal or fulminant hepatitis is an atypical outcome of HBV infection, occurring in less than 1% of icteric cases. Morphologically, the liver shows massive necrosis. Bleeding and hepatic encephalopathy supervene. The pattern of rapid disappearance of HBsAg and HBeAg was previously held to indicate an exaggerated immune response to the infection. Newer evidence suggests that the disease may be related not only to host factors but also to the viral strain.

Patients with fulminant hepatitis B cluster among contacts with blood positive for anti-HBe (Oren et al. 1989; Fagan et al. 1986). Indeed a number of recent studies have shown that HBV clones propagated from the sera of patients with fulminant hepatitis B show pre-C region defects (Kosaka et al. 1991), whereas clones propagated from patients with acute hepatitis B without hepatitic failure serving as controls failed to show such defects.

In some of the cases, the mutants were present in carrier mothers who transmitted the disease to their children (Terazawa et al. 1991). These findings suggest that pre-C-defective mutants have a higher propensity to induce fulminant hepatitis and can be transmitted as a stable strain with altered virulence. This has important implications for the prophylaxis of potential infectivity in anti-HBe-positive children.

The mechanism of liver disease in pre-C-defective mutants is as yet unexplained and may be complex. Experiments in ducks infected with duck HBV or hepG2 cells transfected with mutant e-HBV DNA show that HBeAg expression is not essential for virus viability (Chang et al. 1987; Schlicht et al. 1987). These variants may be cytotoxic, or alternatively, the immunopathologenicity of the host response may be affected. Cytotoxic T cells primed with anti-HBc/anti-HBe might selectively attack hepatocytes harbouring defective mutants because of the lack of blocking HBeAg.

HBcAg and HBeAg are cross-reactive at the T-cell level, but not at the B-cell level, different peptides leading to anti-HBc and anti-HBe responses. T cells but not B cells are made tolerant by HBeAg neonatal mice, and T-cell tolerance elicited by HBeAg also extends to HBcAg-specific T cells. The combination of T-cell tolerance to nucleocapsid antigens in the presence of B-cell response to HBcAg may be important (Milich et al. 1989, 1990).

It is also feasible that lack of serum HBeAg may simply represent a phenotypic feature associated with a particular viral strain responsible for severe disease; this possibility requires exclusion by experiments in which complete sequencing of the isolated genomes is undertaken. It is uncertain whether pre-C peptides are expressed in these patients, which could have a functional or cytopathic significance.

Hepatocellular Carcinoma

A large body of scientific evidence has confirmed a relationship between chronic HBV infection and HCC, but the mechanism of oncogenesis is unknown. Integration of the HBV genome into hepatocyte DNA may be an important initiating factor. Experimental evidence suggests that HBV may play an indirect role in inducing malignant transformation, and thus development of HCC, by insertional mutagenesis, production of novel fusion genes, chromosomal deletions, translocations, loss of tumour suppressor alleles or oncogene activation.

PCR has been used to detect HBV DNA in a proportion of patients with HCC with antibody to hepatitis B or lacking serological evidence of HBV infection. These genomes were apparently transcriptionally active as RNA could also be detected. The role of the viral DNA in the pathogenesis of these tumours is not known, however.

Several recent reports have described a specific allelic deletion from chromosome 17p and a mutation in the P53 gene (a putative tumour suppressor gene) in South African and Chinese patients with HCC. The mutation resembles that seen in aflatoxin-induced experimental carcinogenesis (Hosono et al. 1991; Bressac et al. 1991; Hsu et al. 1991).

Conclusions

PCR methodology has already altered our perceptions of the natural history of chronic hepatitis B infection and is likely to shed further light on this complex disease. The technique is proving useful in the diagnosis of hepatitis B and a valuable research tool to advance our understanding of HBV variants, vaccine escape mutants and HCC. Future studies using this methodology, will, however, require a critical evaluation of the results and methods and demand rigid controls and more precise quantitation to obtain consistent results. The probes and primers used will require a degree of standardisation (Kaneko and Miller 1990a).

At present, the assay does not replace serological testing but is a useful adjunct to the presently available assays. The feasibility of PCR for blood donor screening remains doubtful. It should be remembered that HBV mutations are being studied in isolation, and the effects upon viral replication or epidemiological selection are as yet uncertain. Several major implications for the virulence, chronicity and prevention of hepatitis B have already been uncoverered, however.

References

Akahane Y, Yamanaka T, Suzuki H, Sugai Y, Tsuda F, Yotsumoto S, Omi S, Okamoto H, Miyakawa Y, Mayumi M (1990) Chronic active hepatitis with hepatitis B virus DNA and antibody against e antigen in the serum. Disturbed synthesis and secretion of e antigen from hepatocytes due to a point mutation in the precore region. Gastroenterology 99:1113–1119

Bonino F, Hoyer B, Nelson J, Engle R, Verme G, Gerin J (1981) Hepatitis B virus DNA in the sera of HBsAg carriers: a marker of active hepatitis B virus replication in the liver. Hepatology 1:386–391

Brechot C (1990) Polymerase chain reaction. A new tool for the study of viral infections in hepatology. J Hepatol 11:124–129

Brechot C, Hadchouel M, Scotto J, Degos F, Charnay P, Trepo C, Tiollais P (1981a) Detection of hepatitis B virus DNA in liver and serum: a direct appraisal of the chronic carrier state. Lancet 2:765–768

Brechot C, Hadchouel M, Scotto J, Fonck M, Potet F, Vyas GN, Tiollais P (1981b) State of hepatitis B virus DNA in hepatocytes of patients with hepatitis B surface antigen-positive and -negative liver diseases. Proc Natl Acad Sci USA 78:3906–3910

Brechot C, Nalpas B, Courouce AM, Duhamel G, Callard P, Carnot F, Tiollais P, Berthelot P (1982) Evidence that hepatitis B virus has a role in liver-cell carcinoma in alcoholic liver disease. N Engl J Med 306:1384–1387

Brechot C, Degos F, Lugassy C, Thiers V, Zafrani S, Franco D, Bismuth H, Trepo C, Benhamou JP, Wands J et al. (1985) Hepatitis B virus DNA in patients with chronic liver disease and negative tests for hepatitis B surface antigen. N Engl J Med 312:270–276

Bressac B, Kew M, Wands J, Ozturk M (1991) Selective G to T mutations of p53 gene in hepatocellular carcinoma from southern Africa. Nature 350:429–431

Brunetto MR, Stemler M, Bonino F, Schodel F, Oliveri F, Rizzetto M, Verme G, Will H (1990) A new hepatitis B virus strain in patients with severe anti-HBe positive chronic hepatitis B. J Hepatol 10:258–261

Brunetto MR, Giarin MM, Oliveri F, Chiaberge E, Baldi M, Alfarano A, Serra A, Saracco G, Verme G, Will H, Bonino F (1991) Wild-type and e antigen-minus hepatitis B viruses and course of chronic hepatitis. Proc Natl Acad Sci USA 88:4186–4190

Carman WF, Hadziyannis S, McGarvey MJ et al. (1989) Mutation preventing formation of hepatitis B e antigen in patients with chronic hepatitis B infection. Lancet 2:588–590

Carman WF, Zanetti AR, Karayiannis P, Waters J, Manzillo G, Tanzi E, Zuckerman AJ, Thomas HC (1990) Vaccine-induced escape mutant of hepatitis B virus. Lancet 336:325–329

Chang C, Enders G, Sprengel R, Peters N, Varmus HE, Ganem D (1987) Expression of the precore region of an avian hepatitis B virus is not required for viral replication. J Virol 61:3322–3325

Coursaget P, Yvonnet B, Bourdil C, Mevelec MN, Adamowicz P, Barres JL, Chotard J, N'Doye R, Diop-Mar I, Chiron JP (1987) HBsAG positive reactivity in man not due to hepatitis B virus. Lancet 2:1354–1358

Fagan EA, Smith PM, Davison F, Williams R (1986) Fulminant hepatitis B in successive female sexual partners of two anti-HBe-positive males. Lancet 2:538–540

Feray C, Zignego AL, Samuel D, Bismuth A, Reynes M, Tiollais P, Bismuth H, Brechot C (1990) Persistent hepatitis B virus infection of mononuclear blood cells without concomitant liver infection. The liver transplantation model. Transplantation 49:1155–1158

Fiordalisi G, Cariani E, Mantero G, Zanetti A, Tanzi E, Chiaramonte M, Primi D (1990) High genomic variability in the pre-C region of hepatitis B virus in anti-HBe, HBV DNA-positive chronic hepatitis. J Med Virol 31:297–300

Gerken G, Paterlini P, Manns M, Housset C, Terre S, Dienes H-P, Hess G, Gerlich WH, Berthelot P, Zum Büschenfelde K-HM, Brechot C (1991) Assay of hepatitis B virus DNA by polymerase chain reaction and its relationship to pre-S- and S-encoded viral surface antigens. Hepatology 13:158–166

Hadziyannis SJ, Lieberman HM, Karvountzis GG, Shafritz DA (1983) Analysis of liver disease, nuclear HBcAg, viral replication and hepatitis B virus DNA in liver and serum of HBeAg versus anti-HBe positive carriers of hepatitis B virus. Hepatology 3:656–662

Hosono S, Lee CS, Chou MJ, Yang CS, Shih CH (1991) Molecular analysis of the p53 alleles in primary hepatocellular carcinomas and cell lines. Oncogene 6:237–243

Hsu IC, Metcalf RA, Sun T, Welsh JA, Wang NJ, Harris CC (1991) Mutational hotspot in the p53 gene in human hepatocellular carcinomas. Nature 350:427–428

Jackson JB (1991) Polymerase chain reaction assay for detection of hepatitis B virus. Am J Clin Pathol 95:442–444

Kaneko S, Miller RH (1990) Characterization of primers for optimal amplification of hepatitis B virus DNA in the polymerase chain reaction assay. J Virol Methods 29:225–229

Kaneko S, Miller RH, Feinstone SM, Unoura M, Kobayashi K, Hattori N, Purcell RH (1989) Detection of serum hepatitis B virus DNA in patients with chronic hepatitis using the polymerase chain reaction assay. Proc Natl Acad Sci USA 86:312–316

Kaneko S, Miller RH, Di Bisceglie AM, Feinstone SM, Hoofnagle JH, Purcell RH (1990) Detection of hepatitis B virus DNA in serum by polymerase chain reaction. Gastroenterology 99:799–804

Kosaka Y, Takase K, Kojima M, Shimizu M, Inoue K, Yoshiba M, Tanaka S, Akahane Y, Okamoto H, Tsuda F et al. (1991) Fulminant hepatitis B: induction by hepatitis B virus mutants defective in the precore region and incapable of encoding e antigen. Gastroenterology 100:1087–1094

Krogsgaard K, Kryger P, Aldershvile J, Andersson P, Brechot C, Copenhagen Hepatitis Acute Programme (1985) Hepatitis B virus DNA in serum from patients with acute hepatitis B. Hepatology 5:10–3

Krogsgaard K, Wantzin P, Aldershvile J, Kryger P, Andersson P, Nielsen JO (1986) Hepatitis B virus DNA in hepatitis B surface antigen-positive blood donors: relation to the hepatitis B e system and outcome in recipients. J Infect Dis 153:298–303

Lampertico P, Malter JS, Colombo M, Gerber MA (1990) Detection of hepatitis B virus DNA in formalin-fixed, paraffin-embedded liver tissue by the polymerase chain reaction. Am J Pathol 137:253–258

Li J, Tong S, Vitvitski L, Zoulim F, Tr'po C (1990) Rapid detection and further characterization of infection with hepatitis B virus variants containing a stop codon in the distal pre-C region. J Gen Virol 71:1993–1998

Liang TJ, Isselbacher KJ, Wands JR (1989) Rapid identification of low level hepatitis B-related viral genome in serum. J Clin Invest 84:1367–1371

Liang TJ, Blum HE, Wands JR (1990) Characterization and biological properties of a hepatitis B virus isolated from a patient without hepatitis B virus serologic markers. Hepatology 12:204–212

Liaw YF, Sheen IS, Chen TJ, Chu CM, Pao CC (1991) Incidence, determinants and significance of delayed clearance of serum HBsAg in chronic hepatitis B virus infection: a prospective study. Hepatology 13:627–631

Lin HJ, Lai CL, Lau JY, Chung HT, Lauder IJ, Fong MW (1990) Evidence for intrafamilial transmission of hepatitis B virus from sequence analysis of mutant HBV DNAs in two Chinese families. Lancet 336:208–212

Marcellin P, Giostra E, Martinot-Peignoux M, Loriot MA, Jaegle ML, Wolf P, Degott C, Degos F, Benhamou JP (1991) Redevelopment of hepatitis B surface antigen after renal transplantation. Gastroenterology 100:1432–1434

McMahon G, McCarthy L, Dottavio D, Ostberg L (1991) Surface antigen and polymerase gene variation in hepatitis B virus isolates from a monoclonal antibody-treated liver transplant patient. In: Hollinger BF, Lemon SM, Margolis H (eds) Viral hepatitis and liver disease. Williams and Wilkins, Baltimore, pp 219–221

Milich DR, Jones JE, McLachlan A, Houghten R, Thornton GB, Hughes JL (1989) Distinction between immunogenicity and tolerogenicity among HBcAg T cell determinants. Influence of peptide–MHC interaction. J Immunol 143:3148–3156

Milich DR, Jones JE, Hughes JL, Price J, Raney AK, McLachlan A (1990) Is a function of the secreted hepatitis B e antigen to induce immunologic tolerance in utero? Proc Natl Acad Sci USA 87:6599–6603

Mitsuda T, Yokota S, Mori T, Ibe M, Ookawa N, Shimizu H, Aihara Y, Yoshida N, Kosuge K, Matsuyama S (1989) Demonstration of mother-to-infant transmission of hepatitis B virus by means of polymerase chain reaction. Lancet 2:886–888

Muller R, Gubernatis G, Farle M, Niehoff G, Klein H, Wittekind C, Tusch G, Lautz H-U, Boker K, Stangel W, Pichlmayr R (1991) Liver transplantation in HBs antigen (HBsAg) carriers. Prevention of hepatitis B virus (HBV) recurrence by passive immunization. J Hepatol 13:90–96

Neurath AR, Kent SBH, Strick N et al. (1985) Hepatitis B virus contains pre-S gene-encoded domains. Nature 315:154–156

Okamoto H, Imai M, Tsuda F, Tanaka T, Miyakawa Y, Mayumi M (1987) Point mutation in the S gene of hepatitis B virus for a d/y or w/r subtypic change in two blood donors carrying a surface antigen of compound subtype adyr or adwr. J Virol 61:3030–3034

Okamoto H, Yotsumoto S, Akahane Y, Yamanaka T, Miyazaki Y, Sugai Y, Tsuda F, Tanaka T, Miyakawa Y, Mayumi M (1990) Hepatitis B viruses with precore region defects prevail in persistently infected hosts along with seroconversion to the antibody against e antigen. J Virol 64:1298–1303

Oren I, Hershow RC, Ben-Porath E, Krivoy N, Goldstein N, Rishpon S, Shouval D, Hadler SC, Alter MJ, Maynard JE et al. (1989) A common-source outbreak of fulminant hepatitis B in a hospital. Ann Intern Med 110:691–698

Ou J, Laub O, Rutter W (1986) Hepatitis B virus gene function: the precore region targets the core antigen to cellular membranes and causes the secretion of the e antigen. Proc Natl Acad Sci USA 83:1578–1582

Peters M, Vierling J, Gershwin ME, Milich D, Chisari FV, Hoofnagle JH (1991) Immunology and the liver. Hepatology 13:977–994

Raimondo G, Rodino G, Smedile V, Brancatelli S, Villari D, Longo G, Squadrito G (1990) Hepatitis B virus (HBV) markers and HBV-DNA in serum and liver tissue of patients with acute exacerbation of chronic type B hepatitis. J Hepatol 10:271–273

Samuel D, Bismuth A, Mathieu D, Arulnaden J-L, Reynes M, Benhamou J-P, Brechot C, Bismuth H (1991) Passive immunoprophylaxis after liver transplantation in HBsAg-positive patients. Lancet 337:813–815

Schlicht H-J, von Brunn A, Theilmann L (1991) Antibodies in anti-HBe-positive patient sera bind to an HBe protein expressed on the cell surface of human hepatoma cells: implications for virus clearance. Hepatology 13:57–61

Schlicht H-J, Salfeld J, Schaller H (1987) The duck hepatitis B virus pre-C region encodes a signal sequence which is essential for synthesis and secretion of processed core proteins but not for virus formation. J Virol 61:3701–3709

Schlicht HJ, Esser A, Bartenschlager R, Galle P, Kuhn C, Nassal M, Niepman M, Radziwill G, Schaller H (1989) Unravelling the life cycle of hepatitis B viruses. In: Bannasch P, Keppler D, Weber G (eds) Liver cell carcinoma. Kluwer, Dordrecht, pp 79–91

Seibl R, Holtke HJ, Ruger R, Meindl A, Zachau HG, Rasshofer R, Roggendorf M, Wolf H, Arnold N, Wienberg J, et al. (1990) Non-radioactive labeling and detection of nucleic acids: III. Applications of the digoxigenin system. Biol Chem Hoppe Seyler 371:939–951

Shafritz DA, Shouval D, Sherman HI, Hadziyannis SJ, Kew MC (1981) Integration of hepatitis B virus DNA into the genome of liver cells in chronic liver disease and hepatocellular carcinoma. Studies in percutaneous liver biopsies and post-mortem tissue specimens. N Engl J Med 305:1067–1073

Shih L-N, Sheu J-C, Wang J-T, Huang G-T, Yang P-M, Lee H-S, Sung J-L, Wang T-H, Chen D-S (1990) Serum hepatitis B virus DNA in healthy HBsAg-negative Chinese adults evaluated by polymerase chain reaction. J Med Virol 32:257–260

Stahl S, MacKay P, Magazin M, Bruce SA, Murray K (1982) Hepatitis B virus core antigen: synthesis in *Escherichia coli* and application in diagnosis. Proc Natl Acad Sci USA 79:1606–1610

Takeda K, Akahane Y, Suzuki H, Okamoto H, Tsuda F, Miyakawa Y, Mayumi M (1990) Defects in the precore region of the HBV genome in patients with chronic hepatitis B after sustained seroconversion from HBeAg to anti-HBe induced spontaneously or with interferon therapy. Hepatology 12:1284–1289

Terazawa S, Kojima M, Yamanaka T, Yotsumoto S, Okamoto H, Tsuda F, Miyakawa Y, Mayumi M (1991) Hepatitis B virus mutants with precore-region defects in two babies with fulminant hepatitis and their mothers positive for antibody to hepatitis B e antigen. Pediatr Res 29:5–9

Thiers V, Nakajima E, Kremsdorf D, Mack D, Schellekens H, Driss F, Goudeau A, Wands J, Sninsky J, Tiollais P et al. (1988) Transmission of hepatitis B from hepatitis-B-seronegative subjects. Lancet 2:1273–1276

Tiollais P, Pourcel C, Dejean A (1985) The hepatitis B virus. Nature 317:489–495

Tong SP, Li JS, Vitvitski L, Trepo C (1990) Active hepatitis B virus replication in the presence of anti-HBe is associated with viral variants containing an inactive pre-C region. Virology 176:596–603

Wands JR, Fujita YK, Isselbacher KJ, Degott C, Schellekens H, Dazza MC, Thiers V, Tiollais P, Brechot C (1986) Identification and transmission of hepatitis B virus-related variants. Proc Natl Acad Sci USA 83:6608–6612

Wang J-T, Wang T-H, Sheu J-C, Shih L-N, Lin J-T, Chen D-S (1991) Detection of hepatitis B virus DNA by polymerase chain reaction in plasma of volunteer blood donors negative for hepatitis B surface antigen. J Infect Dis 163:397–399

Yokosuka O, Omata M, Hosoda K, Tada M, Ehata T, Ohto M (1991) Detection and direct sequencing of hepatitis B virus genome by DNA amplification method. Gastroenterology 100:175–181

Zeldis JB, Lee JH, Mamish D, Finegold DJ, Sircar R, Ling Q, Knudsen PJ, Kuramoto IK, Mimms LT (1989) Direct method for detecting small quantities of hepatitis B virus DNA in serum and plasma using the polymerase chain reaction. J Clin Invest 84:1503–1508

Chapter 8 Application of Polymerase Chain Reaction to Hepatitis C Virus Research and Diagnostics

Amy J. Weiner, Venkatakrishna Shyamala, J. Eric Hall, Michael Houghton, and Jang Han

Summary

Rapid advances in virology during the 1980s were due, in part, to the development of powerful molecular biological technologies including lambda GT11 immunoscreening (Young and Davis, 1983) and the polymerase chain reaction (PCR) (Saiki et al. 1985). The hepatitis C virus (HCV), which was cloned using lambda GT11 immunoscreening, is the only virus which was discovered by cloning and characterizing the viral genome prior to biochemical isolation and/or visualization of the virus or viral antigens by microscopy. The application of PCR to HCV enabled researchers to overcome some of the technical limitations associated with the relatively low titers of HCV in blood or liver tissue and the absence of an in vitro tissue culture system for the cultivation of HCV. PCR based diagnostics, the rapid accumulation of HCV sequence information and the availability of cDNA/PCR generated cDNA clones for the expression of putative viral antigens has contributed greatly to our understanding of the structure and biology of HCV.

Introduction

The hepatitis C virus (HCV) was discovered in 1987 (Choo et al. 1989; Houghton et al. 1989) and has since been shown to be the major etiologic agent for posttransfusion and community-acquired non-A, non-B hepatitis (NANBH) (Choo et al. 1990; Kuo et al. 1989). HCV has also been implicated in other forms of liver disease such as hepatocellular carcinoma (HCC) and cryptogenic liver disease (for a review of the literature, see Choo et al. 1990; Houghton et al. 1991 b; Yoneyama et al. 1990). Approximately 150000 cases of NANBH are reported annually in the USA, approximately 50% of which become chronic infections which may progress to cirrhosis and/or HCC (Alter 1985; Dienstag and Alter 1986).

The genome is a single stranded RNA molecule of plus-strand polarity, which appears to encode a polyprotein precursor of 3010–3011 amino acids (Choo et al. 1989, 1991; Kato et al. 1990 a; Takamizawa et al. 1991). Comparison of

Chiron Corporation, 4560 Horton St., Emeryville, CA 94608, USA

the nucleotide and deduced amino acid sequence of the HCV genome with those of other viruses suggests that it is most closely related to the pesti- and flaviviruses. Processing of the polyprotein by cellular and putative viral proteases is believed to generate the putative viral structural (Core, E1) or nonstructural (NS2-NS5) proteins. It is unknown whether the E2/NS1 gene, located immediately downstream from the E1 gene, encodes a structural protein or nonstructural protein corresponding to the gp 53/55 of the pestiviruses or the NS1 protein of the flaviviruses (for a review of the literature, see Choo et al. 1990; Houghton et al. 1991 b).

Historically, HCV was believed to be present at very low titers (10^2-10^4 chimpanzee infectious doses, CID/ml) in the blood and liver of infected individuals as determined by chimpanzee infectivity studies (Bradley and Maynard 1986). These findings appeared to be supported by the inability to reliably detect viral antigens in liver sections by classic immunofluorescence assays using serum from convalescent patients (Bradley and Maynard 1986). The breakthrough in the molecular identification of the HCV genome came when Choo et al. discovered that HCV-specific antibodies in patient sera could bind to recombinant HCV polypeptides expressed in bacterial cells transfected with bacteriophage lambda λGT11 containing cDNA inserts from the HCV genome (Choo et al. 1989). Subsequent to finding the initial HCV cDNA clone, 5-1-1, the remainder of the HCV genome was obtained by identifying overlapping cDNA clones in the original λGT11 cDNA library (Choo et al. 1989). The process of cloning the ~ 9.4-Kb HCV genome by constructing λGT11 libraries from patient or chimpanzee plasma, which contains typically approximately 5 fg/ml – 50 pg/ml HCV RNA, is both time consuming and technically challenging. Of the three full length genomes published to date, only the prototype (HCV-1) sequence is derived from a single individual (Choo et al. 1991). HCV-J (Kato et al. 1990a) and HCV-BK (Takamizawa et al. 1991) are composite cDNA sequences from HCV RNA contained in plasma pools of 9 and 50 individuals, respectively. The advent of polymerase chain reaction (PCR) technology (for a review of the literature, see Erlich et al. 1991), first described in 1985 (Saika et al. 1985), could not have come at a more opportune time for HCV research and diagnostics, since detecting and cloning low levels of cDNA synthesized from the HCV RNA present in small volume samples by conventional techniques is difficult if not impossible. This chapter describes the current status of developments in HCV research and diagnostics which have been obtained using PCR and a discussion of PCR technology as it applies to the hepatitis C virus.

Structure and Heterogeneity of the Hepatitis C Virus Genome

The availability of cDNA and PCR primers from the prototype HCV sequence and subsequently sequenced portions of the HCV genomes from related isolates has contributed substantially to the rapidly increasing number of partial and complete HCV sequences published to date (see Houghton et al. 1991 b for review; Cha et al. 1991 a, b; Chen et al. 1991; Cuypers et al. 1991; Fuchs et al. 1991; Hosada et al. 1991; Li et al. 1991). Three points are evident from the comparison of HCV nucleotide and deduced amino acid sequences obtained by both PCR and non-PCR methods: (a) there are at least three major groups of related HCV genotypes (groups I–III; see Houghton et al. 1991 b for review), (b) HCV appears to be evolving at rates typical for RNA viruses (Chambers et al. 1990; Holland et al. 1982), and the family of related HCVs exists in a quasi-species distribution (Martell et al. 1992), and (c) within the viral genome, there are subregions of increased nucleotide and amino acid sequence heterogeneity including a hypervariable domain (E2 HV) located at the N-terminus of the E2/NS1 protein which is found in all viral isolates described so far (Weiner et al. 1991 a; Cha et al. 1991 a; Hijikata et al. 1991; Kremsdorf et al. 1991; Ogata et al. 1991). Similar hypervariable domains have not been reported for pesti- or flaviviruses. Group-specific variable subregions are also evident (Weiner et al. 1991 a; Cha et al. 1991 a; Hijikata et al. 1991; Kremsdorf et al. 1991; Ogata et al. 1991). The amino acid sequence variation observed in the E2 HV domain may result from immune selection and could be significant with respect to protective immunity (Weiner et al. 1992).

The structure of the 5′- and 3′-termini of the HCV genome were defined using PCR (Han et al. 1991), although it is also possible to obtain them by identifying overlapping clones in cDNA libraries (Kato et al. 1990a; Takamizawa et al. 1991). Previous studies indicated that the HCV genome is retained on oligo dT cellulose and suggested that HCV had a polyadenylated 3′-end (Choo et al. 1989). Cloning the 3′-end of the HCV genome was facilitated by PCR using one primer containing an oligo dT tract and a second primer from an HCV cDNA clone close to the putative 3′-end of the genome (Han et al. 1991). The 5′-terminus was also extended by PCR using an specific HCV primer for cDNA synthesis followed by dA tailing and amplification using the HCV-specific primer and a nonspecific primer sequence attached to the 5′-end of the oligo dT primer used in second-strand synthesis (Han et al. 1991). Similar techniques have been described for obtaining the 5′- and 3′-ends of cellular mRNAs (Frohman et al. 1988). As a result of these experiments, Han et al. provided more evidence that the 3′-terminal region of the HCV genomes contains a poly A tract and showed that the 5′-terminal region (5′-TR) of HCV (a) is very highly conserved among isolates (Choo et al. 1991; Cha et al. 1991 b; Okamoto et al. 1990; Takeuchi et al. 1990) and is therefore the most desirable region from which to obtain cDNA and PCR primers, (b) contains several small open reading frames (ORF) upstream from the putative initiator methionine (Choo et al. 1991), (c) has approximately 50% nucleotide homology with the corresponding region of the pestiviruses (Choo et al. 1990), and (d) has a predictable terminal hairpin structure. The 5′- and 3′-subgenomic HCV RNA molecules of unknown function were also identified in this study. Two other groups reported HCV genomes containing a heterogenous 3′-terminus and a poly T tract rather than a poly A tract (Kato et al. 1990a; Takamizawa et al. 1991).

Virus-Host Interactions and Virus Characterization

Very little is known about the mechanism(s) of HCV pathogenesis. Although it is believed that HCV is cytolytic, as are some flaviviruses, it has never been determined whether liver disease in people with HCV infections results from direct damage to hepatocytes caused by viral replication and protein synthesis or to indirect processes such as host immunologic response(s) to infected cells displaying viral antigens, as is thought to be the case for HBV (Bianchi 1981; Thomas et al. 1984). Data from one PCR study, which showed that the peak of viremia overlapped with the peak of liver damage as judged by a peak of elevated alanine amino transferase (ALT) levels during the acute phase of disease (Shimizu et al. 1990), is consistent with the notion that HCV may be cytolytic. In depth case studies using quantitative PCR in conjunction with better markers for cell death may provide more insight into the issue of cyto-pathogenicity in the future.

Viral persistence and chronic liver disease frequently characterize HCV infections (Bradley et al. 1981, 1985; Bradley and Maynard 1986). Van der Poel et al. and Farci et al. have suggested that the incidence of persistent HCV infections may be higher than previously thought since some individuals, who were classified as having acute, resolved hepatitis on the basis of normal ALT values and lack of disease symptoms, had detectable levels of HCV RNA in their plasma (Farci et al. 1990; van der Poel et al. 1991 a). Liver damage can reoccur up to 3 years following normalization of ALT values in apparently acute resolved NANBH cases. Recent findings showing that E2 HV variants can be found associated with different episodes of disease in the same individual suggest that the rapid evolution of escape mutants may account for the high levels of chronicity found in HCV infections (Weiner et al. 1992).

Currently, there is neither a tissue culture nor transfection system for the propagation of HCV which would be potentially useful for infectivity and neutralization assays, developing potential attenuated vaccine strains, and studying viral replication and the biogenesis of viral proteins. Viral RNA has been detected in monocytes (macrophages), B cells, and to a lesser extent T cells obtained by panning peripheral blood mononuclear cells (PBMCs) from chronically infected individuals using cell type-specific antibodies attached to magnetic beads (Weiner and Steimer, unpublished data). Although minus-stranded RNA was identified in the panned PBMCs, it was also found in the plasma of some of the individuals from which the blood cells were derived. The concentration of the plus- and minus-stranded RNA in the cells relative to the amount observed in the plasma suggests, but does prove, that HCV may replicate in these cells (Weiner and Steimer, unpublished data). In contrast, Takehara et al. found minus-stranded HCV RNA in the liver but not the plasma or PBMCs from HCV-infected individuals (Takehara et al. 1991). PBMCs may serve as a reservoir for HCV in infected individuals; however, the significance of these findings remains to be determined. All efforts to cultivate HCV in PBMCs and related cell lines have been unsuccessful to date. Extensive research in the area of cell culture will most likely continue using PCR and radioimmune precipitation of viral antigens with HCV-specific antibodies as assays for viral replication.

HCV has never been visualized under the electron microscope either directly or in immune complexes, nor has the virus been purified in sufficient quantities to study its structure or composition. Some of its physiochemical properties such as the approximate size, buoyant density, and structural features have been reported using infectivity in chimpanzees as the assay (Bradley et al. 1983, 1991; He et al. 1987). More recently, Yuasa et al. applied a cDNA/PCR assay to trace the purification of HCV and to estimate the size of HCV as not larger than approximately 35 nm using a pore filtration purification scheme (Yuasa et al. 1991).

Transmission and Therapy

One important issue in HCV research concerns the route(s) of viral transmission. Epidemiological studies indicate that approximately 6% of HCV cases are transmitted through blood and blood products (parenteral transmission), approximately 42% through intravenous drug use and approximately 6%, 3%, 2%, and 0.6% through sexual exposure, household exposure, occupational exposure, and hemodialysis, respectively (Alter et al. 1990). Therefore, approximately 50% of individuals with no parenteral source of infection can develop NANBH (Alter 1991). Recent PCR studies indicate that HCV is transmitted from mother to child (vertical transmission) (Thaler et al. 1991). Although HCV may also be transmitted through bodily fluids such as saliva (Abe and Inchauspe 1991; Takamatsu et al. 1990; Dusheiko et al. 1990), unlike other parenterally transmitted viruses such as the human immunodeficiency virus, HIV, there is no evidence for HCV RNA in semen. The contribution of these sources of infection to the total number of cases of HCV is unknown.

HCV cDNA/PCR assays have also been useful in determining the efficacy of drug treatment programs and will serve the same function in the development of vaccines. Kanai et al. and Chayama et al. demonstrated that treatment with α-interferon, one of the few drugs approved for chronic hepatitis, can reduce the amount of HCV in the plasma of people who respond to treatment (Kanai et al. 1990; Chayama et al. 1991; Brillanti et al. 1991). Saracco et al. has confirmed those results and also showed that viremia correlated with patient response to interferon treatment, in that HCV RNA was not reduced in individuals who failed to respond to interferon treatment and was initially reduced but rebounded in individuals who had a relapse of liver disease after the end of treatment (Saracco et al. 1991). Long-term PCR studies should help to determine accurately the number of individuals who have true resolution of hepatitis after treatment with interferon.

Hepatitis C Virus Diagnostics

Studies documenting the transmission of HCV from blood donors to recipients as well as chimpanzee transmission studies have clearly shown that antibodies to HCV antigens correlate well with infectivity (Chao et al. 1990; Houghton et al. 1991 b; van der Poel et al. 1989, 1990, 1991 b). Using cDNA/PCR assays, several groups have shown that HCV-specific antibody-reactive patients and blood donors frequently have HCV RNA in their livers and/or plasma (sera) (Cristiano et al. 1991; Fang et al. 1991; Garson et al. 1990a, b, c; Kato et al. 1990 b; Kaneko et al. 1990; Ohkoshi et al. 1990; Schlauder et al. 1991; Shibata et al. 1990; Simmonds et al. 1990; Tanaka et al. 1991; Ulrich et al. 1990; Weiner et al. 1990 a, b). In some cases, however, C100-3-specific antibody-reactive individuals may consistently lack HCV RNA in serum for several years (Farci et al. 1990). Whether HCV in the liver or other cell types could have contributed to the prolonged antibody response was not determined. In one study, two patients who reportedly had HCV RNA in their livers and were nonreactive for C100-3-specific antibodies (Weiner et al. 1990 a) were later determined to be C33c-specific antibody positive, suggesting that improved immunodiagnostics will reduce the number of viremic, antibody-negative individuals.

As with all antibody assays for viral infections, there is a window from the time of infection to the time of seroconversion which is estimated at an average of 23.9 weeks (range of 7.4–30 weeks including one sample at 53.4 weeks) and 15.3 weeks (range 7.4–30.6) for the IgG class of antibodies to C100-3 (Ortho ELISA) and C25 (C100, C33C, and C22 antigens combined), respectively (Chien et al., in preparation; also see Alter et al. 1989 for seroconversion to C100-3-specific antibodies). HCV RNA has not only been detected in the acute phase plasma of patients and chimpanzees who had not yet seroconverted to HCV-specific antibodies (Farci et al. 1990; Weiner et al. 1990 a; Shibata et al. 1990) but at very early times in the course of HCV infection. Experimentally infected chimpanzees were shown to have HCV in their plasma by 3–4 days postinnoculation (Shimizu et al. 1990), and patients had detectable levels of HCV in their plasma within 1 week of transfusion (Farci et al. 1990). PCR is especially useful in situations where seroconversion to HCV-specific antibodies may be inadequate or insufficient, as in newborn babies or in immunocompromised individuals. Thaler et al. showed that 8 babies of 10 HCV antibody-positive mothers had detectable levels of HCV (Thaler et al. 1991). Lack of seroconversion in some babies may correlate with immune tolerance. Other direct tests for HCV antigens such as immunofluorescence (IFA) have been reported (Infantolino et al. 1990; Krawczynski et al. 1991); however, liver biopsies samples are not as easily obtained as plasma (sera) and the immunological tools required for the assay are currently more difficult to create than PCR primers and probes.

Since PCR is an important test for HCV, reports in the literature often compare PCR data with immunoassay results to draw conclusions about the efficacy and veracity of HCV immunodiagnostics. The interpretation of PCR data in this manner may be misleading because PCR and immunodiagnostic tests measure different molecules. When both indirect and direct tests for HCV are employed, it becomes clear that not all samples which contain antibodies to HCV had circulating virus in the blood (Farci et al. 1990; Garson et al. 1990c; Ohkoshi et al. 1990; Shimizu et al. 1990; Weiner et al. 1990a), and conversely, not all individuals who had circulating virus had developed antibodies to all or a subset of HCV antigens (Farci et al. 1990; Kato et al. 1990b; Ohkoshi et al. 1990; Okamoto et al. 1990; Shimizu et al. 1990; Thaler et al. 1991; Weiner et al. 1990a). Several interpretations for these results are possible. Cases in which individuals are seropositive but lack detectable RNA in plasma could be due to (a) absence of virus in the plasma (sera) but not in the liver or other, as yet unknown tissues, (b) resolved hepatitis as measured by the absence of HCV RNA and normalized ALT levels, (c) extremely low levels of virus in the plasma, (d) sampling error due to fluctuations in virus concentration in the plasma, and (e) inadequate storage conditions for the preservation of intact virus particles. Alternatively, cases in which individuals lack detectable levels of antibodies but have detectable levels of HCV RNA may be due to (a) sampling prior to seroconversion, (b) sampling error due to fluctuating antibody levels, (c) limited number of antigens tested, and (d) lack of response to specific antigens.

Conditions which lead to and precautions to avoid false-positive PCR results have been discussed extensively in the literature (Erlich et al. 1991). Suboptimal collection and storage protocols of plasma or sera may lead to false-negative PCR results. Cuypers et al. suggest that plasma (sera) should be separated from whole blood within 2–3 h after collection since cDNA/PCR products from RNA extracted from plasma (sera) derived from whole blood stored for 8 h or more at room temperature were reduced by at least one order of magnitude (van der Poel 1991a). Storage of whole blood in the presence of the anticoagulant EDTA at room temperature typically resulted in a significant reduction of PCR products within 1 day and a complete loss of detectable PCR products after 4 days, using cDNA/PCR primers from the NS3 region of the HCV genome. Under the same circumstances, PCR products generated with the 5′-TR primers of Han et al. decreased after approximately 3–4 days, with a loss of detectable PCR products between 8 and 14 days (T. Cuypers, personal communication). These data indicate that the 5′-TR of HCV is more stable than downstream regions of the genome, possibly due to the secondary structure in the RNA, but that collection and storage conditions can also influence the yield of cDNA/PCR products from the 5′-TR of HCV.

Quantitative Polymerase Chain Reaction

The efficiency of PCR primers and the purity of the RNA sample are two major factors which can influence the sensitivity, reproducibility, and quantitation of PCR. Primer efficiency may be a function of nucleotide sequence, nucleotide sequence heterogeneities between the primers and the target sequence, the size and structure of the target sequence, and possibly unknown factors. After the entire HCV nucleotide sequence was elucidated, Han et al. and Okayama et al. discovered that the 5'-TR (also referred to as 5'-NC and 5'-UT in the literature) of the genome was highly conserved among all reported HCV sequences, and therefore, the problem of failing to detect HCV in clinical samples due to sequence heterogeneities in the PCR primers could be eliminated. We prefer the "TR" nomenclature since this region of the HCV genome contains 4 small ORFS some of which may be translated in vivo. Although not all PCR primers in the 5'-TR region have equal efficiency, certain primer pairs (Han et al. 1991) or nested primer sets (Cristiano et al. 1991; Carson et al. 1990 b; Okamoto et al. 1990) have been demonstrated to be between 100- and 500-fold more sensitive than previously described primers from the NS3/4 region (Cristiano et al. 1991; Weiner et al. 1990 a, b; Farci et al. 1990) and at least as great, if not greater, than nested PCB primers from other regions of the HCV genome described by others (Cristiano et al. 1991; Ulrich et al. 1990). Primers from other regions of the genome have been described by many groups, although these may not work efficiently on divergent genotypes (Kaneko et al. 1990) and therefore may require either multiple sets of primers (Cristiano et al. 1991; Ulrich et al. 1990) or primers with nucleotide degeneracies (A.J. Weiner, unpublished data). (Note: Probes and primers from the 5'-TR region should be sufficiently mismatched with pestiviral genomes so as to avoid detection of distantly related viruses in the HCV cDNA/PCR assay.) The purity of the template RNA is significant since nonnucleic acid components of plasma, sera, or tissues which can inhibit cDNA synthesis may contaminate RNA preparations. Several RNA extraction methods have also been described, and although in our hands the proteinase K/SDS method appears to give the most consistent yield of RNA, the purity of RNA from guanidinium extraction methods is superior (A.J. Weiner and J.E. Hall, unpublished data; also see Cristiano et al. 1991).

Comparison of cDNA/PCR products derived from RNA extracted from dilutions of titered chimpanzee plasma in parallel with experimental samples have been used to quantitate PCR products (Cristiano et al. 1991; Farci et al. 1990; Fong et al. 1991; Simmonds et al. 1990; Weiner et al. 1990 a). Such methods are semiquantitative since they fail to serve as a control for the extraction efficiency of the template RNA or cDNA synthesis, and they rely on infectivity in chimpanzees which may be differentially susceptible to the same infectious inocula. An alternative method for quantitating cDNA/PCR products involves introducing an in vitro synthesized HCV RNA target sequence of known concentration into the experimental sample (Shyamala and Han, manuscript in preparation). The synthetic RNA, which competes with the native HCV template for the primers and reactants in the PCR reaction and should be present in relatively low copy number, serves as an internal control (IC) for RNA extraction, cDNA, and PCR efficiency. The IC molecule used in the experiment shown in Figs. 1 and 2 was made by inserting a 60-bp foreign DNA sequence into the authentic HCV PCR fragment produced by PCR

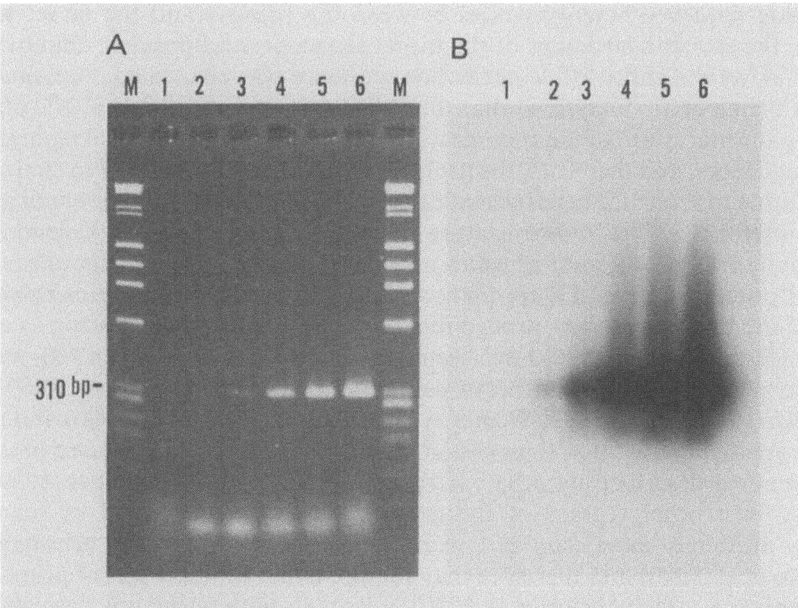

Fig. 1A, B. Determination of the minimum number of detectable hepatitis C virus (HCV) RNA molecules. Internal control (IC) HCV RNA was serially diluted using MS2 RNA (Boehringer Mannheim) as a carrier. Then, $10-10^5$ RNA molecules were reverse transcribed using primer JHC51 (Cha et al. 1991a; Han et al. 1991) in the single-strand cDNA synthesis kit from Pharmacia and amplified in a final volume of 50 µl containing primer JHC 93 (Cha et al. 1991a; Han et al. 1991). A 10-µl sample of the amplification reaction mix was electrophoresed on 1% agarose gel (**A**) and also blotted onto nitrocellulose for Southern hybridization (Southern 1975) (**B**). *Lane 1* contains cDNA/PCR products from IC RNA amplified without reverse transcription. *Lanes 2–6* contain PCR products amplified from cDNA templates derived from $10-10^5$ molecules of IC RNA. *M* denotes molecular weight standards (*Hin*dIII-digested phage λ and *Hae*III-digested ΦX174 DNA). The autoradiogram was exposed overnight at $-20\,°C$

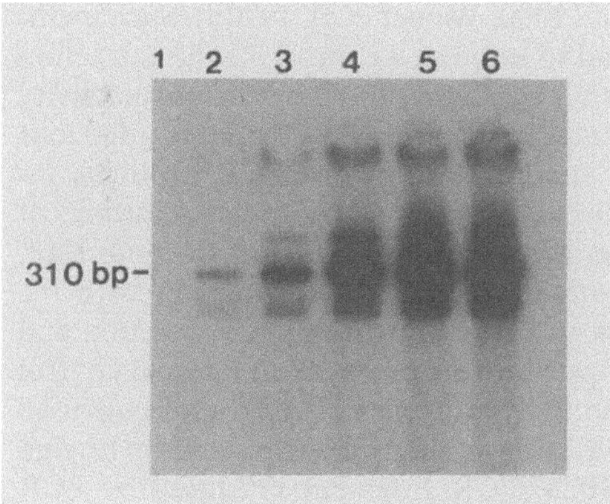

Fig. 2. Solution hybridization of HCV cDNA/PCR products to an oligonucleotide probe. An identical amount of cDNA/PCR product from each lane shown in Fig. 1 was hybridized with ^{32}P-labeled 30-mer oligonucleotide probe in $1 \times$ SSC (0.1 M NaCl, 0.015 M sodium citrate), and 0.1 M EDTA at 55 °C for 20 min and electrophoresed on a 6% neutral polyacrylamide gel. The gel was autoradiographed overnight at $-20\,°C$. For lanes and markers, see Fig. 1

using the 5′-TR primer sequences JHC51 and 93 (Cha et al. 1991 b; Han et al. 1991) in order to distinguish the IC molecule from the native PCR product by gel electrophoresis. The bacteriophage T7 promoter sequence was introduced upstream from the 5′-end of the IC DNA to facilitate in vitro RNA synthesis.

To determine the minimum number of IC RNA molecules which could be detected in the cDNA/PCR assay described by Shyamala and Han, serial dilution samples of the IC RNA template representing $10-10^5$ molecules were subject to reverse transcription and PCR. Figure 1 shows that cDNA/PCR products from as few as 10 IC RNA molecules could be visualized on a 1% neutral agarose gel stained with ethidium bromide (A, lane 2). Specificity of the product was established by Southern blot (Southern 1975) hybridization using a ^{32}P-labeled oligonucleotide probe (B, lanes 1–6), which is internal to the PCR primer sequences. PCR products from lanes 1–6 were also subject to solution hybridization using a ^{32}P-labeled probe (2.5×10^5 cpm/0.7 pmol) and electrophoresed on a 6% polyacrylamide gel which was autoradiographed overnight (Fig. 2). Figure 2 shows that the solution hybridization method is at least as sensitive as Southern blot hybridization and has the advantages (1) of being a rapid assay (about 3 h) and (2) avoiding potential quantitation problems due to inefficient transfer of PCR products onto nitrocellulose or unequal blotting efficiencies. To demonstrate the utility of this approach to quantitative PCR, a known number of in vitro transcribed HCV IC RNA molecules were added to a constant volume of titered chimpanzee plasma (10^6 CID/ml) (Bradley and Maynard 1986), and the amount cDNA/PCR product generated from the IC RNA was compared with the cDNA/PCR product amplified from the native HCV template in the same sample. Shyamala and Han thus determined that the titered chimpanzee plasma actually contained 3×10^7 molecules/ml of HCV RNA which included the 5′-TR region of the HCV genome (Shyamala and Han, manuscript in preparation).

Summary

The low quantities of hepatitis C virus (HCV) in the plasma an livers of infected individuals hampered the discovery of the virus for several years and continued to make certain areas of HCV research difficult to pursue. PCR technology overcame this limitation once the nucleotide sequence of the prototype HCV genome was determined and provided an important tool for HCV research. For diagnostic purposes, cDNA/PCR primers from the 5′-TR of HCV have several advantages over other HCV primers since this region of the genome (a) lacks significant sequence heterogeneity, (b) appears to have increased stability and can contribute to the potential for detecting HCV in plasma (sera) samples collected and stored under suboptimal conditions, and (c) contains nucleotide sequences which serve as excellent templates for efficient cDNA synthesis and PCR amplification. HCV cDNA/PCR assays have been particularly useful in identifying and monitoring viral infections in the absence of HCV-specific antibodies. PCR will undoubtedly continue to play an important role in HCV research related to viral transmission, tissue culture propagation and purification of the virus, viral sequence heterogeneity, and the evaluation of drug therapy and the efficacy of vaccines.

Acknowledgements. We wish to thank Peter Anderson for preparation of the manuscript and D.F.G. for continued support and encouragement.

References

Abe K, Inchauspe G (1991) Transmission of hepatitis C by saliva. Lancet 337:248

Alter HJ (1985) In: Infection, immunity and blood transfusion. Liss, New York, pp 47–61

Alter MJ (1991) Inapparent transmission of hepatitis C: footprints in the sand. Hepatology 14:389–391

Alter HJ, Purcell RH, Smith JW, Melpoler JC, Houghton M, Choo Q-L, Kuo C (1989) Detection of antibody to hepatitis C virus in prospectively followed transfusion recipients with acute and chronic non-A, non-B hepatitis. N Engl J Med 321:1494–1539

Alter MJ, Hadler SC, Judson FN, Mares A, Alexander J, Hu PY, Miller JK, Moyer LA, Fields HA, Bradley DW, Margolis HS (1990) Risk factors for acute non-A, non-B hepatitis in the United States and association with hepatitis C virus infection. JAMA 264:2231–2235

Bianchi L (1981) The immunology of acute type B hepatitis. Springer Semin Immunopathol 3:421–438

Bradley DW, Maynard JE (1986) Etiology and natural history of posttransfusion and enterically-transmitted non-A, non-B hepatitis. Semin Liver Dis 6:56–66

Bradley DW, Maynard JE, Krawczynski KZ, Popper H, Cook EH, Gravelle CR, Ebert JW (1981) Non-A, non-B hepatitis in chimpanzees infected with a factor VIII agent: evidence of persistent hepatic disease. In: Szmuness W, Alter HJ, Maynard JE (eds) Proceedings of the third international symposium on viral hepatitis, New York. Franklin Institute, New York

Bradley DW, Maynard JE, Popper H, Cook EH, Ebert JW, MacCaustland KA, Schable CA, Fields HA (1983) Posttransfusion non-A, non-B hepatitis: physicochemical properties of two distinct agents. J Infect Dis 148:254–265

Bradley DW, McCaustland KA, Cook EH, Schable CA, Ebert JW, Maynard JE (1985) Posttransfusion non-A, non-B hepatitis in chimpanzees. Gastroenterology 88:773–779

Bradley D, McCaustland K, Krawczynski K, Spelbring J, Humphrey C, Cook EH (1991) Hepatitis C virus: buoyant density of the factor VIII-derived isolate in sucrose. J Med Virol 34:206–208

Brillanti S, Garson JA, Tuke TW, Ring C, Briggs M, Masci C, Miglioli M, Barbara L, Tedder RS (1991) Effect of alpha-interferon therapy on hepatitis-C viraemia in community-acquired chronic non-A hepatitis, non-B hepatitis: a quantitative polymerase chain reaction study. J Med Virol 34:136–141

Cha T-A, Beall E, Irvine B, Kolberg J, Kuo G, Urdea MS (1991a) Two hepatitis C viral genotypes occur throughout the world (submitted)

Cha T-A, Kolberg J, Irvine B, Stempien M, Beall E, Yano M, Choo Q-L, Houghton M, Kuo G, Han JH, Urdea MS (1991b) Use of a signature nucleotide sequence of the hepatitis C virus for the detection of viral RNA in human serum and plasma. J Clin Microbiol 29 (in press)

Chambers TJ, Hahn CS, Galler R, Rice CM (1990) Flavivirus genome organization, expression, and replication. Annu Rev Microbiol 44:649–688

Chayama K, Saitoh S, Arase Y, Ikeda K, Matsumoto T, Sakai Y, Kobayashi M, Unakami M, Morinaga T, Kumada H (1991) Effect of interferon administration on serum hepatitis-C virus RNA in patients with chronic hepatitis C. Hepatology 13:1040–1043

Chen P-J, Lin M-H, Tu S-J, Chen D-S (1991) Isolation of a complementary DNA fragment of hepatitis C virus in Taiwan revealed significant sequence variations compared with other isolates. Hepatology 14:73–78

Chien DY, Choo Q-L, Tabrizi A, Kuo C, McFarland J, Shuster JR, Moyer DL, Han JH, Alter HJ, Fong CT, Houghton M, Kuo G, Designed chimeric protein for the serodiagnosis of hepatitis C virus infection: significant improvements in HCV antibody detection as compared with the first generation HCV C100-3 ELISA and the synthetic peptide EIA test (manuscript in preparation)

Choo Q-L, Kuo G, Weiner AJ, Overby LR, Bradley DW, Houghton M (1989) Isolation of a cDNA derived from a blood-borne non-A, non-B viral hepatitis genome. Science 244:359–362

Choo Q-L, Weiner AJ, Overby LR, Kuo G, Houghton M (1990) Hepatitis C virus: the major causative agent of viral non-A, non-B hepatitis. Br Med Bull 46:423–441

Choo Q-L, Richman KH, Han JH, Berger K, Lee C, Dong C, Gallegos C, Coit D, Medina-Selby A, Barr PJ, Weiner AJ, Bradley DW, Kuo G, Houghton M (1991) Genetic organization and diversity of the hepatitis C virus. Proc Natl Acad Sci USA 88:2451–2455

Cristiano K, Di Bisceglie AM, Hoofnagle JH, Feinstone SM (1991) Hepatitis C viral RNA in serum of patients with chronic non-A, non-B hepatitis: detection by the polymerase chain reaction using multiple primer sets. Hepatology 14:51–55

Cuypers HTM, Winkel IN, van der Poel CL, Reesink HW, Lelie PN, Houghton M, Weiner M (1991) Analysis of genomic variability of hepatitis-C virus. Hepatology (in press)

Dienstag JL, Alter HJ (1986) Semin Liver Dis 6:67–81

Dusheiko GM, Smith M, Schever PJ (1990) Hepatitis transmitted by human bite. Lancet 336:503–504

Erlich HA, Gelfand D, Sninsky JJ (1991) Recent advances in the polymerase chain reaction. Science 252:1643–1651

Farci P, Alter HJ, Wong D, Miller RH, Smith JW, Jett B, Purcell RH (1990) A long-term study of hepatitis C virus replication in non-A, non-B hepatitis. N Engl J Med 325:98–103

Fong T-L, Shindo M, Feinstone SM, Hoofnagle JH, Di Bisceglie AM (1991) Detection of replicative intermediates of hepatitis C viral RNA in liver and serum of patients with chronic hepatitis C. J Clin Invest 88:1058–1060

Frohman MA, Dush MK, Martin GR (1988) Rapid production of full-length cDNAs from rare transcripts: amplification using a single gene-specific oligonucleotide primer. Proc Natl Acad Sci USA 85:8998–9002

Fuchs K, Motz M, Schreier E, Zachoval R, Deinhardt F, Roggendorf M (1991) Characterization of nucleotide sequences from European hepatitis-C virus isolates. Gene 103:163–169

Garson JA, Preston FE, Makris M, Tuke P, Ring C, Machin SJ, Tedder RS (1990a) Detection by PCR of hepatitis C virus in factor VIII concentrates. Lancet 335:1473

Garson JA, Ring C, Tuke P, Tedder RS (1990b) Enhanced detection by PCR of hepatitis C virus RNA. Lancet 335:878–879

Garson JA, Tedder RS, Briggs M, Tuke P, Glazebrook JA, Trute A, Parker D, Barbara JAJ, Contreras M, Aloysius S (1990c) Detection of hepatitis C viral sequences in blood donations by "nested" polymerase chain reaction and prediction of infectivity. Lancet 335:1419–1432

Han JH, Shyamala V, Richman KH, Brauer MJ, Irvine B, Urdea MS, Tekamp-Olson P, Kuo G, Choo Q-L, Houghton M (1991) Characterization of the terminal regions of hepatitis C viral RNA: identification of conserved sequences in the 5c untranslated region and poly(A) tails at the 3c end. Proc Natl Acad Sci USA 88:1711–1715

He L-F, Alling D, Popkin T, Shapiro M, Alter HJ, Purcell RH (1987) Determining the size of non-A, non-B hepatitis virus by filtration. J Infect Dis 156:636–640

Hijikata M, Kato N, Ootsuyama Y, Nakagawa M, Ohkoshi S, Shimotohno K (1991) Hypervariable regions in the putative glycoprotein of hepatitis C virus. Biochem Biophys Res Commun 175:220–228

Holland J, Spindler K, Horodyski F, Grabau E, Nichol S, VandelPol S (1982) Science 215:1577–1585

Hosoda K, Yokosuka O, Omata M, Kato N, Ohto M (1991) Detection of partial sequencing of hepatitis-C virus RNA in the liver. Gastroenterology 101:766–771

Houghton M, Choo Q-L, Kuo G (1989) European Patent Application 0318216

Houghton M, Weiner A, Han J, Kuo G, Choo Q-L (1991 b) Molecular biology of the hepatitis C viruses: implications for diagnosis, development and control of viral disease. Hepatology 14:381–388

Infantolino D, Bonino F, Zanetti AR, Lesniewski RR, Barbazza R, Chiaramonte M (1990) Localization of hepatitis C virus (HCV) antigen by immunohistochemistry on fixed-embedded liver tissue. Ital J Gastroenterol 22:198–199

Kanai K, Iwata K, Nakao K, Kako M, Okamoto H (1990) Suppression of hepatitis C virus RNA by interferon-a. Lancet 336:245

Kaneko S, Unoura M, Kobayashi K, Kuno K, Murakami S, Hattori N (1990) Detection of serum hepatitis C virus RNA. Lancet 335:976

Kato N, Hijikata M, Oootsuyama Y, Nakagawa M, Ohkoshi S, Sugimura T, Shimotohno K (1990a) Molecular cloning of the human hepatitis C virus genome from Japanese patients with non-A, non-B hepatitis. Proc Natl Acad Sci USA 87:9524–9528

Kato N, Yokosuka O, Omata M, Hosoda K, Ohto M (1990b) Detection of hepatitis C virus ribonucleic acid in the serum by amplification with polymerase chain reaction. J Clin Invest 86:1764–1767

Krawczynski K, Beach M, Bradley D, Kuo G, DiBisceglie A, Houghton M, Reyes G, Kim J, Alter M (1991) Hepatitis C virus antigen in hepatocytes: immunomorphologic detection and identification (submitted)

Kremsdorf D, Porchon C, Kim JP, Reyes GR, Bréchot C (1991) Nucleotide sequence analysis of a French hepatitis C virus (HCV) isolate. Gene (in press)

Kuo G, Choo Q-L, Alter AJ, Gitnick GI, Redeker AG, Purcell RH, Miyamura T, Dienstag JL, Alter MJ, Stevens CE, Tegtmeier GE, Bonino F, Colombo M, Lee W-S, Kuo C, Berger K, Shuster JR, Overby LR, Bradley DW, Houghton M (1989) An assay for circulating antibodies to a major etiologic virus of human non-A, non-B hepatitis. Science 244:363–364

Li JS, Tong SP, Vitvitski L, Trépo C (1991) Seminar of the evidence for two major subtypes of HCV in France and close relatedness of the predominant one with the prototype virus. In: Genetic heterogeneity of hepatitis viruses: clinical implications. Sestriere, Turin, April 1991

Martell M, Esteban JI, Quer J, Genesca J, Weiner A, Esteban R, Guardia J, Gomez J (1992) HCV circulates as a population of different but closely related genomes ("Quasispecies" nature of HCV-genome distribution) J Virol 66:3225–3229

Ogata N, Alter HJ, Miller RH, Purcell RH (1991) Nucleotide sequence and mutation rate of the H strain of hepatitis C virus. Proc Natl Acad Sci USA 88:3392–3396

Ohkoshi S, Kato N, Kinoshita T, Hijikata M, Ohtsuyama Y, Okazaki N, Ohkura H, Hirohashi S, Honma A, Ozaki T, Yoshikawa A, Kojima H, Asakura H, Shimotohno K (1990) Detection of hepatitis C virus RNA in sera and liver tissues of non-A, non-B hepatitis patients using the polymerase in chain reaction. Jpn J Cancer Res 81:862–865

Okamoto H, Okada S, Sugiyama Y, Yotsumoto S, Tanaka T, Yoshizawa H, Tsuda F, Miyakawa Y, Mayumi M (1990) The 5c-terminal sequence of the hepatitis C virus genome. Jpn J Exp Med 60:167–177

Saiki RK, Scharf S, Faloona F, Mullis KB, Horn GT, Erlich HA, Arnheim N (1985) Enzymatic amplification of b-globin genomic sequences and restriction site analysis for diagnosis of sickle cell anemia. Science 230:1350–1354

Saracco G, Abate M, Brunetto MR, Baldi M, Calvo PL, Manzini P, Rizzetto M, Verme G, Weiner AJ, Polito A, Chien D, Kuo G, Houghton M, Bonino F (1991) Patients with a complete remission of chronic non-A, non-B hepatitis after interferon therapy remain carriers of hepatitis C virus (submitted)

Schlauder GG, Leverenz GJ, Amann CW, Lesniewski RR, Peterson DA (1991) Detection of the hepatitis C virus genome in acute and chronic experimental infection in chimpanzees. J Clin Microbiol 29:2175–2179

Shibata M, Kudo T, Kajiyama M, Takahashi T, Matsushima M, Kimura H, Kuzushima K, Morishima T (1990) Hepatitis C viral sequences in sera during the acute phase of infection. Am J Med 89:830–832

Shimizu YK, Weiner AJ, Rosenblatt J, Wong DC, Shapiro M, Popkin T, Houghton M, Alter HJ, Purcell RH (1990) Early events in hepatitis C virus infection of chimpanzees. Proc Natl Acad Sci USA 87:6441–6444

Shyamala V, Han JH, A simple strategy for the detection and quantitation of HCV RNA in infectious sera by PCR amplification (manuscript in preparation)

Simmonds P, Zhang LQ, Watson HG, Rebus S, Ferguson ED, Balfe P, Leadbetter GH, Yap PL, Peutherer JF, Ludlam CA (1990) Hepatitis C quantification and sequencing in blood products, haemophiliacs, and drug users. Lancet 336:1469–1471

Southern E (1975) Detection of specific sequences among DNA fragments separated by gel electrophoresis. J Mol Biol 98:503

Takamatsu K, Koyanagi Y, Okita K, Yamamoto N (1990) Hepatitis C virus RNA in saliva. Lancet 336:1515

Takamizawa A, Mori C, Fuke I, Manabe S, Murakami S, Fujita J, Onishi E, Andoh T, Yoshida I, Okayama H (1991) Structure and organization of the hepatitis C virus genome isolated from human carriers. J Virol 65:1105–1113

Takehara T, Hayashi N, Mita E, Hagiwara H, Ueda K, Katayama K, Kasahara A, Fusamoto H, Kamada T (1991) Detection of minus strand of hepatitis C virus RNA by reverse transcriptase and polymerase chain reaction. Implication for hepatitis C virus replication in infected tissue

Takeuchi K, Kubo Y, Boonmar S, Watanabe Y, Katayama T, Choo Q-L, Kuo G, Houghton M, Saito I, Miyamura T (1990) The putative nucleocapsid and envelope protein genes of hepatitis C virus determined by comparison of the nucleotide sequences of two isolates derived from an experimentally infected chimpanzee and healthy human carriers. J Gen Virol 71:3027–3033

Tanaka E, Kiyosawa K, Sodeyama T, Nakano Y, Yoshizawa K, Hayata T, Shimizu S, Nakatsuji Y, Koike Y, Furuta S (1991) Significance of antibody to hepatitis C virus in Japanese patient with viral hepatitis: relationship between anti-HCV antibody and the prognosis of non-A, non-B post-transfusion hepatitis. J Med Virol 33:117–122

Thaler MM, Park CK, Landers DV, Wara DW, Houghton M, Veereman-Wauters G, Sweet RL, Han JH (1991) Vertical transmission of hepatitis C virus. Lancet 338:17–18

Thomas HC, Pignatelli M, Goodall A, Waters J, Karayiannis P, Brown D (1984) Immunologic mechanisms of cell lysis in hepatitis B virus infection. In: Viral hepatitis and liver disease, pp 167–177

Ulrich PP, Romeo JM, Lane PK, Kelly I, Daniel LJ, Vyas GN (1990) Detection, semiquantitation, and genetic variation in hepatitis C virus sequences amplified from the plasma of blood donors with elevated alanine aminotransferase. J Clin Invest 86:1609–1614

van der Poel CL, Reesink HW, Lelie PN, Leentvaar-Kuypers A, Choo Q-L, Kuo G, Houghton M (1989) Anti-hepatitis C antibodies and non-A, non-B post-transfusion hepatitis in the Netherlands. Lancet 2:297–298

van der Poel CL, Reesink HW, Schaasberg W, Leentvaar-Kuypers A, Bakker E, Exel-Oehlers PJ, Lelie PN (1990) Infectivity of blood seropositive for hepatitis C virus antibodies. Lancet 335:558–560

van der Poel CL, Cuypers HTM, Reesink HW, Weiner AJ, Quan S, Di Nello R, Van Boven JJP, Winkel I, Mulder-Folkerts D, Exel-Oehlers PJ, Schaasberg W, Leentvaar-Kuypers A, Polito A, Houghton M, Lelie PN (1991a) Confirmation of hepatitis C virus infection by new four-antigen recombinant immunoblot assay. Lancet 337:317–319 van der Poel CL, Cuypers HTM, Reesink HW, Choo Q-L, Kuo G, Han J, Quan S, Polito A, Verstraten JW, van de Wouw JCAM, Schaasberg W, Houghton M, Lelie PN (1991 b) Risk factors in HCV infected blood donors (submitted)

Weiner AJ, Kuo G, Bradley DW, Bonino F, Saracco G, Lee C, Rosenblatt J, Choo Q-L, Houghton M (1990a) Detection of hepatitis C viral sequences in non-A, non-B hepatitis. Lancet 335:1–3

Weiner AJ, Truett MA, Rosenblatt J, Han J, Quan S, Polito AJ, Kuo G, Choo Q-L, Houghton M, Agius C, Page E, Nelles MJ (1990b) HCV testing in low-risk population. Lancet 336:695

Weiner AJ, Brauer MJ, Rosenblatt J, Richman KH, Tung J, Crawford K, Bonino F, Saracco G, Choo Q-L, Houghton M, Han JH (1991a) Variable and hypervariable domains are found in the regions of HCV corresponding to the flavivirus envelope and NS1 proteins and the pestivirus envelope glycoproteins. Virology 180:842–848

Weiner AJ, Geysen HM, Christopherson C, Hall JE, Mason TJ, Saracco G, Bonino F, Crawford K, Marion CD, Crawford KA, Barr PJ, Miyamura T, McHutchinson J, Houghton M (1992) Evidence for immune selection of hepatitis C virus (HCV) putative envelope glycoprotein variants: potential role in chronic HCV infections. PNAS 89:3468–3472

Yoneyama T, Takeuchi K, Watanabe Y, Harada H, Ohbayashi A, Tanaka Y, Yuasa T, Saito I, Miyamura T (1990) Detection of hepatitis C virus cDNA sequence by the polymerase chain reaction in hepatocellular carcinoma tissues. Jpn J Med Sci Biol 43:88–94

Young RA, Davis RW (1983) Efficient isolation of genes using antibody probes. PNAS USA 80:1194–1198

Yuasa T, Ishikawa G, Manabe S, Sekiguchi S, Takeuchi K, Miyamura T (1991) The particle size of hepatitis-C virus estimated by filtration through microporous regenerated cellulose fibre. J Gen Virol 72:2021–2024

Chapter 9 Polymerase Chain Reaction for Hepatitis Delta Virus RNA Identification and Characterization *

Anna Linda Zignego [1,2], Paul Deny [1], Paolo Gentilini [2], and Christian Brechot [1,3]

Summary

Hepatitis delta virus (HDV) is a defective virus dependent on hepatitis B virus (HBV) for infection. Superinfection by HDV on a HBV chronic carrier state is generally associated with an exacerbation of the underlying liver disease. Several HDV RNA sequences are now available from different viral strains which show a relatively high conservation of sequences, although some differences are shown in some parts of the viral genome. The delta antigen (HDAg) is the only identified viral protein. The diagnosis of delta hepatitis is generally based on the detection of anti-HD antibodies (total and IgM) in serum and detection of HDAg in the liver; this antigen is detectable in serum only for a short time at the onset of acute infection. The use of molecular hybridization techniques for HDV RNA identification has recently proved to be a powerful tool in the appraisal of HDV infections, mainly in patients with acute hepatitis and in immunocompromised subjects, where the pattern of serological HDV markers may be modified. Among these techniques, the polymerase chain reaction (PCR) appears to be very sensitive in detecting HDV RNA sequences, both in liver and in serum samples. This approach is most useful when only small amounts of serum HDV RNA are present, for example in the later stages of chronic infection, in IFN-treated patients, or in otherwise inadequately preserved sera. PCR is also useful in interpreting doubtful Slot test results which could correspond either to a negative or to a weak positive. In addition, HDV PCR is a means for cloning and sequencing and for analyzing HDV genetic variability. We analyzed, in a relatively short time, the whole genome of HDV from the liver of an infected woodchuck after several passages, then compared selected parts of the HDV genome with two unrelated HDV strains isolated from humans. PCR allowed, in these circumstances, a better appraisal in the implications of HDV genome variability.

---

[1] Hybridotest, Institut Pasteur, rue du Dr. Roux 28, F-75015, Paris, France
[2] Institut of Internal Medicine, University of Florence, Florence, Italy
[3] INSERM U75 CHU Necker, Service d'Hepatologie, Hôpital Laennec, F-75015 Paris, France
* Anna Linda Zignego was supported by the National Research Council, targeted project "Prevention and Control Disease Factors", subproject "SP3", contract no. 91.00168

Introduction

Hepatitis delta virus (HDV), first identified by Rizzetto et al. (1977), is considered to be a defective virus since it is only infectious in patients who are carriers of hepatitis B virus (HBV). Indeed, infection with HDV can occur either simultaneously with acute HBV infection (coinfection) or, may be superimposed on a chronic HBV infection (superinfection). In the case of coinfection, the acute disease is usually self-limiting, although a fulminant course may also occur. Superinfection by HDV on a HBV chronic carrier state is generally associated with an exacerbation of the underlying HBV infection, leading to chronic active hepatitis, cirrhosis, and, in some cases, fulminant hepatitis (Rizzetto 1983).

HDV has been successfully transmitted in chimpanzees and woodchucks, always in the context of an hepadnavirus infection (Rizzetto et al. 1980a). HBV or woodchuck hepatitis virus (WHV) provide HDV with the envelope needed for infectivity under natural circumstances. However, in infected cells (normal woodchuck hepatocytes), the genome does not seem to depend on the presence of the helper hepadnavirus (Taylor et al. 1987). Furthermore, introduction of the HDV cDNA or RNA into cells by transfection or using the fusion liposome technique is sufficient to induce viral replication (Kuo et al. 1989; Glenn et al. 1990).

In infectious sera, the virus particle has a diameter of about 36 nm and a chimeric structure; the envelope is derived from the helper hepadnavirus, whereas the core consists of phosphoproteins bearing the delta antigenicity and a single-stranded, circular RNA molecule of about 1700 nucleotides in length (1679–1683) (Fig. 1) (Bonino et al. 1981; Rizzetto 1983). This very small genome has a high C-G content and a 70% self-complementarity of bases; the molecule forms an almost unbranched rod structure which is particularly resistant to nuclease digestion (Chen et al. 1986).

Several sequences are now available from HDV, isolated either from human serum or after transmission to woodchucks or chimpanzees (Wang et al. 1986; Kuo et al. 1988; Makino et al. 1987; Kos et al. 1986; Saldanha et al. 1990; Dennistom et al. 1986). The different strains show a relatively high conservation of sequences, although several differences are evident in some parts of the viral genome. Most of these sequences were obtained using classic cDNA cloning.

The proposed mechanisms for replication of HDV RNA utilize a rolling circle model, like plant viroids. In fact, in infected livers (Chen et al. 1986; Negro et al. 1989), dimeric or trimeric molecules are shown. From these replicative intermediates, single genomic or antigenomic molecules are self-cleaved, thus closely resembling the self-cleavage of viroids or group I introns.

It has been suggested that hepatocyte injury resulting from infection with HDV may be caused by a direct virus cytotoxicity in contrast to the immune-mediated injury associated with HBV. This hypothesis has been supported by the absence of a lymphocytic infiltrate in acute HDV hepatitis and by the observation of a temporal relation between the hepatitis and the expression and replication of the virus (Popper et al. 1983; Verme et al. 1990). Further-

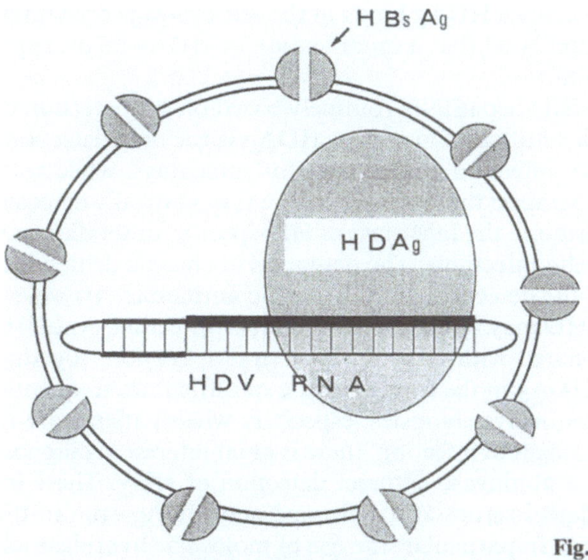

Fig. 1. Structure of hepatitis delta virus (HDV) particle

more, recently performed in vitro studies seem to indicate that the delta antigen (HDAg) is specifically cytotoxic to infected hepatocytes (Cole et al. 1991). On the other hand, in vivo observations of prolonged HDV replication without hepatic damage in immunocomprised patients suggest that, at least in some instances, the host response may be important in determining the hepatocellular injury (Reynes et al. 1989; Zignego et al. 1990 b, c).

HDAg, at present the only identified viral protein detected in infected cells, is located in the nucleus, probably both in the nucleoplasmic area and in the nucleoli (Kuo et al. 1989). This viral protein is phosphorylated on the serine residues with an RNA-binding site in the central region of the HDV genome (Chang et al. 1988; MacNaughton et al. 1990). It acts in *trans* to facilitate genome replication (Kuo et al. 1989). The ORF encoding HDAg has been located on the antigenomic sense RNA and called ORF 5 (Wang et al. 1986), ORF 2 (Makino et al. 1987), or ORF G (Kuo et al. 1988). A clonal heterogeneity yields a potential stop codon in the C-terminal part of the ORF; thus, the size of the protein(s) can be either 195 (Kuo et al. 1988; Wang et al. 1986) or 214 amino acids (Makino et al. 1987; Wang et al. 1986). From different cDNA clones obtained from the same isolate, both proteins may in fact be expressed (Wang et al. 1986). Their respective roles have been analysed in cell culture after transfection of HDV cDNA (Chao et al. 1990). The small protein (195 amino acids) is sufficient to induce and support HDV RNA replication (Kuo et al. 1989; Chao et al. 1990). Only this protein seems to be synthesized during the early phase of HDV infection. On the contrary, the larger protein appears to act as a repressor of replication. This suggested the hypothesis that enhanced expression of the large delta protein might lead to persistent viral infection (Chao et al. 1990). Since HDAg binds to the self-cleavage domain of HDV RNA, it has been speculated that a modification of HDAg might regulate HDV RNA production.

The diagnosis of acute delta hepatitis is routinely based on the detection of HDAg or IgM HD-specific antibodies in serum. HDAg is the only detectable HDV marker in early HDV infection and persists for some days, while anti-HD appears after its disapperance; thus, neither can be systematically detected in acute infections. Furthermore, the IgM class of HD-specific antibodies may also be found in chronic delta infections. The diagnosis of chronic delta infection is generally based on the detection of HD-specific antibodies. However, the detection of total anti-HD may reflect a resolved HDV infection. Anti-HD of the IgM class are associated with ongoing HDV multiplication, together with the identification of HDAg in the liver. Because of difficulties frequently encountered in the analysis of liver biopsies, especially when patients are in acute or chronic advanced stages of infection, there is great interest in diagnostic methods which enable a noninvasive, direct detection of HDV. These include techniques able to detect serum HDV genome and HDAg even in the advanced stages of infection. In particular, the use of molecular hybridization techniques for HDV RNA determination has recently been shown to be a powerful tool in the appraisal of HDV infections. For instance, HDV RNA detection in serum by the slot-hybridization method has been employed as a diagnostic test in acute infections. In addition, hybridization is a useful technique in immunocompromised patients, whose pattern of serological HDV markers may be modified (Zignego et al. 1990 b, c). However, standard hybridization procedures used so far for HDV RNA determination may not be sufficiently sensitive to detect very low levels of replication. This might be important in the first phase of acute infection or in the late stages of chronic infection. Consequently, a rather large percentage of subjects with chronic hepatitis have been found to be HDV RNA-negative in serum although the liver HDAg test result was positive (Smedile et al. 1986; Rasshofer et al. 1987; Saldanha et al. 1989).

The polymerase chain reaction (PCR) is considered a sensitive method in the identification of HDV RNA. We have developed and will report the method here, its clinical applications, and its use for HDV sequence characterization.

Methods

Polymerase Chain Reaction

To amplify the HDV genome, a first step of reverse transcription on a purified RNA is necessary to synthesize HDV cDNA. This synthesized cDNA can then be amplified by PCR.

RNA Purification. For the successful amplification of HDV RNA, it is important to obtain a well-purified and undegraded product. In our experience, RNA extraction from liver tissue using either hot phenol (Farza et al. 1987) or guanidinium methods (Chirgwin et al. 1979) proved to be very effective. For serum samples, a simple and rapid technique can be used as follows: Digestion of 200 µl of serum for 12 h at 37 °C in 2 volumes of a lysis buffer containing 0.4 *M* NaCl, 40 m*M* ethylene diamine tetra-acetic acid (EDTA), 3% sodium dodecyl sulfate (SDS), and 500 µg/ml proteinase K followed by total RNA extraction with phenol-chloroform, precipitation with 1 volume of propan-2-ol, and redissolution in 50 µl of diethylpyrocarbonate-treated water (DEPC H_2O) Zignego et al. 1990 a). Whatever method is used, it is always important to wash precipitated RNA carefully in 70% ethanol.

Oligonucleotide Primers. We have designed several sets of primers which encompass the complete HDV genome. Figure 2 shows some which have proved effective in different combinations. The nucleotide sequences of these primers have been selected according to available HDV sequences and the general criteria useful in choosing of PCR primers (Saiki 1989; Zignego et al. 1990 a). As Fig. 2 shows, the entire genome may be amplified by using different combinations of primers. It is clear, therefore, that when using the following method, the secondary structures of HDV RNA do not prevent the synthesis of cDNA and thus do not impair the amplification experiment.

HDV RNA Reverse Transcriptase-Polymerase Chain Reaction. Due to the RNA nature of the HDV genome, its amplification by PCR must be preceded by complementary DNA (cDNA) synthesis in a reverse transcriptase (RT) reaction. Commercial kits using random priming can be used for this, but we generally obtain better results with the same antigenomic HDV-specific primer (downstream primer) utilized in the PCR reaction (i.e., 2A, 3A, etc.). We usually employ heat denaturation by putting the RNA solution together with the antisense primer (50 pmol) and 1 m*M* dNTPs at 95 °C. After 2–3 min, tubes are frozen in cold ethanol (-70 °C), the cDNA mix [RNasin: 20 U, RT buffer 5 × : 4 µl, MuMLV (BRL, Pharmacia): 200 U/µg of liver RNA or 200 µl of serum] is added to DEPC H_2O in a final volume of 20 µl, and the cDNA reaction is incubated for 30 min at 45 °C and 5 min at 95 °C, then stored for a few minutes at $+4$ °C.

PCR-amplified HDV cDNA is generated by adding the corresponding genomic delta oligonucleotide primer as the upstream primer (i.e., 2S, 3S, etc.; 50 pmol) and Taq polymerase (Perkin-Elmer Cetus) (1 U/µg of liver RNA or 200 µl of serum) in a final volume of 100 µl and a series of 35 cycles in a Perkin-Elmer Cetus automatic thermal cycler. The first denaturation is performed at 94 °C for 5 min and for 1 min in subsequent cycles and annealing and elongation at 55 °C and at 72 °C for 1 min, respectively; the 35th cycle has a 5 min elongation step to fully extend incomplete DNA fragments.

It is also possible to perform a double or "nested" PCR that consists of the reamplification of a first product by using a second set of primers inside the

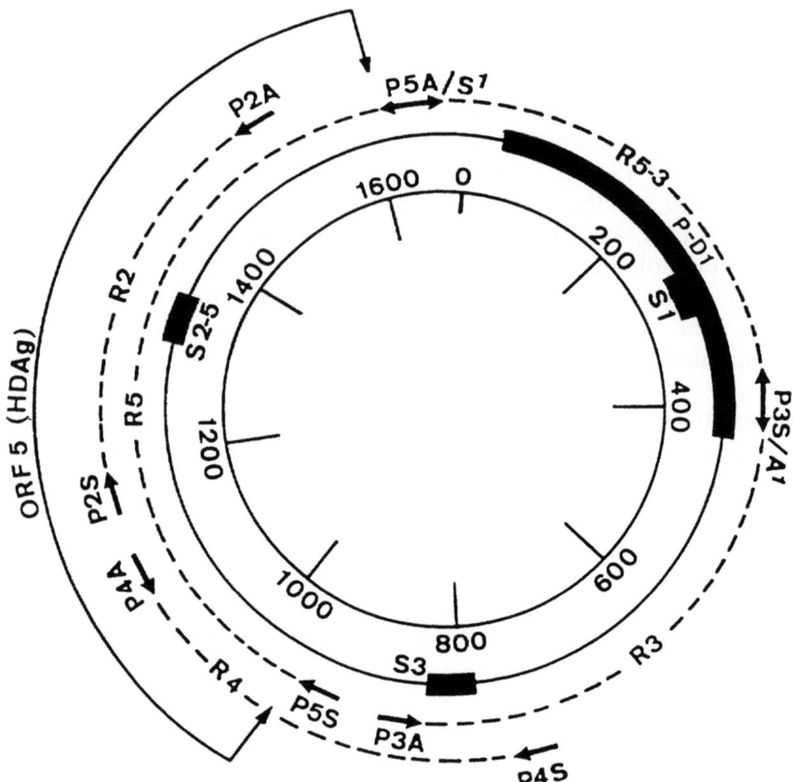


```
Position Name Restriction site Nucleotide sequence 5' ➤ 3'

1525    2A      PstI        AGAAGTTAGAGGAACTGC
867     3A      EcoRI       CGACCTGGGCATCCGAAGGAGGACG
1206    4A      EcoRI       GAGGTTGACCGAGGAAGACGAGAGA
1640    5A      BamHI       AGAAAAGAGTAAGAGTACTGAGGAC
1167    2S      HindIII     TATTCTTCTTTCCCTTCT
391     3S      HindIII     GGAGACCGAAGCGAGGAGGAAAGTA
743     4S      HindIII     CGTCCCCTCGGTAATGGCGAATGGG
929     5S      HindIII     CAAACCTGTGAGTGGAAACCCGCTT
1309    2-5                 TTGTCGGTGAATCCTCCCCTGAGAGGCCTCTTCCTAGGTC
801     3-4                 CGAGAGAAAAGTGGCTCTCCCTTGGCCATCCGAGTGGA
```

Fig. 2. Hepatitis delta virus (HDV) genetic map with the different primes and probes. Sequences coding for delta antigen are indicated by the *curved arrow*, the position and nucleotide sequence of primers (*arrows*) and probes S2−5 and S3(3−4) (*black boxes*) are shown. The nucleotide sequences of primers 5S′ and 3A′ are complementary to those of primers 5A and 3S, respectively. R2, 3, 4, and 3−5: five of the HDV regions that are amplifiable with the primers shown

first one (i.e., the 5S-5A primer pairs for the first one and the 2S-2A pairs for the second one). It would increase both the sensitivity and specificity of the reaction because to obtain a product of the right size, four specific primer hybridizations with corresponding HDV sequences occur. On the other hand, the risk of contamination increases significantly because of the greater amount of amplification products and working steps involved. One way to avoid this is to put a thick layer of mineral oil over the first reaction mix. After a first series of thermal cycles, the second PCR is then obtained by putting the second reaction mix on top of the layer of oil and by centrifuging the tube in order to mix all the reaction components together. To analyse the amplified cDNA we usually subject to electrophoresis one-tenth of the reaction mix on a 1.5% – 1.9% ethidium bromide agarose gel (Sigma) in TRIS-borate EDTA (TBE). The gel is subsequently transferred to a nylon membrane (Gene Screen plus, NEN). The filters are prehybridized in 3 sodium chloride citrate (SCC, 150 mmol/l sodium chloride, 15 mmol/l sodium citrate), $5 \times$ Denhardt's solution (0.02% ficoll, 0.02% polyvinylpyrrolidone, 0.02% bovine serum albumin), 0.5% SDS, and 30% formamide and then incubated at 42 °C overnight with the oligonucleotide probes radiolabelled with γ-AT^{32}P using the polynucleotide kinase (Boehringer Mannheim or Pharmacia). Oligonucleotides used as probes are specific for the amplified fragments and correspond to genomic internal sequences (Fig. 2). After incubation, filters are washed in $2 \times$ SSC and 1% SDS for 15 min at room temperature and in $0.2 \times$ SSC and 0.1% SDS for 10 min at 55 °C. The autoradiograph is exposed for 3 h and 16 h with a single intensification screen.

PCR for Cloning and Sequencing of Amplified cDNA

In order to facilitate HDV RNA cloning, restriction sites, absent from the HDV sequences to be amplified, are included at the 5′-ends of the primers (Fig. 2). After amplification, the HDV cDNA is purified from the agarose gel, i.e., by using a NA 45 membrane, extracted with phenol and chloroform, and precipitated with ethanol. A Centricon step (Amicon) is sometimes used after chloroform extraction. Then, purified HDV cDNA sequences may be cloned in traditional ways, i.e., in M13 mp18/mp19 (Appligene) by using the specific restriction sites at the 5′-end of the primers. The nucleotide sequence may be obtained by using commercially available kits, such as Sequenase (USB).

Otherwise, in association with cloning experiments, it is possible to perform direct sequencing, i.e., with Sequenase, using antisense HDV primers (2.5 μM) as internal primers. We generally prefer to perform both direct and clonage-mediated sequencing of the amplification products because the former is often more difficult to interpret (less clear in reading) and cannot show the eventual coexistence in the same biological sample of different genomic variants. Moreover, it is worth pointing out that the cloning of a small, very homogeneous, and abundant cDNA, as obtained by PCR, is much easier and more rapid than cloning by traditional methods.

Pitfalls in the Procedure

Presence of PCR Inhibitors in Serum. In serial dilution experiments performed with infected sera, an initial dilution of 10^1 to 10^2 in DEPC-treated water improved efficiency and was sometimes essential for obtaining amplification itself. This has been interpreted as an effect of poorly specified reaction inhibitors present in the serum samples (Zignego et al. 1990a) and was observed also in experiments of HBV DNA amplification (Gerken et al. 1990).

Secondary Structures of HDV RNA. The denaturation of RNA is common to all cDNA synthesis methods both in cloning and amplification experiments and may be accomplished using either chemical or physical methods, but its degree varies in different systems. For example, in human immunodeficiency (HIV) RNA, HBV RNA, and HCV RNA reverse transcription, a physical denaturation at $60°-65\,°C$ is generally used, whereas with HDV RNA, due to its abundant secondary structures and internal base pairing, a more drastic hot denaturation is required. This is done, as shown above, by heating the RNA dilution at $95\,°C$ for $2-3$ min and then cooling it at $-70\,°C$ (cold ethanol).

False-Positive Results. It is important to be very careful and make numerous specificity controls when using PCR for HDV identification in clinical samples; this is a general problem in PCR methodology, directly related to the risk of minute contamination by amplification products. The main precautions to take are the physical separation of the various steps involved in the method (RNA extraction, cDNA-PCR reactions, and product analysis), the use of disposable pipettes, new disposable containers, and various specificity controls in each experiment. The latter may be represented by (1) the negative results observed after PCR analysis of RNA extracted from normal human liver or serum (Fig. 3, lanes *c-*) and baker's yeast (Fig. 3, lanes *ry*), DNA from lambda bacteriophage and various plasmids, the reaction mixture without the reverse transcriptase (Fig. 3, lanes *-rt*), and simple water (Fig. 3, lane *R*); (2) the positive results obtained using RNA extracted from liver and/or serum samples known to contain HDV RNA sequences, (3) the size of the fragments amplified corresponding to that expected from the genetic map and their hybridization to a ^{32}P-labelled internal oligonucleotide probe (an example is shown in Fig. 3, where the products of amplification A1 and A2 from regions 5 and 3, respectively, have the expected sizes of 716 and 478 bp, respectively; similarly, in Fig. 4, the amplification products of region 2 are, as predicted, 359 bp long); (4) the repetition of the PCR assay and the use of more than one primer set for each sample analyzed, that is, one positive or negative result must be observed at least twice in different experiments.

Assessment of the Sensitivity of the Assay for the detection of HDV sequences was carried out in experiments using serial dilutions of RNA extracted from serum and liver samples. We were able to obtain a band in Southern blot-PCR analysis with only 0.1 pg of infected liver or 8 pl of infected serum (Fig. 4B).

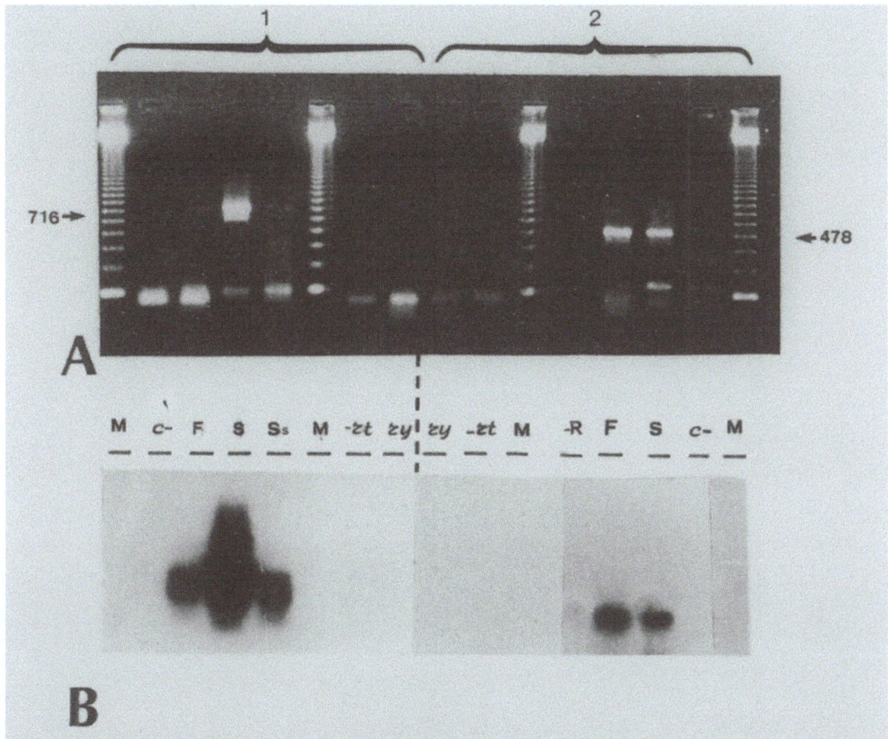

Fig. 3A, B. Amplification of HDV RNA sequences after ethidium bromide staining (**A**) and hybridization with ^{32}P-labelled oligonucleotide probes (**B**). *Panel 1*, amplification with primers 5A, 5S. *Panel 2*, amplification with primers 3A, 3S. *Lanes M*, reference molecular weight markers (123 bp ladder); *lanes c−*, negative control (normal liver RNA); *lanes F*, HDV + woodchucks' liver; *lanes S*, serum of a patient with delta chronic active hepatitis; *lane Ss*, HDV + chimpanzee's serum; *lanes −rt*, the same reaction mix used for *S* without reverse transcriptase; *lanes ry*, baker's yeast RNA; *lane −R*, distilled water

Parallel analysis of the same sample by the slot blot technique (Fig. 4C) showed that the sensitivity of HDV RT-PCR was 10^4–10^6 greater than that of slot blot analysis, which was unable to detect HDV sequences clearly when using less than 1 ng of liver RNA and 8 µl of serum (Fig. 4A, B).

Applications

RT-PCR can be used for very sensitive and specific HDV RNA detection and the rapid generation of HDV probes as well as for HDV RNA cloning and sequencing.

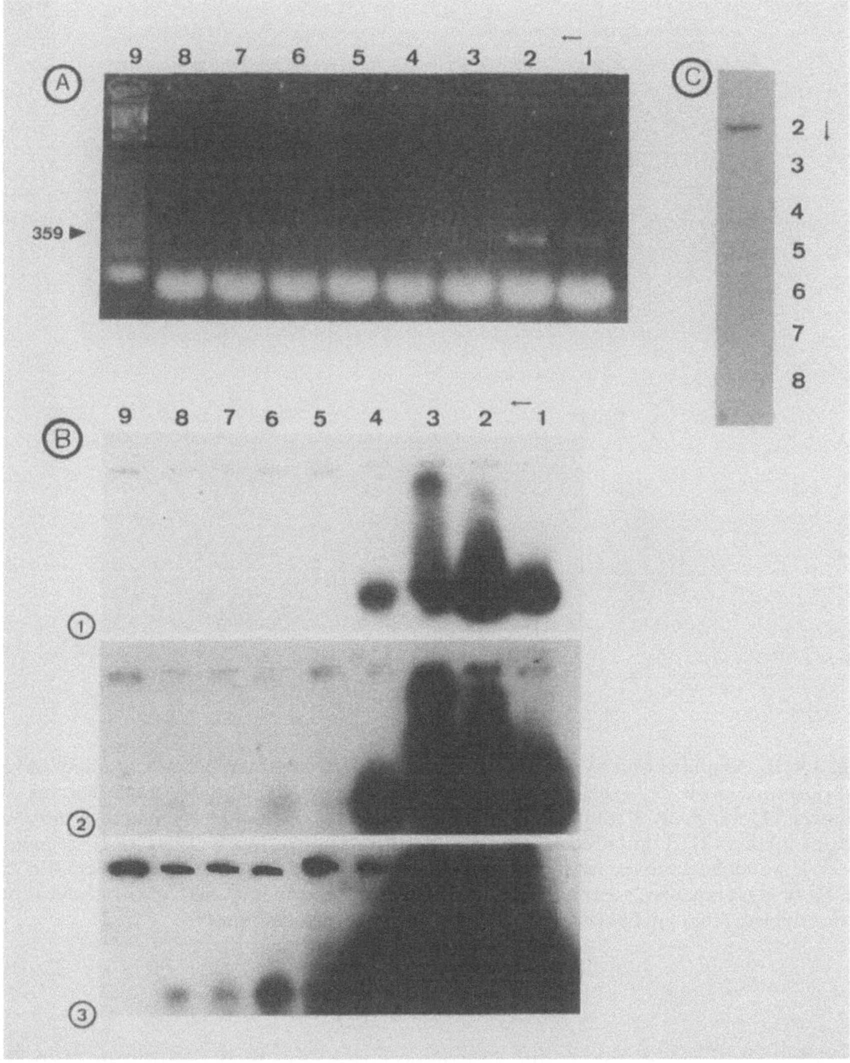

Fig. 4A–C. Comparative analysis of slot blot and RT-PCR for HDV RNA detection. Tenfold serial dilutions of the serum from a patient with delta chronic active hepatitis. In **A** and **B**, lanes 1, 2, 3, 4, 5, 6, 7, and 8 correspond, respectively, to 80 µl, 8 µl, 800 nl, 80 nl, 800 pl, and 8 pl of the initial serum sample. Amplification was performed with primers 2A, 2S. **A** Ethidium bromide staining. **B**, Hybridization with ^{32}P-labelled oligonucleotide probe S2; 1, 2, and 3, 2 h, 1 day, and 3 days exposure. **C** Slot blot analysis of the same dilutions

PCR for HDV RNA Detection

This approach is most useful when the amount of serum HDV RNA is small, that is, in the first phase of acute infection or in the later stages of chronic infection as well as in scarcely viremic sera from interferon (IFN)-treated patients or in otherwise inadequately preserved sera.

Patients (Table 1). We analyzed 13 patients with delta chronic active (CAH) hepatitis whose serum samples had been stored lyophilized and 10 subjects after 6 months of IFN-α treatment for a delta CAH who were HDV RNA positive using the slot blot analysis and whose serum samples had been frozen at $-70\,°C$ (A. L. Zignego et al. 1991, unpublished observation).

Table 1. Serum hepatitis delta virus (HDV) RNA detection by polymerase chain reaction (PCR) analysis of serum samples negative or doubtful with slot blot

Patients	n	Slot blot	RT-PCR[+]
CAH D (lyophilized sera)	13	$-$ 10	4
		$+/-$ 3	1
CAH D	10	$-$ 8	3
IFN treated (frozen sera)		$+/-$ 2	1

CAH D, delta chronic active hepatitis; IFN, interferon; RT, revere transcriptase

Results. PCR analysis showed serum HDV RNA in 4 of 10 lyophilized serum samples that appeared HDV RNA negative and in 1 of 3 which were of doubtful interpretation ($\pm$) with slot blot (light signal after 3 days of exposure) and in 3 of 8 patients negative, and 1 of 2 subjects $\pm$, on the HDV RNA slot blot analysis. HDV PCR positive and negative results correspond to permanently high or normalized ALT values (Table 1).

In summary, by PCR analysis we were able to identify HDV sequences in the serum of patients affected by chronic D hepatitis who had undergone IFN treatment and in lyophilized sera that were negative by the slot test. It was also useful in interpreting $\pm$ slot test results, such as the presence of a shadow at the background noise limit, especially frequent after long exposure (3 days), which could correspond either to a negative or a weak positive. Furthermore, it enabled us to follow directly the effectiveness of antiviral treatment in inhibiting viral replication.

PCR in HDV RNA Cloning and Sequencing

The difficulty in HDV RNA cloning from infected samples makes approaches aiming to obtain different HDV probes or to compare HDV sequences from different isolates laborious. As a result, PCR is an extremely useful means for cloning and sequencing and for effectively analyzing HDV genetic variability.

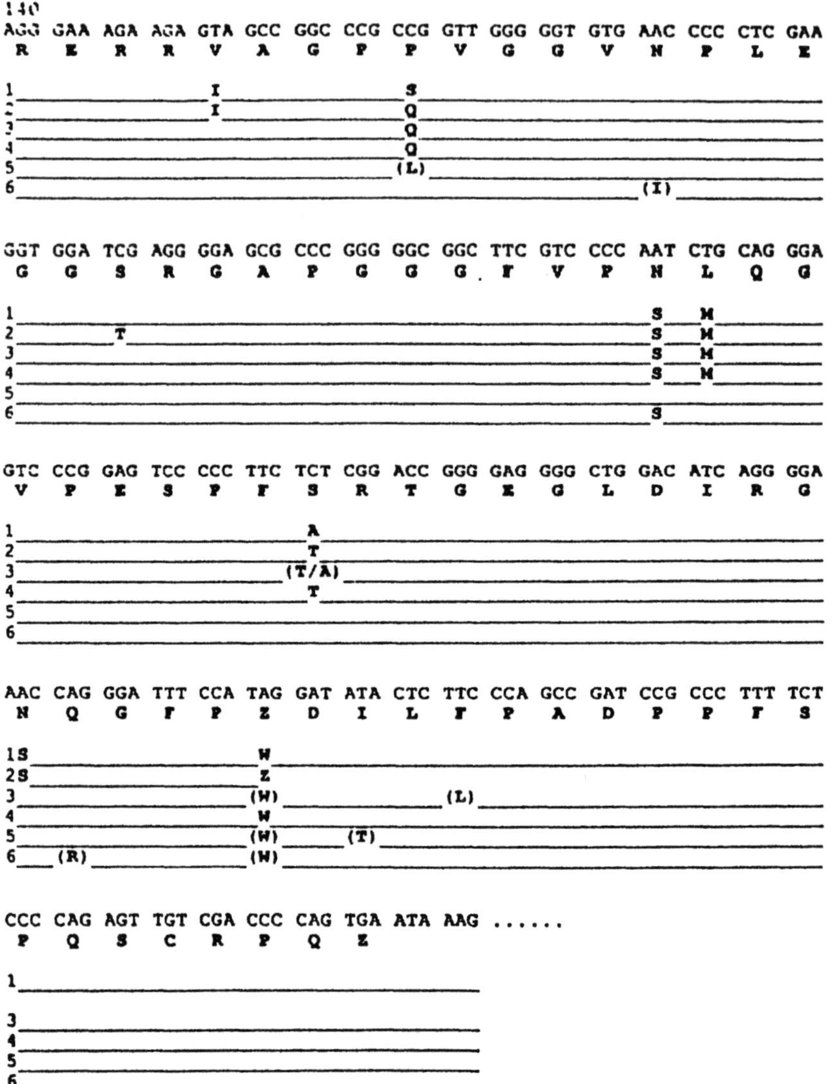

Fig. 5. Analysis of woodchuck delta antigen sequence from amino acid 140 to the C-terminal end. *Lines 1–6* represent comparison between different HDV isolates. *Line 1*, first human HDV isolate (Makino et al. 1987); *line 2*, human HDV isolate (Saldanha et al. 1990); *line 3*, isolate from a chronically infected patient; *line 4*, isolate from an acutely infected patient; *line 5*, woodchuck isolate; *line 6*, first chimpanzee isolate (Wang et al. 1986). Possible heterogeneity due to nucleotide variability between different clones is shown in *parentheses*

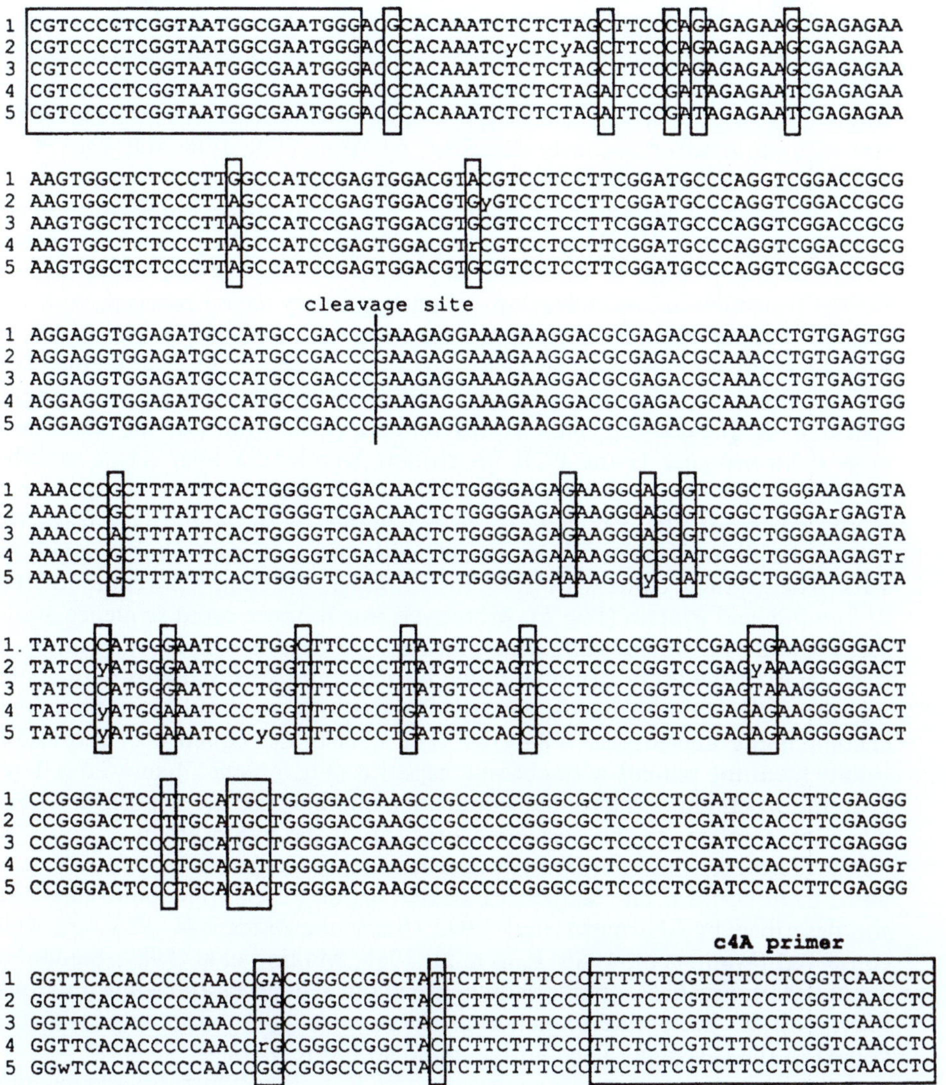

Fig. 6. HDV nucleotide sequences of woodchuck and human isolates and comparison with available sequences. The sequence from nucleotide 743 to 1206 is shown. The *vertical line* corresponds to the antigenomic RNA self-cleavage position (900–901). *Lines 1–5* are as described in Fig. 5. The *lower-case letters* indicate ambiguities in the sequence using the convention *y* represents C or T, *r* represents A or G, and *w* represents A or T

In using PCR, we were able to analyze in a relatively short time the whole genome of HDV belonging to the liver of an infected woodchuck after several passages and to compare selected parts of the HDV genome with two unrelated HDV strains isolated from humans. In the former case, the total genome length was 1679 nucleotides; only point mutations were observed in comparison with sequences previously described by Wang et al. 1986 and Kuo et al. 1988. The different isolates whose sequences were compared were derived from the same human serum transmitted first to chimpanzees (Rizzetto et al. 1980 b) and then to WHV-infected woodchucks. Our viral preparation came from the fifth passage in woodchucks (Ponzetto et al. 1984). During experimental transmission, modifications of pathogenicity were observed, with increased severity of the liver disease (Ponzetto et al. 1984; 1987). The fact that we found only minimal sequence variation might suggest that host factors play a predominant role in this enhanced pathogenicity. In addition, one might speculate on the consequences of different expression rates of the small and large delta proteins. Using PCR for cloning woodchuck liver RNA, we observed a UAG stop codon in all the clones obtained from the antigenomic RNA, which potentially coded for the 195 amino acid protein. In contrast, we found two different genomic molecules with a T to C mutation at position 1012. The genomic clone with the C mutation has the coding potential for the 214 amino acid protein (Fig. 5). Moreover, due to more rapid sequence analysis allowed by a previous amplification step by PCR, we could further analyze this region of the HDV genome in two unrelated infected patients with acute and chronic HDV hepatitis. Several point mutations were evident in the region encompassing nucleotides 768–1181. The nucleotide sequence of the HDV isolate from the patient with chronic hepatitis (Fig. 6, lane 3) showed a T to C mutation similar to that of the woodchuck number 5 isolate at position 1012, which corresponds to the possible expression of the two proteins.

Furthermore, the comparative sequence analysis that we performed enabled us to confirm that sequences located around the antigenomic cleavage site described by Sharmeen et al. 1988 (from nucleotide 844–933) are well conserved (Wang et al. 1986; Kuo et al. 1988; Makino et al. 1987; Saldanha aet al. 1990). Fig. 6 is a comparison between our sequences and some previous ones. This 90 nucleotide sequence conservation might reflect sequences involved in the cleavage of the antigenomic strand in vivo.

In conclusion, these results suggest that PCR, followed by rapid sequencing of the amplified products, will allow a better appraisal in the implications of HDV genome variability.

References

Bonino F, Hoyer B, Ford E, Shih JW, Purcell RH, Gerin JL (1981) The delta agent: HBsAg particles with delta antigen and RNA in the serum of HBV carrier. Hepatology 1:127–131

Chang MF, Baker SC, Soe LH, Kamahora T, Makino S, Govindarajan S, Lai MMC (1988) Human hepatitis delta antigen is a nuclear phosphoprotein with RNA-binding activity. J Virol 62:2403–2410

Chao M, Hsieh SY, Taylor J (1990) Role of two forms of hepatitis delta virus antigen: evidence for a mechanism of self limiting genome replication. J Virol 64:5066–5069

Chen PJ, Kalpana G, Goldberg J, Mason W, Werner B, Gerin J, Taylor J (1986) Structure and replication of the genome of the hepatitis delta virus. Proc Natl Acad Sci USA 83:8774–8778

Cole SM, Gowans EJ, Macnaughton TB, De La M, Hall P, Burrell CJ (1991) Direct evidence for cytotoxicity associated with expression of hepatitis delta virus antigen. Hepatology 13:645–851

Dennistom KJ, Hoyer BH, Smedile A, Wells FV, Nelson J, Gerin JL (1986) Cloned fragment of the hepatitis delta virus RNA genome: sequence and diagnostic applications. Science 232:873–875

Farza H, Salmon AM, Hadchouel M, Moreau JL, Babinet C, Tiollais P, Porcell C (1987) Hepatitis B surface antigen gene expression is regulated by sex steroids and glucocorticoids in transgenic mice. Proc Natl Acad Sci USA 84:1187–1191

Gerken G, Paterlini P, Manns M, Housset C, Terre S, Dienes HP, Hess G, Gerlich WH, Berthelot P, Meyer zum Buschenfelde KH, Brechot C (1990) Assay of hepatitis B virus DNA by polymerase chain reaction and its relationship to Pre-S and S-encoded viral surface antigens. Hepatology 13:158–166

Glenn J, Taylor J, White J (1990) In vitro-synthesized delta virus RNA initiates genome replication in cultured cells. J Virol 164:3104–3107

Kos A, Dijkema R, Arnberg AC, Wan der Meide PH, Schellekens H (1986) The hepatitis delta virus possesses a circular RNA. Nature 323:558–560

Kuo M, Goldberg J, Coates L, Mason W, Gerin J, Taylor J (1988) Molecular cloning of hepatitis delta virus RNA from an infected woodchuck liver: sequence, structure and applications. J Virol 62:1855–1861

Kuo MY-P, Chao M, Taylor J (1989) Initiation of replication of human hepatitis delta virus genome from cloned DNA: role of delta antigen. J Virol 63:1945–1950

Macnaughton TB, Gowans EJ, Reinboth B, Jilbert AR, Burrel CJ (1990) Stable expression of hepatitis delta virus antigen in a eukaryotic cell line. J Gen Virol 71:1339–1345

Makino S, Chang MF, Shieh CK, Kamahora T, Vannier DM, Govindarajan S, Lai MM (1987) Molecular cloning and sequencing of a human hepatitis delta virus RNA. Nature 329:343–346

Negro F, Korba BE, Forzani B, Barondy BM, Brown TL, Gerin JL, Ponzetto A (1989) Hepatitis delta virus (HDV) and woodchuck hepatitis virus (WHV) nucleic acids in tissues of HDV-infected chronic WHV carrier woodchuck. J Virol 63:1612–1618

Ponzetto A, Cote PJ, Popper H, Hoyer BH, London WT, Ford EL, Bonino F, Purcell RH, Gerin JL (1984) Transmission of the hepatitis B virus-associated delta agent to the eastern woodchuck. Proc Natl Acad Sci USA 81:2208–2212

Ponzetto A, Forzani B, Smedile A, Hele C, Avanzini L, Novara R, Canese M (1987) Acute and chronic delta infection in the woodchuck. Progr Clin Biol Res 234:37–46

Popper H, Thung SN, Gerber MA, Hadler SC, De Monzon M, Ponzetto A, Anzola E, Rivera D, Mondolsi A, Bracho A, Francis DP, Gerin JL, Maynard JE, Purcell RH (1983) Histologic studies of severe delta agent infection in Venezuelan Indians. Hepatology 3:906–912

Rasshofer R, Buti M, Esteban R, Jardi R, Roggerdorf M (1987) Demonstration of hepatitis D virus RNA in patients with chronic active hepatitis. J Infect Dis 157:191–195

Reynes M, Zignego AL, Samuel D, Fabiani B, Gugenheim J, Tricottet V, Brechot C, Bismuth H (1989) Graft hepatitis delta virus reinfection after orthotopic liver transplantation in HDV cirrhosis. Transplant Proc 21:2424–2425

Rizzetto M (1983) The delta agent. Hepatology 3:729–737

Rizzetto M, Canese MG, Arico S, Crivelli O, Bonino F, Trepo CG, Verme G (1977) Immunofluorescence detection of a new antigen antibody system (delta/anti-delta) associated to the hepatitis B virus in the liver and in the serum of HBsAg carriers. GUT 18:997–1003

Rizzetto M, Canese M, Gerin JL, London WT, Sly DL, Purcell RH (1980a) Transmission of the hepatitis B virus associated delta antigen to chimpanzees. J Infect Dis 141:590–602

Rizzetto M, Hoyer B, Canese M, Shih JW-K, Purcell RH, Gerin J (1980b) Delta agent: association of delta antigen with hepatitis B surface antigen and RNA in serum of delta-infect chimpanzees. Proc Natl Acad Sci USA 77:6124–6128

Saiki RK (1989) The design and optimization of the PCR. In: Erlich HA (ed) PCR technology: principles and applications for DNA amplification. Stockton, New York, pp 7–16

Saldanha J, di Blasi F, Blas C, Velosa J, Ramalho RM, di Marco V, Mora L, de Moura MC, Carreno V, Craxi A (1989) Detection of hepatitis delta virus RNA in chronic liver disease. Hepatology 9:23–28

Saldanha JA, Thomas HC, Monjardino JP (1990) Cloning and sequencing of RNA of hepatitis delta virus isolated from human serum. J Gen Virol 71:1603–1606

Sharmeen L, Kuo M, Dinter-Gottelieb G, Taylor J (1988) Antigenomic RNA of hepatitis delta virus can undergo self cleavage. J Virol 62:2674–2679

Smedile A, Rizzetto M, Denninston K, Bonino F, Wells F, Verme G, Consolo F, Hoyer B, Purcell RH, Gerin JL (1986) Type D hepatitis: the clinical significance of hepatitis D virus RNA in serum as detected by a hybridization-based assay. Hepatology 6:1297–1302

Taylor J, Mason W, Summers J, Goldberg J, Aldrich C, Coates L, Gerin J, Gowans E (1987) Replication of human hepatitis delta virus in primary cultures of woodchuck hepatocytes. J Virol 61:2891–2895

Verme G, Rocca G, Rizzi R, Mollo F, David E, Solcia E, Sessa F (1990) Histopathology of chronic delta hepatitis. In: Verma G, Bonino F, Rizzetto M (eds) Viral hepatitis and delta infection. Liss, New York, pp 169–176

Wang K-S, Choo Q-L, Weiner AJ, Ou J-H, Najarian RC, Thayer RM, Mullenbach GT, Denniston KJ, Gerin JL, Hougton M (1986) Structure, sequence and expression of the hepatitis delta viral genome. Nature 323:508–514, 328:456

Zignego AL, Deny P, Feray C, Ponzetto A, Gentilini P, Tiollais P, Brechot C (1990a) Amplification of hepatitis delta virus RNA sequences by polymerase chain reaction: a tool for viral detection and cloning. Mol Cell Probes 4:43–51

Zignego AL, Dubois F, Samuel D, Reynes M, Gentilini P, Bysmuth A, Gaudeau A, Benhamou JP, Tiollais P, Bismuth H, Brechot C (1990b) Serum hepatitis delta RNA in patients with delta hepatitis and in liver graft recipients. J Hepatol 11:102–110

Zignego AL, Samuel D, Dubois F, Reynes M, Gentilini P, Gaudeau A, Benhamou JP, Tiollais P, Bismuth H, Brechot C (1990c) Liver transplant recipients as a model for the study of HDV infection natural history. In: Dianzani MU, Gentilini P (eds) Chronic liver damage. Excerpta Medica, Amsterdam, pp 137–144

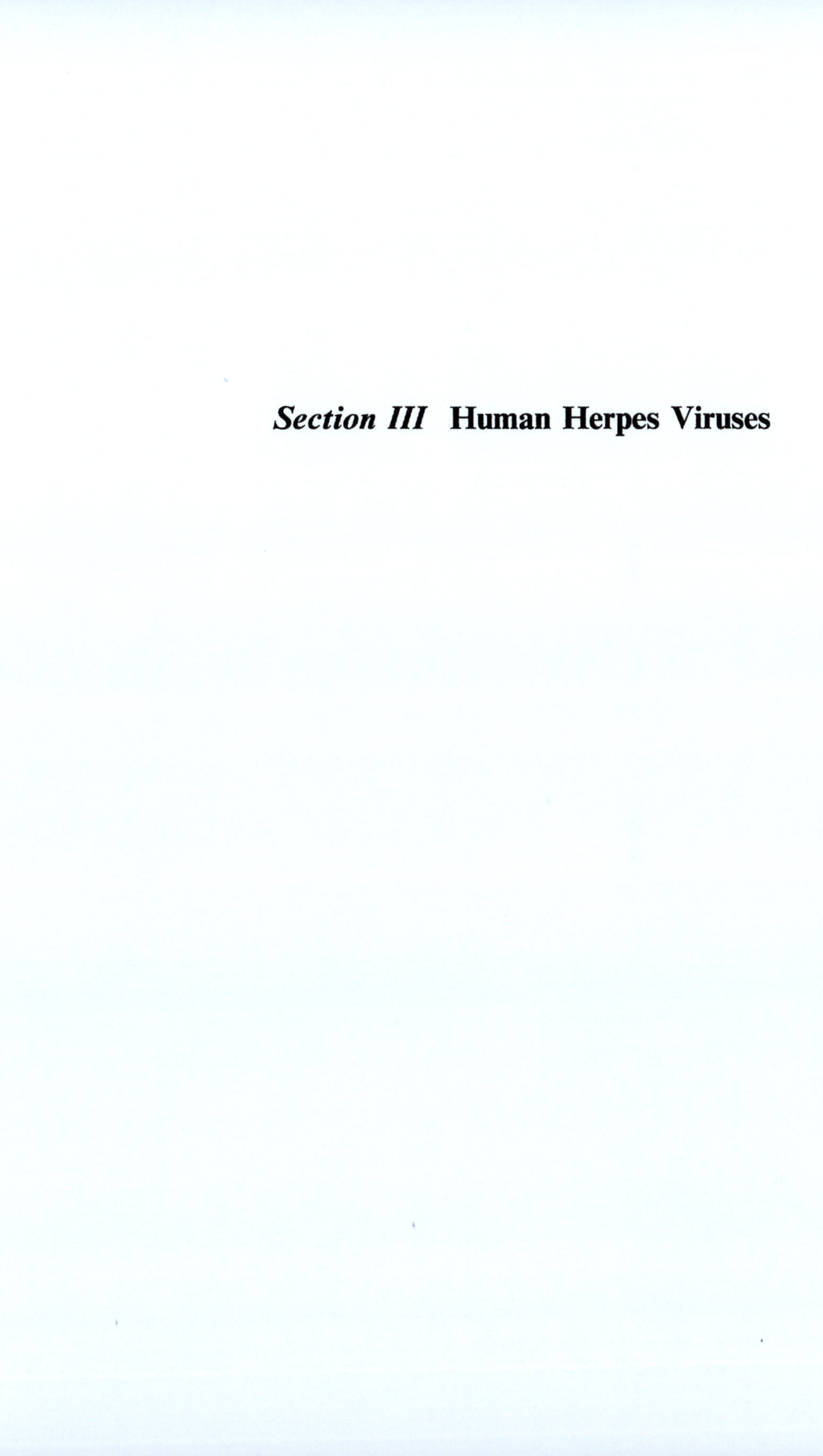

Section III Human Herpes Viruses

Chapter 10 Polymerase Chain Reaction Detection of Herpes Simplex Virus in Cerebrospinal Fluid

R. H. Boerman[1], A. C. B. Peters[1], E. P. J. Arnoldus[1,2], A. K. Raap[2], A. M. van Loon[3], B. R. Bloem[1], and M. van der Ploeg[2]

Summary

Herpes simplex encephalitis (HSE) is a serious disease with significant morbidity and mortality if left untreated. It can be treated with acyclovir, but for this to be effective it must be instituted early during the course of disease. Noninvasive, early diagnosis, however, is presumptive at best, and many patients are treated without a definitive diagnosis.

We have developed a polymerase chain reaction (PCR) for the detection of HSV DNA in the cerebrospinal fluid (CSF) of experimentally infected mice. PCR was compared with viral isolation techniques and immunofluorescence on CSF cells during the course of disease of experimentally infected mice. It was found to be more sensitive than these other methods, and HSV DNA could be detected during the earliest stages of central nervous system infection. The assay has also been used for the detection of HSV DNA in CSF samples from patients with HSE. The results are discussed and compared with those published elsewhere.

Introduction

Herpes simplex virus (HSV) can infect any part of the nervous system, causing HSV encephalitis (HSE), brain stem encephalitis, myelitis, meningitis, radiculitis, and cranial nerve infections (Boerman 1991). Apart from HSV encephalitis, these diseases are very rare indeed. As delay in instituting treatment may result in serious morbidity, the possibility of a HSV infection should always be considered if signs of an inflammation of the nervous system are present.

HSE is the usual manifestation of HSV infection of the CNS and occurs in all age groups. However, HSE in neonates differs in several respects from the illness in children and adults (Whitley 1990). Diagnosis is less difficult, as it

---

[1] Department of Neurology, University Hospital Leiden, The Netherlands
[2] Department of Cytochemistry and Cytometry, State University Leiden, The Netherlands
[3] Department of Virology, Rijksinstituut voor de Volksgezondheid en Milieuhygiëne, Bilthoven, The Netherlands

usually arises in the setting of more or less disseminated disease and as the virus can often be isolated from the cerebrospinal fluid (CSF). The major diagnostic problem of HSE is presented by the disease of children and adults. This is the most common form of sporadic fatal encephalitis in the USA and Western Europe (Baringer 1978; Booss and Esiri 1986), and it is associated with significant mortality (50% – 70%) and morbidity (50% of survivors) in untreated cases (Whitley et al. 1981). These appalling figures have improved since effective antiviral therapy with acyclovir became available (Sköldenberg et al. 1984; Whitley et al. 1986).

Early diagnosis of HSE is, however, still difficult. Although most patients present with a combination of fever, an organic psychosyndrome, impaired consciousness or focal neurological deficit, the clinical signs and symptoms are variable (Barringer 1978; Spaar 1976; Boerman 1991). In contrast to many other viral encephalitides, additional epidemiological information is not helpful in diagnosis due to its worldwide, sporadic occurrence in all age groups. No predisposing conditions or diseases are known, and no association with an immunosuppressive state has been found in contrast to other herpetic diseases. The CSF is rarely normal if neurological signs have appeared, but the alterations are not specific (Koskiniemi et al. 1984). Imaging techniques often do not show any changes during the early stages of disease and findings at later stages lack specificity. Overdependence on the CT scan in the diagnosis of HSE can even result in delays in diagnosis (Whitley et al. 1982). Although often neglected, the EEG frequently shows focal abnormalities in HSE (Nahmias et al. 1982), and it should be included in any battery of tests performed prior to the decision to biopsy or start antiviral therapy (Booss and Esiri 1986).

Viral diagnostic studies in HSE are often uninformative for two reasons. First of all, HSV can only rarely be cultured from the CSF in patients with HSE. This is quite remarkable as HSV is easy to culture from lesions in other herpetic diseases. Secondly, HSV remains latent in trigeminal and other sensory ganglia and may be reactivated by many intercurrent diseases unrelated to HSE. Reactivation may even occur in the absence of overt herpetic skin or mucosal disease, but with the production of an HSV-specific immune response. Consequently, peripheral isolation of HSV and the detection of HSV-specific antibodies during the acute phase are meaningless. Paired sera and CSF samples are required in order to demonstrate both active infection and intrathecal production of antibodies and thus clinch the diagnosis of HSE (Barringer 1978). However, by the time serological diagnosis has been established, the patient will either have died or have recovered. Thus, brain biopsy remains the only conclusive way of proving the diagnosis within the first days of disease and may have to be performed in selected cases (Fishman 1987; Hanley et al. 1987). Clearly, this situation is unsatisfactory, and a rapid, non-invasive diagnostic method is needed.

Material and Methods

Experimental Approach

The central nervous system (CNS) has been referred to as an immunologically privileged site because of the relative isolation of brain and spinal cord from the systemic vascular and immune systems by the blood-brain barrier (Medawar 1948). Therefore, inflammation of CNS tissue may not be reflected in the blood. By contrast, the CSF is in intimate contact with the parenchyma of the brain and spinal cord and is considered to be an excellent indicator of immune processes occurring within the CNS itself (Trotter and Brooks 1980). As CSF is the only intrathecal material readily obtainable from patients, the obvious approach is to develop a rapid diagnostic method which can be applied to it.

The developments in molecular biology and the advent of effective antiviral therapy have given new impetus to diagnostic virology. Detection of the virus by culture on cell monolayers has been the gold standard of diagnostic virology, and its techniques are still being improved. Centrifuging samples on monolayers, in combination with the detection of viral antigen, has resulted in both increased sensitivity and speed of detection of HSV and other viruses (Salmon et al. 1986; Gleaves et al. 1985; Boerman 1991). New, sensitive serological methods have been developed for the detection of a specific intrathecal immune response against HSV at an early stage of disease. Some of these tests even render a correction for leakage of the blood-brain barrier superfluous. The major advances have been in the detection of viral components (Boerman 1991). HSV replicates very inefficiently, producing only a few infective viral particles in relation to the total amount of viral nucleic acid and protein produced (Richman et al. 1984; Watson et al. 1963). Viral DNA in particular offers the additional advantages of stability of its primary structure and the possibility of in vitro amplification.

Except in rare instances (Bergström et al. 1989; Brihaye 1970; Zurukzoglu 1937), asymptomatic reactivation of HSV within the intrathecal compartment has not been found. Consequently, the intrathecal detection of HSV or of a specific intrathecal immune response against HSV proves the diagnosis of HSV infection of the CNS.

Developing a diagnostic method for HSE using CSF from patients confronts one with several problems. First of all, it is difficult to obtain a sufficient number of specimens due to the low incidence and because of the practice of taking only a few samples from each patient. Second, the variability of samples obtained from different patients is large. CSF is obtained at different intervals after the onset of disease, and the day of onset may be impossible to determine. Processing is frequently insufficient because samples are often obtained outside office hours or may have to be transported to facilities outside the hospital. Third, CSF findings in patients with HSE are subject to large variations, even if obtained at comparable intervals after clinical onset of HSE (Koskiniemi et al. 1984). The reason may be that CSF alterations do not reflect the clinical onset and progression of disease but rather the histopathological process of viral infection and inflammation. However, correlation of CSF findings and histopathological examination of the whole area of inflamed brain tissue is rarely possible, and for obvious reasons samples of brain tissue prior to the onset of neurological symptoms cannot be obtained.

Hence, we started a research program using an animal model. Animal models of HSE provide the ideal system for comparing and developing diagnostic tests. Large amounts of identical individuals are available, all inoculated by the same route at a fixed time and with an inoculum of known strength and virus strain. The clinical signs, CSF, histopathological and immunohistochemical progression of the disease can be studied simultaneously in all subjects. Animal models of HSE have been well studied, and the histopathological progression of inflammation is similar to that in HSE (Spaar 1976). The mouse is the animal best studied with regard to experimental nervous system infections with HSV and can be inoculated by various routes (Boerman 1991). Although the exact route of infection in HSE has not been established, reactivation from the site of latency in the trigeminal ganglia and spread along nervous pathways, in particular the trigeminal and olfactory nerves, appears the most likely explanation. In this study mice were inoculated by the corneal route and used to compare viral culture techniques and methods for the detection of viral antigen and DNA in samples of CSF.

Virus and Animals

Following ether anesthesia both corneas were scarified, 10 µl of a virus suspension containing 10^7 median tissue culture infective dose ($TCID_{50}$) HSV-1 (strain McKrae) was placed in the conjunctival sacs, and the eyelids were closed and gently rubbed. Control animals were inoculated with Eagle's MEM using the same technique. Animals were sacrificed every 12–24 h for a period of 5–6 days. A number of infected mice were held in reserve in case a mouse died before the preset time of sacrifice. Clinical signs (keratoconjunctivitis, paralysis, apathy, and ruffled fur) were scored for each mouse.

Nervous System

Brains were dissected, weighed, placed in GLY medium (gelatin, lactalbumin hydrolysate, yeast extract supplemented with gentamicin and vancomycin) and stored at $-70\,^\circ$C. Frozen brains were thawed and cultured within 1 week (Boerman et al. 1988). If definite cytopathological effect (CPE) had developed, the log $TCID_{50}$ per gram of brain tissue was calculated.

For histology, the animals were perfused using 4% formaldehyde in PBS. Heads were transected at the base of the skull and split in the midsagittal plane and fixed in formalin (Boerman 1991; Boerman et al. 1991a).

In order to investigate all sites of inflammation and possible pathways of HSV from the site of inoculation to the brain, decalcified head preparations with the brains in situ were used (Boerman 1991). Sections were collected on polylysine-treated slides and stained with hematoxylin (Harris)/eosin. Parenchymal and meningeal inflammation were scored. For immunocytochemistry, adjacent sections were incubated with polyclonal rabbit HSV-specific antibody (Dakopatts) and peroxidase-conjugated swine anti-rabbit immunoglobulin (Dakopatts). Peroxidase staining was performed with 3,3-diaminobenzidine (DAB) (Merck, BRD). The slides were counterstained with hematoxylin (Harris) and mounted. Sections from uninfected mice subjected to the entire procedure and sections from infected animals omitting the first layer of antibody were used as controls. Areas of tissue damage were identified using a standard atlas of the mouse brain (Sidman et al. 1971) as reference. A histological examination was also performed in order to relate viral spread within the CNS to the results of detection of HSV in the CSF. This investigation was limited, because the experiments had shown the viral spread to the CSF compartment to be correlated to spread in the trigeminal complex only. Moreover, only rapid viral culture (spin-amplified culture with immunofluorescence, SAC/IF) and the polymerase chain reaction (PCR) were compared, as these methods had shown sufficient sensitivity to warrant further study (Boerman et al. 1989). Histological and immunocytochemical investigations were carried out as described above.

Cerebrospinal Fluid

Cisterna magna puncture according to the method of Fleming et al. (1983) was used to obtain CSF, and the volume acquired was measured by comparison with a calibrated pipette. Inspection of the brain after dissection was performed to check whether direct puncture of brain tissue had occurred. Due to the small volume of mouse CSF (5–15 µl), only one analytical procedure could be performed on a sample.

The development of inflammation within the meningeal compartment was studied by cytological methods (Boerman 1991) and related to the results of detection of HSV-DNA by PCR (Boerman et al. 1989; Boerman 1991). The sensitivity of PCR was compared with viral culture by routine techniques (Boerman 1988) and SAC/IF (Boerman et al. 1989, 1991 a), radioactive dot-blot hybridization (Boerman et al. 1989), and immunofluorescence (Boerman et al. 1988) by both in vitro and in vivo experiments.

Polymerase Chain Reaction

For the PCR (Boerman et al. 1989), the procedure as described by Saiki (Saiki et al. 1988) was used, employing primers complementary to the HSV-TK gene from a standard map (Kit et al. 1983) at positions 664–683 (p219) and 769–787 (p220):

primer p219, (5′-3′) CTGCGGGGTTTATATACACGG
primer p220, (3′-5′) TACTGAATCACCGTCCACGA

The CSF samples were heated at 100 °C for 15 min, quenched on ice and centrifuged at 10000 × *g* for 1 min. Some 32 of the following incubation cycles were performed with the aid of a laboratory robot arm (P & P Elektronik; Nürnberg, FRG): 1 min at 98 °C followed by 3 min at 68 °C. The amplified samples (20%) were electrophoresed, photographed under short-wave UV illumination, and transferred to Zeta-probe (BioRad Laboratories, Calif.) by Southern blotting using 0.4 *N* NaOH as transfer solution. Hybridization using a ^{32}P-end-labeled 40-nucleotide sequence contained with the amplified DNA fragment as probe was carried out as described elsewhere (Saiki et al. 1985). PCR was performed using 1 ng HSV DNA as a positive control, 1 ng of mouse and calf thymus DNA as negative controls.

The sensitivity of PCR was determined by amplification of serial dilutions of a virus suspension containing 1 µg/ml total HSV-1 (strain McKrae) DNA.

Human Material

CSF specimens were obtained from 7 patients with HSE. Diagnosis was confirmed by a significant, fourfold HSV antibody titer rise in the CSF in 6 patients and by brain biopsy in 1 patient. As a control group, we examined CSF specimens from 4 patients with other causes of meningoencephalitis. Specimens used in this study were taken from days 2 to 29 after onset of illness. The onset of neurological illness was noted.

Diagnostic Methods

The PCR was carried out as stated above. The previously described antibody-capture ELISA (van Loon et al. 1981, 1985 b, 1989) was used for the detection of HSV-specific IgA and IgG antibodies in CSF specimens.

For virus isolation, 10% wt/vol homogenates of brain biopsy or autopsy specimens were inoculated onto duplicate cultures of primary monkey kidney cells and human diploid fibroblasts (Flow 2002; Flow Laboratories). Levels of antibodies were further determined by the complement fixation microtechnique of Casey (1965).

Results

Studies in the Animal Model

Histopathology and Immunohistochemistry

Two major pathways for the spread of HSV were found: the trigeminal nerve and the lacrimal duct. HSV immediately enters nerve fibers at the site of inoculation and spreads along branches of the ophthalmic nerve to the trigeminal ganglion and its nuclei within the brain stem. Other cranial nerves were also infected, but HSV does not reach the CNS via these pathways. The virus spread along the lacrimal duct, via the nasal mucosa to the olfactory nerves. The olfactory bulbs and meningeal surface surrounding the olfactory filia were only slightly infected (Boerman et al. 1991 b). This pathway does not appear important for the induction of experimental encephalitis, but it is an important route with respect to the pathogenesis of HSE as it affords HSV access to the parts of the brain involved in human HSE. In general, the spread of HSV along nervous pathways is consistent with the results of other investigators (Kristensson et al. 1978; Dyson et al. 1987) and other models of experimental HSE. Primary neural spread is a pivotal feature of experimental HSE and of great importance in understanding the process of spread of HSV to the CSF.

Within the CNS, nerve cells in the trigeminal spinal nuclei and the oligodendrocytes in the trigeminal tract are the CNS elements first infected. During the early phase of cellular infection, the infected cell does not show morphological alterations, but it does contain viral antigen. At a later stage, intranuclear inclusion bodies appear, and the infected cells collapse and die. Neuronal infection is accompanied by infection and necrosis of glia cells surrounding the infected neurons and the development of an inflammatory exudate. A typical multifocal infection is seen with patches of infected cells in the midst of noninfected nervous tissue. Spread to nervous structures within the brain stem but not directly connected with the trigeminal system is very limited (Boerman et al. 1991 a). Outside the brain stem, cellular infection is also found in the olfactory system and parts of the parietal meninges, but it is much more limited at these sites, and ballooning degeneration of CNS cells is not observed. The histopathological and immunohistochemical observations made in the mouse model would be best explained by intercellular spread of HSV through the neurons of the trigeminal system, accompanied by infection of surrounding cells in direct contact with infected nerve cell bodies or axons. Other investigations have also shown the very limited extracellular dissemination of HSV within the CNS of the mouse (Cook and Stevens 1973) and the preferential spread through axonal processes (Ugolini et al. 1989). In fact, the spread through neurons is such a dominant mechanism that it has been used as a neuronal tracer in animal studies (Ugolini et al. 1989).

The observations cited above must be considered as evidence that extracellular spread of HSV to the CSF is of minor importance. HSV could also be directly shed into the CSF from infected meningeal cells covering the brainstem. No correlation between detection of virus in the CSF and the presence or absence of HSV in the meningeal surface has, however, been found (Boerman et al. 1991 a). HSV has been shown to infect the ependyma of the fourth ventricle. This has been linked to positive viral culture of CSF samples in patients with HSE (Spaar 1976). In contrast, we could not find a correlation with the detection of HSV in CSF (Boerman et al. 1991 c). Ependymal infection was also an inconsistent finding occurring at a late stage after inoculation Third, HSV might disseminate to the CSF by infecting leukocytes. Infection and replication of HSV in mouse leukocytes has been observed (Lopez and Dudas 1979) and is consistent with our results of detection of viral antigen in CSF cells from infected mice (Boerman et al. 1988, 1991 b).

Thus, it would appear that shedding of HSV into the CSF is determined by two processes: neuronal, intercellular spread of HSV with very limited and late extracellular spread, and direct contact between infected cells and inflammatory cells for cellular transport of HSV to the CSF.

CSF

At the height of disease, HSV could be cultured from 20% of CSF samples by routine viral culture (RVC) and in up to 100% by SAC/IF (Table 1) (Boerman et al. 1989, 1991 a). Similar differences in sensitivity were obtained by titrations of virus stock. Positive results on CSF samples could only be obtained if conditions for processing were optimal with rapid inoculation of the CSF. Because little CPE was present in the monolayers, very low numbers of infectious virions are present in the CSF of infected mice, even at the height of infection. The sensitivity of SAC/IF compares favorably with PCR (Boerman et al. 1989, 1991 a), but the correlation with viral spread within brain tissue is less convincing (Boerman et al. 1991 a). Nevertheless, SAC/IF is far more sensitive than RVC, and improved viral culture techniques cannot be considered obsolete in the diagnosis of viral encephalitis.

HSV antigen detection in CSF cells by immunofluorescence (IF) has been claimed in proven cases of HSE early in the course of disease (Dayan and Stokes 1973; Taber et al. 1976; Peters and Versteeg 1980). However, the reliability and applicability of IF have remained controversial (Nahmias et al. 1982). We have demonstrated that IF is a rapid, specific, and sensitive tech-

Table 1. Comparison of virus detection methods in cerebrospinal fluid (CSF)

	Viral cultures		Animal experiments
	A	B	C
PCR	(0.01)	1	21/31 (68%)
RVC	1	–	3/15 (20%)
SAC/IF	0.03	–	12/22 (55%)
IF	–	–	18/24 (75%)
DBA	(10)	500	1/40 (2.5%)

PCR, polymerase chain reaction; RVC, routine viral culture; SAC/IF, spin-amplified viral culture with immunofluorescence; IF, indirect immunofluorescence on CSF cells; DBA, radioactive dot-blot hybridization

Column A, minimal amount of infectious viral particles detected, expressed in $logTCID_{50}$ (RVC, SAC/IF) or approximate equivalent (DBA, PCR)

Column B, minimal amount of total HSV-1 DNA detected (fg)

Column C, number (and percentage) of positive CSF samples per total number of infected mice

nique for the detection of HSV-1 antigen in CSF cells of experimentally infected mice (Boerman et al. 1988). In this model, IF is also more sensitive than RVC for the detection of HSV in CSF, but optimal results require special microscopic techniques and a rigorous controlled procedure (Boerman et al. 1988). Unambiguous positive results were not obtained prior to day 5 p.i., and faint signals were present at day 4 p.i. (Boerman et al. 1991 b). The late emergence of HSV-specific fluorescence can be accounted for by the evolution of the inflammatory process. Meningitis and CSF pleocytosis are present as soon as HSV has spread to CNS cells, but perivascular infiltrations do not appear prior to day 4 p.i. It is only at this stage that inflammatory cells come into contact with infected CNS cells and may become infected. Accordingly, the results of antigen detection in CSF cells nicely fit the histopathological and immunohistochemical observations of predominant if not exclusive intercellular spread of HSV in the CNS.

Various techniques for the detection of viral DNA were compared, including radioactive in situ hybridization on CSF cells, radioactive dot-blot hybridization (DBA), and the PCR. The PCR proved to be a very sensitive and specific technique for the detection of HSV-DNA in CSF (Boerman et al. 1989, 1991 a). HSV DNA was first detected early on day 4 p.i. in a few samples. At this stage, infected neurons were found in the trigeminal spinal nucleus, and there was infiltration of leukocytes into the brainstem (Boerman et al. 1989, 1991 a). The correspondence of detection of HSV DNA in CSF with the spread of HSV DNA in brain tissue is important (Boerman et al. 1989, 1991 a). Apparently, HSV DNA can be detected in the CSF as soon as neurons within the nervous system are infected. Viral DNA can be detected by the immunoperoxidase technique in cells only a few hours after infection. As disturbance of function of infected neurons is expected to occur at about the same time, this would also be the moment of the appearance of symptoms referable to the CNS. If these observations can be extrapolated to HSE, it would mean the detection of HSV DNA in CSF as soon as symptoms of CNS disease appear. Spread to higher-order neurons and increase in the severity of inflammation are even related to detection of HSV DNA in all specimens (Boerman et al. 1991 a). However, the concentrations of both infectious virus and viral components in the CSF are very low, barely exceeding the limit of detection of PCR and SAC/IF (Boerman et al. 1989, 1991 a).

Studies on Human Material

HSV-DNA was detected in CSF of 4 patients with HSE during the second week after onset of illness (Table 2). During the first and after the second week, no HSV DNA could be detected in the CSF (Fig. 1). No signal was obtained in specimens from control patients and total human DNA.

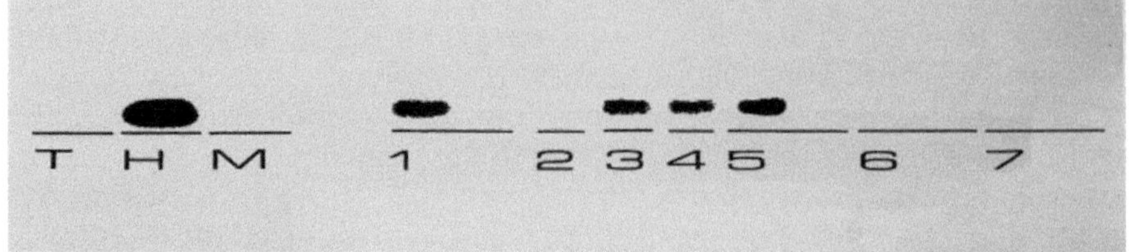

Fig. 1. Results of polymerase chain reaction, PCR (nonautoradiographic Southern blot hybridization). *Lane T*, total human DNA; *lane H*, herpes simplex virus HSV DNA; *lane M*, PCR reaction mix; *lanes 1–7*, samples from patients 1–7 (see Table 1). Results of investigation of two samples from patients 1, 5, 6, and 7 are shown: first sample on the *left*, second sample on the *right*

Table 2. Results of ELISA (van Loon et al. 1989) and PCR in comparison with conventional serological diagnosis

Patient	Days after onset of illness	ELISA HSV IgG	PCR	CF
1	6	ND	pos	neg
	20	ND	neg	1:32
2	3	pos	ND	ND
	27	pos	neg	ND
3	10	neg	pos	neg
	17	pos	ND	1:4
4	7	pos	pos	neg
	16	pos	ND	1:4
5	10	pos	pos	neg
	22	pos	neg	1:16
6	2	neg	neg	neg
	9	neg	neg	1:1
7	10	pos	neg	1:4
	29	pos	neg	ND

CF, Complement fixation; ND, not done; ELISA, enzyme-linked immunosorbent assay; PCR, polymerase chain reaction

Discussion

Two conclusions drawn from the animal experiments are relevant to CSF diagnosis of HSE. Firstly, HSV is present in the CSF of infected mice in very low amounts due to the peculiarities of viral spread within the CNS. Secondly, the best techniques for the detection of HSV in the CSF are SAC/IF and viral DNA amplification techniques.

Virus isolation by RVC from the CSF of patients with HSE enjoys little success. This may be partly due to delays in transport of samples (Nahmias et al. 1982). However, HSV can often be cultured from the CSF of patients with HSV meningitis (Sköldenberg et al. 1975) and of 50% of babies with encephalitis (Whitley 1990). As HSV virions are not deactivated in the CSF, the best explanation for the low success rate of RVC in HSE would be limited shedding of infectious virions into the CSF. Since the histopathological evolution of the disease in mice and other animals is similar to the findings in the human CNS (Spaar 1976), it is reasonable to assume that the mode of spread of HSV to the CSF also occurs in animal models, resulting in such limited viral spread to the CSF. SAC/IF has proven to be more sensitive than RVC. The sensitivity could be further improved by increasing the speed of centrifugation (Tenser 1978) and perhaps by combining this with other techniques such as nonradioactive in situ hybridization. These improvements in culture techniques are not required for routine diagnosis of HSV infections and have limited applicability in a routine diagnostic laboratory. However, HSE is an important disease, and the effort should be made to improve sensitivity of viral culture as this may lead to rapid diagnosis of other viral encephalitides. Viral culture does require rapid processing of CSF samples. As the incidence of HSE is rather low, it was not possible to obtain fresh samples for viral culture from patients in our hospital.

We have studied DNA detection techniques for the diagnosis of HSE. HSV DNA was detected by DBA in the CSF from 1 patient with brain stem HSE (Boerman et al. 1987). This finding appears to be an exception, however, with respect to the large amount of viral DNA present in the CSF.

PCR is by far the most sensitive technique. We were able to detect HSV DNA in 4 samples from 7 patients with serologically proven and 1 patient with biopsy-proven HSE. All positive results were obtained during the second week after onset of illness and within one week after the onset of neurological symptoms (Table 2). This confirms the applicability of PCR for the noninvasive diagnosis of HSE (Rowley et al. 1990; Aurelius et al. 1991; Puchhammer-Stöckl et al. 1990; Klapper et al. 1990).

Various techniques have been used for the preparation of DNA from CSF for PCR. Most have used CSF samples stored at from $-30°$ to $-40°C$ (Aurelius et al. 1991; Puchhammer-Stöckl et al. 1990; Klapper et al. 1990). Centrifugation was either not performed (Aurelius et al. 1991) or prior to extraction of DNA (Rowley et al. 1990; Klapper et al. 1990). The volume of the CSF samples varied between 5 (Boerman 1991) and 100 μl (Puchhammer-Stöckl et al. 1990; Aurelius et al. 1991), and DNA was extracted by simple boiling and ethanol precipitation (Aurelius et al. 1991) or phenol/chloroform extraction either with (Rowley et al. 1990; Puchhammer-Stöckl et al. 1990) or without proteinase K (Klapper et al. 1990). However, the use of more elaborate techniques for DNA preparation do not appear to improve the sensitivity of PCR. The technique used by Aurelius et al. (1991) is most applicable: storage of samples at $-30°C$ without centrifugation; boiling of 100 μl of CSF for 15 min and precipitation by addition of 250 μl 95% ethanol and incubation at $-30\,^{\circ}$C for 10 min. This method of storage is very convenient in a clinical setting, because samples are often taken outside office hours or in hospitals without adequate facilities for viral diagnosis.

Primer and probe sequences have been selected from the TK gene (Klapper et al. 1990; Boerman 1991), the glycoprotein D gene (Aurelius et al. 1991), and the UL 42 region (Puchhammer-Stöckl et al. 1990). The choice of these primers does not affect the sensitivity of the procedure, although it may be further enhanced by choosing primers within the repeated regions, as these are represented more than once in the HSV genome. Specificity is affected by homology with other herpesviruses, but information regarding this source of error can only be found in one study (Aurelius et al. 1991). In the future, the use of sets of different primers may allow for the simultaneous typing of the HSV strain (Kimura et al. 1990) and the detection of mutant virus strains that are less sensitive to the effects of acyclovir.

The study of Aurelius et al. (1991) was rigorously checked by control specimens and internal contamination controls. This is most important, as HSV is ubiquitous, and most adults harbor and excrete the virus throughout life (Anonymous 1991). Thus, all persons handling the material are potential sources of contamination. This liability to false-positive results of the PCR technique in general and regarding the detection of HSV in particular remains its main drawback. The use of a nested PCR (Aurelius et al. 1991; Porter-Jordan et al. 1990) may partly circumvent this problem, but the utmost care should be taken in order to obtain reliable results for the reasons outlined above.

The virologist's search for a reliable, noninvasive method for the early diagnosis of HSE is over: In the largest and most important study of noninvasive diagnosis of HSE by PCR (Aurelius et al. 1991), HSV DNA could be detected in CSF samples of 42 of 43 patients with proven HSE. Positive results could be obtained as early as a few days after the onset of neurological symptoms, and all controls remained negative.

References

Anonymous (1991) Acute diagnosis of herpes simplex encephalitis. Lancet 337:205–206

Aurelius E, Johansson B, Sköldenberg B, Staland A, Forsgen M (1991) Rapid diagnosis of herpes simplex encephalitis by nested polymerase chain reaction assay of cerebrospinal fluid. Lancet 337:189–192

Baringer JR (1978) Herpes simplex infections of the nervous system. In: Vinken P, Bruyn GW (eds) Infections of the nervous system. Elsevier, Amsterdam, pp 145–159 (Handbook of clinical neurology, vol 34, part 2)

Bergström T, Andersen O, Vahlne A (1989) Isolation of herpes simplex virus type 1 during first attack of multiple sclerosis. Ann Neurol 26:283–285

Boerman RH (1991) Herpes simplex encephalitis. An experimental and clinical study on early noninvasive diagnosis. Thesis, Leiden University

Boerman RH, Raap AK, van der Ploeg, Peters ACB, Rothbart PH (1987) Rapid diagnosis of HSV-1 encephalitis by radioactive genome detection in cerebrospinal fluid. Lancet ii:94

Boerman RH, Peters ACB, Verhey M, Raap AK, van Ekdom LTS, van Eer A, van der Ploeg M (1988) Immunofluorescence detection of herpes simplex virus type 1 antigen in cerebrospinal fluid cells of experimentally infected mice. J Neurol Sci 87:245–254

Boerman RH, Arnoldus EPJ, Raap AK, Bloem BR, Verhey M, van Gemert G, Peters ACB, van der Ploeg M (1989) Polymerase chain reaction and viral culture techniques to detect HSV in small volumes of cerebrospinal fluid. An experimental mouse encephalitis study. J Virol Methods 25:189–198

Boerman RH, Arnoldus EPJ, Peters ACB, Bloem BR, Raap AK, van der Ploeg M (1991a) Polymerase chain reaction for early detection of HSV-DNA in cerebrospinal fluid. An experimental mouse encephalitis study. J Med Virol 33:53–55

Boerman RH, Bloem BR, Peters ACB, Raap AK, van der Ploeg M (1991b) Spread of herpes simplex virus to the cerebrospinal fluid and the meninges in experimental mouse encephalitis. Acta Neuropathol (Berl) (accepted for publication)

Boerman RH, Mitro A, Bloem BR, Arnoldus EPJ, Raap AK, Peters ACB, van der Ploeg M (1991c) Detection of herpes simplex virus in ependyma of experimentally infected mice. Acta Virol (Praha) (accepted for publication)

Booss J, Esiri MM (1986) Sporadic encephalitis I. In: Viral encephalitis: pathology, diagnosis and management. Blackwell Scientific, Oxford, pp 55–94

Brihaye J (1970) Etude des encéphalites herpétiques et des encéphalites necrosantes aigues. Acta Neurol Psych Belg 59:1–114

Casey HL (1965) Adaptation of the LBCF microtechnique. In: Standardized diagnostic complement fixation and adaptation to microtest. US Public Health Service, Washington (US Public Health Service monograph no. 74), pp 31–34

Cook ML, Stevens JG (1973) Pathogenesis of herpetic neuritis and ganglionitis in mice-evidence for intraaxonal transport of infection. Infect Immun 7:272–288

Dayan AD, Stokes MI (1973) Rapid diagnosis of encephalitis by immunofluorescent examination of cerebrospinal fluid cells. Lancet i:177–179

Dyson H, Shimeld C, Hill TJ, Blyth WA, Easty DL (1987) Spread of herpes simplex virus within ocular nerves of the mouse: demonstration of viral antigen in whole mounts of eye tissue. J Gen Virol 68:2989–2995

Fishman RA (1987) No, brain biopsy need not be done in every patient suspected of having herpes simplex encephalitis. Arch Neurol 44:1291–1292

Fleming JO, Ting JYP, Stohlman SA, Weiner LP (1983) Improvements in obtaining and characterizing mouse cerebrospinal fluid. J Neuroimmunol 4:129–140

Gleaves CA, Wilson DJ, Wold AD, Smith TF (1985) Detection and serotyping of herpes simplex virus in MRC-5 cells by use of centrifugation and monoclonal antibodies 16 h postinoculation. J Clin Microbiol 21:29–32

Hanley DF, Johnson RT, Whitley RJ (1987) Yes, brain biopsy should be a prerequisite for herpes simplex encephalitis treatment. Arch Neurol 44:1289–1290

Johnson GD, Davidson RS, McNamee KC, Russel G, Goodwin D, Holborrow EJ (1982) Fading of immunofluorescence during microscopy: a study of the phenomenon and its remedy. J Immunol Methods 55:231–242

Kimura H, Shibata M, Kuzushima K, Nishikawa K, Nishiyama Y, Morishima T (1990) Detection and direct typing of herpes simplex virus by polymerase chain reaction. Med Microbiol Immunol (Berl) 179:177–184

Kit S, Kit M, Qavi H, Trkula D, Otsuka H (1983) Nucleotide sequence of the herpes simplex type 2 (HSV-2) thymidine kinase gene and predicted amino acid sequence of thymidine kinase polypeptide and its comparison with the HSV-1 thymidine kinase gene. Biochim Biophys Acta 741:158–170

Klapper PE, Cleator GM, Dennett C, Lewis AG (1990) Diagnosis of herpes encephalitis via Southern blotting of cerebrospinal fluid DNA amplified by polymerase chain reaction. J Med Virol 32:261–264

Koskiniemi M, Vaheri A, Taskinen E (1984) Cerebrospinal fluid alterations in herpes simplex virus encephalitis. Rev Infect Dis 6:609–618

Kristensson K, Vahlne A, Persson LA, Lycke E (1978) Neural spread of herpes simplex virus types 1 and 2 in mice after corneal or subcutaneous (footpad) inoculation. J Neurol Sci 35:331–340

Leestma JE, Bornstein MB, Sheppard DR, Feldman LA (1969) Ultrastructural aspects of herpes simplex infection in organized cultures of mammalian nervous tissue. Lab Invest 20:70–78

Lopez C, Dudas G (1979) Replication of herpes simplex virus type 1 in macrophages of resistant and susceptible mice. Infect Immun 23:432–437

Maniatis T, Fritsch EF, Sambrook J (1982) Molecular cloning: a laboratory manual. Cold Spring Harbor Laboratory, Cold Spring Harbor

Medawar PB (1948) Immunity to homologous grafted skin, part 3. Fate of skin homographs transplanted to the brain, to subcutaneous tissue, and to the anterior chamber of the eye. Br J Exp Pathol 299:58–69

Nahmias AJ, Whitley AN, Visintine AN, Takei Y, Alford CA Jr, Collaborative Antiviral Study Group (1982) Herpes simplex encephalitis: laboratory evaluations and their diagnostic significance. J Infect Dis 145:829–936

Nii S, Kamahora J (1963) Location of herpetic viral antigen in interphase cells. Biken J 6:145–154

Peden K, Mounts P, Hayward GS (1982) Homology between mammalian cell DNA sequences and human herpesvirus genomes detected by a hybridization procedure with high-complexity probe. Cell 31:71–80

Peters ACB, Versteeg J (1980) Immunofluorescence of CSF cells in viral neurological diseases. In: Den Hartog-Jager WA (ed) Color atlas of CSF cytopathology. Elsevier, Amsterdam

Porter-Jordan K, Rosenberg EI, Keiser JF, Gross JD, Ross AM, Nasim S, Garrett CT (1990) Nested polymerase chain reaction assay for the detection of cytomegalovirus overcomes false positives caused by contamination with fragmented DNA. J Med Virol 30:85–91

Puchhammer-Stöckl E, Popow-Kraupp T, Heinz FX, Mandl CW, Kunz C (1990) Establishment of PCR for the early diagnosis of herpes simplex encephalitis. J Med Virol 32:77–82

Reed LJ, Muench H (1938) A simple method for estimating fifty percent endpoints. Am J Hyg 27:493–497

Richman DD, Cleveland PH, Redfield DC, Oxman MN, Wahl GM (1984) Rapid viral diagnosis. J Infect Dis 149:298–310

Roizman B (1961) Virus infections of cells in mitosis: I. Observations on the recruitment of cells in karyokinesis into giant cells inducted by herpes simplex virus and bearing on the site of antigen formation. Virology 13:387–401

Rowley AH, Whitley RJ, Lakeman FD, Wolinsky SM (1990) Rapid detection of herpes-simplex-virus DNA in cerebrospinal fluid of patients with herpes simplex virus encephalitis. Lancet 335:440–441

Saiki RK, Scharf S, Faloona F, Mullis KB, Korn GT, Erlich HA, Arnheim N (1985) Enzymatic amplification of globin genomic sequences and restriction site analysis for diagnosis of sickle cell anemia. Science 230:1350–1354

Saiki RK, Gelfland DH, Stoffel S (1988) Primer-directed enzymatic amplification of DNA with a thermostable DNA polymerase. Science 239:487–491

Salmon VC, Turner RB, Speranza MJ, Overall JC Jr (1986) Rapid detection of herpes simplex virus in clinical specimens by centrifugation and immunoperoxidase staining. J Clin Microbiol 23:683–686

Schuster V, Matz B, Wiegand H et al. (1986) Detection of herpes simplex virus and adenovirus DNA by dot-blot hybridization using in vitro synthesized RNA transcripts. J Virol Methods 13:291–299

Sidman RL, Angevine JB, Pierce ET (1971) Atlas of the mouse brain and spinal cord. Harvard University Press, Cambridge

Sköldenberg B, Jansson S, Wolorit S (1975) HSV type 1 and acute . . . meningitis. Scand J Infect Dis 7:227–232

Sköldenberg B, Alestig K, Forkman A, Lövgren K, Norrby R, Stiernstedt G, Forsgren M, Bergstrom T, Dahlqvist E, Frydén A, Olding-Stenkvist E, Uhnoo I, De Vahl K (1984) Acyclovir versus vidarabine in herpes simplex encephalitis. Lancet 2:707–711

Spaar F-W (1976) Die menschliche Herpes-Simplex-Encephalitis und -Meningitis. Eine klinisch-neuropathologische Untersuchung. Fischer, Stuttgart

Taber LH, Brasier F, Couch RB, Greenberg SB, Jones D, Knight V (1976) Diagnosis of herpes simplex virus infection by immunofluorescence. J Clin Microbiol 3:309–312

Tenser RB (1978) Ultracentrifugal inoculation of herpes simplex virus. Infect Immun 21:281–285

Trotter JL, Brooks BR (1980) Pathophysiology of cerebrospinal fluid immunoglobulins. In: Wood JH (ed) Neurobiology of cerebrospinal fluid. Plenum, New York, pp 465–486

Ugolini G, Kuypers HGJM, Strick PL (1989) Transneuronal transfer of herpes virus from peripheral nerves to cortex and brainstem. Science 243:89–91

van den Berg FM, van Ooyen AJJ, Volkers H, Walboomers JMM (1984) Heterogeneity in subregions of the terminal repeats of herpes simplex virus type 2 DNA. Intervirology 21:96–103

van Loon AM, van der Logt JTM, van der Veen J (1981) Diagnosis of herpes encephalitis by ELISA. Lancet 2:1228–1229

van Loon AM, van der Logt JTM, Heessen FWA, van der Veen J (1985a) Quantitation of immunoglobulin E antibody to cytomegalovirus by antibody capture enzyme-linked immunosorbent assay. J Clin Microbiol 21:558–561

von Loon AM, van der Logt JTM, Heessen FWA, van der Veen J (1985b) Use of enzyme-labelled antigen for the detection of immunoglobulin M and A antibody to herpes simplex virus in serum and cerebrospinal fluid. J Med Virol 15:183–195

van Loon AM, van der Logt JTM, Heessen FWA, Postma B, Peeters MF (1989) Diagnosis of herpes simplex virus encephalitis by detection of virus-specific immunoglobulins A and G in serum and cerebrospinal fluid by using an antibody-capture enzyme-linked immunosorbent assay. J Clin Microbiol 27:1983–1987

Watson DH, Russell WC, Wildy P (1963) Electron microscopic particle count on herpes virus using the phosphotungstate negative staining technique. Virology 19:250–260

Whitley RJ (1990) Herpes simplex virus infections. In: Remington JS, Klein JO (eds) Infections diseases of the fetus and newborn infant. Saunders, Philadelphia, pp 282–305

Whitley RJ, Soong SJ, Hirsch MS, Karchmer AW, Dolin R (1981) Herpes simplex encephalitis. Vidarabine and diagnostic problems. N Engl J Med 304:313–318

Whitley RJ, Soong SJ, Linneman C Jr, Liu C, Pazin G, Alford CA (1982) Herpes simplex encephalitis. Clinical assessment. JAMA 247:317–320

Whitley RJ, Alford CA, Hirsch MS et al. (1986) Vidarabine vs acyclovir therapy in herpes encephalitis. N Engl J Med 314:144–149

Zurukzoglu S (1937) Über das Vorkommen von Herpesvirus in Liquor Cerebrospinalis. Zentralbl Bakteriol [Orig] 139:86–92

Chapter 11 Polymerase Chain Reaction Diagnosis of Varicella Zoster Virus*

Ravi Mahalingam[1], Randall Cohrs[1], Aud N. Dueland[1], and Donald H. Gilden[1,2]

Summary

Primary infection of humans by varicella zoster virus (VZV) results in chickenpox (varicella). Virus establishes latency in sensory ganglia and reactivates decades later to produce shingles (zoster). Disseminated zoster and postherpetic neuralgia are the two major complications of VZV reactivation in the elderly. There is increasing evidence that VZV can reactivate subclinically (zoster without skin rash), particularly in immunocompromised patients. Thus, the clinical import of rapid diagnosis and use of antiviral drugs in these patients is clear. Although most clinical assays are restricted by their dependence on an adequate host immune response, and by their limited sensitivity and specificity, PCR technology enables the detection of minute amounts of a specific viral DNA sequence in a large excess of background DNA. The availability of the complete DNA sequence of the viral genome has facilitated the diagnosis of VZV by PCR during acute and latent infections. VZV DNA has already been detected by PCR in DNA isolated from human blood MNCs, throat swabs, vesicles and crusts obtained at different times during viral infection, and in latently infected human ganglia. In addition, PCR has also been used to study the early site of viral replication and the mode of transmission of VZV infection. Lastly, the use of nested-primers and quantitative PCR will enhance the sensitivity of detection and permit determination of virus burden in biological specimens. These findings will aid in studies of VZV pathogenesis.

Introduction

Varicella zoster virus (VZV) causes childhood chickenpox (varicella), becomes latent in dorsal root ganglia, and reactivates decades later to produce shingles

Departments of [1]Neurology and [2]Microbiology and Immunology, University of Colorado School of Medicine, Denver CO 80262, USA

* Supported in part by grants AG06127, AG07347, and NS07321 from the National Institutes of Health, Bethesda, Md., and by a grant from the Roy and Beatrice Backus Foundation. A.N.D. is the recipient of an advanced postdoctoral fellowship (FG861-A-1) from the National Multiple Sclerosis Society

(zoster). Rash is characterized by vesicles on an erythematous base. Skin and mucosal lesions of varicella are generalized, while those of zoster are restricted to 1–3 dermatomes. Both disorders are acquired by household or direct contact with skin lesions or respiratory secretions of infected humans (Ross 1962; Gordon and Meader 1929; Evans 1940; LeClair et al. 1980; Gustafson et al. 1982; Brunell 1989).

Zoster, a common problem in elderly and immunocompromised individuals (Gallagher and Merigan 1979), is often complicated by postherpetic neuralgia (pain which persists for months to years after rash), by disseminated zoster or encephalitis or both, and less often by granulomatous arteritis. Rapid clinical diagnosis of VZV infection and any attendant complications is essential since antiviral treatment does exist.

Because conventional approaches to the diagnosis of VZV are time-consuming and not always precise, they are of limited value to the clinician. Nevertheless, before discussing the polymerase chain reaction (PCR), we briefly review the standard diagnostic techniques to offer some perspective on their usefulness and some speculations regarding their potential application in conjunction with PCR in the diagnosis and study of the pathogenesis of VZV-induced disease. These techniques include histologic and ultrastructural examination, attempts to isolate virus from infected tissue, and serologic assays that measure the humoral or cell-mediated immune response to VZV.

Histopathological diagnosis of VZV infection uses Giemsa-stained smears of vesicle scrapings to detect multinucleated giant cells and Cowdry A intranuclear inclusions characteristic of herpesviruses (Taylor-Robinson and Caunt 1972). Oral mucosa smears and cells in sputum from VZV-infected individuals have also been shown to display intranuclear inclusions (Cooke 1960, 1963; Williams and Capers 1959).

Electron microscopy (EM) study may reveal herpes virions in infected tissue. Monocytes obtained during the early stages of zoster have been shown to contain herpesvirus particles (Twomey et al. 1974). However, EM is cumbersome and does not distinguish the different herpesviruses; its diagnostic usefulness today is to complement viral and immunocytochemical analyses of herpesvirus-infected tissue.

VZV can occasionally be isolated by cocultivation of infected tissue with human or primate cells. It has been isolated from blood mononuclear cells (MNCs) of 10 of 12 (83%) otherwise healthy immunocompetent children 4–5 days before the onset of varicella, and 100% of children 1–2 days before rash, as well as on the first day of rash (Asano et al. 1985; Ozaki et al. 1986). These findings indicate that the greatest magnitude of blood-borne virus dissemination occurs just before the onset of disease. In those studies, the cytopathic effect characteristic of VZV was seen only after a second subculture in human embryonic lung cells. VZV has also been isolated from the blood of immunosuppressed adult zoster patients (Gold 1966; Feldman and Epp 1976; Gershon et al. 1978; Myers 1979), but not from the blood of healthy individuals with zoster or immunosuppressed patients without zoster. Isolation of VZV from the pharynx, a putative site of virus infection and replication shortly after infection, is usually difficult. One study describes the isolation of VZV from the pharynx in 5 of 117 (4.3%) healthy children within 3 days after varicella infection (Ozaki et al. 1989). The cell-associated nature of VZV, together with its inability to produce disease in rodents, has made isolation of the virus difficult and limited to tissue culture.

Serological assays may show a significant four-fold rise or fall in antibody to VZV. They are used in conjunction with histological and viral isolation studies and are particularly important when vesicle fluid or other infected tissue is not available. Historically, agar-gel diffusion and countercurrent immunoelectrophoresis have been used to analyze the moist tops of lesions, crusts, or scabs (Uduman et al. 1972; Frey et al. 1981); however, this technique cannot be used on dried smears of vesicle fluid (Macrae et al. 1969). Complement-fixing (CF) antibodies to VZV-specific proteins have been detected 5–7 days after primary varicella and within 1–2 days after zoster (Taylor-Robinson and Caunt 1972). The neutralizing antibody titer to VZV is usually less than 1:40 after chickenpox but increases and persists after shingles (Caunt and Shaw 1969). The sensitivity of detection of VZV-specific antibody can be increased by the addition of human-specific immunoglobulin to virus-serum mixtures (Asano et al. 1982; Takahashi 1983). Also useful is the fluorescent antibody to VZV-induced membrane antigen (FAMA) test (Williams et al. 1974; Brunell et al. 1975; Gershon and Krugman 1975), which is more sensitive than CF and is used not only in acute infection, but also in evaluating the humoral immune status of children at risk for varicella (Gershon et al. 1974). Currently, the most widely used antibody test to diagnose VZV infection is the enzyme-linked immunoabsorbant assay (ELISA). However, in our laboratory, VZV antibody has been repeatedly detected by immunoprecipitation in serum negative by ELISA. For a detailed review of these serological techniques, their sensitivities, and specificities, the reader is referred to Gilden et al. (1991 b). Finally, in any serological assay, cross-reactivity among the human herpesviruses, leading to misdiagnosis, is a potential problem.

Recently, nucleic acid hybridization has been used to detect VZV in infected tissue. Viral nucleic acid can be detected on Southern blots of DNA or by in situ hybridization of cells fixed on glass slides to VZV-specific probes. Nucleic acid hybridization has detected VZV DNA in the brain of patients with VZV encephalitis (Ryder et al. 1986; Gilden et al. 1988 b), in nerve roots of a patient with VZV meningoradiculitis (Dueland et al. 1991), and in blood mononuclear cells of patients with varicella and zoster (Gilden et al. 1988 a; Koropchak et al. 1989). Furthermore, nucleic acid hybridization has detected latent VZV infection in normal human ganglia (Hyman et al. 1983; Gilden et al. 1983, 1987; Croen et al. 1988). An advantage of in situ hybridization is the ability to identify the cell containing VZV. However, its sensitivity compared with Southern blot hybridization remains to be determined.

PCR technology has provided a means to examine virus at the molecular level with a specificity and precision not previously attainable. However, there have been very few reports on the use of PCR in VZV diagnosis (Kido et al. 1991; Koropchak et al. 1991; Mahalingam et al. 1990, 1992; Ozaki et al. 1991;

Puchhammer-Stockl et al. 1991; Devlin et al. 1992). We have used it to verify that VZV DNA is latent in human trigeminal and thoracic ganglia and to show that more than one region of the viral genome is present during latency (Mahalingam et al. 1990, 1992). This chapter will focus on the contributions of PCR in the evaluation of disease produced by VZV, including the potential use of PCR to detect subclinical reactivation of VZV.

Experimental Approach

Polymerase Chain Reaction

Application of the PCR technique results in the exponential amplification of nucleic acid sequences in vitro (Mullis and Faloona 1987). A pair of synthetic oligonucleotides (primers) are used, whose sequences flank the segment of target DNA. One of the primers is complementary to the sense strand and the other, to the antisense strand of the target DNA. The primers are oriented so that the extension product serves as the template for further amplification. The extension products of the primers annealing to the target DNA template result in "long products" which contain 3'-ends of various lengths. The amplification involves repeated cycles of heat denaturation (94°–98°C) of the target DNA, annealing (37°–65°C) of the primers to their complementary sequences, and extension (72°C) of the annealed primers with DNA polymerase in the presence of dNTPs. After three cycles, "short products" begin to appear and accumulate geometrically with each cycle. Each "short product" is identical double-stranded DNA, the ends of which are bound by the two primers (Eisenstein 1990).

In theory, a single copy of target DNA can be amplified to 10^6 copies in a few hours. An essential component of PCR is a heat-stable DNA polymerase. PCR can also be used with RNA templates by an initial reverse transcription step to convert RNA to cDNA.

Factors which influence the efficiency of PCR amplification include the sequence specificity of the primers, the $G+C$ content of the primers and the amplified segment, the concentrations of the DNA polymerase, dNTPs, and magnesium in the reaction mixture, and the number of cycles (Innis and Gelfand 1990; Saiki 1990; Williams and Kwok 1991).

Varicella Zoster Virus DNA and Primer Selection

The availability of the complete nucleotide sequence of the VZV genome (Davison and Scott 1986) has enabled the use of PCR in diagnosis. The VZV genome consists of a linear, double-stranded, DNA molecule 124 kilobase pairs (kbp) in size (Davison and Scott 1986) with a molar $G+C$ ratio of 46%–47% (Ludwig et al. 1972; Gilden et al. 1982a) and with large and small segments of DNA each bounded by inverted repeats (Gilden et al. 1982 b). Since the $G+C$ content of the inverted repeats is greater than that of the unique segments (Davison and Scott 1986), amplification of DNA sequences from the inverted repeats may require a higher temperature for denaturation and annealing.

The size of the oligonucleotide primers used in PCR ranges from 20 to 30 bases. Primers are selected to have an average $G+C$ content of 50%, without secondary structures and without complementarity, especially at the 3'-terminus (Saiki 1990; Williams and Kwok 1991). We have found the DNASIS program (Hitachi Software Engineering Co.) to be useful in selecting primers. Since VZV is usually present in small quantities in biological specimens and is highly cell-associated, the primers are chosen from a region of the viral genome which is not homologous to cell DNA. Primers can also be selected to identify different strains of VZV based on minor sequence differences (Hondo and Yogo 1988; Hondo et al. 1989; Dohner et al. 1988; Vlazny and Hyman 1985). Inclusion of uninfected cell DNA as a control in PCR is useful.

Oligonucleotide primers are commercially available or can also be generated using a DNA synthesizer. The primers, obtained as dry pellets, are redissolved in TE buffer (10 mM TRIS-HCl, 1 mM EDTA, pH 8.0), aliquoted, lyophilized, and stored as dry pellets at $-20\,°C$. Tables 1 and 2 list the sequences and location on the VZV genome of primers used in our laboratory.

Sample Preparation

Biological specimens used for VZV diagnosis include blood MNCs (Koropchak et al. 1991; Devlin et al. 1992), skin vesicles and crusts (Kido et al. 1991; Koropchak et al. 1991), throat swabs (Ozaki et al. 1991; Kido et al. 1991; Koropchak et al. 1991), cerebrospinal fluid (CSF) (Puchhammer-Stockl et al. 1991), and human tissues (Mahalingam et al. 1990). Target DNA is obtained by lysis of the cells with detergent in the presence of proteinase K and heating to 94 °C for 10 min before use in PCR. Phenol extraction of DNA obtained from CSF is sometimes required for efficient amplification.

MNCs are obtained from heparinized blood, and heparin may inhibit PCR (Holodniy et al. 1991). Therefore, the DNA is treated with heparinase (Beutler et al. 1990). Porphyrin compounds derived from blood heme can also inhibit Taq polymerase (Higuchi 1989) but can be removed from blood by digestion of the DNA with a restriction enzyme followed by denaturation and gel filtration through Sephadex G50 (de Franchis et al. 1988). Alternatively, blood samples can be collected in anticoagulants such as acid citrate dextrose or EDTA (Holodniy et al. 1991).

Protocol for Varicella Zoster Virus Detection

In a typical PCR, 1 µg of DNA isolated from the specimen or 1 ng or uninfected or virus-infected cell DNA or an equivalent volume of sterile water is used

Table 1. Primer pairs for varicella zoster virus (VZV) DNA amplification by polymerase chain reaction (PCR)

Gene and primer pair	Sequence (5′–3′)	Length of PCR product (bp)	Location on VZV genome[a]
Gene 28			
Primer 1	CGGAACTTCTTTTCCATTACAGTA	249	50391–50415
Primer 2	TAAAATGGCGATCAGAACGGGGTT		50616–50640
Gene 29			
Primer 1	TACGGGTCTTGCCGGAGCTGGTAT	272	51066–51090
Primer 2	AATGCCGTGACCACCAAGTATAAT		51314–51338
Gene 37			
Primer 1	ATGTTTGCGCTAGTTTTAGCGGTG	250	66074–66098
Primer 2	GGGTTAACCTTAACAATAACCAAG		66300–66324
Gene 40			
Primer 1	ATGACAACGGTTTCATGTCCCGCT	250	71540–71564
Primer 2	TCTAGAAAACGCACAAAGTTTAAT		71766–71790
Gene 62			
Primer 1	GCAAGACGTTTGGTCTTACGAATC	227	107835–107859
Primer 2	GAGTTTGTTTCGTCTTCATCCTCT		108062–108086
ORI[b]			
Primer 1	GTGGGGGGGGTGAAAAAGGGGGGGGG	325	110034–110058
Primer 2	TCACGTCAAATCGATTTTAAAAAG		110335–110359
Gene 63			
Primer 1	GCACTGGAATGTGACGTATCTGAT	283	111031–111055
Primer 2	GCTGTATATTCCGCGGTTTCTGCA		111290–111314

[a] Davison and Scoot 1986
[b] Denotes origin of VZV replication

Table 2. Internal oligonucleotide sequences used to detect varicella zoster virus (VZV)

Primer	Sequence (5′–3′)	Location on VZV genome[a]
Gene 28	TTCTCTGTTACTACCGCGCC	50544–50564
Gene 29	ACTCACTACCAGTCATTTCT	51098–51118
Gene 37	TATGTCTCTGAAATTAGAAGCCTT	66207–66231
Gene 40	TAACAGCCTTTACTCGGCTC	71707–71727
Gene 62	GGACCTGCTGCCTGTAGTTT	107975–107995
ORI[b]	ATGTCTGTGGTGTACGCCAATCGG	110077–110101
Gene 63	TGGTGAAGACGATAGCGACGATGA	111061–111085

[a] Davison and Scoot 1986
[b] Denotes origin of VZV replication

as a template in a final 100-µl reaction mixture containing 0.010 M TRIS-HCl, pH 8.3, 0.050 M KCl, 0–5 mM $MgCl_2$, and 0.01% gelatin, dATP, dTTP, dCTP, and dGTP (each nucleotide at a final concentration of 200 µM), and 1 µM of each primer. Reduction of the final primer concentration to 250 nM decreases the formation of primer dimers during amplification, particularly when more than 35 cycles are used. Samples are heated to 95 °C for 5 min before adding 2.5 U of Taq DNA polymerase.

An automated DNA thermal cycler is set for a denaturation step of 2 min at 94 °C, an annealing step of 2 min at 45 °C, and an elongation step of 3 min at 72 °C with an automatic extension of the elongation step by 15 s per cycle. The total number of cycles is usually 35. The optimum concentration of $MgCl_2$ in the reaction buffer required for efficient amplification using a particular set of primers is determined in a pilot experiment using a range (0–5 mM) of concentrations of $MgCl_2$ with DNA from virus-infected cells. The reproducibility of amplification is enhanced by filling any unused holes in the automated thermal cycler with the same tubes containing TE buffer (10 mM EDTA, 1 mM TRIS-HCl, pH 8.0) and oil.

Detection of Virus-Specific Sequences in PCR-Amplified Products

The amplified products are usually separated by gel electrophoresis, stained with ethidium bromide, visualized under ultraviolet light followed by Southern transfer and hybridization to radioactive (Mahalingam et al. 1990, 1992) or nonradioactive internal oligonucleotide probes. Other methods for the detection of PCR products include: hybridization to an internal oligonucleotide followed by restriction enzyme digestion and gel electrophoresis (Kwok et al. 1989), by dot or slot-blot analysis (Ting and Manos 1990; Puchhammer-Stockl et al. 1991), by reverse dot-blot hybridization (Saiki et al. 1989).

In PCR amplification of VZV DNA from human specimens, we noted that visualization (under ultraviolet light) of amplification products separated on agarose gels does not always reveal the products, and Southern hybridization is required using radiolabeled probe for detection. Ozaki et al. (1991) used primers specific for VZV gene 68 (glycoprotein I) and human placental DNA mixed with a recombinant clone containing the target sequence to show that the virus DNA could be detected by ethidium bromide staining at a level of 5×10^3 copies and by Southern blot hybridization at a level of 50 copies. We have found oligomer hybridization with radioactive probes to be very useful in the detection of VZV DNA by PCR. Amplified DNA fragments are separated by electrophoresis on an agarose gel. After staining with ethidium bromide and visualization under a ultraviolet light, the DNA fragments in the gel are transferred to Zeta-probe (BIO-RAD) membranes, hybridized to ^{32}P-end-labeled oligonucleotide probes (prepared from synthetic oligonucleotides representing a region internal to the amplified fragment) and detected by autoradiography. The size of the amplified product is determined by comparison with standard molecular weight markers.

The use of "nested" primers in PCR has been shown not only to increase the product yield but also to reduce background (Mullis and Faloona 1987). In this procedure, a very small portion of the amplified product obtained using the first set of primers is reamplified with a second set of primers whose sequences are enclosed by the first set. The second amplification results in significant quantities of the product since the first amplification enriches its substrate. Although this procedure increases the likelihood of contamination, it enhances the sensitivity of detection and reduces the background arising from nonspecific primer extension. We have found nested primers to be useful in VZV PCR (Mahalingam, unpublished observation).

Coamplification using primer pairs that are specific for different regions of the VZV genome from the same target DNA sample (multiplex PCR) is also possible. Each primer pair may require a specific concentration of $MgCl_2$ for efficient amplification. Therefore, the primer pairs selected must have the same optimal $MgCl_2$ concentration. Multiplex PCR is potentially useful to analyze biological specimens for the presence of the complete or partial VZV genome.

Results and Clinical Applications

Due to the extraordinary sensitivity of PCR, the contamination of a reaction either with products from an earlier amplification or with material from an exogenous source is a potential problem. Therefore, extreme care must be taken to avoid "carryover" of reaction products. Guidelines have been published to prevent such problems (Kwok and Higuchi 1989; Kwok 1990). False-negative PCRs are not uncommon and are usually controlled by amplifying a cellular gene from the same sample (Mahalingam et al. 1990).

Application of PCR technology to VZV has resulted in the detection of the virus DNA in blood MNCs (Koropchak et al. 1991; Devlin et al. 1992). MNCs obtained within 24 h after the onset of varicella exanthem from 8 of 12 (67%) patients (both adults and children) were shown to be positive by PCR using primers specific for VZV gene 31 (glycoprotein II) (Koropchak et al. 1991). Recently, we have accumulated PCR data showing that VZV DNA can be detected in blood MNCs in the absence of rash in elderly patients (Devlin et al. 1992). We are currently using quantitative PCR to determine whether there is a difference in virus burden in MNCs of elderly patients with and without postherpetic neuralgia.

Kido et al. (1991) used primers specific for VZV gene 68 (glycoprotein I) and detected VZV DNA in 100% of 20 vesicle samples collected from 18 patients with clinical chickenpox. One other report demonstrates the detection of VZV DNA in 21 of 28 (75%) lesion samples using primers specific for VZV gene 31 (Koropchak et al. 1991).

Koropchak et al. (1991) report that only 1 of 30 (3.3%) throat swab samples, obtained from 30 children within a day after rash, contained VZV DNA. They speculate that most individuals who are symptomatic with varicella may not harbor virus in the oropharynx. In contrast, results of PCR analysis by Ozaki et al. (1991) of 81 throat swabs from 18 patients taken at different times around the appearance of disease show that the detection rate during the first 3 days after clinical onset of disease was 100% (Fig. 1). These authors used primers representing more than one region of the VZV genome and detected VZV DNA from 15 days before until 11 days after the appearance of rash. They detected virus DNA in 11 of 42 (26.2%) samples during the incubation period and in 35 of 39 (89.7%) samples after the clinical onset. The rate of detection did not correlate with the maximum body temperature during the disease. Based on their data, the authors concluded that the pharynx is the probable site of early VZV replication. Finally, 5 of 6 (83%) crust samples from 18 patients obtained during convalescent varicella were shown to contain VZV DNA by PCR using three different primer pairs (Kido et al. 1991).

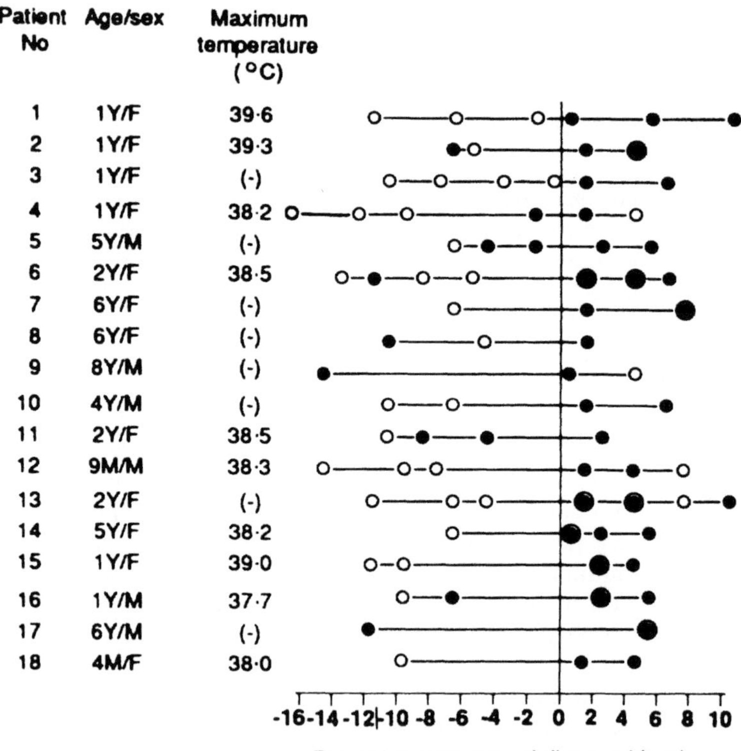

Fig. 1. Results of polymerase chain reaction analysis of throat swab samples obtained from children at different times around the onset of clinical varicella. Not detected (*open circle*), detected by hybridization using internal oligonucleotides (*small solid circle*), and detected before hybridization by direct gel electrophoresis and ethidium bromide staining as well as by hybridization (*large solid circle*); (−) denotes body temperature less than 37.5°C during the disease; *Y* denotes years, and *M* denotes months. (Modified from and printed with permission from Kido et al. 1991)

VZV has been detected by PCR in CSF from 3 of 5 (60%) children with postvaricella cerebellitis and in all of 7 (100%) herpes zoster patients with various neurological features (Puchhammer-Stockl et al. 1991). In these studies, positive PCR results were obtained as early as 3 days after the onset of zoster rash. Furthermore, PCR detected VZV DNA in the CSF from 4 patients with meningitis, 2 of whom had facial palsy. Two of the 4 patients did not have rash but were seropositive for VZV.

The detection of latent VZV DNA in human ganglia by Southern blot hybridization (Gilden et al. 1983) was confirmed by PCR using primers specific for the VZV origin of replication, gene 29 (major DNA binding protein), gene 40 (capsid protein), and gene 28 (DNA polymerase) (Mahalingam et al. 1990, 1992). Examination of 24 trigeminal and 61 thoracic ganglia from 23 individuals revealed latent VZV DNA in trigeminal ganglia from 13 of 15 (87%) subjects and in thoracic ganglia from 9 of 17 (53%) subjects. Most recently, when more than 6–7 thoracic ganglia from an individual were pooled, PCR detected VZV DNA in more than 90% of adults (Mahalingam, unpublished results).

The PCR technique is also useful in the detection of subclinical reactivation of VZV (Puchhammer-Stockl et al. 1991). Subclinical reactivation of VZV has been reported in various settings (Ljungman et al. 1986; Dueland et al. 1991; Gilden et al. 1991 b). Puchhammer-Stockl et al. (1991) detected VZV DNA by PCR in CSF from two patients with meningitis without rash, one of whom had facial palsy. In the case of fatal zoster meningoradiculitis reported by Dueland et al. (1991) and in the patients with preherpetic neuralgia described by Gilden et al. (1991 a), PCR might have been useful in diagnosing VZV-induced neurologic disease without zosteriform eruption. Finally, the combined use of PCR and in situ hybridization will help to determine the mononuclear cell subtypes in which VZV persists. Similarly, PCR in concert with pulsed-field gel electrophoresis can be applied to studies of VZV DNA configuration in latently infected ganglia.

PCR analysis can be extended to detect viral RNA in infected tissue. This would be particularly useful for studies of latency in which the extent of VZV transcription in ganglia is unknown. Usually, RNA PCR involves the selection of primers on the exons of a spliced transcript to overcome the background amplification arising from trace amounts of contaminant DNA (Gilliland et al. 1990). Since there is no known spliced transcript in VZV, complete removal of viral DNA by DNAase is mandatory before VZV RNA PCR can be performed.

Acknowledgements. We thank Mary Wellish for technical help and Mary E. Devlin and Marina Hoffman for editorial review and Marcia Conner for preparing this chapter for publication.

References

Asano Y, Albrecht P, Stagno S, Takahashi M (1982) Potentiation of neutralization of varicella-zoster virus by antibody to immunoglobulin. J Infect Dis 146:524–529

Asano Y, Itakura N, Hiroishi Y, Hirose S, Nagai T, Ozaki T, Yazaki T, Yamanishi K, Takahashi M (1985) Viremia is present in incubation period in nonimmunocompromised children with varicella. J Pediatr 106:69–71

Beutler E, Gelbart T, Kuhl W (1990) Interference of heparin with the polymerase chain reaction. Biotechniques 9:166

Brunell PA (1989) Transmission of chickenpox in a school setting prior to the observed exanthem. Am J Dis Child 143:1451–1452

Brunell PA, Gershon AA, Uduman SA, Steinberg S (1975) Varicella-zoster imunoglobulins during varicella, latency and zoster. J Infect Dis 132:49–54

Caunt AE, Shaw DG (1969) Neutralization test with varicella-zoster virus. J Hyg 67:343–352

Cooke BED (1960) Epithelial smears in diagnosis of herpes simplex and herpes zoster affecting the oral mucosa. Br Dent J 109:83–96

Cooke BED (1963) Exfoliative cytology in evaluating oral lesions. J Dent Res (Suppl) 42:343–347

Croen KD, Ostrove JM, Dragovic LJ, Straus SE (1988) Patterns of gene expression and sites of latency in human nerve ganglia are different from varicella-zoster and herpes simplex viruses. Proc Natl Acad Sci USA 85:9773–9777

Cruickshank JG, Bedson HS, Watson DH (1966) Electron microscopy in the rapid diagnosis of smallpox. Lancet 2:527–530

Davison AJ, Scott JE (1986) The complete DNA sequence of varicella-zoster virus. J Gen Virol 67:1759–1816

de Francis R, Cross NCP, Foulkes NS, Cox TM (1988) A potent inhibitor of Taq polymerase copurifies with human genomic DNA. Nucleic Acids Res 16:10355

Devlin ME, Gilden DH, Mahalingam R, Dueland AN, Cohrs R (1992) Peripheral blood mononuclear cells of the elderly contain varicella-zoster virus DNA. J Infect Dis 165:619–620

Dohner DE, Adams SG, Gelb LD (1988) Varicella-zoster virus DNA from persistently infected cells contains novel tandem duplications. J Gen Virol 69:2229–2249

Dueland AN, Devlin M, Martin JR, Mahalingam R, Cohrs R, Manz H, Trombley I, Gilden D (1991) Fatal varicella zoster virus meningoradiculitis without skin involvement. Ann Neurol 29:569–572

Eisenstein BI (1990) The polymerase chain reaction: a new method of using molecular genetics for medical diagnosis. N Engl J Med 322:178–183

Evans P (1940) An epidemic of chickenpox. Lancet 2:339–340

Feldman S, Epp E (1976) Isolation of varicella-zoster virus from blood. J Pediatr 88:265–267

Frey HM, Steinberg SP, Gershon AA (1981) Rapid diagnosis of varicella-zoster virus infections by countercurrent immunoelectrophoresis. J Infect Dis 143:274–280

Gallagher JG, Merigan TC (1979) Prolonged herpeszoster infection associated with immunosuppressive therapy. Ann Intern Med 91:842–846

Gershon AA, Krugman S (1975) Seroepidemiologic survey of varicella: value of specific fluorescent antibody test. Pediatrics 56:1005–1008

Gershon AA, Steinberg S, Brunell PA (1974) Zoster immune globulin: a further assessment. N Engl J Med 290:243–245

Gershon AA, Steinberg S, Silber R (1978) Varicella-zoster viremia. J Pediatr 92:1033–1034

Gilden DH, Shtram Y, Friedmann A, Wellish M, Devlin M, Cohen A, Fraser N, Becker Y (1982a) Extraction of cell-associated varicella-zoster virus DNA with triton X-100-NaCl. J Virol Methods 4:263–275

Gilden DH, Shtram Y, Friedmann A, Wellish M, Devlin M, Fraser N, Becker Y (1982b) The internal organization of the varicella-zoster virus genome. J Gen Virol 60:371–374

Gilden DH, Vafai A, Shtram Y, Becker Y, Devlin M, Wellish M (1983) Varicella-zoster virus DNA in human sensory ganglia. Nature 306:478–480

Gilden DH, Rozemann Y, Murray R, Devlin M, Vafai A (1987) Detection of varicella-zoster virus nucleic acid in neurons of normal human thoracic ganglia. Ann Neurol 22:377–380

Gilden DH, Devlin M, Wellish M, Mahalingam R, Huff C, Hayward A, Vafai A (1988a) Persistence of varicella-zoster virus DNA in blood mononuclear cells of patients with varicella or zoster. Virus Genes 2:299–305

Gilden DH, Murray RS, Wellish M, Kleinschmidt-DeMasters BK, Vafai A (1988b) Chronic progressive varicella-zoster virus encephalitis in an AIDS patient. Neurology 38:1150–1153

Gilden DH, Dueland AN, Cohrs R, Martin JR, Kleinschmidt-DeMasters BK, Mahalingam R (1991a) Preherpetic neuralgia. Neurology 41:1215–1218

Gilden DH, Mahalingam R, Dueland AN, Cohrs R (1991b) Herpes zoster: pathogenesis and latency. In: Melnick J (ed) Progress in medical virology, vol 39. Karger, Basel

Gilliland G, Perrin S, Blanchard K, Bunn HF (1990) Analysis of cytokine mRNA and DNA: detection and quantitation by competitive polymerase chain reaction. Proc Natl Acad Sci USA 87:2725–2729

Gold E (1966) Serologic and virus-isolation studies of patients with varicella or herpes-zoster infection. N Engl J Med 274:181–185

Gordon JE, Meader FM (1929) The period of infectivity and serum prevention of chickenpox. JAMA 93:2013–2015

Gustafson TL, Lavely GB, Brawner ER Jr, Hutcheson RH Jr, Wright PF, Schaffner W (1982) An outbreak of airborne nosocomial varicella. Pediatrics 70:550–556

Higuchi R (1989) Simple and rapid preparation of samples for PCR. In: Erlich HA (ed) PCR technology: principles and applications for DNA amplification. Stockton, New York, p 31

Holodniy M, Kim S, Katzenstein D, Konrad M, Groves E, Merigan TC (1991) Inhibition of human immunodeficiency virus gene amplification by heparin. J Clin Microbiol 29:676–679

Hondo R, Yogo Y (1988) Strain variation of R5 direct repeats in the right-hand portion of the long unique segment of varicella-zoster virus DNA. J Virol 62:2916–2921

Hondo R, Yogo Y, Yoshida M, Fujima A, Itoh S (1989) Distribution of varicella-zoster virus strains carrying a *Pst*I-site-less mutation in Japan and DNA change responsible for the mutation. Jpn J Exp Med 59:233–237

Hyman RW, Ecker JR, Tenser RB (1983) Varicella-zoster virus RNA in human trigeminal ganglia. Lancet 2:814–816

Innis MA, Gelfand DH (1990) Optimization of PCRs. In: Innis MA, Gelfand DH, Sninsky JJ, White TJ (eds) PCR protocols: a guide to methods and applications. Academic, San Diego, p 3

Kido S, Ozaki T, Asada H, Higashi K, Kondo K, Hayakawa Y, Morishima T, Takahashi M, Yamanishi K (1991) Detection of varicella-zoster virus (VZV) DNA in clinical samples from patients with VZV by the polymerase chain reaction. J Clin Microbiol 29:76–79

Koropchak CM, Solem SM, Diaz PS, Arvin AM (1989) Investigation of varicella-zoster virus infection of lymphocytes by in situ hybridization. J Virol 63:2392–2395

Koropchak CM, Graham G, Palmer J, Winsberg M, Ting SF, Wallace M, Prober CG, Arvin AM (1991) Investigation of varicella-zoster virus infection by polymerase chain reaction in the immunocompetent host with acute varicella. J Infect Dis 163:1016–1022

Kwok S (1990) Procedures to minimize PCR-product carry-over. In: Innis MA, Gelfand DH, Sninsky JJ, White TJ (eds) PCR protocols: a guide to methods and applications. Academic, San Diego, p 142

Kwok S, Higuchi R (1989) Avoiding false positives with PCR. Nature 339:237

Kwok S, Mack DH, Sninsky JJ, Ehrlich GD, Poiesz BJ, Dock NC, Alter HJ, Mildvan D, Grieco MH (1989) Diagnosis of human immunodeficiency virus in seropositive individuals: enzymatic amplification of HIV viral sequences in peripheral blood mononuclear cells. In: Luciw PA, Steimer KS (eds) HIV detection by genetic engineering methods. Dekker, New York, p 243

Leclair JM, Zaia JA, Levin MJ, Congdon RG, Goldmann DA (1980) Airborne transmission of chickenpox in a hospital. N Engl J Med 302:450–453

Ljungman P, Lonnqvist B, Gahrton G, Ringden O, Sundqvist VA, Wahren B (1986) Clinical and subclinical reactivations of varicella-zoster virus in immunocompromised patients. J Infect Dis 153:840–847

Ludwig H, Haines HG, Biswal N, Benyesh-Melnick M (1972) The characterization of varicella-zoster virus DNA. J Gen Virol 14:111–114

MaCrae AD, Field AM, McDonald JR, Meurisse EV, Porter AA (1969) Laboratory differential diagnosis of vesicular skin rashes. Lancet 2:313–316

Mahalingam R, Wellish M, Wolf W, Dueland AN, Cohrs R, Vafai A, Gilden DH (1990) Latent varicella-zoster viral DNA in human trigeminal and thoracic ganglia. N Engl J Med 323:627–631

Mahalingam R, Wellish M, Dueland AN, Cohrs R, Gilden DH (1992) Localization of herpes simplex virus type 1 and varicella-zoster virus in human ganglia. Ann Neurol 31:444–448

Mullis KB, Faloona FA (1987) Specific synthesis of DNA in vitro via a polymerase-catalyzed chain reaction. Methods Enzymol 155:335

Myers MG (1979) Viremia caused by varicella-zoster virus: association with malignant progressive varicella. J Infect Dis 140:229–233

Ozaki T, Ichikawa T, Matsui Y, Kondo H, Nagai T, Asano Y, Yamanishi K, Takahashi M (1986) Lymphocyte-associated viremia in varicella. J Med Virol 19:249–253

Ozaki T, Matsui Y, Asano Y, Okuno T, Yamanishi K, Takahashi M (1989) Study of virus isolation from pharyngeal swabs in children with varicella. Am J Dis Child 143:1448–1450

Ozaki T, Miwata H, Matsui Y, Kido S, Yamanishi K (1991) Varicella zoster virus DNA in throat swabs. Arch Dis Child 66:333–334

Puchhammer-Stockl E, Popow-Kraupp T, Heinz FX, Mandl CW, Kunz C (1991) Detection of varicella-zoster virus DNA by polymerase chain reaction in the cerebrospinal fluid of patients suffering from neurological complications associated with chickenpox or herpes zoster. J Clin Microbiol 29:1513–1516

Ross AH (1962) Modification of chickenpox in family contacts by administration of gamma globulin. N Engl J Med 267:369–376

Ryder JW, Croen K, Kleinschmidt-DeMasters BK, Ostrove JM, Straus SE, Cohn DL (1986) Progressive encephalitis three months after resolution of cutaneous zoster in a patient with AIDS. Ann Neurol 19:182–188

Saiki RK (1990) Amplification of genomic DNA. In: Innis MA, Gelfand DH, Sninsky JJ, White TJ (eds) PCR protocols: a guide to methods and applications. Academic, San Diego, p 13

Saiki RK, Walsh PS, Levenson CH, Erlich HA (1989) Genetic analysis of amplified DNA with immobilized sequence-specific oligonucleotide probes. Proc Natl Acad Sci USA 86:6230–6234

Takahashi M (1983) Chickenpox virus. Adv Virus Res 28:285–356

Taylor-Robinson D, Caunt AE (1972) Varicella virus. Springer, Berlin Heidelberg New York, pp 13–17 (Virology monographs, vol 12)

Ting Y, Manos M (1990) Detection and typing of genital human papillomaviruses. In: Innis MA, Gelfand DH, Sninsky JJ, White TJ (eds) PCR protocols: A guide to methods and applications. Academic, San Diego, p 356

Twomey JJ, Gyorkey F, Norris SM (1974) The monocyte disorder with herpes zoster. J Lab Clin Med 83:768–777

Uduman SA, Gershon AA, Brunell PA (1972) Rapid diagnosis of varicella-zoster infections by agar-gel diffusion. J Infect Dis 126:193–195

Vlazny DA, Hyman RW (1985) Errant processing and structural alterations of genomes present in a varicella-zoster virus vaccine. J Virol 56:92–101

Williams B, Capers TH (1959) The demonstration of intranuclear inclusion bodies in sputum from a patient with varicella pneumonia. Am J Med 27:836–839

Williams SD, Kwok S (1991) Polymerase chain reaction: applications for viral detection. In: Lennette EH (ed) Laboratory diagnosis of viral infection, 3rd edn. Dekker, Washington, DC, pp 147–158

Williams V, Gershon A, Brunell PA (1974) Serologic response to varicella-zoster membrane antigens measured by direct immunofluorescence. J Infect Dis 30:669–672

Chapter 12 **Detection of Human Cytomegalovirus by Polymerase Chain Reaction**

Mark F. Mangano [1,2], Richard L. Hodinka [2,3], and Jordan G. Spivack [1,2]

Summary

Human cytomegalovirus (HCMV) can produce life-threatening disease in immunosuppressed AIDS patients and organ transplant recipients. Congenital HCMV is the leading infectious cause of birth defects, including blindness, deafness, mental retardation, and death. HCMV can be transmitted in utero, during delivery, or postnatally through physical contact, blood transfusion, and organ transplantation. There are few treatments for HCMV disease. However, ganciclovir (DHPG) has been beneficial in HCMV-infected transplant patients.

The diagnosis of HCMV infection requires laboratory confirmation. Currently, a combination of routine virus culture, shell vial assay (which combines culture and immunofluorescence), cytopathology, and serology are used. A more rapid and sensitive technique, such as PCR, might allow for earlier DHPG treatment in transplant and other high-risk patients and might improve the monitoring of disease before and after antiviral therapy. In this chapter, PCR detection of HCMV DNA in urine specimens is compared with routine viral culture and the shell vial assay. There is a high correlation between these techniques (83%). The PCR is sensitive and specific for detecting HCMV DNA, since 10 template molecules could be detected and other herpesvirus genomes were not amplified.

Introduction

Human cytomegalovirus (HCMV) infections can be life-threatening in immunosuppressed children and adults. HCMV infections occur in one-half to two-thirds of all transplant recipients and contribute to failure of the transplanted organ and morbidity or mortality in the recipient (Rubin 1989).

[1]   The Wistar Institute, 3601 Spruce Street, Philadelphia, PA 19104, USA
[2]   Division of Infectious Diseases and Immunology, Children's Hospital of Philadelphia, Philadelphia, PA 19104, USA
[3]   Departments of Pathology and Pediatrics, Children's Hospital of Philadelphia and the University of Pennsylvania, Philadelphia, PA 19104, USA

HCMV can be transmitted by transplanted organs, such as liver (Paya et al. 1989), kidney (Naraqi et al. 1978), or heart (Preiksaitis et al. 1983), and by blood or bone marrow transfusions (Mirkovic et al. 1971; Yeager et al. 1981). Asymptomatic shedding of HCMV is common in urine (Chou and Merigan 1983), saliva, and breast milk (Hayes et al. 1972; Stagno et al. 1980).

Congenital HCMV infections occur in 1% of all newborns (reviewed in Ho 1982) and are the leading infectious cause of birth defects, including blindness, deafness, and mental retardation (Stagno et al. 1982). About 10% of HCMV-infected children are symptomatic at birth, and another 5%–10% develop sequellae (Ho 1982; Yow 1989). It is estimated that congenital HCMV infections cause brain damage in 3000–6000 infants each year in the USA (Stagno and Whitley 1985). While the effects of HCMV infections are already of great concern, they may become even a larger problem in the future because of the increasing numbers of AIDS patients, organ transplant recipients, and children in day-care centers.

HCMV infections are characterized by relatively slow growth, nuclear and cytoplasmic inclusions, and strong species-specificity (reviewed in Gold and Nankervis 1989). The standard for laboratory diagnosis of HCMV is a combination of (a) direct examination of tissues or white blood cells for cytomegalic inclusions, HCMV antigens, or nucleic acids, (b) conventional virus culture (c) the shell vial culture amplification assay, and (d) serology. The shell vial assay, which combines low speed centrifugation and immunofluorescent detection of HCMV early nuclear antigen, is considerably faster (24 h) than routine culture (up to 30 days) (Gleaves et al. 1984; Thiele et al. 1987). While these techniques are adequate to confirm the majority of HCMV infections, the culture and shell vial assays can be discordant. In three studies comparing the techniques, 17%–43% (average 28%) of the HCMV-positive samples were positive by conventional culture only, 21%–38% (average 30%) of the HCMV-positive samples were positive by shell vial only, and the remaining 36%–46% (average 42%) of the HCMV-positive samples were positive by both techniques (Paya et al. 1988; Rabella and Drew 1990; Schirm et al. 1987).

The HCMV genome is linear, doubled-stranded DNA of about 230 kb with the potential to encode more than 200 genes (Chee et al. 1990). A wide variety of HCMV-specific PCR primers can be designed because the entire genome of strain AD169 has been sequenced (Chee et al. 1990). HCMV DNA has been detected by PCR in peripheral blood cells (Cassol et al. 1989; Hsia et al. 1989; Stanier et al. 1989), urine (Cassol et al. 1989; Demmler et al. 1988; Hsia et al. 1989; Olive et al. 1989), and a variety of tissues from AIDS patients (Shibata et al. 1988; Shibata and Klatt 1989). In this chapter, PCR amplification of HCMV DNA from urine specimens is compared with routine viral culture and the shell vial assay. There is a high correspondence between these techniques, which indicates that PCR is both sensitive and specific for detecting HCMV DNA.

There are few treatments available for HCMV or other herpesvirus infections. Dihydroxypropoxymethyl guanine (DHPG, ganciclovir) treatment has helped immunocompromised patients and might be beneficial for serious infections (Erice et al. 1987; Laskin et al. 1987). PCR detection of HCMV might be valuable for life-threatening infections (a) when time is a factor in the decision to use DHPG, (b) if detection of HCMV DNA in blood by PCR prior to the appearance of infectious virus indicates that HCMV disease will develop in transplant recipients, or (c) when isolation of HCMV from tissues, such as white blood cell buffy coats, has been difficult.

Materials and Methods

Clinical Samples. The 29 coded urine samples were obtained from a group of both pediatric and adult transplant and HIV-infected patients. Specimens were transported on ice within 2–4 h to the clinical virology laboratory at the Children's Hospital of Philadelphia (CHOP). After the pH of each sample was neutralized with 0.1 *N* NaOH or 0.1 *N* HCl, the samples were treated for 30 min at 4 °C with a mixture of vancomycin (20 µg/ml), gentamicin (50 µg/ml), and amphotericin B (10 µg/ml). All samples were tested by routine viral culture, shell vial assay, and PCR. Based upon culture data, samples were selected for PCR to insure that about half of the samples would be culture positive.

Conventional Virus Isolation. Two tubes of confluent monolayers of human embryonic lung fibroblasts (MRC-5) were inoculated with 0.2 ml of the urine specimens and inspected daily for 4 weeks for HCMV cytopathic effects (CPE). The cells were refed twice weekly with fresh media containing 2% fetal bovine serum (FBS) and antibiotics. Suspected HCMV isolates were confirmed by indirect immunofluorescence using reagents from Syva Corp. (Palo Alto, CA).

Spin-Amplification Shell Vial Assay. Two shell vials containing 12-mm coverslips with MRC-5 monolayers were inoculated with 0.2 ml of each specimen and centrifuged at $700 \times g$ for 40 min at 25 °C. To each vial 2 ml Eagle's MEM containing 2% FBS and antibiotics were added, and they were incubated at 37 °C for 16–24 h. The coverslips were washed twice with PBS, fixed at room temperature for 10 min with acetone, and washed with PBS. Indirect immunofluorescence was performed with the E-13 mouse anti-HCMV MAb (MAb 810; Chemicon, El Segundo, Calif.) as the primary antibody, followed by biotin-labeled goat anti-mouse antibody (Kirkegaard & Perry Labs, Gaithersberg, M.) and avidin conjugated with fluorescein isothiocyanate (Organon Teknika Corp., West Chester, PA). The cells were counterstained with Evans blue, mounted on glass slides, and observed under a fluorescence microscope. Uninfected and HCMV-infected (strain AD169) monolayers were included as controls.

Polymerase Chain Reaction. Due to the high sensitivities of the PCR, many precautions were taken to minimize contamination and false-positive results.

Disposable tubes, pipettes, and positive displacement micropipettes were used throughout these studies. All PCR templates were prepared in a room physically separate from the main lab. All PCR experiments contained positive controls (an HCMV-infected MRC-5 cell lysate or culture-positive pancreatic tissue) and negative controls (no DNA template and salmon sperm DNA template). The negative controls were set up last, so that if the solutions or pipettes were contaminated with HCMV DNA it would be detected in the negative control. All clinical samples were heated in a boiling water bath for 5 min to inactive nucleases, proteinases, and infectious agents. Primers and probes were synthesized at the Wistar DNA Synthesis Facility (Table 1). Taq DNA polymerase and 10 × Taq buffer (500 mM KCl, 100 mM TRIS-HCl, pH 9.0, 15 mM MgCl$_2$, 0.1% gelatin (w/v), 1% Triton X-100) were purchased from Promega. The 100-µl reaction mixtures consist of: 5–20 µl urine sample, 10 µl 10 × buffer, 100–500 µM primer (each), 200 µM dNTP (each), and 2.5 U Taq. Optimal denaturation, annealling, and extension conditions were established empirically for each pair of primers using an HCMV-infected cell lysate and were 94°C for 1 min, 68°C for 2 min, and 72°C for 3 min, for 30 cycles.

Southern Blots, Hybridization, and Washing. PCR products were resolved by electrophoresis on 1.5% agarose gels and transferred to Gene Screen Plus (Dupont) by Southern blotting, as recommended by the manufacturer. After transfer the filters were soaked for 60 s in 0.4 M NaOH, rinsed in 2 × SSC/ 0.2 M TRIS-HCl, pH 7.4, and air-dried. Prehybridization was for 30 min at 37°C in 5 × SSC, 10 × Denhardt's (0.2% polyvinylpyrrolidone w/v, 0.2% bovine serum albumin, 0.2% Ficoll), sodium dodecylsulfate (SDS), and 200 µg/ml salmon sperm DNA. End-labeled oligonucleotide probes (Table 1) were added at 5–10 ng/ml, and the filters were hybridized for 3–4 h at 37°C. Filters were rinsed at room temperature and washed at 37°C, 50°C, and 65°C,

Table 1. Human cytomegalovirus polymerase chain reaction primers and oligonucleotide probes

Primer/ probe	Gene	Sequence	%GC	Restriction site	Fragment size DNA	RNA	Reference
FP	IE1	GTCCTCTGCCAAGAGAAAGATGGAC	52%	–			Akrigg
RP	IE1	ACATCTTTCTCGGGGGTTCTCGTTGC	52%	–			et al.
Probe	IE1	GACCAAGGCCACGACGTTCC TGCAGACTATGTTGAGGAAG	55%	*Pst* I	351	237	1985
FP	MLA	CGCTGATCTTGGTATCGCAGTACAC	52%	–			Rueger
RP	MLA	ACGTTGATGCTGGGGATGTTCAGCA	52%	–			et al.
Probe	MLA	GAGCCCATGTCGATCTATGT GTACGCGCTGCCGCTCAAGA	58%	*Pst* I	223	184	1987

FP, forward primer; RP, reverse primer; IE 1, immediate early gene 1 (IE 1 exons 2 and 3; UL123); MLA, major late antigen (MLA; pp65; UL83)

for 15 min each, in 5 × SSC/1% SDS. After a final wash at 65 °C in 2.5 × SSC/ 1% SDS, the filters were air-dried and autoradiographed with XAR film (Kodak) at − 70 °C with an intensifying screen.

Preparation of Probes. Oligonucleotide probes were end-labeled with $\gamma[^{32}P]$-ATP with polynucleotide kinase (BRL) and separated from unincorporated nucleotides with push columns (Stratagene).

Results

Specific PCR Amplification of HCMV DNA

The PCR primer pairs used in this work (Table 1) correspond to the immediate early gene 1 (IE1; UL123) and the major late antigen (MLA; pp65; UL83). The primers satisfy the following criteria: (a) approximately 50% GC content; (b) lack of self-homology or complementarity; (c) exact alignment in only a single portion of the HCMV genome; (d) sequence conservation between Towne, AD169, and Davis strains; (e) lack of complementarity to herpes simplex virus (HSV), varicella zoster virus (VZV), Epstein-Barr virus (EBV), human herpes virus 6 (HHV-6), and, for the available sequence data (GenBank), human DNA; (f) primer pairs span an intron so that the products amplified from DNA and RNA are different sizes; and (g) each set of primers yields amplified DNA products of different sizes so that the primers could be used in combination. Oligonucleotides were also synthesized as internal probes to hybridize to amplified PCR products. Each probe contains a restriction site for an enzyme that cuts once in the amplified product. After enzyme digestion, these probes hybridized to both restriction fragments. Thus, the specificity of PCR products could be confirmed by their size, specific hybridization to internal probes, and restriction enzyme digestion.

Specificity

The PCR primers chosen specifically yielded amplification products of appropriate size with HCMV-infected cell lysates, but not with HSV-1-, HSV-2-, HHV-6-, VZV-, or EBV-infected cell lysates (data not shown). The specificity of the amplified products was confirmed by Southern blot analysis with an internal probe and by digestion with restriction enzymes that cut once within the PCR product (see Table 1).

Sensitivity

Both IE1 and MLA primers yielded PCR products from infected cell supernatants that were detectable on ethidium bromide-stained gels (data not shown).

Less than 1 plaque-forming unit (PFU) of HCMV could be detected per PCR reaction in tissue culture homogenates (data not shown). Due to the large ratio of virus particles to PFUs it is difficult to relate viral infectivity directly to the number of HCMV DNA molecules. In reconstruction experiments with known amounts of plasmid DNA in a background of salmon sperm DNA, 10 molecules of target HCMV sequences could be detected by hybridization with Southern blots of PCR products (Fig. 1). This is similar to the sensitivity achieved by other investigators (Cassol et al. 1989; Olive et al. 1989).

PCR Correspondence with Culture and Shell Vial

We have blindly tested 29 coded urine samples from CHOP by PCR (Table 2). Of these, 13 were positive by PCR (LA and IE primers), culture, and shell vial, and 11 were negative by all three procedures. Thus, the PCR results corresponded with both the culture and shell vial procedures in 24/29 (83%) of the

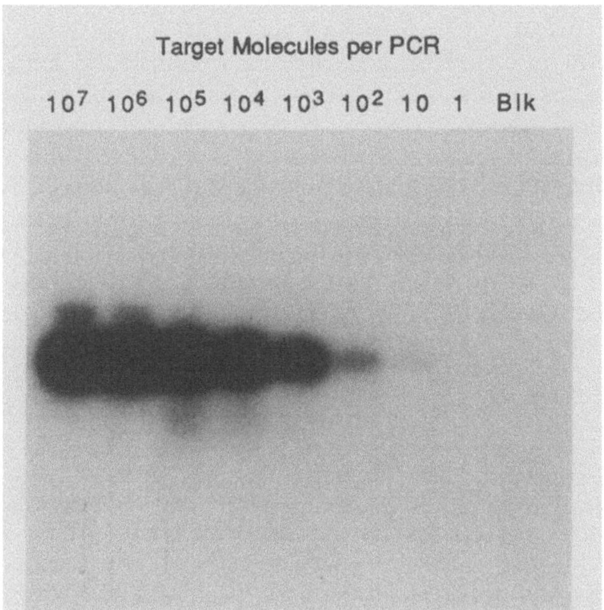

Fig. 1. Sensitivity of polymerase chain reaction (PCR) for human cytomegalovirus (HCMV) DNA. A cosmid clone containing the target sequence for the HCMV major late antigen (MLA) was used (Fleckenstein et al. 1982; Rueger et al. 1987). Based upon the cosmid size and the DNA concentration, PCR reactions were set up containing the indicated number of template molecules with the HCMV MLA primers. The reaction products were resolved by agarose gel electrophoresis, Southern blotted, hybridized with the internal oligonucleotide probe, washed, and autoradiographed. A faint band is detectable in the sample containing 10 molecules of template DNA

Table 2. Comparison of human cytomegalovirus (HCMV), polymerase chain reaction (PCR) with viral culture and shell vial assays for evaluation of urine specimens

PCR	Culture		Shell vial	
	+	−	+	−
+	13	4	14	3
−	0	12	1	11

samples. The 13/13 (100%) culture-positive samples and 14/15 (93%) shell vial-positive samples were PCR-positive. Conventional viral culture and the shell vial assay were in agreement for 27/29 (93%) of the samples, which is a higher correlation than reported by other groups (Paya et al. 1988; Rabella and Drew 1990; Schirm et al. 1987). Two samples that were culture-negative were shell vial-positive. One of these was PCR-positive, the other PCR-negative. The three remaining samples were PCR-positive, but culture- and shell vial-negative. Either PCR detected HCMV DNA that was not infectious, or the PCR results were artifactual or due to contamination. We favor the first possibility because the PCR controls were negative, and the results were confirmed by hybridization with internal probes (Table 1) and restriction enzyme digestion of the PCR product (data not shown).

Discussion

HCMV DNA was detected in human urine samples by PCR, standard viral culture, and shell vial assay. For 24 of 29 samples (83%) all three techniques were in agreement. Three samples were PCR-positive by Southern blot hybridization, yet culture- and shell vial-negative. It is not known if the results for these samples represent an increased sensitivity of PCR, the presence of noninfectious viral DNA, or false-positives due to DNA contamination. Nevertheless, the correlation of the PCR technique with culture techniques was high. All culture-positive samples were PCR-positive, and the PCR products were clearly visible on ethidium bromide-stained agarose gels. The time required from start to finish for PCR and gel electrophoresis is less than 6 h and could probably be even further reduced, which makes PCR a rapid method for detecting HCMV DNA in clinical material.

There are two important issues that need to be resolved before PCR detection of HCMV DNA is used clinically to diagnose HCMV: (a) Is it of predictive value to know when a patient is PCR-positive for HCMV DNA? (b) Is PCR practical in a clinical laboratory? If patients become PCR-positive before becoming culture or shell vial-positive, then this would be valuable in making an earlier decision to use DHPG or to reduce immunosuppressive therapy. PCR might also be useful in evaluating the effect of such changes in treatment on the course of HCMV disease and viral clearance. However, if HCMV PCR positivity is not an indicator for the development of HCMV disease, then PCR will be of less clinical significance. In a heart transplant study, HCMV PCR positivity did not always correlate with HCMV disease (Gerna et al. 1991). Much more work needs to be done in the clinical setting to define the significance of PCR results for the management of HCMV disease in immunosuppressed children and adults. There is no indication for PCR diagnosis of HCMV infection in immunocompetent individuals.

PCR is an extremely sensitive procedure, which can be a double-edged sword. The sensitivity allows the detection of a few molecules of template DNA but is exquisitely sensitive to contamination. This limitation can be dealt with in a research lab where sample processing PCR, and viral culture can be performed in separate rooms on different days with a number of precautions and controls. In a clinical laboratory, this may be more difficult because of the large volume of samples, most of which are unknowns. To reduce the number of false-positive results, clinical labs may need additional space, personnel, or training. When advances in technology make PCR automated, less expensive, and less susceptible to contamination, its use should become more widespread in clinical laboratories.

The applications of PCR that may have the greatest impact for HCMV in the near future are in research. By employing several primer sets and by restriction enzyme analysis of PCR products, clinical isolates can be distinguished (Chou 1990). This would be valuable in epidemiological studies of HCMV transmission and pathogenesis. The variability in restriction pattern of clinical isolates might also be helpful in differentiating between authentic PCR signals and lab contaminants. Another research area is the study of latency. HCMV, like all herpesviruses, establishes life-long latent infections. Latent infections can reactivate to produce serious and fatal disease in immunocompromised patients. PCR should be valuable in determining which tissues harbor latent HCMV DNA and in the study of the mechanism of HCMV latency.

References

Akrigg A, Wilkinson GW, Oram JD (1985) The structure of the major immediate early gene of human cytomegalovirus strain AD169. Virus Res 2:107–121

Cassol SA, Poon M-C, Pal R, Naylor MJ, Culver-James J, Bowen TJ, Russell JA, Krawetz SA, Pon RT, Hoar DI (1989) Primer-mediated enzymatic amplification of cytomegalovirus (CMV) DNA. J Clin Invest 83:1109–1115

Chee MS, Bankier AT, Beck S, Bohni R, Brown CM, Cerny R, Horsnell T, Hutchinson CA III, Kouzarides T, Martignetti JA, Preddie E, Satchwell SC, Tomlinson PA, Weston KM, Barell BC (1990) Analysis of the protein-coding content of the sequence of human cytomegalovirus strain AD169. Curr Top Microbiol Immunol 154:125–170

Chou S (1990) Differentiation of cytomegalovirus strains by restriction analysis of DNA sequences amplified from clinical specimens. J Infect Dis 162:738–742

Chou S, Merigan TC (1983) Rapid detection and quantitation of human cytomegalovirus in urine through DNA hybridization. N Engl J Med 308:921–925

Demmler GJ, Buffone GJ, Schimbor CM, May RA (1988) Detection of cytomegalovirus in urine from newborns by using polymerase chain reaction DNA amplification. J Infect Dis 158:1177–1184

Erice A, Jordan MC, Chace BA, Fletcher C, Chinnock BJ, Balfour HH (1987) Ganciclovir treatment of cytomegalovirus disease in transplant recipients and other immunocompromised hosts. JAMA 257:3082–3087

Fleckenstein B, Muller I, Collins J (1982) Cloning of the complete human cytomegalovirus genome in cosmids. Gene 18:39–46

Gerna G, Zipeto D, Parea M, Revello MG, Silini E, Percivalle E, Zavattoni M, Grossi P, Milanesi G (1991) Monitoring of human cytomegalovirus infections and ganciclovir treatment in heart transplant recipients by determination of viremia, antigenemia, and DNAemia. J Infect Dis 164:488–498

Gleaves CA, Smith TF, Shuster EA, Pearson GR (1984) Rapid detection of cytomegalovirus in MRC-5 cells inoculated with urine specimens by using low speed centrifugation and monoclonal antibody to an early antigen. J Clin Microbiol 19:917–919

Gold E, Nankervis GA (1989) Cytomegalovirus. In: Evans AS (ed) Viral infections of humans. Plenum, New York, pp 143–162

Hayes KD, Danks M, Gibas H, Jack I (1972) Cytomegalovirus in human milk. N Engl J Med 287:177–178

Ho M (1982) Epidemiology of cytomegalovirus infection in man. In: Ho M (ed) Cytomegalovirus: biology and infection. Plenum, New York, pp 79–104

Hsia K, Spector DH, Lawrie J, Spector SA (1989) Enzymatic amplification of human cytomegalovirus sequences by polymerase chain reaction. J Clin Microbiol 27:1802–1809

Laskin OL, Stahl-Bayliss CM, Kalman CM, Rosecan LR (1987) Use of ganciclovir to treat serious cytomegalovirus infections in patients with AIDS. J Infect Dis 155:323–327

Mirkovic R, Werch J, South MA, Benyesh-Melnick M (1971) Incidence of cytomegaloviremia in blood bank donors and in infants with congenital cytomegalic inclusion disease. Infect Immun 3:45–50

Naraqi S, Jackson GG, Jonasson O, Rubenis M (1978) Search for latent cytomegalovirus in renal allografts. Infect Immun 19:699–703

Olive DM, Simsek M, Al-Mufti S (1989) Polymerase chain reaction for detection of human cytomegalovirus. J Clin Microbiol 27:1238–1242

Paya CV, Wold AD, Smith TF (1988) Detection of cytomegalovirus from blood leukocytes separated by Sephacell-MN and Ficoll-Paque/Macrodex methods. J Clin Microbiol 26:2031–2033

Paya CV, Hermans PE, Wiesner RH, Lugwig J, Smith TF, Rakela J, Krom RAF (1989) Cytomegalovirus hepatitis in liver transplantation: prospective analysis of 93 consecutive orthotopic liver transplantations. J Infect Dis 160:752–758

Preiksaitis JK, Rosno S, Grumet C, Merigan TC (1983) Infection due to herpesviruses in cardiac transplant recipients: role of the donor heart and immunosuppressive therapy. J Infect Dis 147:974–981

Rabella N, Drew L (1990) Comparison of conventional and shell vial cultures for detecting cytomegalovirus infection. J Clin Microbiol 28:806–807

Rubin RH (1989) The indirect effects of cytomegalovirus infection on the outcome of organ transplantation. JAMA 261:3607–3609

Rueger B, Klages S, Walla B, Albrecht J, Fleckenstein B, Tomlinson P, Barrell B (1987) Primary structure and transcription of the genes coding for the two virion phosphoproteins pp65 and pp71 of human cytomegalovirus. J Virol 61:446–453

Schirm J, Timmerije W, van der Bij W, The TH, Wilterdink JB, Tegzess AM, van Son WJ, Schroder FP (1987) Rapid detection of infectious cytomegalovirus in blood with the aid of monoclonal antibodies. J Med Virol 23:31–40

Shibata D, Klatt EC (1989) Analysis of human immunodeficiency virus and cytomegalovirus infection by polymerase chain reaction in the acquired immunodeficiency syndrome. Arch Pathol Lab Med 113:1239–1244

Shibata D, Martin WJ, Appleman MD, Cuasey DM, Leedom JM, Arnheim N (1988) Detection of cytomegalovirus DNA in peripheral blood of patients infected with human immunodeficiency virus. J Infect Dis 158:1185–1192

Stagno S, Whitley RJ (1985) Herpesvirus infections of pregnancy: I. Cytomegalovirus and Epstein-Barr virus infections. N Engl J Med 313:1270–1274

Stagno S, Reynolds DW, Pass RF, Alford CA (1980) Breast milk and risk of cytomegalovirus infection. N Engl J Med 302:1073–1076

Stagno S, Pass RF, Dworsky ME, Henderson RE, Moore EG, Walton PD, Alford CA (1982) Congenital cytomegalovirus infection: the relative importance of primary and recurrent maternal infection. N Engl J Med 306:945–949

Stanier P, Taylor DL, Kitchen AD, Wales N, Tryhorn Y, Tyms AS (1989) Persistence of cytomegalovirus in mononuclear cells in peripheral blood and from blood donors. Br Med J 299:897–898

Thiele GM, Bicak MS, Young A, Kinsey J, White RT, Purtillo DT (1987) Rapid detection of cytomegalovirus by tissue culture, centrifugation, and immunofluorescence with a monoclonal antibody to an early nuclear antigen. J Virol Methods 16:327–338

Yeager AS, Grumet FC, Hafleigh EB, Arvin AM, Bradley JS, Prober CG (1981) Prevention of transfusion-acquired cytomegalovirus infections in newborn infants. J Pediatr 98:281–287

Yow MD (1989) Congenital cytomegalovirus disease: a NOW problem. J Infect Dis 159:163–167

Chapter 13 Semiquantitative Analysis of Epstein-Barr Virus DNA by Polymerase Chain Reaction in Clinical Samples of Lymphoproliferative Disorders *

Hans Knecht[1], Roland Sahli[2], David J.L. Joske[1], Edith Bachmann[1], Fedor Bachmann[1], Daniel Hayoz[1], Bernhard F. Odermatt[4], and Phil Shaw[3]

Summary

Epstein-Barr virus (EBV) DNA was detected and semiquantitatively analyzed in clinical samples from patients with Hodgkin's disease (HD), angioimmunoblastic lymphadenopathy (AILD), hairy cell leukemia (HCL), T-lymphoblastic lymphoma (TLL), familial gastric lymphoma, lymphocytic/mixed thymoma, and reactive lymph node hyperplasia (HR). High numbers of viral DNA copies were detected in HD and AILD, whereas no or very few viral DNA copies were found in HCL and thymoma. One case of TLL and two cases of HR contained relatively high numbers of EBV DNA copies. In the case of TLL, the viral genomes probably originated from accessory B cells. The high amount of EBV DNA identified in the HD and AILD cases is not associated with the Bam W/Bam Z rearrangement known to disrupt viral latency. Technical details of amplification, oligonucleotide hybridization, and semiquantitative EBV DNA analysis are described. A sensitive detection system of the Bam W/Bam Z rearrangement is also presented.

Introduction

Epstein-Barr virus (EBV) is a lymphotropic human herpesvirus endemic in all populations. EBV is the causative agent of infectious mononucleosis and oligoclonal B-cell lymphomas of immunocompromised hosts (Cleary et al. 1988; Fischer et al. 1991). It is also closely associated with endemic Burkitt's lymphoma (Lenoir and Bornkamm 1987; Rowe and Gregory 1989) and nasopharyngeal carcinoma (Desgranges et al. 1975; Tugwood et al. 1987). In the human host infectious EBV (virions) are continuously shed at very low

[1] Department of Medicine and Institutes of [2] Microbiology and [3] Pathology, University Hospital CHUV, Lausanne, and [4] Department of Pathology, University Hospital, Zürich, Switzerland

* This study was supported in part by grants no. 31-26389.89 and no. 31-30865.91 from the Swiss National Foundation. D. J. L. J. is a recipient of the PF Sobotka postgraduate research fellowship of the University of Western Australia.

rates from infected nasopharyngeal epithelial cells and latently infect B cells which are consequently immortalized (Thorley-Lawson 1988). In an immunocompetent host these cells are rapidly controlled, mainly by HLA class I-restricted cytotoxic T cells, whereas in the immunocompromised patient they may continuously proliferate (Wallace et al. 1982; Wallace and Murray 1989).

EBV infection of human B lymphocytes and epithelial cells is mediated by the C3d (CD21) receptor (Fingeroth et al. 1984). Binding of the EBV envelope glycoprotein gp350/220 to the B lymphocyte plasma membrane CD21 is followed by crosslinking, capping, and endocytosis of the virus (Tanner et al. 1987). Developing T cells, when expressing CD21 (Thore et al. 1979), may also become infected (Jones et al. 1988). The double-stranded EBV DNA genome is about 172 kilobases (kb) in length and has been completely sequenced (Baer et al. 1984). The EBV DNA is linear in the infective form, viz. virion (Pritchett et al. 1976), but is a covalently closed circular duplex DNA (CCC) episome in latently infected cells (Lindahl et al. 1976). The complex mechanisms leading to EBV DNA circularization and establishment of latency in newly infected B cells have recently been elucidated (Hurley and Thorley-Lawson 1988). In most latently infected cell lines EBV DNA is present in about 5–500 episomal CCC copies per cell (Sudgen et al. 1979; Hurley and Thorley-Lawson 1988; Thorley-Lawson 1988). However, there is one Burkitts lymphoma cell line, Namalwa, in which two linear EBV copies are integrated into the DNA of the unique chromosome 1 without episomal (CCC) DNA (Henderson et al. 1983; Matsuo et al. 1984; Lawrence et al. 1988). Further information of the molecular biology of EBV is contained in a review recently published by Kieff and Liebowitz (1991).

In clinical medicine the participation of EBV in the pathogenesis of disorders other than those cited above has been suspected for a long time, but methods to detect EBV were relatively insensitive (Southern blotting) or fastidious (cell culture). However, the recent development of the polymerase chain reaction (PCR) makes it possible to detect one unique DNA sequence among 10^5 human diploid genomes (Saiki et al. 1988), and as a result, the number of clinical entities closely associated with EBV infection continues to increase (Saito et al. 1989; Herbst et al. 1990; Katzenstein and Peiper 1990; Knecht et al. 1990, 1991).

In this article we focus on the detection and semiquantitative analysis of EBV genomes in clinical samples from different diagnostic entities. We present a sensitive system to detect the EBV Bam W/Bam Z rearrangement known to disrupt viral latency and which occurs in the P3J-HR-1 cell line clone HH543-5 (Countryman and Miller 1985; Countryman et al. 1987; Jenson and Miller 1988).

Materials and Methods

Clinical Samples. Genomic DNA from fresh frozen tissue biopsies or peripheral blood mononuclear cells (PBMC) was purified under sterile conditions according to Reymond (1987). All glass and plastic ware for DNA extraction were autoclaved prior to use.

Polymerase Chain Reaction. Screening for the presence of viral DNA was performed with 1 µg genomic DNA in a 50-µl reaction mixture containing 1 µM each of two oligonucleotide primers, 400 µM of each dNTP, 16.6 mM $(NH_4)_2SO_4$, 67,7 mM TRIS-HCl, pH 8.8, 6.7 mM $MgCl_2$, 10 mM β-mercaptoethanol, 170 µg/ml bovine serum albumin (BSA), and 1.5 U Taq polymerase (*Ampli Taq*; Perkin Elmer Cetus, Norwalk, Conn.). The reaction mixture was prepared in Eppendorf tubes with tips stuffed with cottonwool to avoid contamination by aerosols and overlayed with 50 µl paraffin oil. All plastic ware was autoclaved prior to use. The PCR was run with a HYBAID intelligent heating block, model no. IHB101 (Hybaid Limited, Teddington, UK) in a room distinct from the one in which DNA samples were prepared. Temperature steps (timed only after the target temperature has been reached) were controlled by a thermocouple externally mounted in an Eppendorf tube containing identical volumes of reaction mixture and paraffin oil.

Cycling. Two standard cycling protocols are shown in Table 1. For cases without specific amplification product after 30 cycles, an additional 25–30 cycles were run after addition of 1.5 U Taq polymerase. Increasing the annealing temperature by 3–4 C° for cycles 30–60 lowered the background DNA without affecting sensitivity (Knecht et al. 1990).

Gels. A 10-µl sample of the amplification product was electrophoresed through a 3% NuSieve GTG agarose gel (FMC Bio Products, Rockland, M.). DNA was revealed by ethidium bromide staining.

Oligonucleotide Primers and Internal Controls. The sequences of five pairs of EBV oligonucleotide primers are given in Table 2. Specificity of the amplified

Table 1. Cycling protocols for the Bam W and the BMRF 1 set of primers

Set of primers	Precycling (1 round)		→	Amplification (30 rounds)		→	Postcycle (1 round)		→	Amplification (25–30 rounds)	
Bam W	98 °C	5 min		93 °C	1 min					93 °C	1 min
	72 °C	5 min[a]		56 °C	1 min					60 °C	1 min
↓	56 °C	2 min	↓	72 °C	1.5 min		72 °C	5 min[b]	↓	72 °C	1.5 min[c]
	72 °C	3 min									
BMRF 1	98 °C	5 min		93 °C	1 min					93 °C	1 min
	72 °C	5 min[a]		60 °C	1 min					63 °C	1 min
	60 °C	2 min		72 °C	1.5 min		72 °C	5 min[b]		72 °C	1.5 min[c]
	72 °C	3 min									

[a] Add Taq polymerase
[b] Take out 10 µl of solution and add new Taq polymerase or stop the amplification
[c] After last round to be held for 3 min at 72 °C

Table 2. Sequences of Epstein-Barr virus (EBV) primers (A, B) and internal controls (C)

Region of EBV genome	T_m[a] (°C)	Oligonucleotide sequences	Size of DNA (bp)	Reference and genomic location
BamW	56		121	Cheung and Kieff (1982)
A	to	5'-CCAGAGGTAAGTGGACTT-3'		Bam HI-W 1399–1416
B	60	5'-GACCGGTGCCTTCTTAGG-3'		1520–1503
BMRF 1	60		159	Pfitzner et al. (1987)
A	to	5'-TTAGCGTGCCAATCTTGAGG-3'		Bam HI-M 413–432
B	63	5'-TTATCTTCTGGCTCAGAGGC-3'		572–553
C		5'-CAGGTCTGGCATCATAGCTGT-3'		438–458
Eco RI D	60		593	Fennewald et al. (1984)
A	to	5'-CATGTCATAGGCTTGCTGAC-3'		*Eco*RI D_{het} 1340–1359
B	63	5'-AAGAAGGCCAGAGGAATGTG-3'		1933–1914
C			378 + 215	*Sma*I site at 1554
Bam WZ 1	59		180	Patton et al. (1990)
A	to	5'-AATAGACAGCCCAGTTGAAA-3'		Bam HI-WZ 1469–1488
B	63	5'-GTCCAGCGCGTTTACGTAAG-3'		1649–1630
Bam WZ 3	60		322	Jenson and Miller (1988)
A	to	5'-GTGGCTCATGCATAGTTTCC-3'		Bam HI-WZ 1398–1417
B	63	5'-GAGTGGGCTTGTTTGTGACT-3'		1720–1701
C[b]		5'-GTCCAGCGCGTTTACGTAAG-3'		1649–1630

[a] T_m (optimal annealing temperature) is initially calculated as 4 °C (G + C) plus 2 °C (A + T) after Itakura et al. (1984) and adjusted to the highest level of specificity and sensitivity
[b] C for Bam WZ 3 is identical to primer B of Bam WZ 1

products obtained with the BMRF 1 set of primers was confirmed by hybridization with a ^{32}P-end-labelled 21-bp internal probe (BMRF 1 C in Table 2). The specificity of the Bam W/Bam Z rearranged amplification product (Bam WZ 3 oligonucleotides) was assessed by hybridization of the ^{32}P-labelled Bam WZ 1 B primer. The Bam WZ 1 set of primers can also be used as nested primers because they are located within the DNA sequence delimited by the Bam WZ 3 set of primers. In the Eco RI D amplification system specificity of the amplified DNA was confirmed by presence of a specific *Sma*I restriction site.

Southern Blotting. For hybridization, 10 µl of each amplification reaction were electrophoresed in a 3% NuSieve GTG agarose gel (FMC Bio Products, Rockland, Me.), stained with ethidium bromide, and photographed. Southern transfern to GeneScreen filters (Du Pont de Nemours, Dreieich, FRG) was performed overnight in 25 mM NaH$_2$PO$_4$, pH 6.5. Prehybridization was for 2 h in a solution containing 6 × SSC, 50 mM NaH$_2$PO$_4$ (pH 6.5), 1 mM EDTA (pH 8), 0.5% SDS, 100 µg/ml salmon sperm DNA, and 0.25% nonfat dry milk followed by hybridization for 8–12 h (same solution). Concentration of the ^{32}P-end-labelled internal probe was 4 × 10^5 cpm/ml hybridization solution.

Controls. Human fibroblast DNA and human placental DNA were used as negative controls. Positive controls for EBV were DNA from 4 human lymphoblastoid cell lines derived from cord blood lymphocytes infected with 4 different clinical samples, genomic DNA from an affected lymph node in a person with florid EBV infection, and genomic DNA from Namalwa, Raji, and Jijoye cell lines. Positive control for the Bam WZ rearrangement was genomic DNA from the P3J-HR-1 cell line, clone HH543-5 (this clone is a kind gift from Dr. G. Miller, Yale University, New Haven, Conn.). To exclude negative results due to inhibitors of PCR amplification, controls with HLA-DQα primers were included.

Sensitivity. The sensitivity of detecting EBV genomes by gel electrophoresis was two initial viral copies per 10^5 human diploid genomes (about 1 µg of placenta or fibroblast DNA) with the BMRF 1 set of primers using 60 amplification cycles (Fig. 1). With the Bam W set of primers and identical amounts of target DNA, a specific amplification product was already identified after 55 amplification cycles. However, generation of background DNA was considerably higher in the Bam W system. Using the Bam WZ 3 set of primers and serial dilutions of genomic DNA from the HH543-5 cell line, specific amplification products were still identified by gel electrophoresis when 1 pg of HH543-5 DNA (0.1 cell) mixed with 1 µg of placenta DNA (10^5 cells) served as input DNA. This result indicates both a high sensitivity of the Bam WZ 3 primer system and the presence of at least 10 rearranged (Bam W/Bam Z) EBV genomes in a single HH543-5 cell.

Semiquantitative Analysis of Viral DNA. It has been shown by Lawrence et al. (1988) that two EBV genomes are closely integrated at a known site of the unique chromosome 1 of Namalwa cells. Thus, a single Namalwa cell harbors 2 EBV DNA copies. We have shown (Knecht et al. 1991) that detection of the EBV genome with the BMRF 1 set of primers (only one BMRF 1 copy per EBV genome) is still possible when the DNA equivalent of one Namalwa cell (about 10 pg) mixed with 1 µg placenta or fibroblast DNA serves as input. Increasing the amount of Namalwa DNA gradually by a factor of 10 yields correspondingly more specific amplification product as determined by both gel electrophoresis and subsequent blotting followed by oligonucleotide hybridization. In EBV DNA-positive samples (with 1 µg of input DNA) graduated dilution (factor of 10) of input DNA therefore allows determination of the greatest dilution still giving a specific amplification product after 60 cycles of PCR with the BMRF 1 set of primers. This last positive dilution (theoretically corresponding to a minimum of 2, maximum of 19 EBV copies, practically to about 10 viral copies) allows us to calculate approximately the number of EBV copies in the initial DNA sample (10 multiplied by the appropriate dilution factor).

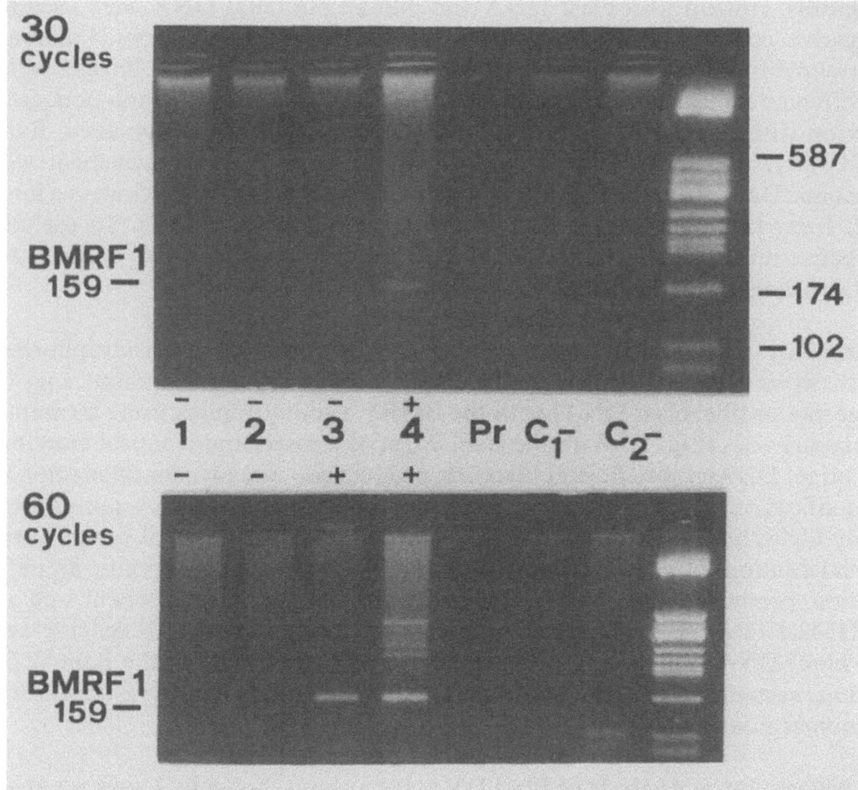

Fig. 1. Sensitivity of polymerase chain reaction (PCR) in detecting Epstein-Barr virus (EBV) DNA with the BMRF 1 set of primers. *Lanes 1–4* are DNA samples from patients with reactive lymph node hyperplasia. C_1^- (fibroblast DNA) and C_2^- (placenta DNA) are negative controls. PCR was run with 1 µg of input DNA except the *Pr* lane (primers only). After 30 cycles a specific amplification product is seen only in lane 4, whereas after 30 additional cyles a specific band is also detected in lane 3. End-point dilution analysis showed lane 3 to contain about 10, lane 4 to contain about 1000 EBV DNA copies per 10^5 diploid genomes

Results

Semiquantitative Analysis of EBV. Genomic DNA from 96 patients with various disorders was screened with the Bam W and BMRF 1 sets of primers for the presence of EBV genomes and further semiquantitatively analyzed using the BMRF 1 set of primers (Table 3). A high number of EBV genomes was found in samples from patients with Hodgkin's disease (HD) (Fig. 2) and AILD. PBMC from seven additional patients with chronic immune stimulation were negative for EBV DNA except two severe hemophiliacs (HIV-negative) receiving regularly Factor VIII products; PBMC from both patients harbored a few EBV genomes (<10 per 10^5 diploid genomes).

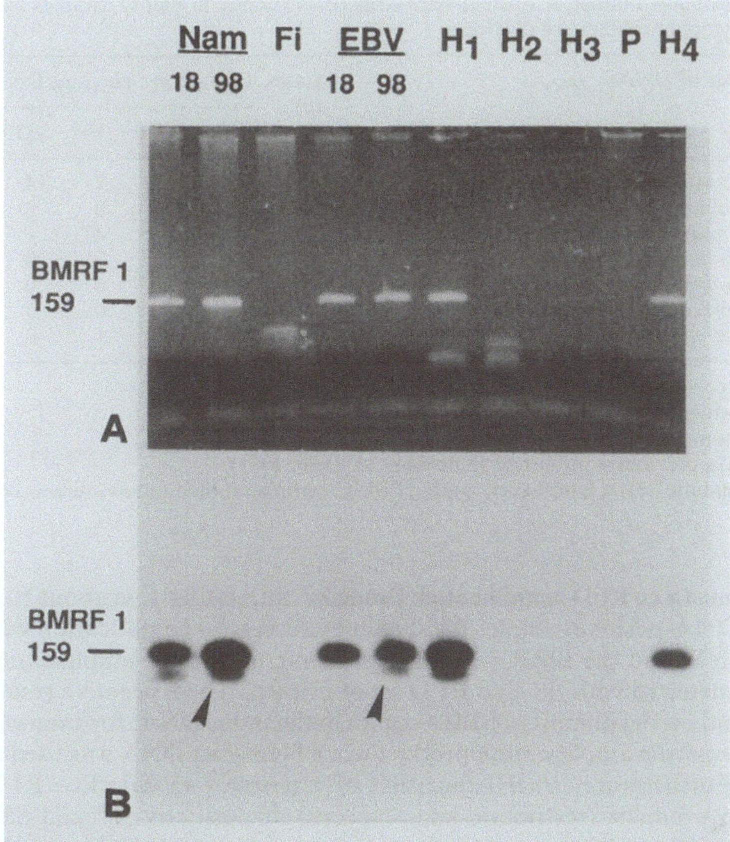

Fig. 2 A, B. Graduated dilution analysis of Epstein-Barr virus (EBV) genomes and identification of single-stranded DNA (ssDNA) using polymerase chain reaction (PCR) with the BMRF 1 set of primers followed by hybridization with an internal oligonucleotide probe. A 10-ng sample of positive controls *Nam* (Namalwa DNA) and *EBV* (DNA from a EBV-infected lymphoblastoid cell line) was subjected to 30 amplification cycles. A 10-ng sample of negative controls *Fi* (fibroblast DNA) and *P* (placenta DNA) was subjected to 60 amplification cycles. A 100-pg sample of genomic DNA from Hodgkin's disease cases H_1-H_4 was also run for 60 cycles. **A** Gel electrophoretic profile of the amplification products. No specific amplification product is detectable in the negative controls Fi and P and H_2 and H_3. A specific amplification band is seen in the positive controls Nam and EBV as well as in H_1 and H_4. The numbers *18* and *98* above lanes 1 and 4, or 2 and 5, indicate that the first half of the reaction was kept at room temperature, the second half heated at 98°C for 3 min shortly before being loaded. **B** Southern blot of the gel shown in **A**. Hybridization with an end-labeled internal oligonucleotide (BMRF 1 C in Table 2) shows specific radiolabelling in positive controls and H_1 and H_4. The adjacent band in H_1 represents ssDNA as proven by generation or increase of corresponding bands when the reaction was heated before being loaded (*arrows*). Note that the identification of EBV genomes in only 100 pg of template DNA in H_1 and H_4 indicates at least 10^5 EBV genomes per 1 µg genomic DNA in these cases

Table 3. Approximate number of Epstein-Barr virus (EBV) copies in biopsy samples from patients with lymphoproliferative disorders

Disorder (source of DNA)	Cases (n)	EBV copies per 1 µg DNA [a]			
		neg	$1-10^2$	10^3	$\geq 10^4$
Hodgkin's disease (lymph node)	50 [b]	11	11	2	14
Hairy cell leukemia (spleen)	6	3	3	0	0
AILD (lymph node)	7	2	0	4 [c]	1
Reactive lymph node hyperplasia	17	12	3	2	0
Familial gastric lymphoma (tumor)	2	1	1	0	0
Lymphocytic/mixed thymoma (tumor)	8	5	3	0	0
T lymphoblastic lymphoma (1 PBMC, 5 lymph nodes)	6	5	0	1	0

[a] 1 µg DNA corresponds to 10^5 diploid human genomes
[b] 12 EBV-positive cases were not further analyzed
[c] In 1 case no more DNA left for further analysis
Note: Some data previously published (Knecht et al. 1990, 1991)
AILD, angioimmunoblastic lymphadenopathy; PBMC, peripheral blood mononuclear cells

Polymorphism of Eco RI D Amplification Products. Interestingly, in about 10% of the EBV DNA-positive samples (as shown by successful amplification with both the Bam W and the BMRF 1 sets of primers), no specific amplification product was detected with the Eco RI D set of primers. These negative results did not depend on the quantity of EBV copies in the input DNA; for example, there was no specific amplification product when Namalwa DNA was used as a template. Furthermore, *Hae*III digestion of a series of specific Eco RI D amplification products (restriction sites expected at positions 312 and 588) revealed at least three different restriction patterns suggesting that this genomic region of the virus might be subject to a high spontaneous mutation rate.

Search for Autonomous EBV Proliferation. In the 16 cases of HD and the five cases of AILD with 10^3 or more EBV copies per 1 µg of DNA, we looked for the presence of the Bam W/Bam Z rearrangement of the EBV genome. This rearrangement could not be detected in any of these cases (Fig. 3).

Discussion

Successful detection of EBV DNA by Southern blotting requires at least 2×10^4 viral DNA copies (corresponding to 2×10^5 Bam W sequences) in the transferred DNA sample when the Bam W sequence is used as a radiolabeled probe (Staal et al. 1989). In a PCR system based on 30 amplification cycles, considerably fewer viral targets (approximately 10^3 viral copies in the input DNA) will suffice to produce a clearly visible amplification product on ethidium bromide-stained gels (Herbst et al. 1990). Thus, 30 cycles of PCR immedi-

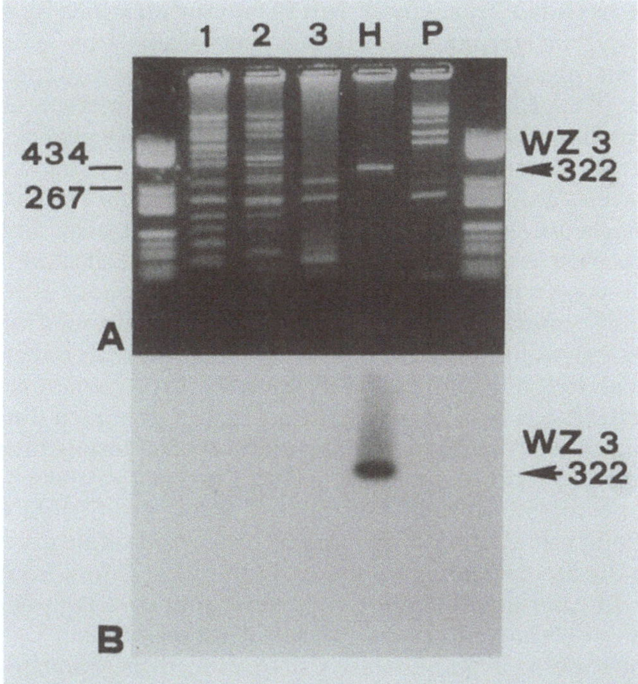

Fig. 3A, B. Search for the BamW/BamZ rearrangement in Hodgkin's disease (HD) cases (*1–3*) with a high content of EBV genomes (approximate number of EBV copies per 1 µg of genomic DNA: 10^4 in lane 1, 10^5 in lane 2, 10^4 in lane 3). Positive control *H* is genomic DNA from the cell line HH543-5. Negative control *P* is placenta DNA. In all, 55 amplification cycles were performed using the BamWZ3 set of primers and 1 µg of input DNA (except H where amplification was stopped after 30 cycles). **A** A specific amplification product is detectable in H. In HD cases 1 and 2 small bands of nearly specific length (such variations in length could result from different junction sites of the BamW and Z fragments) are visible besides multiple extraneous bands. **B** Southern blot of the gel shown in **A**. Hybridization with a radiolabeled internal oligonucleotide probe (Bam WZ1 B in Table 2) shows radiolabelling specific for the amplified Bam W/Bam Z rearrangement present in H. Absence of hybridization in HD cases and P proves the nonspecific nature of all other bands observed in **A**

ately improve the sensitivity of detecting EBV DNA significantly as suggested by 58% (Herbst et al. 1990) compared with 29% of EBV DNA-positive HD cases revealed by Southern blotting (Staal et al. 1989). However, increasing the number of amplification cycles up to 60 further improves the sensitivity by a factor of at least 10^3 and allows identification of as few as two EBV genomes per 10^5 human diploid genomes (Saito et al. 1989; Knecht et al. 1991). Moreover, using graduated dilutions of Namalwa DNA (only two integrated EBV genomes per cell) as a standard, calculation of the approximate number of EBV copies in a clinical DNA sample is possible (Knecht et al. 1991). This end-point dilution method offers a simple way of semiquantitative EBV genome analysis in clinical samples.

Clonal EBV genomes (Raab-Traub and Flynn 1986) were identified by the Southern blot technique in HD, and EBV DNA was localized by in situ hybridization within Sternberg-Reed cells (Weiss et al. 1989; Anagnostopulos et al. 1989). However, Katz et al. (1989), in a paper dealing with latent and replicative EBV in lymphoproliferative disease, demonstrated both oligoclonal CCC and linear (replicating) EBV DNA in a case of HD. They found no strict correlation between the amount of EBV DNA and the presence of linear viral DNA. This observation and our findings in HD (Knecht et al. 1991) showing no correlation between the number of Sternberg-Reed cells and the amount of EBV DNA suggest that the role of EBV in the pathogenesis of HD remains to be elucidated. Especially in cases with eventual chromosomal integration of EBV DNA – an event probably much more frequent than hitherto recognized (Hurley et al. 1991) – identification of only a few EBV-positive cells is potentially of interest. It has been shown that transfection of human fibroblasts with UV-irradiated EBV DNA-containing plasmid p500 results in enhanced cell transformation when the plasmid is integrated but not when it remains episomal (Vos et al. 1989). Even a few cells with integrated EBV DNA – not detectable by Southern blotting or 30-cycle PCR – could influence the aggressiveness of the disease depending on the location of the chromosomal integration site. Therefore, sensitive detection and semiquantitative analysis of EBV DNA in lymphoproliferative disorders will be of major interest.

In the present report we have additionally semiquantified the number of EBV genomes in AILD (Frizzera et al. 1974), a prelymphoma known to be closely associated with EBV (Knecht et al. 1990). In this disorder, the rearrangement pattern of immunoglobulin and TcR genes (Lipford et al. 1987; Knecht 1989) closely resembles the one described in immunocompromised hosts (Cleary et al. 1988). It is possible that relatively high amounts of EBV DNA, as identified in five of our seven cases and in a further case (White et al. 1989), result from an accumulation of EBV-infected B cells due to impaired HLA class I-restricted cytotoxic T cell elimination or escape from immunosurveillance by spontaneous mutations within the EBV genomic region coding for the late membrane protein (LMP) (Rowe et al. 1987). This region is adjacent to the PCR amplification product obtained with the Eco RI D set of primers.

High amounts of EBV RNA were recently identified in clinically aggressive hairy cell leukemia (HCL) by in situ hybridization (Wolf et al. 1990). We investigated six HCL cases prior to treatment using PCR and identified a few EBV DNA copies in only three of them. In the light of these differences, semiquantitative EBV DNA analysis of prospective HCL cases before and during interferon-α treatment might be of interest since interferon-α and EBV share a receptor on B cells (Delcayre et al. 1991).

Only three out of eight thymomas examined contained a few EBV genomes. This is in contrast to the findings of McGuire et al. (1988), who reported a high incidence and amount of EBV DNA in cases of thymoma and thymic hyperplasia in patients of the Hong Kong area. Our findings argue against a pathogenetic role of EBV in thymomagenesis and suggest that the difference between our results and those of McGuire et al. (1988) are related to geographic distribution of EBV infections.

EBV genomes have recently been identified in T cells of T-cell lymphomas (Jones et al. 1988), large granular lymphocyte lymphocytosis (Kawa-Ha et al. 1989), chronic active EBV infection (Kikuta et al. 1988), benign polyclonal T cell proliferation (Yoneda et al. 1990), and aggressive peripheral T-cell lymphomas (Su et al. 1991). In all these cases, infection of T cells presumably occurred early during T-cell ontogeny via C3 (CD21) receptors (Thore et al. 1979). Su et al. (1991) were not able to demonstrate EBV DNA in their cases of T-lymphoblastic lymphoma, whereas in our series one out of six cases contained EBV DNA. However, the approximate number of 10^3 EBV genomes per 10^5 diploid genomes is an argument against their presence in the T lymphoblasts, since in case of EBV-infected T lymphoblasts, many more viral copies should have been identified. Since the DNA was extracted from a lymph node sample containing some nonmalignant residual B cells, it is more likely that the EBV DNA originated from these cells, eventually proliferating in the absence of T-cell-driven elimination.

We found that about 10% of the EBV DNA-positive samples among the population analyzed in this paper was negative when tested with the Eco RI D set of primers. This negative result was independent of the number of EBV targets in the input DNA, and in the positive samples the amplified sequences showed a *Hae*III restriction fragment length polymorphism. Furthermore, the sequence of primer Eco RI D A as published by Bankier et al. (1983) differs from that used by us (Fennewald et al. 1984) by an additional G between positions 1355 and 1356. These observations are consistent with a considerable heterogeneity in this region of the viral genome. In fact, the Eco RI D set of primers is located within the transcribed sequence of LMP, and heterogeneity of LMP was observed in transformed B-cell lines (Rowe et al. 1987). Taken together, these findings favor the hypothesis that small changes in the coding sequence of LMP may occur during latent viral replication and so contribute to the escape of EBV-infected B cells from recognition by HLA class I-restricted cytotoxic T cells.

Miller and coworkers identified a subclone of a Burkitt's lymphoma cell line, P3J-HR-1 HH 543-5, characterized by abundant infectious EBV (Countryman and Miller 1985; Countryman et al. 1987; Jenson and Miller 1988). They observed the juxtaposition of the Bam W and the Bam Z fragment as the event leading to the disruption of viral latency. Autonomous viral replication due to defective virus containing the Bam W/Bam Z rearrangement was recently identified in two out of ten oral hairy leukoplakia samples from AIDS patients (Patton et al. 1990). We have looked for the Bam W/Bam Z rearrangement in DNA samples from HD and AILD patients with a relatively high number of EBV genomes. As demonstrated, the negative results were due to a real absence of the Bam W/Bam Z rearrangement in the 21 samples examined because the sensitivity of detection was one positive cell within 10^5 diploid human genomes.

References

Anagnostopoulos I, Herbst H, Niedobitek G, Stein H (1989) Demonstration of monoclonal EBV genomes in Hodgkin's disease and Ki-1-positive anaplastic large cell lymphoma by combined Southern blot and in situ hybridization. Blood 74:810–816

Baer R, Bankier AT, Biggin MD, Deininger PL, Farrell PJ, Gibson TJ, Hatfull G, Hudson GS, Satchwell SC, Séguin C, Tuffnell PS, Barrell BG (1984) DNA sequence and expression of the B95-8 Epstein-Barr virus genome. Nature 310:207–211

Bankier AT, Deininger PL, Satchwell SC, Baer R, Farrell PJ, Barrell GB (1983) DNA sequence analysis of the *Eco*RI Dhet fragment of B95-8 Epstein-Barr virus containing the terminal repeat sequences. Mol Biol Med 1:425–445

Cheung A, Kieff E (1982) Long internal direct repeat in Epstein-Barr virus DNA. J Virol 44:286–294

Cleary ML, Nalesnik MA, Shearer WT, Sklar J (1988) Clonal analysis of transplant associated lymphoproliferations based on the structure of the genomic termini of the Epstein-Barr virus. Blood 72:349–352

Countryman J, Miller G (1985) Activation of expression of latent Epstein-Barr herpesvirus after gene transfer with a small cloned subfragment of heterogeneous viral DNA. Proc Natl Acad Sci USA 82:4085–4089

Countryman J, Jenson H, Seibl R, Wolf H, Miller G (1987) Polymorphic proteins encoded within BZLF1 of defective and standard Epstein-Barr viruses disrupt latency. J Virol 61:3672–3679

Delcayre AX, Salas F, Mathur S, Kovats K, Lotz M, Lernhardt W (1991) Epstein Barr virus/complement C3d receptor is an interferon α receptor. EMBO J 10:919–926

Desgranges C, Wolf H, De-Thé G, Shanmugaratnam K, Cammoun N, Ellouz R, Klein G, Lennert K, Munoz N, zur Hausen H (1975) Nasopharyngeal carcinoma: X. Presence of Epstein-Barr genomes in separated epithelial cells of tumours in patients from Singapore, Tunisia and Kenya. Int J Cancer 16:7–15

Fennewald S, van Santen V, Kieff E (1984) Nucleotide sequence of a mRNA transcribed in latent growth-transforming virus infection indicates that it may encode a membrane protein. J Virol 51:411–419

Fingeroth JD, Weiss JJ, Tedder TF, Strominger JL, Biro PA, Fearson DT (1984) Epstein-Barr virus receptor of human B lymphocytes is the C3d receptor CR2. Proc Natl Acad Sci USA 81:4510–4516

Fischer A, Blanche S, Le Bidois J, Bordigoni P, Garnier JL, Niaudet P, Morinet F, Le Deist F, Fischer AM, Griscelli C, Hirn M (1991) Anti-B-cell monoclonal antibodies in the treatment of severe B-cell lymphoproliferative syndrome following bone marrow and organ transplantation. N Engl J Med 324:1451–1456

Frizzera G, Moran EM, Rappaport H (1974) Angio-immunoblastic lymphadenopathy with dysproteinaemia. Lancet i:1070–1073

Henderson A, Ripley S, Heller M, Kieff E (1983) Chromosome site for Epstein-Barr virus DNA in a Burkitt tumor cell line and in lymphocytes growth-transformed in vitro. Proc Natl Acad Sci USA 80:1987–1991

Herbst N, Niedobitek G, Kneba M, Hummel M, Finn T, Anagnostopoulos I, Bergholz M, Krieger G, Stein H (1990) High incidence of Epstein-Barr virus genomes in Hodgkin's disease. Am J Pathol 137:13–18

Hurley EA, Thorley-Lawson DA (1988) B cell activation and the establishment of Epstein-Barr virus latency. J Exp Med 168:2059–2075

Hurley EA, Agger S, McNeil JA, Lawrence JB, Calendar A, Lenoir G, Thorley-Lawson DA (1991) When Epstein-Barr virus persistently infects B-cell lines, it frequently integrates. J Virol 65:1245–1254

Itakura K, Rossi JJ, Wallace RB (1984) Synthesis and use of synthetic oligonucleotides. Annu Rev Biochem 53:323–356

Jenson HB, Miller G (1988) Polymorphisms of the region of the Epstein-Barr virus genome which disrupts latency. Virology 165:549–564

Jones JF, Shurin S, Abramowsky C, Tubbs RR, Sciotto CG, Wahl R, Sands J, Gottman D, Katz BZ, Sklar J (1988) T-cell lymphomas containing Epstein-Barr viral DNA in patients with chronic Epstein-Barr virus infections. N Engl J Med 318:733–741

Katz BZ, Raab-Traub N, Miller G (1989) Latent and replicating forms of Epstein-Barr virus DNA in lymphomas and lymphoproliferative disorders. J Infect Dis 160:589–598

Katzenstein AL, Peiper S (1990) Detection of Epstein-Barr virus genomes in lymphomatoid granulomatosis: analysis of 29 cases using the polymerase chain reaction technique. Mod Pathol 3:435–441

Kawa-Ha K, Ishihara S, Ninomiya T, Yumura-Yagi K, Hara J, Murayama F, Tawa A, Hirai K (1989) CD-3 negative lymphoproliferative disease of granular lymphocytes containing Epstein-Barr viral DNA. J Clin Invest 84:51–55

Kieff E, Liebowitz D (1991) Epstein-Barr virus and its replication. In: Fields BN, Knipe DM (eds) Fundamental virology, 2nd edn. Raven, New York, pp 897–928

Kikuta H, Taguchi Y, Tomizawa K, Ishizaka A, Sakiyama Y, Matsumoto S, Imai S, Kinoshita T, Koizumi S, Osato T, Kobayashi I, Hamada I, Hirai K (1988) Epstein-Barr virus genome-positive lymphocytes in a boy with chronic active EBV infection associated with Kawasaki-like disease. Nature 333:455–457

Knecht H (1989) Angioimmunoblastic lymphadenopathy; ten years' experience and state of current knowledge. Semin Hematol 26:208–215

Knecht H, Sahli R, Shaw P, Meyer C, Bachmann E, Odermatt BF, Bachmann F (1990) Detection of Epstein-Barr virus DNA by polymerase chain reaction in lymph node biopsies from patients with angioimmunoblastic lymphadenopathy. Br J Haematol 75:610–614

Knecht H, Odermatt BF, Bachmann E, Teixeira S, Sahli R, Hayoz D, Heitz P, Bachmann F (1991) Frequent detection of Epstein-Barr virus DNA by the polymerase chain reaction in lymph node biopsies from patients with Hodgkin's disease without genomic evidence of B or T cell clonality. Blood 78:760–767

Lawrence JB, Villnave CA, Singer RH (1988) Sensitive, high-resolution chromatin and chromosome mapping in situ: presence and orientation of two closely integrated copies of EBV in a lymphoma line. Cell 52:51–61

Lenoir GM, Bornkamm GW (1987) Burkitt's lymphoma, a human cancer model for the study of the multistep development of cancer: proposal of a new scenario. In: Klein G (ed) Advances in viral oncology, vol 7. Raven, New York, pp 173–206

Lindahl T, Adams A, Byursell G, Bornkamm GW, Kaschka-Dierich C, Jehn U (1976) Covalently closed circular duplex DNA of Epstein-Barr virus in a human lymphoid cell line. J Mol Biol 102:511–530

Lipford EH, Smith HR, Pittaluga S, Jaffe ES, Steinberg AD, Cossman J (1987) Clonality of angioimmunoblastic lymphadenopathy and implications for its evolution to malignant lymphoma. J Clin Invest 79:637–642

Matsuo T, Heller M, Petti L, O'Shiro E, Kieff E (1984) Persistence of the entire Epstein-Barr virus genome integrated into human lymphocyte DNA. Science 226:1322–1324

McGuire LJ, Huang DP, Teoh R, Arnold M, Wong K, Lee JCK (1988) Epstein-Barr virus genome in thymoma and thymic lymphoid hyperplasia. Am J Pathol 131:385–390

Patton DF, Shirley P, Raab-Traub N, Resnick L, Sixbey JW (1990) Defective viral DNA in Epstein-Barr virus-associated oral hairy leukoplakia. J Virol 64:397–400

Pfitzner AJ, Strominger JL, Speck SH (1987) Characterization of a cDNA clone corresponding to a transcript from the Epstein-Barr virus Bam HI M fragment: evidence of overlapping mRNAs. J Virol 61:2943–2946

Pritchett R, Pedersen M, Kieff E (1976) Complexity of EBV homologous DNA in continuous lymphoblastoid cell lines. Virology 74:227–231

Raab-Traub N, Flynn K (1986) The structure of the termini of the Epstein-Barr virus as a marker of clonal cellular proliferation. Cell 47:883–889

Reymond CD (1987) A rapid method for the preparation of multiple samples of eucaryotic DNA. Nucleic Acids Res 15:8118

Rowe M, Gregory C (1989) Epstein-Barr virus and Burkitt's lymphoma. In: Klein G (ed) Advances in viral oncology, vol. 8. Raven, New York, pp 237–259

Rowe M, Evans HS, Young LS, Hennessy K, Kieff E, Rickinson AB (1987) Monoclonal antibodies to the latent membrane protein of Epstein-Barr virus reveal heterogeneity of the protein and inducible expression in virus-transformed cells. J Gen Virol 68:1575–1586

Saiki RK, Gelfand DH, Stoffel S, Scharf SJ, Higuchi R, Horn GT, Mullis KB, Erlich HA (1988) Primer-directed enzymatic amplification of DNA with a thermostable DNA polymerase. Science 239:487–491

Saito I, Servenius B, Compton T, Fox RI (1989) Detection of Epstein-Barr virus DNA by polymerase chain reaction in blood and tissue biopsies from patients with Sjögren's syndrome. J Exp Med 169:2191–2198

Staal SP, Ambinder R, Beschorner WE, Hayward GS, Mann R (1989) A survey of Epstein-Barr virus DNA in lymphoid tissue. Am J Clin Pathol 91:1–5

Su IJ, Hsieh HC, Lin KH, Uen WC, Kao CL, Chen CJ, Cheng AL, Kadin ME, Chen JY (1991) Aggressive peripheral T-cell lymphomas containing Epstein-Barr viral DNA: a clinicopathologic and molecular analysis. Blood 77:799–808

Sudgen B, Phelps M, Domoradzki J (1979) Epstein-Barr virus DNA is amplified in transformed lymphocytes. J Virol 31:590–595

Tanner J, Weis J, Fearon D, Whang Y, Kieff E (1987) Epstein-Barr virus gp350/220 binding to the B lymphocyte C3d receptor mediates absorption, capping, and endocytosis. Cell 50:203–213

Thore A, Dosch HM, Gelfand EW (1979) Expression and modulation of C3 receptors during early T-cell ontogeny. Cell Immunol 45:157–159

Thorley-Lawson DA (1988) Basic virological aspects of Epstein-Barr virus infection. Semin Hematol 25:247–260

Tugwood JD, Law WH, O SK, Tsao SY, Martin WM, Shiu W, Desgranges C, Jones PH, Arrand JR (1987) Epstein-Barr virus specific transscription in normal and malignant nasopharyngeal biopsies and in lymphocytes from healthy donors and infectious mononucleosis patients. J Gen Virol 68:1081–1091

Vos JMH, Wauthier EL, Hanawalt PC (1989) DNA damage stimulates human cell transformation by integrative but not episomal Epstein-Barr virus-derived plasmid. Mol Carcinog 2:237–244

Wallace LE, Murray RJ (1989) Immunological aspects of the Epstein-Barr virus system. In: Klein G (ed) Advances in viral oncology, vol 8. Raven, New York, pp 219–236

Wallace LE, Rickinson AB, Rowe M, Epstein MA (1982) Epstein-Barr virus specific cytotoxic T-cell clones restricted through a single HLA antigen. Nature 297:413–415

Weiss LM, Movahed LA, Warnke RA, Sklar J (1989) Detection of Epstein-Barr viral genomes in Reed-Sternberg cells of Hodgkin's disease. N Engl J Med 320:502–506

White Jr AC, Katz BZ, Silbert JA (1989) Association of Epstein-Barr virus with an angioimmunoblastic lymphadenopathy-like lymphoproliferative syndrome. Yale J Biol Med 62:263–269

Wolf BC, Martin AW, Neiman RS, Janckila AJ, Yam LT, Caracansi A, Leav BA, Winpenny R, Schultz DS, Wolfe HJ (1990) The detection of Epstein-Barr virus in hairy cell leukemia cells by in situ hybridization. Am J Pathol 136:717–723

Yoneda N, Tatsumi E, Kawanishi M, Teshigawara K, Masuda S, Yamamura Y, Inui A, Yoshino G, Oimomi M, Baba S, Yamaguchi N (1990) Detection of Epstein Barr virus genome in benign polyclonal proliferative T cells of a young male patient. Blood 76:172–177

Chapter 14 Detection of Human Herpesvirus 6 DNA in Clinical Samples of Patients by Polymerase Chain Reaction Amplification

K. Kondo, C. Tomomori, T. Kondo, T. Mukai, T. Yamamoto, and K. Yamanishi

Summary

The detection of human herpesvirus 6 (HHV-6) DNA in peripheral blood mononuclear cells (PBMC), throat swabs, and cerebrospinal fluid (CSF) was determined using the polymerase chain reaction (PCR). The clinical samples were collected from patients with exanthem subitum (ES) during the acute and convalescent phases of infection, and from healthy children and adults. HHV-6 DNA was detected in mononuclear cells which were collected from ES patients during both acute and convalescent phases, and even from healthy adults. Mononuclear cells of peripheral blood were then separated into adherent and nonadherent cells. While DNA could be detected in nonadherent and adherent mononuclear cells during the acute phase, it was detected predominantly in adherent cells during convalescent phase. Furthermore, viral DNA was found in adherent cells of healthy adults. HHV-6 DNA could also be detected in throat swabs from children and adults having antibody to HHV-6. HHV-6 DNA was detected in CSF of ES patients who had some neurological symptoms at a high frequency. However, there was no detection of HHV-6 DNA in samples from neonates.

Finally, we have developed nested double PCR, which could detect HHV-6 DNA more sensitively. A few copies of HHV-6 DNA could be detected by this method, even in clinical samples.

Introduction

A novel human herpesvirus, now named human herpesvirus 6 (HHV-6), has been isolated independently by several groups from patients with lymphocytic disorders (Salahuddin et al. 1986; Tedder et al. 1987; Downing et al. 1987; Agut et al. 1988; Lopez et al. 1988). Although this virus was initially named "human B lymphotropic," it was later found mainly to infect and replicate in lymphocytes of the T cell lineage (Ablashi et al. 1987; Lusso et al. 1988; Taka-

Department of Virology, Research Institute for Microbial Diseases, Osaka University, 3-1 Yamada-oka, Suita, Osaka 565, Japan

hashi et al. 1989). Characterization of HHV-6 indicated that it was antigenically and genetically distinct from other human herpesviruses (cytomegalovirus
CMV, herpes simplex virus HSV types 1 and 2, varicella-zoster virus VZV, and
Epstein-Barr virus EBV; Josephs et al. 1986; Salahuddin et al. 1986; Lopez
et al. 1988). It was, however, not known at that time whether HHV-6 causes
disease in humans. We first reported HHV-6 as the causal agent of exanthem
subitum (ES) in 1988 (Yamanishi et al. 1988). The clinical appearance of ES
has been well described in pediatric textbooks. The main clinical characteristics are as follows: The onset of this disease is often abrupt, and it lasts for a
few days. The typical eruption appears after the decline of the temperature.
Neurological symptoms such as convulsion may be seen during the febrile
phase of roseola, although encephalitis has been reported rarely. The prognosis is uniformly excellent, even for those cases complicated by convulsive
seizures. We have reported a case of encephalitis following roseola confirmed
by seroconversion to HHV-6 (Ishiguro et al. 1990).

A novel technique for the amplification of DNA or RNA in vitro, the
polymerase chain reaction (PCR), has recently been developed. The PCR can
be used for many purposes, such as the analysis of inherited disorders and the
detection of somatic diseases (Saiki et al. 1985, 1988; Mullis and Faloona
1987), and one of the most important applications has been in the detection of
infectious agents that are present in small numbers in clinical samples. Here we
show the use of the PCR for the detection of HHV-6 in peripheral blood
mononuclear cells (PBMC), throat swabs of children and adults, and cerebrospinal fluid from ES patients. Furthermore, we also describe the nested
double PCR method, which can detect a small amount of template DNA
derived from clinical material.

Materials and Methods

Collection of Clinical Samples

Peripheral blood was collected from patients with ES and healthy adults and
separated to obtain PBMC by Ficoll-Paque centrifugation as described previously (Yamanishi et al. 1988). CSF was obtained from patients with ES who
had been admitted to hospitals and showed neurological symptoms such as
vomiting, convulsion, irritability, and bulging of the anterior fontanel during
the acute phase of the disease. Throat swabs were also collected from ES
patients, healthy adults, and adults and children who had common colds with
fever.

Methods for DNA Extraction

Method 1: Phenol-Chloroform Method. Up to 5×10^6 PBMC or CSF was
incubated for 8–16 h at 60 °C in 0.5 ml NET buffer (final concentrations:

150 m*M* NaCl, 15 m*M* Tris-HCl, 1 m*M* EDTA) with 0.1% sodium dodecylsulfate (SDS) and proteinase K 1.0 mg/ml. Throat swabs were ultracentrifuged at 70 000 rpm (Beckman TL 100) for 2 h. The pellets were suspended in 500 µl NET buffer. The mixture was extracted three times with an equal volume of phenol:chloroform:isoamyl alcohol (25:24:1) and treated with chloroform twice. DNA was precipitated by adding ethanol. Transfer RNA (tRNA) (5 µg) was used as a carrier; between 5 and 20 µg tRNA did not interfere with the sensitivity of the PCR. The DNA was washed three times with 80% ethanol and resuspended in distilled water. Approximately 0.1–5 µg of PBMC DNA was used in the PCR system.

Method 2: K Buffer Method. PBMC or CSF was incubated for 3 h at 56°C in 0.5 ml K buffer (final concentrations: 50 m*M* KCl, 10 m*M* Tris-HCl pH 8.3), 3 m*M* MgCl$_2$, 0.45% NP40, 0.45% Tween 20, proteinase K 1 mg/ml. The samples were then heated to 98°C for 10 min to denature proteinase K and cooled on ice.

Primer Design

The primers which were used in this context were a part of the *Sal*I fragment (approximately 6 kilobases, kb) from the HHV-6 Hashimoto strain. Primer 3 (5'-GTGTTTCCATTGTACTGAAACCGGT-3') and primer B, 5'-TAAA-CATCAATGCGTTGCATACAGT-3' were used as primers at beginning (Kondo et al. 1990). Five other primers shown in Fig. 1 were also designed for nested double PCR. It was recently found from DNA sequence data that this area is in the coding region of a nucleocapsid protein (Lawrence et al. 1990).

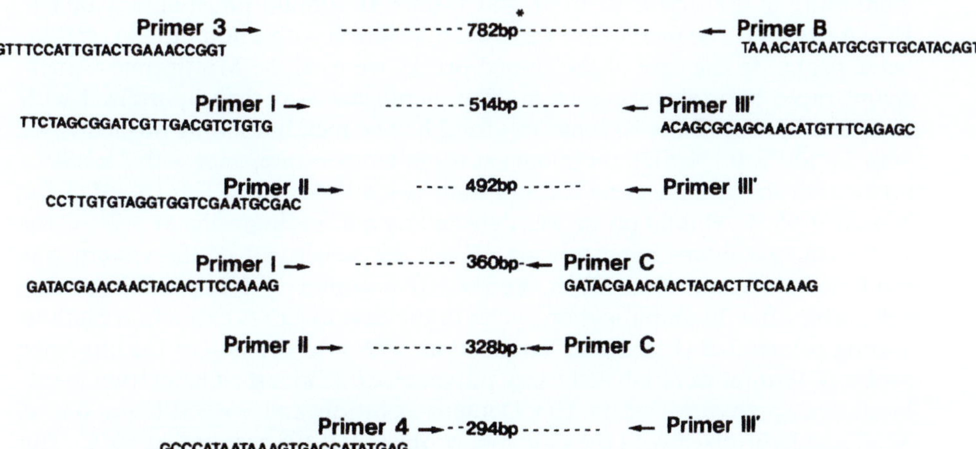

Fig. 1. Primer sets and size of the PCR product. * In the case of 3-B we calculated the size to be 776 bp (Kondo et al. 1990), but in another study it appeared to be 782 bp (Lawrence et al. 1990)

Polymerase Chain Reaction

A total volume of 50 µl of reaction mixture consisting of 50 mM KCl, 10 mM Tris-HCl pH 8.3, 1.5 mM MgCl$_2$, 0.01% (w/v) gelatin, 200 µM each of dGTP, dATP, TTP, dCTP, and 5 U Taq polymerase (Behringer-Mannheim). The sample was first denatured at 94 °C for 2 min and then subjected to 30 amplification cycles of annealing at 62 °C for 2 min, extension at 72 °C for 5 min, and denaturation at 90 °C for 1 min. The primers were used at 1.0 µM each and the reaction was carried out in a DNA thermal cycler (Perkin-Elmer/Cetus). β-Globin primers (Saiki et al. 1985) were used as a positive control in which amplification of viral DNA was not observed.

In the case of nested double PCR, 5 µl of the PCR product after 30 cycles was added to 45 µl of the new reaction mixture (the ingredients were the same as the first PCR reaction mixture). This was then subjected to another 30 amplification cycles.

Detection of Amplified Products

The amplified product was detected by direct gel analysis and Southern blot hybridization. In case of Southern blot hybridization, we used a cloned DNA probe (part of the *Sal*I fragment) or an inner oligomer probe (5'-GAGATG-TACTGGGAGAGTATGTTGGTGAGT-3'). For direct gel analysis, 10 µl of the reaction mixture was run on a 1.5% agarose gel, and DNA was monitored under UV light after staining with ethidium bromide. A band of 782 base pairs (bp) was seen when samples were amplified. DNA was transferred to a Hybond-N$^+$ membrane (Amersham) by alkali blotting. Briefly, we used alkali transfer buffer (0.4 M NaOH), capillary blotting for 3 h, and rinsing of the membrane in 2× SSPE (0.36 M NaCl, 0.02 M sodium phosphate, 0.002 M EDTA pH 7.7). The membrane was then hybridized with a homologous ^{32}P-labeled probe. In the case of the cloned probe, we used the Multiprime (Amersham) rapid hybridization system. The membrane was then hybridized with the cloned probe (2.0 × 10^6 cpm/ml) for 2 h. The membrane was washed twice with 2 × SSPE/0.1% SDS for 20 min at room temperature, once with 2 × SSPE/0.1% SDS for 20 min at 65°C, and then twice in 0.1 × SSPE/0.1% SDS for 20 min at 65 °C. Bound probe was detected by autoradiography at −70°C for 16 h with two intensifying screens. When the sensitivity of the system was estimated using purified DNA, at least five copies of HHV-6 DNA were detectable after 30 amplification cycles in the case of DNA extraction method 1 using primer 3-B (Fig. 1 and Kondo et al. 1991). In the case of the oligomer probe, 5' termini were labeled using polynucleotide kinase. The blotted membrane was prehybridized in 10 × Denhart solution and 6 × SSPE for 6 h at 50°C and hybridized with the oligomer probe (1.0 × 10^6 cpm/ml) at 50°C. The membrane was washed twice with 2 × SSPE/0.1% SDS for 20 min at room temperature and then twice in 2 × SSPE/0.1% SDS for 20 min at 50°C. Bound probe was detected as described above.

Results and Discussion

Sensitivity and Specificity of PCR

The ability of our primers to detect eight different clinical isolates (six strains isolated from patients with ES in Japan, one isolated from a patient who had a renal transplantation, and one isolated from a patient with acquired immune deficiency syndrome in the United States) was evaluated. Samples of 0.05–0.2 µg DNA from infected cells were used as templates in the PCR using the primer-pair 3-B. Each strain gave a positive band on direct gel electrophoresis (Kondo et al. 1990).

The specificity of PCR and the primers for HHV-6 were evaluated using five other human herpesviruses, HSV-1, HSV-2, VZV, CMV, and EBV. DNAs from cells infected with these viruses were extracted and purified by DNA extraction method 1, and DNA samples of 0.01–1 µg were used as templates in the PCR. No amplification was detected by direct gel electrophoresis or Southern blot hybridization. Because the samples might have been contaminated with inhibitors, DNA samples of approximately 1 µg containing 1 ng phage DNA were used as templates, and the PCR was performed in accordance with the instructions provided with the GeneAmp DNA amplification kit (Perkin-Elmer/Cetus). A positive band was observed on direct gel electrophoresis. Next, β-globin DNA from all samples except EBV was also amplified using primers as described by Saiki et al. (1985). A band with a molecular weight of 150 was observed (Kondo et al. 1990).

Detection of HHV-6 DNA in PBMC from Patients with ES

Blood samples were collected during the acute and convalescent phases from five clinically and serologically diagnosed ES patients. All acute phase samples were collected from patients with fever; convalescent phase samples were obtained 1.5–2.5 months after the onset of the illness. Sera were first tested for anti-HHV-6 antibody by an immunofluorescent antibody test. Antibodies were not detected (<1:10) in acute phase sera, but the antibody titer became 1:640 to 1:2560 during the convalescent phase.

PBMC of ES patients were separated into adherent and nonadherent cells as described previously (Kondo et al. 1991). We then attempted to detect HHV-6 DNA in adherent cells (monocytes) and nonadherent cells using the PCR. Using small amounts of DNA and 30 cycles of PCR amplification, the accumulation of amplification product appeared to be an unlimited process, and the product was thought to increase in proportion to the original HHV-6 DNA copy number in the samples (data not shown). HHV-6 DNA could be detected in both adherent and nonadherent cells collected during the acute phase, and predominantly from the adherent cell fraction during the convalescent phase (Fig. 2). Since it was not possible to isolate virus from PBMC during the convalescent phase, HHV-6 might persist in a latent state in mono-

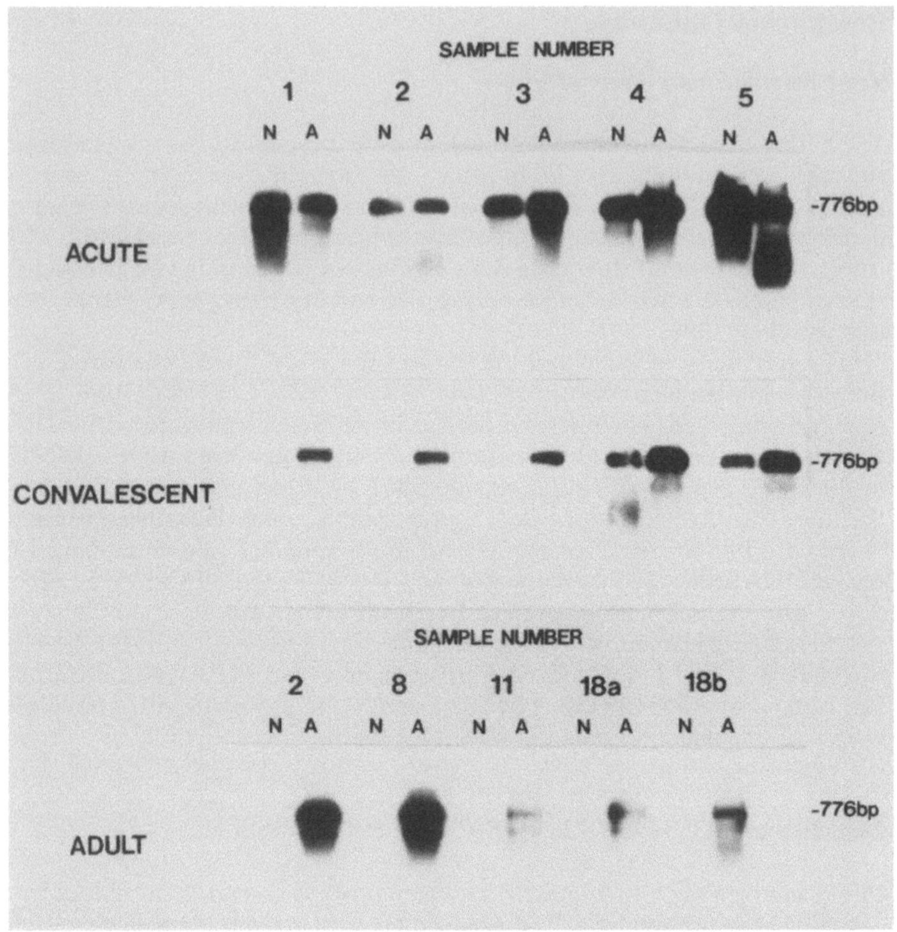

Fig. 2. Detection of HHV-6 DNA in peripheral blood mononuclear cells of acute and convalescent phase ES patients and healthy adults by PCR. *N*, nonadherent cells; *A*, adherent cells

cytes. Viral DNA could also be detected in the monocyte fraction of the peripheral blood of healthy adults by using 45 cycles of PCR amplification, although the detection rate was not so high (Fig. 2).

Detection of HHV-6 DNA in Throat Swabs

Specimens were classified according to the following criteria: group I, 30 healthy adults aged 27–47 years, including 10 mothers of ES infants; group II, 9 adults with common colds when samples were collected; group III, 10 infants aged 5–9 months with ES; group IV, 10 children with fever under aged 1 year;

group V, 29 children with fever aged over 1 year; group VI, 14 healthy neonates, 1 week after birth. The throat swabs of each group were centrifuged as described in "Materials and Methods." DNA was extracted by method 1 and HHV-6 DNA was then amplified by PCR. In group I, HHV-6 DNA was detected in 1 of 30 healthy adults (3.3%) including mothers of ES patients. In group II, HHV-6 DNA was detected in 2 of 9 adults with common colds (22%). In group III, HHV-6 DNA was detected in 2 of 10 ES patients (20%) and in group IV in 3 of 10 children under 1 year with antibody to HHV-6 and fever (30%). The rate of detection of HHV-6 DNA was about 3% (1/29) in children over 1 year with antibody to HHV-6 and with fever (group V). However, no DNA was detected in the 14 neonates within 1 week after birth in group VI (Kido et al. 1990).

Since DNA was detected in children and adults with antibodies to HHV-6, and it is widely known that HHV latently infects humans and can be reactivated, HHV-6 may be reactivated from the latent state and may be transmitted to uninfected individuals through the air.

Detection of HHV-6 DNA in CSF of ES Patients

Specimens of CSF were collected from ES patients with neurological symptoms. When cell counts in the CSF were determined at the time of collection, they were found to be in the normal range in each sample. PCR was attempted after DNA extraction by method 1. A band of 782 bp was detected and could be visualized by ethidium bromide staining in some patients. The presence of HHV-6 gene products was confirmed by Southern blot analysis using a HHV-6-specific probe (data not shown).

These results indicate that HHV-6 genomic material could be detected in CSF of patients with neurological symptoms such as febrile convulsion, vomiting, and bulging of the anterior fontanel. Although we have only attempted to detect DNA in CSF of roseola patients with severe symptoms, we speculate that HHV-6 replicates in the brain of at least some children with roseola and may cause some neurological signs.

Development of Nested Double PCR

Since it takes a long time to purify DNA by method 1, this method does not seem suitable for applicable in clinical laboratories. Therefore, to determine the sensitivity of the PCR in clinical samples, DNA from CSF of two ES patients was prepared by both methods 1 and 2, and the sensitivity was compared by PCR. DNA of patient 1 was clearly detectable by ethidium bromide staining using method 1 (Fig. 3, lane 1), but no bands were seen when DNA was prepared by method 2 (Fig. 3, lanes 3 and 4). When DNA was Southern hybridized, only a faint band could be detected by method 1 in patient 2 (Fig. 3, lane 2). This result clearly shows that DNA from clinical samples is

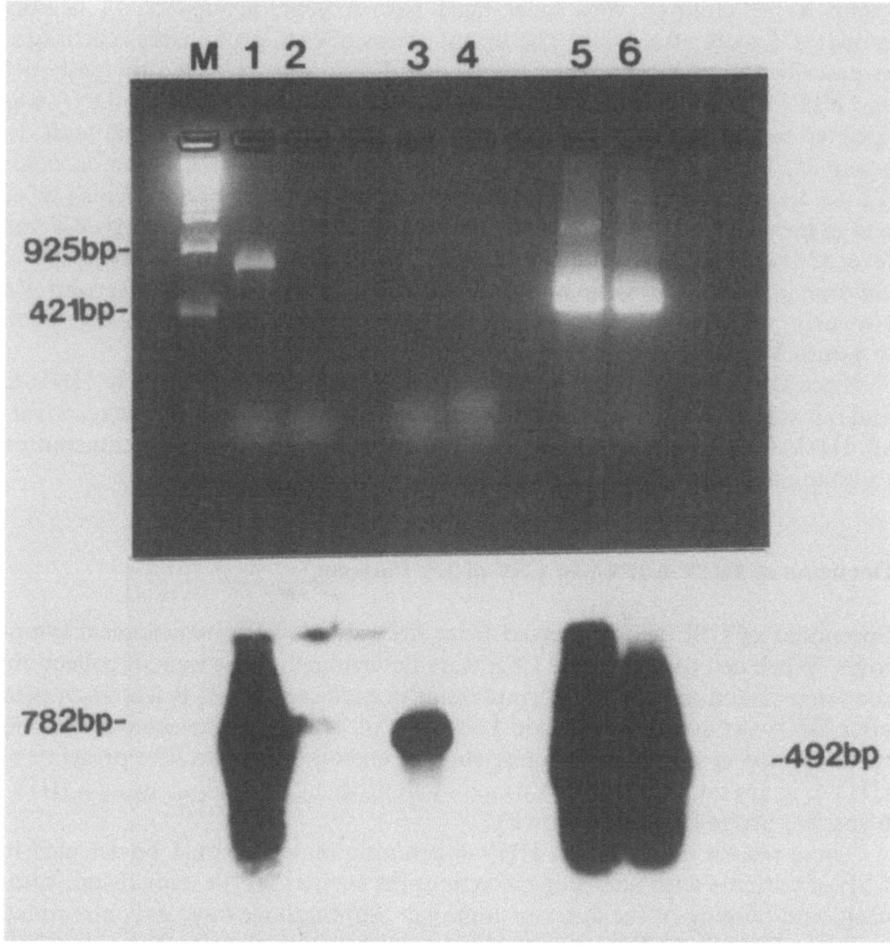

Fig. 3. Comparison of two methods for DNA extraction for PCR. CSF of two patients was prepared to produce DNA samples by method 1 or 2. Amplified product was detected by staining by ethidium bromide and by Southern blot hybridization, as described in "Materials and Methods." *Lanes 1 and 2*, prepared by method 1, *lanes 3 and 4*, prepared by method 2; *lanes 5 and 6*, nested double PCR; *1, 3, and 5*, patient 1; *lanes 2, 4, and 6*, patient 2

detectable by PCR when samples are prepared by method 1, although this method is more troublesome and time-consuming.

In order to have an easier and more sensitive method for use with clinical samples, nested double PCR has been developed using five other primers, as shown Fig. 4. A viral DNA sample with copy numbers of 3×10^3/ml was diluted to 10^{-4} and the sensitivity of PCR using five paired primers as second primers was compared. When the I-C, II-C, and 4-III' pairs were used as the second primers, more than 30 copies of DNA were detectable. When I-III' and II-III' were used as the second paired primers, one more band was detectable

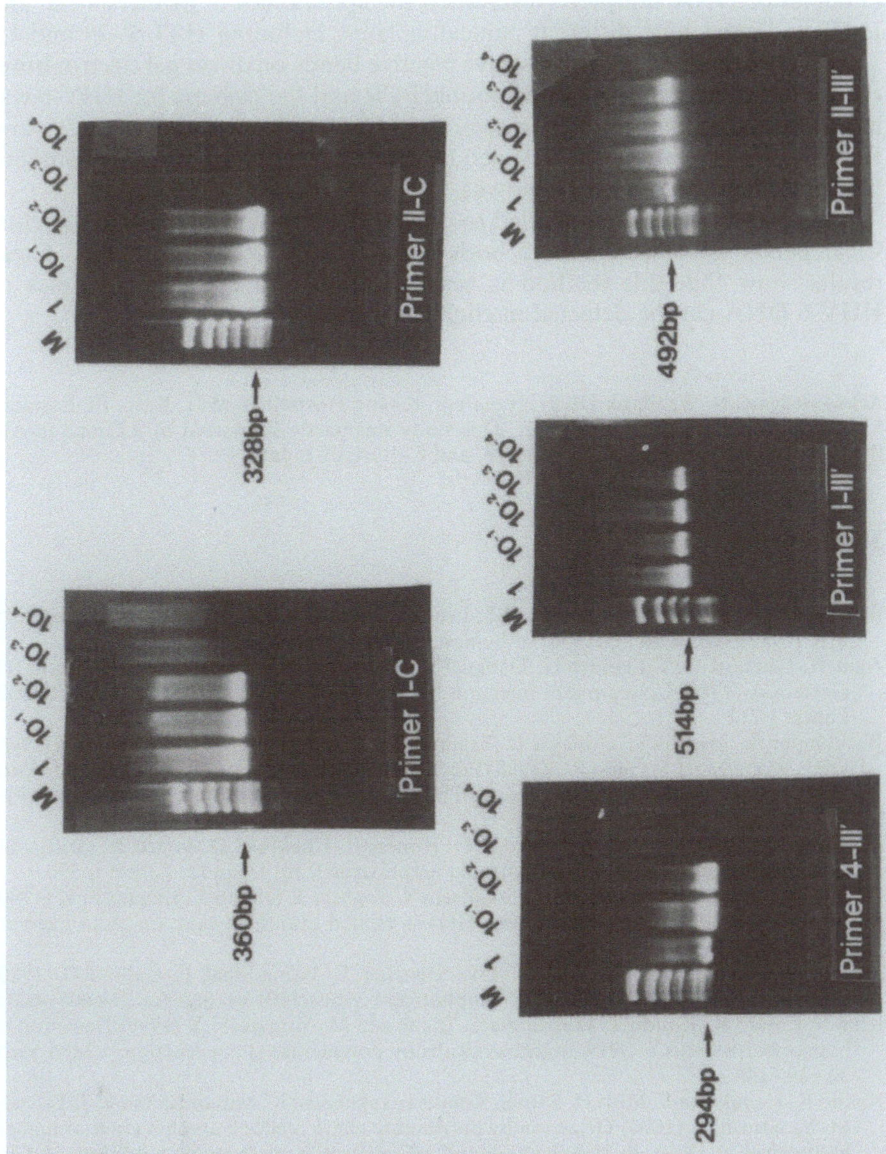

Fig. 4. Analysis of viral DNA after nested double PCR and direct gel electrophoresis. The diluted samples of viral DNA (equivalent to 3×10^3 to 0.3 copies) were amplified by the nested double PCR using five paired primers as the second primers

(3 copies of DNA/sample). Next, when the specificities of primers 4-III' and II-III' to detect nine different clinical isolates, including HST, Z-29 and U-1102, were evaluated, all strains gave positive bands on direct gel electrophoresis. The specificity of this nested double PCR and the primers for HHV-6 was evaluated using five other herpesviruses, HSV-1, HSV-2, VZV, CMV, and EBV. No amplification was detected by direct gel electrophoresis or Southern blot hybridization (data not shown).

This method was then applied to detect DNA in CSF from two patients. Clear bands were detectable in both samples (Fig. 3, lanes 5 und 6). These results show that this method is very sensitive, and even single copies of HHV-6 DNA can be detected in clinical samples.

Acknowledgments. We thank Dr. H. Nagafugi, Kitano Hospital, and H. Kato, PL hospital, for kindly providing clinical material. This study was partly supported by a Grand-in-Aid from the Ministry of Education, Science, and Culture of Japan.

References

Ablashi DV, Salahuddin SZ, Josephs SF, Iwan F, Lusso P, Gallo RC, Hung C, Markham MK (1987) HBLV (or HHV-6) in human cell lines. Natur 329:207

Agut H, Guetard D, Collandre H, Dauguet C, Montagnier L, Miclea JM, Baurmann H, Gessain A (1988) Concomitant infection by human herpesvirus 6, HTLV-1 and HIV-2. Lancet i:712

Buchbinder A, Josephs SF, Ablashi D, Salahuddin SZ, Klotman ME, Manak M, Krueger GRF, Wong-Staal F, Gallo RC (1988) Polymerase chain reaction amplification and in situ hybridization for the detection of human B lymphotropic virus. J Virol Methods 21:191–197

Downing RG, Sewankambo N, Sewadda D, Honess R, Crawford D, Jarrett R, Griffin BE (1987) Isolation of human lymphotropic herpesvirus from Uganda. Lancet ii:390

Ishiguro N, Yamada S, Takahashi T, Takahashi Y, Togashi T, Okuno T, Yamanishi K (1990) Mengino-encephalitis assoicated with HHV-6 related exanthem subitum. Acta Paediatr Scand 79:987–989

Josephs SF, Salahuddin SZ, Ablashi DV, Schacher F, Wong-Staal F, Gallo RC (1986) Genomic analysis of the human B-lymphotropic virus (HBLV). Science 234:601–603

Kido S, Kondo K, Kondo T, Morishima T, Takahashi M, Yamanishi Y (1990) Detection of human herpesvirus 6 DNA in throat swabs by polymerase chain reaction. J Med Virol 32:139–142

Kondo K, Hayakawa Y, Mori H, Sato S, Kondo T, Takahashi K, Minamishima Y, Takahashi M, Yamanishi K (1990) Detection by polymerase chain reaction amplification of human herpesvirus 6 DNA in peripheral blood of patients with exanthem subitum. J Clin Microbiol 28:970–974

Kondo K, Kondo T, Okuno T, Takahashi M, Yamanishi K (1991) Latent infection of human herpesvirus 6 (HHV-6) in human monocytes/macrophages. J Gen Virol 72:1401–1408

Lawrence GL, Chee M, Craxton MA, Gompels UA, Honess RW, Barrel BG (1990) Human herpesvirus 6 is closely related to human cytomegalovirus. J Virol 64:287–299

Lopez C, Pellett P, Stewart J, Goldsmith C, Sanderlin K, Back J, Warfield D, Ferino P (1988) Characteristics of human herpesvirus 6. J Infect Dis 157:1271–1273

Lusso P, Markham PD, Ischachler E, Veronese FM, Salahuddin SZ, Ablashi DV, Pahwa S, Krohn K, Gallo RC (1988) In vitro cellular tropism of human B-lympho-tropic virus (human herpesvirus-6). J Exp Med 167:1659–1670

Mullis KB, Faloona FA (1987) Specific synthesis of DNA in vitro via a polymerase catalyzed chain reaction. Methods Immunol 155:335–350

Saiki RK, Scharf S, Faloona F, Mullis KB, Horn GT, Erich HA, Arnheim N (1985) Enzymic amplification of β-globin genomic sequences and restriction site analysis for diagnosis of sickle cell anemia. Science 230:1350–1354

Saiki RK, Gelfond DH, Stoffel S, Scharf SJ, Higuchi R, Horn GT, Mullis KM, Erlich HA (1988) Primer directed enzymatic amplification of DNA with a thermostable DNA polymerase. Science 239:487–491

Salahuddin SZ, Ablashi DV, Markham PD, Josephs SF, Sturzenegger S, Koplan M, Halligan G, Biberfeld P, Wong-Staal F, Kramarsky B, Gallo RC (1986) Isolation of a new-virus, HBLV, in patients with lymphoproliferative disorders. Science 234:596–600

Takahashi K, Sonoda S, Higashi K, Kondo T, Takahashi H, Takahashi M, Yamanishi K (1989) Predominant CD4 T-lymphocyte tropism of human herpesvirus 6 related virus. J Virol 63:3161–3163

Tedder RS, Briggs M, Cameron CH, Honess R, Robertson D, Whittle H (1987) A novel lymphotropic herpesvirus. Lancet ii:390

Yamanishi K, Okuno T, Shiraki K, Takahashi M, Kondo T, Asano Y, Kurata T (1988) Identification of human herpesvirus-6 as a causal agent for exanthem subitum. Lancet i:1065–1067

Section IV **Human Papova Viruses**

Chapter 15 Human Papillomavirus Infections of the Genital Tract: Clinical Significance and Diagnosis by Polymerase Chain Reaction *

Stina Syrjänen and Kari Syrjänen

Summary

Current data implicate that human papillomavirus (HPV) lesions in the genital tract are frequently associated with precancer lesions and invasive squamous cell carcinomas. Almost 70 different HPV types have been recognized during the past 10 years, and a significant risk for the development of an invasive cancer has been ascribed to infections of the high risk HPV types. On the other hand, spontaneous regression has been histologically confirmed in a significant proportion of genital HPV infections in prospective cohort studies. Of special importance from the epidemiological point of view is the mode of transmission of this virus by sexual contact, thus conferring a potential risk for the development of genital precancer lesions to both sexual partners. Recently, HPV DNA has been detected by different hybridization techniques and PCR in normal squamous epithelium of the genital tract in both sexes, suggesting that these sites of latent HPV infections might act as a reservoir for such infections. In the present review, the clinical significance of genital HPV infections is discussed. The importance of making a distinction between (1) clinically manifest, (2) subclinical, and (3) latent HPV infections is emphasized, and the potentially precancerous nature of the manifest HPV infections is underlined. The applicability and limitations of different diagnostic techniques are discussed with special emphasis on the important role of PCR as the only method capable of disclosing also the subclinical and latent HPV infections. The significance of detecting minute amounts (a few molecules) of HPV DNA is unknown. Many of these PCR-positive results may not represent an infection per se, but rather a tissue surface contamination. Extensive studies are still needed to establish the clinically relevant amounts of viral DNA. Thus, caution should be exercised in labelling as an HPV infection those cases where

Department of Pathology and Kuopio Cancer Research Centre, University of Kuopio, SF-70211 Kuopio, Finland

* Our original studies included in this review have been supported in part by a research grant from the Finnish Cancer Society, a research grant from Sigrid Juselius Foundation, a PHS grant no. 5 RO1 CA 42010-03 awarded by the National Cancer Institute, DHHS, a research grant from the Social Insurance Institution of Finland, a joint research grant from Fabriques de Tabac Reunies S.A., and British-American Tobacco Company Ltd, and by a research contract (#1041051) from the Medical Research Council of the Academy of Finland.

HPV DNA is detected by PCR alone. It seems clear that the examination and treatment of all patients with even a clinically manifest HPV infection will be an overwhelming task. In the light of the epidemiological data and substantiated by the current PCR results, it is equally obvious that patients with subclinical and latent HPV infections cannot be traced and treated by the currently available facilities. It should be borne in mind, however, that it is not the HPV itself which is harmful to the patients, but the precancer (and cancer) lesions that it causes at various anatomic sites. Thus, in countries where nation-wide mass-screening programmes are effective, the means are available to prevent the development of invasive cervical carcinomas by tracing and eradicating the precancer lesions. The old concept on early detection of cervical precancer lesions still remains valid, despite the role of HPV in genital carcinogenesis. In prevention of cervical cancer worldwide, it will be enormously much more effective to first establish covering mass-screening programmes in the high-risk countries than to introduce sophisticated DNA technology (hybridization tests) or PCR amplification procedures to screen large populations for subclinical and latent HPV infections. It will be of major importance to promptly elucidate the risk factors predisposing the clinically manifest HPV lesions for rapid progression. Although some of these factors are well established by now (i.e., lesion grade and HPV type), additional factors involved in the regulation of the viral life cycle within the cell certainly exist which, hopefully, could be used to better predict the disease outcome of the genital HPV infections in the future.

Introduction

Since 1976, it has been well recognized that human papillomavirus (HPV) lesions (i.e., flat, inverted, and exophytic condylomas) in the genital tract are frequently associated with precancerous lesions (intraepithelial neoplasias, CIN, VIN, VAIN, PIN, AIN) and invasive squamous cell carcinomas (K.J. Syrjänen 1986, 1991). Almost 70 different HPV types have been recognized during the past 10 years (de Villiers 1989), and a significant risk for the development of an invasive cancer has been ascribed to infections of the so-called high-risk HPV types (Broker and Botchan 1986; Howley 1983; K.J. Syrjänen et al. 1987). One of the key issues in understanding the epidemiology of HPV infections is the assessment of their natural history, i.e., the disease outcome without therapeutic interventions (K.J. Syrjänen et al. 1987; Roman and Fife 1989; zur Hausen 1989). Evidence on the definite progressive potential of certain HPV lesions has been obtained by prospective cohort studies (K.J. Syrjänen et al. 1985a, b, 1988). On the other hand, a spontaneous regression has been histologically confirmed in a significant proportion of genital HPV infections in such studies (K.J. Syrjänen et al. 1985b, 1988; Kataja et al. 1989, 1990, 1991). Of special importance from the epidemiological point of view is the mode of transmission of this virus by sexual contact, thus conferring a potential risk for the development of genital precancer (and eventually cancer) lesions to both sexual partners (Barrasso et al. 1987; Campion et al. 1985). Recently, however, other possible modes of transmission have been suggested (Sedlacek et al. 1989; Ferenczy et al. 1989; Schneider 1990). Before appropriate preventive measures can be entertained, it is essential to identify the reservoirs of the virus (Barrasso et al. 1987; Campion et al. 1985; S. Syrjänen 1987). During the past few years, several reports have been published in which HPV DNA has been disclosed by different hybridization techniques and polymerase chain reaction (PCR) in normal squamous epithelium of the genital tract in both sexes, suggesting that these sites of latent HPV infections might act as such a reservoir (Meanwell et al. 1987; Fife et al. 1987; Macnab et al. 1986; Wickenden et al. 1985).

In the present review, the clinical significance of genital HPV infections is discussed on the basis of current epidemiological data. The importance of making a distinction between (a) clinically manifest, (b) subclinical, and (c) latent HPV infections is emphasized, and the potentially precancerous nature of the manifest HPV infections is underlined. The applicability and limitations of different diagnostic techniques are shortly discussed with special emphasis on the important role of PCR as the only currently available method capable of disclosing subclinical and latent HPV infections.

Clinical Manifestations of Genital HPV Infections

The extensive literature describing the morphological manifestations of clinically overt HPV infections in the genital tract which has accumulated since the late 1970s is expected to be familiar to the reader, and it is not reviewed in detail in this context (K.J. Syrjänen et al. 1987; K.J. Syrjänen 1991; Schneider 1990). It is generally agreed that the clinically manifest genital HPV lesions include the following: condyloma acuminatum, giant condyloma, flat condyloma, endophytic (inverted) condyloma, bowenoid papulosis, and pigmented papulosis. The accurate diagnosis of these HPV lesions does not usually pose any major problems, provided that proper colposcopic (peniscopic) and pathology facilities are available (Schneider 1990; Syrjänen 1986, 1991).

Subclinical and Latent HPV Infections

During the past couple of years, reports have been published in which HPV DNA has been disclosed in lesions not fulfilling the criteria of the above-listed HPV manifestations or even in the genital mucosa of healthy women (Meanwell et al. 1987; Fife et al. 1987; Macnab et al. 1986; Wickenden et al. 1985). However, contradictory reports questioning the estistence of HPV DNA in histologically normal epithelium have been published as well (Syrjänen and

Syrjänen 1989; Cornelissen et al. 1989; Kulski et al. 1989). For proper evaluation of the epidemiology of HPV infections, it is essential to define their different manifestations accurately. This is especially important because marked inconsistencies and discrepanies exist in the usage of the terms subclinical and latent HPV infections in the current literature (Schneider 1990; Meanwell et al. 1987; Fife et al. 1987; Macnab et al. 1986; Wickenden et al. 1985; Syrjänen and Syrjänen 1989; K.J. Syrjänen 1989, 1990, 1991; Cornelissen et al. 1989; Kulski et al. 1989). The suggested diagnostic criteria of these three categories of HPV infections are summarized in Table 1.

Accordingly, to be called a clinical HPV infection, the lesion should be readily apparent by all clinical diagnostic means, i.e., colposcopy, PAP smear, and biopsy. At the other end of the spectrum, latent HPV infections should not be detectable by any of these clinical diagnostic techniques. Thus, in latent infections, HPV DNA demonstrable by DNA hybridization or PCR should be confined to entirely normal epithelium (K.J. Syrjänen 1989, 1990, 1991). Self-evidently, colposcopy and PAP smear alone are not sufficient diagnostic tools to establish the epothelial normality, which should invariably be based on histopathologic examination of the biopsy (Syrjänen and Syrjänen 1989).

The most problematic are the infections currently referred to as subclinical, a term being ascribed to different issues by different authors (Schneider 1990; Meanwell et al. 1987; Fife et al. 1987; Macnab et al. 1986; Wickenden et al. 1985; Syrjänen and Syrjänen 1989; K.J. Syrjänen 1989; Cornelissen et al. 1989; Kulski et al. 1989). According to our view, however, even the minor epithelial changes not fulfilling the criteria of classic HPV lesions should be called a subclinical infection whenever shown to contain HPV DNA by hybridization or PCR techniques (K.J. Syrjänen 1989, 1990, 1991). In practice, the diagnosis of subclinical HPV infections frequently encounters major problems, while findings classifiable only as suspicious of HPV are present on colposcopy, PAP smear, and biopsy. Suspicious on colposcopy usually refers to faint acetowhite staining of mucosa, whithout the classic colposcopic patterns of leukoplakia, mosaic, or punctation. On PAP smears, the only manifestations might be the weakly dyskeratotic superficial cells presenting with mild nuclear pyknosis, but devoid of the diagnostic koilocytosis (and dyskaryosis). Furthermore, in

Table 1. Different manifestations of human papillomavirus (HPV) infections and their diagnostic criteria

Diagnostic method	Clinical infection	Subclinical infection	Latent infection
Colposcopy	Positive	Suspicious	Negative
Pap smear	Positive	Suspicious	Negative
Histology	Positive	Suspicious	Negative
Hybridization	Positive	Positive	Positive
PCR amplification	Positive	Positive	Positive

PCR, polymerase chain reaction

punch biopsy the slightly acanthotic epithelium might contain superficial dyskeratosis and intermediate cells with slightly vacuolized cytoplasm, but again devoid of any nuclear changes consistent with HPV-induced koilocytosis.

Representative series of both subclinical and latent HPV infections were biopsied from the cervix, vagina, and vulva in 100 women who on a recent mass screening presented with a normal PAP smear (S. Syrjänen et al. 1990). In the majority of these biopsies (shown to contain HPV DNA only by PCR amplification), only subtle morphological changes were present. Without awareness of the DNA data, most of them would have been extremely difficult (if not impossible) to classify as HPV lesions. Undoubtedly, further work is needed to correlate the DNA data with morphology, to establish the histopathological criteria (if any) for subclinical HPV infections, especially in the vagina, introitus, vulva, and male genitalia (Franquemont et al. 1989). According to the experience of the authors, HPV lesions at these sites are much more problematic to assess histologically than those of the uterine cervix. The important epidemiological role of the subclinical and latent HPV infections at these sites is apparent, however (zur Hausen 1989; K.J. Syrjänen 1990, 1991).

Biological Behavior of Genital HPV Infections

Compared with the vast amount of literature which has accumulated on HPV lesions during the past 14 years, the number of adequately conducted prospective follow-up studies is still surprisingly low (K.J. Syrjänen et al. 1985b, 1988; Campion et al. 1986; Nash et al. 1987; Hollingworth and Barton 1988). To elucidate the natural history of genital HPV infections, a prospective follow-up study of women infected with this virus was started in our clinic in 1981. A cohort of 530 women (including cervical, vaginal, and vulvar HPV lesions with and without intraepithelial neoplasia of all grades) has been followed-up for a mean of 71 months (SD$\pm$21), using colposcopy, PAP smears, and punch biopsies, but without instituting any kind of treatment (Kataja et al. 1990, 1991). Using a wide panel of techniques applied to cytological smears and repeated punch biopsies, a substantial amount of new data has been elaborated on the biological behavior of genital HPV infections (K. J. Syrjänen et al. 1985a, 1985b, 1988; K.J. Syrjänen 1989, 1990, 1991). As established by the life-table analysis, progression and regression seem to be significantly related to grade of the lesion in the first biopsy ($P < 0.00001$ and $p = 0.0005$, respectively). Clinical progression was related to the cellular atypia in the first PAP smear ($p = 0.006$, overall) as well as to the type of HPV ($p = 0.0012$). With regard to the role of HPV type as a prognostic indicator, HPV 16-induced CIN lesions seem to run a more aggressive course than those caused by any of the other HPV types analysed (HPV 6, 11, 18, 31, and 33) (Kataja et al. 1989, 1990). When entered into Cox's model, HPV 16 infections showed a fivefold higher risk for progression compared with HPV DNA-negative lesions and those infected by HPV 6 or 11 (Kataja et al. 1991). The data on the prognostic significance of HPV type in invasive cervical carcinomas remain controversial, however (Walker et al. 1989; King et al. 1989). No prognostic influence could be ascribed to any of the HPV types in our recent series of cervical carcinomas (Ji et al. 1991). Furthermore, PAP smear or HPV type were not of predictive value for spontaneous regression, and colposcopic pattern predicted the clinical course very poorly (Kataja et al. 1989, 1990, 1991).

During the 10-year follow-up period, the vast majority of the HPV infections has regressed spontaneously (61%), many seem to persist (25%), and a certain proportion (14%) underwent clinical progression up to the stage of carcinoma in situ (CIS) (K.J. Syrjänen 1990). Quite interestingly, the percentage of spontaneous regressors seems to increase in parallel with the extent of the follow-up time, i.e., from <30% at 35-month follow-up, to 40% at 45-month, to over 55% at 58-month, up to 61% at 70-month (data not published yet). Noteworthy is the fact that during that period, the percentage of clinical progressors remains practically unchanged, i.e., at 14% of these 530 women. In many cases, however, genital HPV infections seem to run an extremely fluctuating course, a transition from a manifest to a subclinical or latent infection being frequently encountered in individual patients when examined at 6-month intervals for prolonged periods (K. J. Syrjänen 1989; K. J. Syrjänen et al. 1990a).

When viewed as a group, several distinct disease profiles can be distinguished among these 530 prospectively followed-up women, with regard to the clinical behavior of their HPV infections. *Early regressors* are women in whom the lesion disappears quickly, i.e., sometimes during the period elapsed from the invitation (i.e., PAP smear diagnostic to HPV) to the first attendance at the clinic, which usually takes 2–3 months. *Persistors* are the patients whose HPV lesion remains morphologically unchanged for a prolonged time period. Quite a substantial percentage of the patients are *fluctuators*, with their lesions alternatively disappearing and reappearing during the follow-up. *Progression*, when established, is usually evident during the first 2 years of the follow-up, as evidenced by the settling of the progression rate at a constant level of 14% in our study after the mean follow-up of about 3 years. *Late regressors* are women in whom the lesion abruptly disappears after a prolonged follow-up period. This is reflected as the continuously increasing rate of regression as a function of the follow-up time. Most unfortunate are the *recurrators*, i.e., women whose lesion was eradicated by cone treatment because of progression into CIS but recurs after a variable period of post-treatment follow-up (K.J. Syrjänen 1990, 1991).

It is clear that this kind of complex clinical behavior for the most part explains the significantly divergent prevalence figures recently reported for HPV infections in different series (ranging from a few percent to 80% and more) (Roman and Fife 1989; Schneider 1990; Meanwell et al. 1987; Fife et al. 1987; Macnab et al. 1986; Wickenden et al. 1985; S. M. Syrjänen et al. 1990). In this context, it is also imperative to realize the fact that these figures are completely dependent on the technique used to analyse the samples for the presence of HPV, whether PAP smear, DNA hybridization, or PCR amplification, as recently emphasized (K. J. Syrjänen 1989, 1990, 1991). Extreme caution should be exercised in interpreting the reports on over 80% detection rates of HPV 16 DNA in 'normal women' by PCR (Tidy et al. 1989). This is because of at least two obvious reasons: (a) The reported 80% prevalence of HPV 16 does not even nearly correspond to the known representation of this single HPV type in clinical lesions, and (b) the 'normal women' studied have been in most cases those with a single normal PAP smear only. As emphasized before, however, a single PAP smear can be normal even in women with CIN, CIS, and cancer lesions (Syrjänen and Syrjänen 1989). The role of PCR will be discussed in more detail in the subsequent sections of this review.

Natural History

The experience gained from our follow-up study was recently summarized by constructing a model to give an overview on the natural history of genital HPV infections (K. J. Syrjänen 1989, 1991). Upon entrance of HPV into the epithelial cells, three different outcomes can be expected: (a) a manifest infection, (b) a subclinical infection, and (c) a latent infection. According to our data, the prevalence of manifest HPV infections in the genital tract among the Finnish female population (age range 25–60 years) currently falls between 2 and 3% (K.J. Syrjänen et al. 1990a). This is based on the figures derived from the nationwide mass-screening program for cervical cancer run in our country since the 1960s. As an example, the prevalence of HPV-induced cytopathic changes in the screened women in Kuopio (approx. 8000 smears examined annually) steadily increased from 1980 through 1987, i.e., 0.05% and 1.72%, respectively. Of interest is the observed decline in these prevalence figures after that year, i.e., to 1.4% in 1988, and to 1.02% in 1989 (K.J. Syrjänen et al. 1990b). The full significance of these data remains to be evaluated, however.

Clinical infections, which thus represent only the peak of an iceberg (2%–3% of all HPV infections), possess the potential for developing in three directions: regression, persistence or progression, as conclusively established by our follow-up program (Kataja et al. 1989, 1990, 1991). Regression seems possible even in lesions diagnosed as CIN I or II, but the eventual regression rate of CIN III lesions is hard to establish for ethical reasons, i.e., all our follow-up women are treated by cone whenever the lesion progresses to CIN III. As an example of a sometimes quite rapid progression of clinical HPV infections is the single (and so far the only) woman in our follow-up series who developed an invasive cancer from CIN I in less than 3 years (K.J. Syrjänen et al. 1985a). As pointed out above, the marked fluctuation in the clinical course of HPV infections observed in individual patients followed up for extended periods is a clear indication of the fact that a transition from a clinical infection to either a subclinical or a latent infection is a frequent event indeed. It should also be emphasized that no reliable follow-up data are available as yet on subclinical and latent HPV infections. Thus, the rate of progression from latent to subclinical and further to clinical infections is not known. More importantly, no firm evidence is available to indicate whether or not a complete clearance of HPV infection is possible. To declare a women HPV-free (healthy) should require that all HPV DNA detection techniques, including PCR, give negative results. In such an assessment, overwhelming problems arise with the sampling, because of the established multifocal nature of HPV infections.

Epidemiology of Genital HPV Lesions

Prevalence and Incidence

To assess the prevalence and incidence figures in the general population, a mass screening was conducted which focused on an unselected cohort of 22-year-old women in Kuopio province (K. J. Syrjänen et al. 1990a). Thus, in 2 successive years, the same cohort of women was screened using routine PAP smears. The prevalence of HVP infection among these women was about 3% in the first year of screening, and about 7% 1 year later. The crude annual incidence was 7.0%. The prelavence figures for those attending both rounds and for those attending only once indicate that the nonattenders had some 25% higher risk of HPV infection than the attenders. Adjusting for this sources of bias increased the annual incidence to 8% at this age group (K. J. Syrjänen et al. 1990a).

Lifetime Risk

Based on the above figures and those derived from a random sample of 2084 (out of 28 861) routine PAP smears examined in our laboratory, the lifetime risk of genital HPV infections was calculated. According to these estimates, up to 79% of Finnish women would contract at least one HPV infection between ages 20 and 79 years (K.J. Syrjänen et al. 1990a). It should be realized, however, that this 79% lifetime risk of contracting an HPV infection, or even the observed 14% of progression rate for clinically manifest HPV infections, by no means reflects the established risk for the development of cervical cancer (i.e., $0.79 \times 0.14 = 11\%$). This is clearly evident from the statistics of cervical cancer in Finland (current incidence $< 5/10^5$). These mass-screening data are in full agreement, however, with the proposed model of the natural history of HPV infections, of which only a fraction exhibits clinically overt infections at any time period (K.J. Syrjänen 1989, 1990, 1991). It should be pointed out that the above figures do not include any estimates on the prevalence and incidence of genital HPV infections in men. This is because no reliable data for the basis of such calculations are available as yet (Oriel 1987).

Use of PCR in the Diagnosis of Genital HPV Infections

The PCR, orginally introduced by Saiki et al. (1985) and subsequently automated by Mullis and Faloona (1987), has emerged as a powerful tool in molecular genetics for the exponential in vitro amplification of specific sequences of interest from minute quantities of DNA. PCR has rapidly established itself as a standard technique also in several laboratories working with HPV diagnosis. Gene detection experiments most dramatically illustrate the extreme sensitivity of PCR, amplifying specific genes from single cells, and a few abnormal cells against a background of normal cells, or latent virus infection of less than 1 in 10 000 cells. However, the major problem with PCR detection is that even if the specificity of the conditions can be established (primers, annealing temperatures) beyond doubt, the sensitivity is so high that there is a major risk of obtaining false-positive results.

During the past few years, PCR has been widely used to analyze the occurrence of HPV infections in different populations (Manos et al. 1990; S. M. Syrjänen et al. 1990; Bauer et al. 1991; Evander et al. 1991). Also, several methodological papers have been published describing highly sensitive PCR-based methods that utilize consensus or general primers to amplify many unidentified HPV types in addition to the known types (Gregoire et al. 1989; Manos et al. 1989; van den Brule et al. 1989, 1990; Snijders et al. 1990; Evander and Wadell 1991). In HPV diagnosis, amplification has been successfully applied to a variety of samples including exfoliative cytology (van den Brule et al. 1990; Bauer et al. 1991), fresh or formalin-fixed, paraffin-embedded materials (Shibata et al. 1988; He et al. 1989; S. M. Syrjänen et al. 1990; Nuovo 1990), urine, and sperm (Le et al. 1988; Kataoka et al. 1991) (Table 2). Several rapid methods to process the samples for PCR analysis have recently been published (van den Brule et al. 1990; Kallio et al. 1991; Saiki 1990). These topics will be shortly discussed in the following paragraphs.

Sample Preparation

A large number of different protocols for the isolation of nucleic acids are available (for review, see Maniatis et al. 1982; Kawasaki 1990). Most of the methods used initially in PCR-based HPV diagnosis produced highly purified DNA. Subsequent works, however, have shown that rapid methods that do not involve extensive purification of the nucleic acids nevertheless yield preparations readily applicable for DNA amplification by PCR (Kawasaki 1990; Saiki 1990).

Table 2. Detection rates of human papillomavirus (HPV) DNA by polymerase chain reaction (PCR) Southern blot, in situ and dot-blot hybridization

Reference	Year	No. of samples	Type of samples	Positive with PCR	Positive with Southern blot	Positive with in situ	Positive with dot-blot
Morris et al.	1988	107	Cervical lavages (50 normal)	32% (TP)	–	–	–
			Cervical lavages (57 abnormal)	100%	–	–	–
Kiyabu et al.	1989	88	Biopsies (squamous Ca)	50% (HPV 16, 18)	–	–	–
Melchers et al.	1989	100	Scrapes (normal)	5%	–	–	–
		80	Scrapes (HPV + smear)	70% (CP)	46%	–	–
Van den Brule et al.	1989	1841	Scrapes (normal)	–	–	–	1.5% (+FISH)
		340	Scrapes (normal)	5% (TP)	–	–	–
		70	Scrapes (dysplasia, Ca)	–	–	–	53% (+FISH)
		120	Scrapes (dysplasia, Ca)	62% (TP)	–	–	–
Low et al.	1990	83	Biopsies (cervical Ca)	93% (HPV 16, 18)	72%	–	–
Nuovo	1990	57	Vulvar and cervical lesions without koilos.	23% (TP)	30%	5%	–
Van den Brule et al.	1990	411	Scrapes (normal)	5% (CP+TP)	–	–	9% (+FISH)
		48	Scrapes (dysplasia)	56%	–	–	–
Manos et al.	1990	144	Biopsies in healthy women	40.1% (CP)	–	–	–
Mayelo et al.	1990	68	Scrapes (normal)	10% (HPV 16)	–	–	–
		58	Scrapes (dysplasia, Ca)	40%	–	–	–
McNicol and Dodd	1990	5	Normal prostate (fixed biopsies)	20% (HPV 16, 18)	–	–	–
		19	Prostate (hyperplasia, Ca.)	95% (HPV 16, 18)	–	–	–
S. M. Syrjänen et al.	1990	109	Cervical, vaginal, vulvar biopsies (PAP, women)	36.4% (TP)	–	3%	–
Evander et al.	1991	12	Cervival cancer	92%	–	–	–
		15	Cervical scrapes (normal)	7%	–	–	–
		25	Scrapes (condyloma, CIN)	68% (CP)	–	–	–
Fujinaga et al.	1991	39	Frozen biopsies (cervical cancer)	85% (TP)	–	–	–
Nishikawa et al.	1991	92	Scrapes (normal)	5% (TP)	–	–	–

Bauer et al.	1991	421	Scrapes (normal)	31%		–	–	7% (ViraPap)
		33	Scrapes (dysplasia, Ca)	73% (CP+TP)		–	–	15%
Gravitt et al.	1991	362	Scrapes (normal)	37% (CP)		–	–	15.2% (ViraPap)
Ji et al.	1991	81	Fixed biopsies (cervical Ca)	37% (TP)		–	–	–
Schiffman et al.	1991	68	Scrapes (normal)	29% (TP)	8%	–	–	–
			Scrapes (CIN, Ca)	71%	51%	–	–	–
Yoshikawa et al.	1991	101	Scrapes and biopsies (CIN, Ca)	95% (CP)	–	–	–	–
		102	Scrapes (normal)	12%	–	–	–	–
Tham et al.	1991	57	Cervical biopsies (HPV lesions)	100% (TP)	74%	–	–	–
Kataoka et al.	1991	66	Scrapes (normal)	12% (TP)	6%	–	–	–
			Scrapes (acetowhite)	26%	8%	–	–	–

TP, type-specific primers (mostly HPV 6, 11, 16, 18); CP, consensus/general primers; Ca, squamous cell carcinoma; –, not done
FISH, filter in situ hybridization; Koilos, koilosyte

Cellular Swabs

In HPV diagnosis, PCR has been most widely applied on scrapings from the uterine cervix, vagina, or vulva (Bauer et al. 1991; van den Brule et al. 1990; Melchers et al. 1989) or cervical lavages (Morris et al. 1988). In addition, urethral scrapings have been used to analyse HPV infections of men (Kataoka et al. 1991). In most methods, the samples are collected with a cotton swab or cytobrush. The specimen is then placed in a 5–10 ml test tube containing buffer (mostly phosphate-buffered saline with antibiotics). The samples should be processed in 1 day when kept at room temperature, while refrigerator temperature allows a longer delay. After washings, the pretreatment of the cells can be performed. Lysing the cells in hypotonic solution at high temperature (i.e., boiling in water) is a quick and effective method of preparing DNA for PCR. The main limitation is that only relatively few cells, usually 10^4 or less, can be used; otherwise, the accumulation of cellular debris will begin to inhibit the reaction (Kawasaki 1990). In addition, cells in complex biological fluids (blood) or cells resistant to lysis (sperm) require additional manipulations (Kawasaki 1990; Saiki 1990). It has been shown that as little as 1% v/v of blood is enough to totally inhibit Taq polymerase (Panaccio and Lew 1991).

Recently, van den Brule and coworkers (1990) described a rapid PCR method applicable to all cervical scrapes to amplify HPV DNA. To optimize the use of PCR directly on crude cell suspensions for HPV detection, they tested several pretreatment methods published earlier. Briefly, cells were subjected to proteinase K incubation (Li et al. 1988; Manos et al. 1989), alkaline denaturation followed by neutralization (Jiwa et al. 1989; Melchers et al. 1989), or common protein fixation methods such as buffered formalin (Shibata et al. 1988), paraformaldehyde, ethanol, and acetone (Mullnik et al. 1989). Their optimal protocol used cell suspensions that had been pelleted and resuspended in phosphate-buffered saline and then frozen at $-40\,°C$. After thawing, the cells were vortexed vigorously. Subsequently, 10 µl of the homogeneous cell suspension was taken up with a disposable pipette, denatured at $+100\,°C$ for 10 min, cooled on ice, and centrifuged for 1 min at $3000 \times g$. The PCR was performed in 50 µl of PCR solution containing 50 mM KCl, 3.5 mM $MgCl_2$, 0.01% gelatin, 200 µM of each dNTP, 25 pmol of general primers, 1U of thermostable DNA polymerase, and 10 µl denatured cell suspension. After initial denaturation ($95\,°C$, 5 min) each cycle consisted of a denaturation step at $95\,°C$ for 1 min, following by a primer annealing step at $40\,°C$ for 2 min, and a chain elongation step of $72\,°C$ for 1.5 min. Finally, 10 µl of the amplification product was analyzed by 1.5% agarose electrophoresis. For type-specific PCR, these authors used a lower $MgCl_2$ concentration (1.5 mM), type-specific primers, and annealing temperature of $55\,°C$ (van den Brule et al. 1990).

This method has also been reproducible in our hands and has consistently resulted in successful amplification of HPV target sequences. With this method, reduced PCR efficiency has not been noted when cervical scrapes contain blood. This is probably because the serum components were lost during the initial pelleting of the cells (van den Brule 1990). The detection level of the method is about 10–100 copies of HPV in a given sample as determined with a serial dilution of SiHa cells (1–10 copies of HPV per genome) in 10^5 human fibroblasts (van den Brule et al. 1990).

Biopsies

Although several rapid methods exist for PCR-based HPV diagnosis performed on scrapings, fresh surgical material has usually been processed according to the classic DNA extraction methods with standard proteinase K, phenol, and chloroform extractions with ethanolic salt precipitations (Maniatis et al. 1982). Currently, we are using the saturated sodium chloride method as a standard to extract DNA from biopsies for different purposes (Miller et al. 1988). The method is easier to perform than the classic one and avoids all toxic reagents. The amount of purified DNA needed for each reaction is usually 0.1–0.05 μg. This method is currently used in our laboratory to detect HPV in all our research material. In our hands, rapid PCR methods have not been as reliable when performed on fresh biopsies as on scrapings. This might be due to the large number of cells in biopsies resulting in accumulation of cellular debris which inhibits the PCR reaction (Kawasaki 1990). However, isolation of clean DNA is impractical for routine tests because of the complexity of the procedure.

Retrospective studies with molecular biology techniques have recently become possible with the advent of DNA extraction methods applicable for formalin-fixed, paraffin-embedded tissues (for review, see Wright and Manos 1990). Such retrospective studies can provide valuable epidemiological data on several infectious diseases including HPV infection. PCR analysis of archival tissues from several years old to over 40 years old has been accomplished successfully with 5- to 10-μm sections (Shibata et al. 1988). Recently, several reports on PCR amplification of DNA from fixed tissues have been published (S.J. Syrjänen et al. 1991; Chang et al. 1990; Wright and Manos 1990).

In our laboratory, we are mostly working with PCR performed on formalin-fixed material to detect HPV. We have recently developed an improved method which can be applied to routine PCR using archival material (Kallio et al. 1991). The method is briefly as follows: One to several 5-μm thick sections are cut to achieve an average surface area of 1 cm^2. Before slicing, the excess paraffin should be cut off from the block. The sections are directly placed into a 500-μl microfuge tube. The paraffin is extracted twice with 500 μl xylene by mixing gently for 5 min at room temperature, followed by washes with absolute ethanol to remove xylene. After removing the ethanol, a few drops of acetone is added to each tube to be evaporated. Onto dried samples, 50 μl of sterile distilled water is added. The pellet is gently resuspended and boiled for 10 min. The tubes are transferred on ice and centrifuged to pellet the undissolved material. Samples are used immediately or stored at −20 °C until used. For PCR, 15 μl of supernatant is used. PCR is done in 50 μl of the reaction mixture described in the paragraph for scrapings (van den Brule et al.

1990). After an initial template thermal denaturation at 95 °C for 5 min, the following cycle profile is used: 30 s at 55 °C (annealing), 1 min at 72 °C (extension), and 30 s at 95 °C (denaturation). At the end of the 35th cycle, the extension step is lengthened by an additional 5 min. Every assay includes negative controls in which PCR is performed with no DNA. In all reactions, an additional pair of primers amplifying β-globin (a single copy gene) should be included to analyze the adequacy of the sample. The PCR product is then analyzed by routine gel electrophoresis with subsequent hybridization (Fig. 1).

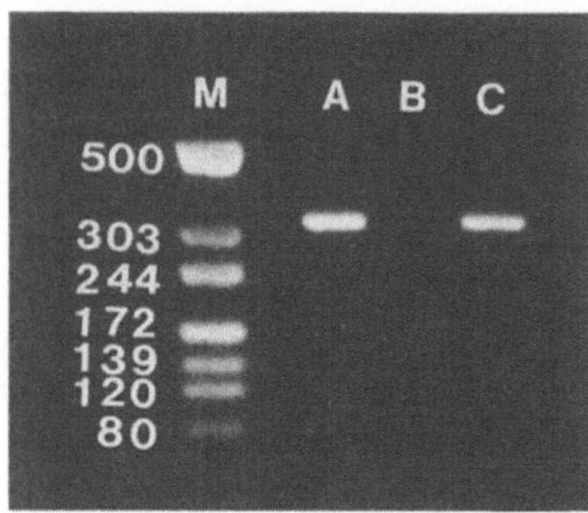

Fig. 1. Formalin-fixed, paraffin-embedded biopsies from the uterine cervix amplified for human papillomavirus (HPV) 16 DNA. The specific polymerase chain reaction (PCR) product is of size 315 bp. *Lanes A* and *C* represent two HPV 16 DNA-positive cases. *Lanes B* and *M* represent a negative control and DNA size marker, respectively

Although several reports favor the isolation of DNA from paraffin blocks with pronase treatment (Wright and Manos 1990), we have found that it results in a significantly lower PCR yield than obtained with the protocol described above. Proteinase K treatment might be needed when large fragments are amplified (Wright and Manos 1990). In the same study, we also examined the effect of tissue debris on amplification by adding the PCR mixture directly to the boiled sample. Although this gave fairly good amplification, in some cases the amplification was suppressed (Kallio et al. 1991). This may be due to fixation effects (Shibata et al. 1988) or inhibitory substances present in the boiled pellet (Wright and Manos 1990). The simplicity and rapidness of our method makes it particularly suitable for routine diagnostic work. This also diminishes the need for manual manipulation and will thus minimize the risks of cross-contamination and carry-over from previous amplifications.

Primer Selection for HPV Diagnosis

Currently, there are at least 65 known HPV types; however, the complete sequence of HPV is known only for 10. Thus, HPV typing with PCR is applicable only for the amplification of the specific target HPV DNA with a known nucleotide sequence. In the earliest reports on PCR-based HPV diagnosis, specific primers for HPV 6, 11, 16, and 18 were mostly used (Shibata et al. 1988; S.M. Syrjänen 1990; Tham et al. 1991). Also, a few studies have analysed HPV 31 and 33 infections with PCR (van den Brule et al. 1989; Evander et al. 1991). The primer selection (HPV 6, 11, 16, 18) is used not only because they present the HPV types with known sequences but they are the most relevant and frequent HPV types in genital infections. The type-specific HPV primers are usually selected from open reading frames (ORFs) E1, E6, and E7. The E6 and E7 regions of the HPV 16 and 18 genomes have been selected for amplification for several reasons. First, these sequences are consistently present in HPV-associated lesions, even when integration of the viral genome into host chromosomes has occurred (Schwarz et al. 1987). Second, expression of the E6 and E7 region appears to be an important component in the process of cellular transformation by these oncogenic types (Schwarz et al. 1987).

Anticontamination Primers

The main disadvantage of PCR is its sensitivity for laboratory infection or contamination, such as virus particles, cloned HPV plasmids, and PCR products themselves. In particular, the last two sources of contamination can cause false-positive results. Van den Brule and coworkers (1989) developed a special strategy to avoid the false-positive results. The use of anticontamination primers ruled out the detection of cloned HPV plasmids. Thus, specific HPV detection is possible even in the presence of contamination with the cloned HPV plasmids. The localization of the anticontamination primers (flanking the plasmid cloning sites, i.e., primers either from L1 or from E5) also minimizes false-negative results caused by integration.

We recently analysed formalin-fixed, paraffin-embedded biopsies from 81 women, treated for an invasive carcinoma of the uterine cervix during 1964–1987, for HPV DNA content by PCR. Amplified HPV DNA was found in 37% of the cases when the type-specific primers were used. However, when we repeated the PCR with the type-specific anticontamination primers, only 24% of the cases were positive. HPV 16 was the most frequent type present in 60% of the cases, followed by HPV 18 in 37%. The low overall HPV-positivity in these samples was probably due to the age of the paraffin blocks and the long fixation time in formalin (Ji et al. 1991).

Consensus General Primers

Several studies have investigated the capability of PCR to detect a broad spectrum of HPV genotypes in order to obtain a powerful tool for the demonstration of unknown HPV types. Matrix comparison of sequenced HPV types 1a, 6b, 15 and 18 revealed that the most conserved regions are localized within the E1 and L1 ORFs (Giri and Danos 1986). By comparing the DNA sequences of genital HPV types 6, 11, 16, 18, and 31, regions of homology 20–25 bp in length have been identified within L1 or E1. Consensus or general primers have been shown to amplify at least 20–30 distinct genital HPV types (Gregoire et al. 1989; Manos et al. 1989; van den Brule et al. 1989, 1990; Snijders et al. 1990; Evander et al. 1991). Several HPV types have been recently identified by using the general primers. Although some of the primers created are working well with different HPV plasmids, nonspecific or poor amplification has been found with human material. In addition, attention should be paid to possible amplification of human DNA. Cross-hybridization has been shown at least between HPV 18 DNA and human DNA. So far, the consensus primers published by Manos et al. (1989) have yielded the best results in our hands.

Prevalence of HPV Infection as Detected by PCR

In various reports, the detection rates of HPV in cervical scrapes from patients with positive cytology have ranged from 17% to 80%. In cervical cancer, HPV DNA has been found in up to 90% of cases. The positivity rate depends on the method used for HPV DNA detection (S. Syrjänen 1990). Recently, several studies comparing different HPV detection techniques have been published, which are summarized in Table 2.

Normal Histology/Cytology and Infection

A number of early PCR papers claimed that HPV DNA can be detected in the uterine cervix of nearly all healthy subjects (Tidy et al. 1989; Ward et al. 1989). Of peculiar interest seemed to be the extremely high rates of cervical infections by the high-risk type HPV 16 (Tidy et al. 1989; Ward et al. 1989). A variant of HPV 16 (16b) was reported to be common in the normal population (Tidy et al. 1989), but this finding was subsequently retracted as a laboratory contamination (Tidy et al. 1989). In more recent studies, however, the figures of HPV prevalence have dramatically declined. Recently, Manos et al. (1990) reported a prevalence of cervical HPV infection ranging from 31% to 44% using the consensus primers. More importantly, the prevalence of HPV 16 ranged from 0% to 22%. They analysed cervical swabs from over 200 women and found no evidence of HPV 16 b.

The explanation for the decline in prevalence figures for HPV infection in the more recent PCR studies probably lies in various forms of accidental contamination or carry-over to which PCR is vulnerable (Fig. 2). Melchers and coworkers (1989) analysed 100 women involved in a triennial checkup program, who had a normal Pap smear result and no history of cervical lesions. HPV was detected in 5% of the women. However, it was not known whether the cervix was clinically or histologically normal in these subjects, because no colposcopy was performed. Bauer and coworkers (1991) reported their results from 467 women visiting a university health service for a routine

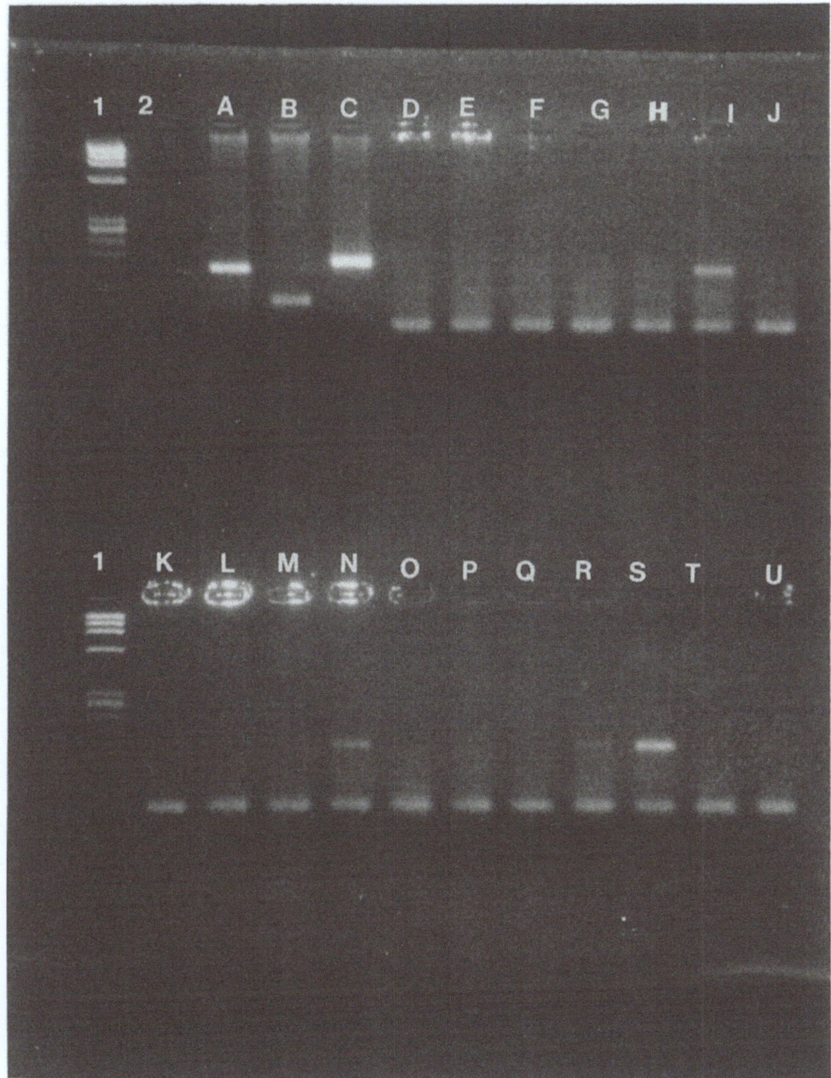

Fig. 2. Amplification of DNA from human tissue with HPV-specific oligonucleotide primers, as shown in the low-melting temperature gel (3%). *Lane A*, a positive amplification control for HPV 6; *lane B*, a positive amplification control for HPV 16 DNA; *lane C*, a positive amplification control for HPV 18; *lane D*, a positive amplification control for HPV 11 DNA; *lane J*, a negative control for HPV 11 DNA amplification, now showing contamination; *Lanes E–I, K–U*, clinical samples (formalin sections) amplified with HPV 11 and 18 primers showing positive amplification results with HPV 11 due to the contamination of PCR buffer with HPV 11 DNA. Clinical samples in lanes I, N, R, and S show positivity for HPV 18

annual gynecologic examination. Using PCR, they found that 46% of the study population was infected with HPV; the ViraPap test showed a prevalence of 11% infected. Of these subjects, normal cytology was found in 421 subjects. HPV DNA was found in 31% and 7% of the normal cytological samples by PCR and ViraPap, respectively. We also performed a study to assess the prevalence of HPV infection in the genital tract of women with normal cytology (S. Syrjänen et al. 1990). A random series of 109 women was re-examined using colposcopy, PAP smear, and punch biopsy taken from the cervix (33 cases), vagina (212 cases), and vulva (20 cases). Changes consistent with HPV infection were seen in 7% of the biopsies under light microscopy. However, only 3% of the biopsies harbored HPV DNA when analysed by in situ hybridization. By PCR, 93 biopsies derived from 40 subjects were analysed, and HPV DNA was found in 35.5%. Only 17 biopsies analysed by PCR were graded as histologically normal, and only 3 of them contained HPV DNA (S. Syrjänen et al. 1990).

Clinical Importance of HPV Infection Detected Only by PCR

This topic can be summarized as recently discussed in Lörincz and Spring (1991). Most authorities agree that PCR is prone to false-positive results. Even if all possibility of laboratory contamination is eliminated, the significance of detecting minute amounts (a few molecules) of HPV DNA is unknown. Many of these PCR-positive results may not represent an infection per se but rather a tissue surface contamination. Extensive studies are still needed to establish the "clinically relevant amounts of viral DNA" (or rather DNA) that are clinically relevant. Caution should be exercised in labelling an HPV infection (especially the genital ones) in those cases where HPV DNA is detected by PCR methods alone.

Conclusions

It seems clear by now that the examination and treatment of all patients with even a clinically manifest HPV infection will be an overwhelming task. In the light of the epidemiological data and substantiated by the current PCR results, it is equally obvious that patients with subclinical and latent HPV infections cannot be traced and treated by the currently available facilities. It should be borne in mind, however, that according to the best available data, it is not the papillomavirus itself which is harmful to the patients, but the precancer (and cancer) lesions that it causes at various anatomic sites. Thus, in countries in which nationwide mass-screening program are effective, the means are available (at least in theory) to prevent the development of invasive cervical carcinomas by tracing and eradicating the precancerous lesions. Accordingly, the old concept of early detection of cervical precancerous lesions still remains valid, despite the role of HPV in genital carcinogenesis. The situation is quite different, however, in the high-risk countries, where such programs do no exist. In the prevention of cervical cancer worldwide, it will be enormously much more effective first to establish mass-screening programs in those high-risk countries than to introduce sophisticated DNA technology (hybridization tests) or PCR amplification procedures to screen the large populations for subclinical and latent HPV infections. Of major importance will be to elucidate promptly the risk factors predisposing the clinically manifest HPV lesions to rapid progression (K.J. Syrjänen et al. 1985a). Although some of these factors are well established by now (i.e., lesion grade and HPV type) (K.J. Syrjänen 1990, 1991), additional factors intimately involved in the regulation of the viral life cycle within the cell certainly exist (zur Hausen 1989), which hopefully could be used to predict better the disease outcome of genital HPV infections.

References

Barrasso R, de Brux J, Croissant O, Orth G (1987) High prevalence of papillomavirus-associated penile intraepithelial neoplasia in sexual partners of women with cervical intraepithelial neoplasia. N Engl J Med 317:916–923

Bauer HM, Ting Y, Greer CE, Chambers JC, Tashiro CJ, Chimera J, Reingold A, Manos MM (1991) Genital human papillomavirus infection in female university students as determined by a PCR-based method. JAMA 265:472–477

Broker TR, Botchan M (1986) Papillomaviruses: retrospectives and prospectives. Cancer Cells 4:17–36

Campion MJ, Singer A, Clarkson PK (1985) Increased risk of cervical neoplasia in consorts of men with penile condylomata acuminata. Lancet i:943–946

Campion MJ, McCane DJ, Cuzick J, Singer A (1986) Progressive potential of mild cervical atypia: progressive cytological, colposcopic, and virological study. Lancet i:238–240

Chang FC, Syrjänen S, Nuutinen J, Karjä J, Syrjänen K (1990) Detection of human papillomavirus (HPV) DNA in oral squamous cell carcinomas by in situ hybridization and polymerase chain reaction. Arch Dermatol Res 282:493–497

Chang F, Janatuinen E, Pikkarainen P, Syrjänen S, Syrjänen K (1991) Esophageal squamous cell papillomas. Failure to detect human papillomavirus (HPV) DNA by in situ hybridization and polymerase chain reaction (PCR). Scand J Gastroenterol 26:535–543

Cornelissen MTE, Tweel JGVD, Struyk AHB, Jebbink MF, Briet M, Noordaa JVD, Schegger JT (1989) Localization of human papillomavirus type 16 DNA using the polymerase chain reaction in the cervix uteri of women with cervical intraepithelial neoplasia. J Gen Virol 70:2555–2562

de Villiers EM (1989) Heterogeneity of the human papillomavirus group. J Virol 63:4898–4903

Evander M, Wadell G (1991) A general primer pair for amplification and detection of genital human papillomavirus types. J Virol Methods 31:239–250

Evander M, Bodén E, Bjersing L, Rylander E, Wadell G (1991) Oligonucleotide primers for DNA amplification of the early regions 1, 6, and 7 from human papillomavirus types 6, 11, 16, 18, 31, and 33. Arch Virol 116:221–233

Ferenczy A, Bergeron C, Richart RM (1989) Human papillomavirus DNA in formites on objects used for the management of patients with genital human papillomavirus infections. Obstet Gynecol 74:950–954

Fife KH, Rogers RE, Zwickl BW (1987) Symptomatic and asymptomatic cervical infections with human papillomavirus during pregnancy. J Infect Dis 156:904–911

Franquemont DW, Ward BE, Andersen WA, Crum CP (1989) Prediction of high-risk cervical papillomavirus infection by biopsy morphology. Am J Clin Pathol 92:577–582

Fujinaga Y, Shimada M, Okazawa F, Fukushima M, Kato I, Fijinaga K (1991) Simultaneous detection and typing genital human papillomavirus DNA using the polymerase chain reaction. J Gen Virol 72:1039–1044

Giri I, Danos O (1986) Papillomavirus genomes. From sequence data to biological properties. Trends Genet 2:227–232

Gravitt P, Hakenewerth A, Stoerker J (1991) A direct comparison of methods proposed for use in widespread screening of human papillomavirus infections. Mol Cell Probes 5:65–72

Gregoire L, Arella M, Campione-Piccarco J, Lancaster WD (1989) Amplification of human papillomavirus DNA sequences by using conserved primers. J Clin Microbiol 27:2660–2665

He Y, Zhang J, Xu Q, Gao J (1989) Detection of human papillomavirus DNA in cervical cancer tissue by the polymerase chain reaction. J Virol Methods 26:17–26

Hollingworth T, Barton S (1988) The natural history of early cervical neoplasia and cervical human papillomavirus infection. Cancer Surv 7:519–527

Howley PM (1983) The molecular biology of papillomavirus transformation. Am J Pathol 113:414–421

Ji HX, Syrjänen S, Klemi P, Chang F, Tosi P, Syrjänen K (1991) Prognostic significance of nuclear DNA content and human papillomavirus (HPV) type in invasive cervical cancer. Polymerase chain reaction (PCR) and flow cytometric analysis. Int J Gynecol Cancer 1:59–67

Jiwa NM, van Gemert GW, Raap AK, Rijke FM, Mulder A, Lens PF, Salimans MMM, Zwaan FE, van Dorp W, van der Ploeg M (1989) Rapid detection of human cytomegalovirus DNA in peripheral blood leucocytes of viremic transplant recipients by the polymerase chain reaction. Transplantation 48:72–76

Kallio P, Syrjänen S, Tervahauta A, Syrjänen K (1991) A simple isolation of DNA from formalin-fixed, paraffin-embedded samples for PCR. J Virol Methods 35:39–47

Kataja V, Syrjänen K, Mäntyjärvi R, Väyrynen M, Syrjänen S, Saarikoski S, Parkkinen S, Yliskoski M, Salonen JT, Castren O (1989) Prospective follow-up of cervical HPV infections. Life table analysis of histopathological, cytological and colposcopic data. Eur J Epidemiol 5:1–8

Kataja V, Syrjänen S, Yliskoski M, Parkkinen S, Saarikoski S, Mäntyjärvi R, Väyrynen M, Salonen JT, Castren O (1990) Prospective follow-up of genital HPV infections. Life-table analysis of HPV typing data. Eur J Epidemiol 6:9–14

Kataja V, Syrjänen K, Syrjänen S, Mäntyjärvi R, Yliskoski M, Saarikoski S, Salonen JT (1992) Prognostic factors in cervical human papillomavirus (HPV) infections. Sex Transm Dis (in press)

Kataoka A, Claesson U, Hansson BG, Eriksson M, Lindh E (1991) Human papillomavirus infection of the male diagnoses by Southern-blot hybridization and polymerase chain reaction: comparison between urethra samples and penile biopsy samples. J Med Virol 33:159–164

Kawasaki ES (1990) Sample preparation from blood, cells and other fluids. In: Innis MA, Gelfand DH, Sninsky JJ, White TJ (eds) PCR protocols: a guide to methods and applications. Academic, San Diego, pp 153–159

King LA, Tase T, Twiggs LB, Okagaki T, Savage JE, Adcock LL, Prem KA, Carson LF (1989) Prognostic significance of the presence of human papillomavirus DNA in patients with invasive carcinoma of the cervix. Cancer 63:879–890

Kiyabu M, Shibata D, Arnheim N, Martin WJ, Fitzgibbons PL (1989) Detection of human papillomavirus in formalin-fixed, invasive squamous carcinomas using the polymerase chain reaction. Am J Surg Pathol 13:221–224

Kulski JK, Pakoczy DP, Sterrett GF, Pixley EC (1989) Human papillomavirus coinfections of the vulva and uterine cervix. J Med Virol 27:244–251

Li H, Gyllensten UB, Cui X, Saiki RK, Erlich HA, Arnheim N (1988) Amplification and analysis of DNA sequences in single human sperm and diploid cells. Nature 335:414–417

Low SH, Thong TW, Ho TH, Lee YS, Morita T, Sing M, Yap EH, Chan YC (1990) Prevalence of human papillomavirus types 16 and 18 in cervical carcinomas: a study by dot-blot and Southern blot hybridization and the polymerase chain reaction. Jpn J Cancer 81:1118–1123

Lundeberg J, Wahlberg J, Uhlén M (1991) Rapid colorimetric quantification of PCR-amplified DNA. Biotechniques 10:68–75

Lörincz A, Spring S (1991) The detection of genital human papillomavirus infection using polymerase chain reaction. JAMA 265/21:2809

Macnab JCM, Walkinshaw SA, Cordiner JW, Clements JB (1986) Human papillomavirus in clinically and histologically normal tissue of patients with genital cancer. N Engl J Med 315:1052–1058

Maniatis T, Fritsch EF, Sambrook J (1982) Molecular cloning: a laboratory manual. Cold Spring Harbor Laboratory, Cold Spring Harbor

Manos MM, Ting Y, Wright DK, Lewis AJ, Broker TR, Wolinsky SM (1989) Use of polymerase chain reaction amplification for the detection of genital human papillomaviruses. Cancer Cells 7:209–214

Manos M, Lee K, Greer C, Waldman J, Kiviat N, Holmes K, Wheeler C (1990) Looking for human papillomavirus type 16 by PCR. Lancet 24:734

Mayelo V, Coursaget P, Fignon A, Lhuintre Y, Lansac J, Anthonioz P (1990) Détection du papillomavirus humain de type 16 dans les cellules épithéliales du col utérin. Presse méd 19:1978–1980

McNicol PJ, Dodd JG (1990) Detection of human papillomavirus DNA in prostate gland tissue by using the polymerase chain reaction amplification assay. J Clin Microbiol 28:409–412

Meanwell CA, Blackledge G, Cox MF, Maitland NJ (1987) HPV 16 DNA in normal and malignant cervical epithelium: implications for the aetiology and behaviour of cervical neoplasia. Lancet i:703–707

Melchers W, van den Brule A, Walboomers J, de Bruin M, Burger M, Herbrink P, Meijer C, Lindeman J, Quint W (1989) Increased detection rate of human papillomavirus in cervical scrapes by the polymerase chain reaction as compared to modified FISH and Southern-blot analysis. J Med Virol 27:329–335

Miller JF, Dower WJ, Tompkins (1988) High-voltage electroporation of bacteria: Genetic transformation of *Campylobacter jejuni* with plasmid DNA. Proc Natl Acad Sci 85:856–871

Morris BJ, Flanagan JL, McKinnon KJ, Nightingale BN (1988) Papillomavirus screening of cervical lavages by polymerase chain reaction. Lancet 10:1368

Mullink H, Walboomers JMM, Tadema TM, Jansen DJ, Meijer CJLM (1989) Combined immuno- und nonradioactive hybridocytochemistry on cells and tissue sections: influence of fixation, enzyme pretreatment, and choice of chromogen on detection of antigen and DNA sequences. J Histochem Cytochem 37:603–609

Mullis KB, Faloona FA (1987) Specific synthesis of DNA in vitro via a polymerase-catalyzed chain reaction. Methods Enzymol 155:335–350

Nash JD, Burke TW, Hoskins WJ (1987) Biologic course of cervical human papillomavirus infection. Obstet Gynecol 69:160–162

Nishikawa A, Fukushima M, Shimada M, Yamakawa Y, Shimano S, Kato I, Fujinaga K (1991) Relatively low prevalence of human papillomavirus 16, 18 and 33 DNA in the normal cervices of Japanese women shown by polymerase chain reaction. Jpn J Cancer 82:532–538

Nuovo GJ (1990) Human papillomavirus DNA in genital tract lesions histologically negative for condylomata. Am J Surg Pathol 14:643–651

Oriel JD (1987) Genital and anal papillomavirus infections in human males. In: Syrjänen KJ, Gissmann L, Koss L (eds) Papillomaviruses and human disease. Springer, Berlin Heidelberg New York, pp 182–196

Panaccio M, Lew A (1991) PCR based diagnosis in the presence of 8% (v/v) blood. Nucleic Acid Res 19:1151

Park JS, Jones RW, McLean MR, Currie JL, Woodruff JD, Shah KV, Kurman RJ (1991) Possible etiologic heterogeneity of vulvar intraepithelial neoplasia. Cancer 67:1599–1607

Roman A, Fife KH (1989) Human papillomaviruses: are we ready to type. Clin Microbiol Rev 2:166–190

Saiki RK (1990) Amplification of genomic DNA. In: Innis MA, Gelfand DH, Sninsky JJ, White TJ (eds) PCR protocols: a guide to methods and applications. Academic, San Diego, pp 153–159

Saiki RK, Scharf S, Faloona F, Mullis KB, Horn GT, Erlich HA, Arnheim N (1985) Enzymatic amplification of β-globulin genomic sequences and restriction site analysis for diagnosis of sickle cell anemia. Science 230:1350–1354

Schiffman MH, Bauer HM, Lorincz AT, Manos MM, Byrne JC, Glass AG, Cadell DM, Howley PM (1991) Comparison of Southern blot hybridization and polymerase chain reaction methods for the detection of human papillomavirus DNA. J Clin Microbiol 29:573–577

Schneider A (1990) Latent and subclinical genital HPV infections. Papillomavirus Rep 1:2–5

Schwarz E, Schneider-Gedicke A, zur Hausen H (1987) Human papillomavirus type-18 transcription in cervical carcinoma cell lines and in human cell hybrids. In: Steinberg BM, Brandsma JL, Taichman LA (eds) Cancer cells: 5. Papillomaviruses. Cold Spring Harbor Laboratory, Cold Spring Harbor, pp 47–53

Sedlacek TV, Lindheim S, Eder C, Hasty L, Woodland M, Ludomirsky A, Rando RF (1989) Mechanism for human papillomavirus transmission at birth. Am J Obstet Gynecol 161:55–59

Shibata D, Arnheim KN, Martin WJ (1988) Detection of human papillomavirus in paraffin-embedded tissue using the polymerase chain reaction. J Exp Med 167:225–230

Snijders PJF, van den Brule AJC, Schrijnemakers HFJ, Snow G, Meijer CJLM, Walboomers JMM (1990) The use of general primers in the polymerase chain reaction permits the detection of a broad spectrum of human papillomavirus genotypes. J Gen Virol 71:173–181

Syrjänen KJ (1986) Human papillomavirus (HPV) infections of the female genital tract and their associations with intraepithelial neoplasia and squamous cell carcinoma. Pathol Annu 21:53–89

Syrjänen KJ (1989) Epidemiology of human papillomavirus (HPV) infections and their associations with genital squamous cell cancer. APMIS 97:957–970

Syrjänen KJ (1990) Natural history of genital HPV infections. Papillomavirus Rep 1(4): 1–5

Syrjänen KJ (1992) Genital human papillomavirus (HPV) infections and their associations with squamous cell cancer: reappraisal of the morphologic, epidemiologic and DNA data. Prog Surg Pathol XII:217–240

Syrjänen KJ, Syrjänen SM (1989) Concept on the existence of human papillomavirus (HPV) DNA in histologically normal squamous epithelium of the genital tract should be re-evaluated. Acta Obstet Gynecol Scand 68:613–617

Syrjänen KJ, de Villiers E-M, Väyrynen M, Mäntyjärvi R, Parkkinen S, Saarikoski S, Castren O (1985a) Cervical papillomavirus infection progressing to invasive cancer in less than three years. Lancet i:510–511

Syrjänen KJ, Väyrynen M, Saarikoski S, Mäntyjärvi R, Parkkinen S, Hippeläinen M, Castren O (1985b) Natural history of cervical human papillomavirus (HPV) infections based on prospective follow-up. Br J Obstet Gynaecol 92:1086–1092

Syrjänen KJ, Gissmann L, Koss LG (eds) (1987) Papillomaviruses and human disease. Springer, Berlin Heidelberg New York

Syrjänen KJ, Mäntyjärvi R, Saarikoski S, Väyrynen M, Syrjänen S, Parkkinen S, Yliskoski M, Saastomoinen J, Castren O (1988) Factors associated with progression of cervical human papillomavirus (HPV) infections into carcinoma in situ during a long-term prospective follow-up. Br J Obstet Gynaecol 95:1096–1102

Syrjänen KJ, Hakama M, Saarikoski S, Väyrynen M, Yliskoski M, Syrjänen S, Kataja V, Castren O (1990a) Prevalence, incidence and estimated life-time risk of cervical human papillomavirus (HPV) infections in nonselected Finnish female population. Sex Transm Dis 17:15–19

Syrjänen KJ, Yliskoski M, Kataja V, Hippeläinen M, Syrjänen S, Saarikoski S, Ryhänen A (1990b) Prevalence of genital human papillomavirus (HPV) infections in a mass-screened Finnish female population aged 20–65 years. Int J STD AIDS 1:410–415

Syrjänen SM (1987) Human papillomavirus infections in the oral cavity. In: Syrjänen KJ, Gissmann L, Koss L (eds) Papillomaviruses and human disease. Springer, Berlin Heidelberg New York, pp 104–137

Syrjänen SM (1990) Basic concepts and practical applications of recombinant DNA techniques in detection of human papillomavirus (HPV) infections. APMIS 98:95–110

Syrjänen SM, Saastamoinen J, Chang F, Ji H, Syrjänen K (1990) Colposcopy, punch biopsy, in situ DNA hybridization, and the polymerase chain reaction in searching for genital human papillomavirus (HPV) infections in women with normal PAP smears. J Med Virol 31:259–266

Syrjänen S, Andersson B, Juntunen L, Syrjänen K (1991) Use of polymerase chain reaction in generation of biotinylated human papillomavirus DNA probes for in situ hybridization. J Virol Meth 31:147–160

Tham KM, Chow VTK, Singh P, Tock EPC, Ching KC, Lim-Tan SK, Sng ITY, Bernard HU (1991) Diagnostic sensitivity of polymerase chain reaction and Southern blot hybridization for the detection of human papillomavirus DNA in biopsy specimens from cervical lesions. Am J Clin Pathol 95:638–646

Tidy J, Farrell PJ (1989) Retraction: human papillomavirus subtype 16b. Lancet ii:1535

Tidy JA, Vousden KH, Farrell PJ (1989) Relation between infection with a subtype of HPV 16 and cervical neoplasia. Lancet ii:1225–1117

van den Brule AJC, Claas ECJ, du Maine M, Melchers WJG, Helmerhorst T, Quint WGV, Lindeman J, Meijer CJLM, Walboomers JMM (1989) Use of anticontamination primers in the polymerase chain reaction for the detection of human papilloma virus genotypes in cervical scrapes and biopsies. J Med Virol 29:20–27

van den Brule AJC, Meijer CJLM, Bakels V, Kenemans P, Walboomers JMM (1990) Rapid detection of human papillomavirus in cervical scrapes by combined general primer-mediated and type-specific polymerase chain reaction. J Clin Microbiol 28:2739–2743

Walker J, Bloss JD, Liao SY, Berman M, Bergen S, Wilczynski SP (1989) Human papillomavirus genotype as a prognostic indicator in carcinoma of the uterine cervix. Obstet Gynecol 74:781–785

Wang AM, Doyle MV, Mark DF (1989) Quantitation of mRNA by the polymerase chain reaction. Biochemistry 86:9717–9721

Ward P, Parry GN, Yule (1989) Human papillomavirus subtype 16a. Lancet ii:170

Wickenden C, Malcolm ADB, Steele A, Coleman DV (1985) Screening for wart virus infection in normal and abnormal cervices by DNA hybridization of cervical scrapes. Lancet i:65–67

Wright DK, Manos MM (1990) Sample preparation from paraffin-embedded tissues. In: Innis MA, Gelfand DH, Sninsky, JJ, White TJ (eds) PCR protocols: a guide to methods and applications. Academic, San Diego, pp 153–159

Yoshikawa H, Kawana T, Kitagawa K, Mizuno M, Yoshikura H, Iwamoto A (1991) Detection and typing of multiple genital human papillomaviruses by DNA amplification with consensus primers. Jpn J Cancer Res 82:524–531

zur Hausen H (1989) Papillomavirus in anogenital cancer: the dilemma of epidemiologic approaches. INCI 81:1681–1682

Chapter 16 Human Papillomavirus Types 6 and 11 in Tumours

R. L. Bryan and J. Crocker

Summary

Human papillomavirus (HPV) types 6 and 11 are known to be associated with tumours of the upper respiratory tract and anogenital region. Their presence has been demonstrated by a variety of techniques, particularly in benign tumours at these sites. The extremely sensitive polymerase chain reaction indicates that these viruses are more widespread than was previously thought; they are frequently associated with malignant tumours and also with tissue showing no histological abnormality.

Introduction

Viral particles were first identified in laryngeal papillomas by means of electron microscopy (Boyle et al. 1973), and the viral antigen was later detected by immunohistochemistry (Braun et al. 1982). In 1983, Gissmann and colleagues demonstrated the recently characterised human papillomavirus (HPV) 6 (Gissmann and zur Hausen 1980) and closely related HPV 11 (Gissmann et al. 1982 b) in these lesions as well as in cervical squamous tumours.

The polymerase chain reaction (PCR) is an extremely sensitive technique which is able to detect a single DNA sequence in 10^5 cells (L.S. Young et al. 1989). In an attempt to clarify the role of HPV 6 and 11 in tumours of the upper respiratory tract, we used the PCR to analyse benign tumours (squamous papillomas and inverted nasal papillomas) and malignant tumours (invasive squamous and verrucous carcinomas). We also analysed biopsies of the pharynx and larynx showing no histological abnormality, for comparison (Bryan et al. 1990).

Department of Histopathology, East Birmingham Hospital, Birmingham B9 5ST, UK

Materials and Methods

The 32 tumours of the upper respiratory tract selected comprised 6 laryngeal squamous papillomas, 8 squamous carcinomas of the pharynx and larynx, 5 verrucous carcinomas from these sites and 13 nasal inverted papillomas. Also, 14 biopsies of histologically normal nasopharyngeal mucosa were analysed.

A representative histological section was chosen in each case, and five 10-μm sections were cut from the corresponding paraffin block. These were cleared in xylene, washed twice in 95% ethanol and dried. Each sample was then boiled in 250 μl water for 30 min. A 35-μl aliquot of the resulting solution was amplified using the PCR. Plasmid-cloned DNA of HPV 6 and 11 was used as positive control material, and lymphoblastoid cell line and fetal liver DNA were used as negative controls. Oligonucleotide primers complementary to sequences in the E6 region of HPV 6 and 11 were used to amplify segments of differing length in the two HPV types (Table 1). The samples for analysis and the control samples were amplified in a 100-μl reaction mixture containing: (a) 16.6 mmol/l ammonium sulphate, 6.7 mmol/l magnesium chloride and 67 mmol/l trometamol, pH 8.8, (b) 1.5 mmol/l of each dNTP, (c) 10 mmol/l 2-mercaptoethanol and 10% v/v dimethyl sulphoxide, (d) 200 μg/l gelatin, (e) 1 μmol/l HPV type-specific oligonucleotide primer and (f) 2 U thermostable DNA polymerase (Taq polymerase). The samples were overlaid with mineral oil, denatured for 7 min at 94°C and subjected to 30 cycles of amplification.

A cycle consisted of primer extension for 2 min at 70°C, denaturation for 1 min at 92°C, reannealing for 2 min at room temperature, followed by primer extension as before.

After amplification, samples were incubated for 10 min at 70°C, and 15 μl of the reaction mixture were electrophoresed on 8% polyacrylamide gels. The reaction products were visualised by ethidium bromide staining. The specificity of the amplified product was confirmed by Southern blotting using ^{32}P-labelled oligonucleotide probes (40-mer) to an internal area of the amplified sequence. The specificity of the amplification technique and primers for HPV 6 and 11 has been demonstrated previously (Young et al. 1989; Griffin et al. 1990).

Table 1. Sequences of oligonucleotide primers

HPV type		Sequences (5′–3′)	Genomic location	Size of amplified product (bp)
6	A	GCTAATTCGGTGCTACCTGT	401–420	140
	B	CTGGACAACATGCATGGAAG	521–540	
11	A	CGCAGAGATATATGCATATG	221–240	90
	B	AGTTCTAAGCAACAGGCACA	291–301	

HPV, human papillomavirus

Discussion

The results of PCR analysis are shown in Table 2. According to our results, benign and malignant tumours of the upper respiratory tract are usually associated with infection by HPV 6, HPV 11 or both. Of interest is the fact that most biopsies of histologically normal upper respiratory tract mucosa were also positive for one or both HPV types; indeed, there was no significant difference in the prevalence of HPV DNA between normal biopsies and those containing tumour (Bryan et al. 1990).

HPVs belong the the genus Papovaviridae. The HPV virion is approximately 55 nm in diameter and shows icosahedral symmetry with 72 capsomeres, comprising 60 hexons and 12 pentons at the vertices (Klug and Finch 1965).

The HPV genome consists of circular, double-stranded DNA of approximate molecular weight 5000 kDa and 8 kB in length. By definition, the DNA of all HPV types is able to cross-hybridise at low stringency. A new HPV type is identified when cross-hybridisation is less than an arbitrary 50% at high stringency. More than 60 types have now been described. Several of the different HPVs are further divisible into subtypes; although these show much cross-hybridisation, they have many different restriction endonuclease cleavage sites resulting from stable point mutations and deletions.

The open reading frames (ORFs) within the viral genome include eight early genes (E1 to E8) which code for proteins before DNA replication occurs. In HPV 6b the largest of these is the E1 ORF which extends for 1947 nucleotides (Schwarz et al. 1983). Two late genes, L1 and L2, encode the major and minor capsid proteins, respectively (Firzlaff et al. 1988). The L1 protein (57 kDa) forms disulphide bridges and contains the papillomavirus common antigens. The L2 protein (78 kDa) is the minor capsid component (Doorbar and Gallimore 1987). All the major ORFs are confined to one DNA strand, and only this strand is expressed as mRNA (Schwarz et al. 1983). The homologies between the different HPV types are not evenly distributed over the

Table 2. Polymerase chain reaction (PCR) analysis of sequamous tumours of the upper respiratory tract

	HPV 6		HPV 11	HPV 6 & HPV 11	Total
	Negative	Positive	Positive	Positive	
Normal nasopharynx	5	3	4	2	14
Inverted papilloma	3	1	2	7	13
Squamous papilloma	1	2	1	2	6
Verrucous carcinoma	1	–	1	3	5
Squamous carcinoma	1	–	3	4	8
Totals	11	6	11	18	46

HPV, human papillomavirus

genome. There is usually great homology within reading frames E1, E2 and L1 and most divergence in E4 and part of L2 (Pfister 1987).

The association of HPV with laryngeal papillomas has been known for some time (Boyle et al. 1973). With the subclassification of HPV into different types came the realisation that certain of these are particularly associated with tumours. Indeed, HPV 11 was first described in laryngeal squamous papilloma (Gissmann et al. 1982b). This virus closely resembled HPV 6, which had been described previously in genital warts (Gissmann and zur Hausen 1980), but it soon became apparent that both viruses are encountered at both sites (Gissmann et al. 1983). These viruses show 25% cross-hybridisation under conditions of high stringency and therefore fulfill the requirements for identification as separate types (Gissmann et al. 1982b). They have a nucleotide homology of 82% (Schwarz et al. 1983). The similarity of these viruses perhaps explains why they are both associated with squamous tumours of the same two anatomical regions.

Laryngeal papillomas are benign exophytic tumours composed of hyperplastic squamous epithelium of mature appearance. They occur mainly on the vocal cords, especially at the commissure, and hence the patient usually presents with hoarseness. They have a tendency to recur and may be single or multiple (papillomatosis). Adult papillomas (occurring in patients over 16 years of age) are more common in men. Symptoms of the juvenile form of laryngeal papillomatosis usually begin before 5 years of age, and this is a comparatively aggressive disease which is more likely than the other types to require tracheostomy. The histological appearances of the papillomas are identical in each variant of the disease.

Results of analysis for HPV 6 and 11 have been variable. Some studies have shown increased evidence of infection in juvenile papilloma when compared with the adult form (Terry et al. 1989), but others have shown no such difference (Quiney et al. 1989). Papillomatosis may more often be associated with HPV infection than single papillomas (Tsutsumi et al. 1989), and mixed infections are frequent. Wright and colleagues (1990) noted no geographical differences in HPV analysis when comparing cases in Australia and Papua New Guinea.

HPV DNA has been detected in squamous papillomas of the respiratory tract which have undergone malignant change, which in some cases follows radiotherapy (McCance 1986). Squamous carcinoma at this site may also arise from nonpapillomatous areas of epithelial dysplasia. Tumours may take the form of conventional invasive squamous carcinoma or of verrucous carcinoma, an exophytic, very well differentiated tumour with blunt 'pushing' rather than infiltrating margins. In our series HPV 6 and 11 were associated with squamous carcinomas of both types.

Inverted papillomas are endophytic tumours of the nasal cavity composed of transitional-type epithelium. They may be erosive and may recur. HPV 6/11 has been demonstrated in these tumours, including cases showing dysplasia (Brandwein et al. 1989), although Judd and colleagues (1991) demonstrated HPV DNA only in the fungiform (exophytic) variant of transitional cell papilloma.

Juvenile laryngeal papillomatosis is thought to result from neonatal infection during passage through an infected birth canal. HPV DNA has been detected in nasopharyngeal aspirates and in amniotic fluid (Sedlacek et al. 1989), supporting the theory of perinatal and possibly in utero infection. The association of HPV with viral warts (condylomata acuminata) has been recognised for some time (Oriel and Almeida 1970). HPV 6 was first characterised in these lesions (Gissmann and zur Hausen 1980) and HPV 11 was later demonstrated in them (Gissmann et al. 1983). HPV 6 appears to be particularly associated with these lesions (Gissmann et al. 1982a, 1983), and combined infections are seen (O'Brien et al. 1989). The infective nature of genital warts has been demonstrated; a significant proportion of sexual partners of patients with condylomata acuminata develop these lesions (Oriel 1971). HPV 6 and 11 DNA has been detected by PCR in the urine of male patients with condylomata, probably within exfoliated cells (Melchers et al. 1989).

Anogenital warts in children may follow sexual abuse and may also result from perinatal transmission of HPV, especially in younger children. In many cases the lesions are negative for HPV 6/11 and positive for those HPV types associated with skin papillomas, notably types 2 and 3. This applies particularly to older children who often have such skin lesions, implying autoinfection (Padel et al. 1990).

Condylomata acuminata are seen in association with non-condylomatous intraepithelial neoplasia of the cervix, vulva, vagina, and penis, including Bowen's disease and bowenoid papulosis, both in the same patient and in sexual partners, who may show either or both types of lesion, suggesting common aetiological factors. Cervical intraepithelial neoplasia (CIN) is the most common and extensively studied form of intraepithelial neoplasia. All are continuous processes of dysplasia, divided arbitrarily into grades I, II and III, which correspond to nuclear atypia with atypical mitotic figures in the lower $\frac{1}{3}$, the lower $\frac{2}{3}$ and the full thickness of the epithelium, respectively. High grades of CIN are associated with a much greater risk of progression to invasive squamous carcinoma than low grades; many cases of the latter probably resolve spontaneously. HPV 6 and 11 show a strong association with low-grade CIN, but high-grade CIN is more often associated with other HPV types, especially 16 and 18 (Crum et al. 1985; McCance 1986; zur Hausen 1987). These correlations are not absolute, and in some cases high-grade CIN is associated with the former and low-grade CIN with the latter. It is dangerous to attempt to predict the clinical outcome of a case of CIN solely in the light of the type of HPV present (Weaver et al. 1990).

Invasive squamous carcinoma of the cervix is also associated with HPV, usually types 16 or 18 as found in high-grade CIN, but tumours containing HPV 6/11 are also seen (Gissmann et al. 1983). The former types are probably associated with a seven- to ten-fold risk of carcinoma compared with the latter (Pfister 1987). Tumours containing HPV 6/11 include verrucous carcinomas of identical histological appearance to those found in the upper aerodigestive tract (Rando et al. 1986). These tumours occur on the genitalia of both sexes.

HPV 6/11 have been detected in squamous tumours at various other locations, including the mouth (de Villiers et al. 1986), the nipple (Kulke et al. 1989), the urethra (Grussendorf-Conen et al. 1987) and the bladder (Kerley et al. 1991). Tumours of the bladder are usually transitional cell carcinomas, and studies have failed to demonstrate HPV DNA there, suggesting that squamous metaplasia may be a prerequisite for infection (Kerley et al. 1991). Against this theory is the finding of HPV 6 DNA in a case of transitional cell carcinoma of the urethra (Mevorach et al. 1990). Squamous tumours at sites away from the upper respiratory tract such as the oesophagus (Loke et al. 1990) and non-squamous tumours such as adenocarcinoma of the cervix (F. I. Young et al. 1991), ovary (McLellan et al. 1990) and prostate (Masood et al. 1991) have proved negative for HPV 6/11 DNA.

The variation in results of HPV DNA analysis obtained by different groups when investigating tumours is at least partly a reflection of the different sensitivities of the detection methods used. Quantitative methods such as DNA in situ hybridization show a marked difference in the concentration of viral DNA, even amongst tumours of the same type such as laryngeal papillomas (Quiney et al. 1989). In some laryngeal papillomas, the concentration is very low, at a level below that of the threshold of detection of some techniques (Abramson et al. 1987). Tumours may be more resistant to treatment if there is a high viral DNA copy number (Quiney et al. 1989).

Theoretically, PCR is 10^5 times as sensitive as DNA in situ hybridisation (L.S. Young et al. 1989), although there may be some reduction in sensitivity when the method is applied to paraffin-embedded material (Griffin et al. 1990). Furthermore, it may appear less sensitive than other methods because of its high specificity, which eliminates the cross-reactions sometimes encountered here (Skyldberg et al. 1991). The fact that PCR has detected HPV DNA in histologically normal tissue not shown to contain virus by other methods suggests that HPV is much more widespread than was previously thought, albeit at a much lower concentration than in most tumours. It may be argued that PCR is simply too sensitive and detects amounts of viral DNA which are of no significance. On the other hand, it seems more likely from studies of respiratory tract papillomatosis (Duggan et al. 1990) and genital condylomata (Del Mistro et al. 1987) that such 'latent' virus acts as a reservoir. Both of these tumours tend to recur, often many years after treatment, and in most cases, the recurrent tumour contains HPV DNA which is identical to that detected in the original lesion. This observation suggests that 'latent' virus is very widespread and is normally kept under control by host defence mechanisms, but occasionally these are overcome, and a neoplasm is formed. Other factors such as smoking or chemical exposure may also be required for this. PCR is the most sensitive method available for the detection of viral DNA, but it has the disadvantage of being only qualitative. When analysing paraffin-embedded material by PCR it is essential to avoid cross-contamination of samples by careful cleaning of the microtome knife and surrounding areas between each cut. It is also advisable to cut at least two empty paraffin blocks between blocks containing samples and include these in the PCR analysis as negative controls. The sensitivity of the PCR should then ideally be combined with a confirmatory and quantitative detection method such as in situ hybridization (Skyldberg et al. 1991).

Papillomaviruses show an obvious affinity for squamous epithelium, including epithelium which has undergone squamous metaplasia. This may indeed be more prone to the development of HPV-associated tumours than native squamous epithelium, as suggested by the much higher frequency of tumours, for example of the cervical transformation zone than of the vulva, vagina and ectocervix (McCance et al. 1987). More than 90% of cervical HPV infections may occur at the transformation zone (Gupta et al. 1989); the metaplastic epithelium at this site is by its very nature unstable.

Histological features associated with subclinical HPV infection of squamous epithelium include acanthosis, lack of cellular glycogenation, hyperkeratosis and parakeratosis. Another feature is the presence within the epithelium of koilocytotic atypia. Koilocytes are squamous epithelial cells showing large clear perinuclear zones and nuclear atypia, often with binucleation (Koss and Durfee 1956). These changes are first seen in the intermediate epithelial layers and later extend to the surface. The association of koilocytes with HPV infection has been proven using an in vivo model. Cervical squamous epithelium infected with HPV from condylomata acuminata was grafted beneath the renal capsules of nude mice, resulting in koilocytosis (Kreider et al. 1985). Many studies have investigated the reliability of koilocytotic atypia in the diagnosis of HPV infection, and this is a controversial subject. Koilocytes appear to be particularly associated with HPV 6 and 11 and also, but probably less often, with 16 and 18, which are more often associated with simple nuclear atypia (zur Hausen 1987; Gupta et al. 1989). Atypical mitotic figures are seen with infections by viruses in both groups (Jenkins et al. 1986). The considerable variation and overlap in histological appearance probably reflects the many possible combinations of infection by multiple HPV types in addition to single virus infections (zur Hausen 1987). The sensitivity and selectivity of the traditional koilocytotic appearance in predicting HPV infection of cells are low. The nuclear atypia seen in koilocytes may correlate better with HPV infection than the cytoplasmic changes (Mittal et al. 1990). Attempts such as the use of a 'wart score' (McLeod 1987) have been made to improve the accuracy of histological prediction of HPV infection.

It is believed that HPV infects the basal cells of squamous epithelium. It probably gains access via small abrasions in the epithelium (Pfister 1987). Viral DNA is detected basally and uniformly throughout all of the other epithelial layers (McCance 1986). As the cells mature and differentiate, they migrate upwards, and it is only in the suprabasal and superficial layers that RNA (Brandwein et al. 1989) and gene products (Sekine et al. 1989) are detectable. It seems that virus maturation is governed by viral promoters and enhancers restricted to mature cells in the superficial epithelium (McCance 1986). Virus replication is therefore restricted to the stratum spinosum and stratum granulosum. These superficial cells are also the source of infection and are protected from an effective host immune response by the lower layers. The strict requirements of these viruses for completion of their cycle have hampered attempts at their in vitro culture (McCance 1986).

HPV infection causes hyperplasia of squamous epithelium. This may result partly from increased cell proliferation in the basal layers, although some workers have found normal rates of mitosis and growth in infected cells after explantation, suggesting that the hyperplasia results from abnormal keratinocyte differentiation and hence delayed exfoliation. Infected cells show abnormal protein production with a reduction in the proportion of cells taking up tritiated thymidine and uridine (Steinberg et al. 1990).

Friedman and Fialkow (1976) showed that genital warts are of multicellular origin by analysis of the glucose-6-phosphate dehydrogenase (G6PD) phenotype of warts from heterozygous women. This may be because of infection of several cells by virus or because of repeated autoinfection by virus shed from the tumour.

Integration of HPV DNA sequences into the host genome may play a role in malignant transformation, and this is seen in infections with HPV types associated with a high malignant potential such as HPV 16, 18, 31 and 33 (McCance 1986). In contrast, the DNA of HPV 6 and 11 apparently remains as an episome within the host cell nucleus, even in malignant tumours (Rando et al. 1986; Kasher and Roman 1988). A different method of oncogenesis may therefore apply in the case of these viruses, although the possibility cannot be excluded that HPV 6 and 11 integrate at less than one copy per cell (Kasher and Roman 1988).

Depending on their location within the viral genome, relatively minor differences in the DNA sequence between different viral types and subtypes may be responsible for major differences in pathogenicity (Pfister 1987). HPV 6vc, which shows high homology with HPV 6b, was first characterised in vulvar verrucous carcinoma (Rando et al. 1986). It shows minor genome alterations in the non-coding upstream regulatory region (URR) with enhancer activity. Such changes in control regions may play a significant role in oncogenesis, probably via increased gene expression (Pfister 1987).

Acknowledgement. We are grateful to Blackwell Scientific Publications for permission to include details of our work.

References

Abramson AL, Steinberg BM, Winkler B (1987) Laryngeal papillomatosis: clinical, histologic and molecular studies. Laryngoscope 97:678–685

Boyle WF, Riggs JL, Oshiro LS, Lennette EH (1973) Electron microscopic identification of papova virus in laryngeal papilloma. Laryngoscope 83:1102–1108

Brandwein M, Steinberg B, Thung S, Biller H, Dilorenzo T, Galli R (1989) Human papillomavirus 6/11 and 16/18 in Schneiderian inverted papillomas. In situ hybridization with human papillomavirus RNA probes. Cancer 63:1708–1713

Braun L, Kashima H, Eggleston J, Shah K (1982) Demonstration of papillomavirus antigen in paraffin section of a laryngeal papilloma. Laryngoscope 92:640–643

Bryan RL, Bevan IS, Crocker J, Young LS (1990) Detection of HPV 6 and 11 in tumours of the upper respiratory tract using the polymerase chain reaction. Clin Otolaryngol 15:177–180

Crum CP, Mitao M, Levine RU, Silverstein S (1985) Cervial papillomaviruses segregate within morphologically distinct precancerous lesions. J Virol 54:675–681

de Villiers EM, Schneider A, Gross G, zur Hausen H (1986) Analysis of benign and malignant urogenital tumours for human papillomavirus infection by labelling cellular DNA. Med Microbiol Immunol (Berl) 174:281–286

Del Mistro A, Braunstein JD, Halwer M, Koss LG (1987) Identification of human papillomavirus types in male urethral condylomata acuminata by in situ hybridisation. Hum Pathol 18:936–940

Doorbar J, Gallimore PH (1987) Identification of the proteins encoded by the L1 and L2 open reading frames of human papillomavirus 1a. J Virol 61:2793–2799

Duggan MA, Lim M, Gill MJ, Inoue M (1990) HPV DNA typing of adult-onset respiratory papillomatosis. Laryngoscope 100:639–642

Firzlaff JM, Kiviat NB, Beckmann AM, Jenison SA, Galloway DA (1988) Detection of human papillomavirus capsid antigens in various squamous epithelial lesions using antibodies directed against the L1 and L2 open reading frames. Virology 164:467–477

Friedman JM, Fialkow PJ (1976) Viral "tumorigenesis" in man: cell markers in condylomata acuminata. Int J Cancer 17:57–61

Gissmann L, zur Hausen H (1980) Partial characterization of viral DNA from human genital warts (condylomata acuminata). Int J Cancer 25:605–609

Gissmann L, de Villiers EM, zur Hausen H (1982a) Analysis of human genital warts (condylomata acuminata) and other genital tumors for human papillomavirus type 6 DNA. Int J Cancer 29:143–146

Gissmann L, Diehl V, Schulz-Coulon HJ, zur Hausen H (1982b) Molecular cloning and characterization of human papillomavirus DNA derived from a laryngeal papilloma. J Virol 44:393–400

Gissmann L, Wolnik L, Ikenberg H, Koldovsky U, Schnurch HG, zur Hausen H (1983) Human papillomavirus types 6 and 11 DNA sequences in genital and laryngeal papillomas and in some cervical cancers. Proc Natl Acad Sci USA 80:560–563

Griffin NR, Bevan IS, Lewis FA, Wells M, Young LS (1990) Demonstration of multiple HPV types in normal cervix and in cervical squamous cell carcinoma using the polymerase chain reaction on paraffin wax embedded material. J Clin Pathol 43:52–56

Grussendorf-Conen EI, Deutz FJ, de Villiers EM (1987) Detection of human papillomavirus-6 in primary carcinoma of the urethra in men. Cancer 60:1832–1835

Gupta JW, Saito K, Saito A, Fu YS, Shah KV (1989) Human papillomaviruses and the pathogenesis of cervical neoplasia. A study by in situ hybridization. Cancer 64:2104–2110

Jenkins D, Tay SK, McCance DJ, Campion MJ, Clarkson PK, Singer A (1986) Histological and immunocytochemical study of cervical intraepithelial neoplasia (CIN) with associated HPV 6 and HPV 16 infections. J Clin Pathol 39:1177–1180

Judd R, Zaki SR, Coffield LM, Evatt BL (1991) Sinonasal papillomas and human papillomavirus: human papillomavirus 11 detected in fungiform Schneiderian papillomas by in situ hybridization and the polymerase chain reaction. Hum Pathol 22:550–556

Kasher MS, Roman A (1988) Characterization of human papillomavirus type 6b DNA isolated from an invasive squamous carcinoma of the vulva. Virology 165:225–233

Kerley SW, Persons DL, Fishback JL (1991) Human papillomavirus and carcinoma of the urinary bladder. Mod Pathol 4:316–319

Klug A, Finch JT (1965) Structure of viruses of the papilloma-polyoma type: 1. Human wart virus. J Mol Biol 11:403–423

Koss LG, Durfee GR (1956) Unusual patterns of squamous epithelium of the uterine cervix: cytologic and pathologic study of koilocytotic atypia. Ann NY Acad Sci 63:1245–1261

Kreider JW, Howett MK, Wolfe SA, Bartlett GL, Zaino RJ, Sedlacek TV, Mortel R (1985) Morphological transformation in vivo of human uterine cervix with papillomavirus from condylomata acuminata. Nature 317:639–641

Kulke R, Gross GE, Pfister H (1989) Duplication of enhancer sequences in human papillomavirus 6 from condylomas of the mamilla. Virology 173:284–290

Loke SL, Ma L, Wong M, Srivastava G, Lo I, Bird CC (1990) Human papillomavirus in oesophageal squamous cell carcinoma. J Clin Pathol 43:909–912

Masood S, Rhatigan RM, Powell S, Thompson J, Rodenroth N (1991) Human papillomavirus in prostatic cancer: no evidence found by in situ DNA hybridization. South Med J 84:235–236

McCance DJ (1986) Human papillomaviruses and cancer. Biochim Biophys Acta 823:195–205

McLellan R, Buscema J, Guerrero E, Shah KV, Woodruff JD, Currie JL (1990) Investigation of ovarian neoplasia of low malignant potential for human papillomavirus. Gynecol Oncol 38:383–385

McLeod K (1987) Prediction of human papilloma virus antigen in cervical squamous epithelium by koilocyte nuclear morphology and "wart scores": confirmation by immunoperoxidase. J Clin Pathol 40:323–328

Melchers WJG, Schift R, Stolz E, Lindeman J, Quint WGV (1989) Human papillomavirus detection in urine samples from male patients by the polymerase chain reaction. J Clin Microbiol 27:1711–1714

Mevorach RA, Cos LR, di-Sant'Agnese PA, Stoler M (1990) Human papillomavirus type 6 in grade I transitional cell carcinoma of the urethra. J Urol 143:126–128

Mittal KR, Chan W, Demopoulos RI (1990) Sensitivity and specificity of various morphological features of cervical condylomas. An in situ hybridization study. Arch Pathol Lab Med 114:1038–1041

O'Brien WM, Jenson AB, Lancaster WD, Maxted WC (1989) Human papillomavirus typing of penile condyloma. J Urol 141:863–865

Oriel JD (1971) Natural history of genital warts. Br J Vener Dis 47:1–13

Oriel JD, Almeida JD (1970) Demonstration of virus particles in human genital warts. Br J Vener Dis 46:37–42

Padel AF, Venning VA, Evans MF, Quantrill AM, Fleming KA (1990) Human papillomaviruses in anogenital warts in children: typing by in situ hybridisation. Br Med J 300:1491–1494

Pfister H (1987) Human papillomaviruses and genital cancer. Adv Cancer Res 48:113–147

Quiney RE, Wells M, Lewis FA, Terry RM, Michaels L, Croft CB (1989) Laryngeal papillomatosis; correlation between severity of disease and presence of HPV 6 and 11 detected by in situ DNA hybridisation. J Clin Pathol 42:694–698

Rando RF, Groff DE, Chirikjian JG, Lancaster WD (1986) Isolation and characterization of a novel human papillomavirus type 6 DNA from an invasive vulvar carcinoma. J Virol 57:353–356

Schwarz E, Durst M, Demankowski C, Lattermann O, Zech R, Wolfspender E, Suhai S, zur Hausen H (1983) DNA sequence and genome organization of genital human papillomavirus type 6b. EMBO J 2:2341–2348

Sedlacek TV, Lindheim S, Eder C, Hasty L, Woodland M, Ludomirsky A, Rando RF (1989) Mechanism for human papillomavirus transmission at birth. Am J Obstet Gynecol 161:55–59

Sekine H, Fuse A, Inaba N, Takamizawa H, Simizu B (1989) Detection of the human papillomavirus 6b E2 gene product in genital condyloma and laryngeal papilloma tissues. Virology 170:92–98

Skyldberg B, Kalantari M, Karki M, Johansson B, Hagmar B, Walaas L (1991) Detection of human papillomavirus infection in tissue blocks by in situ hybridization as compared with a polymerase chain reaction procedure. Hum Pathol 22:578–582

Steinberg BM, Meade R, Kalinowski S, Abramson AL (1990) Abnormal differentiation of human papillomavirus-induced laryngeal papillomas. Arch Otolaryngol Head Neck Surg 116:1167–1171

Terry RM, Lewis FA, Robertson S, Blythe D, Wells M (1989) Juvenile and adult laryngeal papillomata: classification by in situ hybridization for human papillomavirus. Clin Otolaryngol 14:135–139

Tsutsumi K, Nakajima T, Gotoh M, Shimosato Y, Tsunokawa Y, Terada M, Ebihara S, Ono I (1989) In situ hybridization and immunohistochemical study of human papillomavirus infection in adult laryngeal papillomas. Laryngoscope 99:80–85

Weaver MG, Abdul-Karim FW, Dale G, Sorensen K, Huang YT (1990) Outcome in mild and moderate cervical dysplasias related to the presence of specific human papillomavirus types. Mod Pathol 3:679–683

Wright RG, Murthy DP, Gupta AC, Cox N, Cooke RA (1990) Comparative in situ hybridisation of juvenile laryngeal papillomatosis in Papua New Guinea and Australia. J Clin Pathol 43:1023–1025

Young FI, Ward LM, Brown LJR (1991) Absence of human papilloma virus in cervical adenocarcinoma determined by in situ hybridisation. J Clin Pathol 44:340–341

Young LS, Bevan IS, Johnson MA, Blomfield PI, Bromidge T, Maitland NJ, Woodman CBJ (1989) The polymerase chain reaction: a new epidemiological tool for investigating cervical human papillomavirus infection. Br Med J 298:14–18

zur Hausen H (1987) Papillomaviruses in human cancer. Cancer 59:1692–1696

Chapter 17 Detection of JC and BK Viruses in Pathological Specimens by Polymerase Chain Reaction*

Ray R. Arthur

Summary

Detection of the human polyomaviruses JC and BK by the polymerase chain reaction (PCR) has been used to diagnose the neurological disease, progressive multifocal leukoencephalopathy, and to study viruria in immunocompromised and nonimmunosuppressed individuals. The universal nature of PCR technology now makes it possible for many laboratories to diagnose infections produced by these viruses.

Introduction

The human polyomaviruses JC and BK are distributed worldwide, and they infect a high proportion of the population (Brown et al. 1975). JC virus (JVC) is the etiologic agent of progressive multifocal leukoencephalopathy (PML) (Walker and Padgett 1983). Once a rare disease, the incidence of PML has increased substantially because it is a complication of AIDS; estimates of the frequency of PML in this population range from 1% to 4% (Gillespie et al. 1991; Krupp et al. 1985; Snider et al. 1983). Nearly all cases of PML occur in subjects with conditions known to impair T-cell function (Padgett and Walker 1983) and are the result of reactivation of latent infections, presumably in the kidney. The virus reaches the brain by the hematogenous route and lytically infects the myelin-producing oligodendrocytes. Patients with this progressively fatal disease commonly experience limb weakness and cognitive abnormalities; mean survival from the onset of symptoms is 4 months, and nearly 75% die by 6 months (Berger et al. 1987; Padgett and Walker 1983).

Hemorrhagic cystitis in recipients of bone marrow transplants is the most frequent pathological condition associated with BK virus (BKV) (Arthur and Shah 1989). Unlike JCV, BKV does not appear to be neurotropic, and infection and pathology are limited to the urinary tract. BKV has also been associ-

Department of Immunology and Infectious Diseases, School of Hygiene and Public Health, Johns Hopkins University, 615 N. Wolfe Street, Baltimore, MD 21205, USA
* These studies were supported by Public Health Service grant 1P50 AI28748-01 from the National Institute of Allergy and Infectious Diseases.

ated with ureteral stenosis in renal allograft recipients (Gardner et al. 1984). Viruria with JC and BK occurs frequently in patients with conditions that impair immunity, e.g., pregnancy, malignancy, immunosuppressive therapy, tissue and organ transplantation, but it also develops in individuals without any evidence of disease (reviewed in Arthur and Shah 1989).

The primary infections with JCV and BKV occur during childhood and adolescence and produce viremia that results in infection of the kidneys. Persistence in renal tissue of both JCV and BKV has been documented by examining autopsy tissue from immunocompetent subjects (Chesters et al. 1983; Heritage et al. 1981). Symptoms accompanying primary infections, if present, are mild and nonspecific. As described above, the illnesses associated with JCV and BKV are more evident in immunodeficient hosts.

The detection of JCV and BKV in clinical specimens has primarily depended on nonculture techniques because permissive cell lines for the primary isolation of JCV are not widely available, and prolonged periods of incubation (weeks to months) are required to obtain positive JCV and BKV cultures (Hogan et al. 1984). Diagnostic techniques have included cytologic and histologic examination of specimens for characteristic cytoplasmic viral inclusions, electron microscopy study, immunologic techniques to detect viral antigens, and most recently, nucleic acid hybridization techniques for the detection of viral genomes (Arthur and Shah 1988). Serologic techniques are of little diagnostic value since high proportions of the general population are infected, nearly 100% and 70% for BKV and JCV, respectively (Shah et al. 1973; Padgett and Walker 1973), and disease is usually the result of viral reactivation.

We recently described the application of the polymerase chain reaction (PCR) in the detection of polyomaviruses in urine and brain tissue (Arthur et al. 1989). We took advantage of the 75% nucleotide sequence homology between JCV and BKV to select a single pair of primers capable of amplifying both viruses. However, the region flanked by these primers was unique for each virus, and the amplification products could be identified by *Bam*HI cleavage patterns and by hybridization with JCV- and BKV-specific oligonucleotide probes. This sensitive, specific, and rapid assay represents a notable improvement over previously described techniques for diagnosing polyomavirus infections, because it can detect as few as 10–100 genome copies, can differentiate JCV and BKV, and has a short turn-around time (<24 h) between the time of specimen receipt and the final diagnosis.

This chapter describes our additional investigations with the PCR method since our initial report and emphasizes the diagnosis of PML. A preliminary report of viruria in nonimmunocompromised individuals is given, and these findings are contrasted to those in immunosuppressed patients. Finally, a brief account is given of applications of PCR for characterizing the genomes of JCV and BKV variants.

Materials and Methods

Study Subjects and Specimens

In the PML studies, fresh-frozen and paraffin-preserved tissues from HIV-positive subjects with confirmed or suspected PML were examined. Fresh tissues included brain biopsies from two cases, autopsy brain tissue from a third case, and brain, spleen, kidney, adrenal gland, liver, bone marrow, and cerebrospinal fluid (CSF) from a fourth case. The diagnosis of PML was established by histological findings in brain tissue that were consistent with PML or by detection of polyomavirus antigen or virus particles. In addition, brain tissue specimens collected from six suspected PML cases by computed tomography(CT)-guided stereotactic needle biopsy and embedded in paraffin were tested (Silver et al. 1991). Antemortem CSF specimens from eight confirmed PML cases were also examined.

A total of 210 urine samples was collected sequentially from 30 healthy, HIV-seronegative, homosexual males. From each subject, six specimens were collected at 6-month intervals, and a seventh specimen was obtained 5.5 years after the first specimen. The subjects were part of a larger population of men whose ages ranged from 18 to 69 years (median age of 33 years) and who are being studied to ascertain the relationship between HIV infection and polyomavirus shedding. A total of 68 urine samples from immunocompromised patients, predominantly bone marrow transplant recipients, was tested. These specimens were submitted to the Microbiology Laboratories of Johns Hopkins Hospital specifically for BKV and JCV testing and were primarily from patients with hemorrhagic cystitis.

All specimens except the needle biopsies of the brain were stored at $-70\,°C$ prior to testing. DNA was extracted from all tissue and bone marrow specimens. Tissues ($25-30\ mm^3$) were minced, suspended in TE ($50\ mM$ TRIS-HCl, $1\ mM$ EDTA, ethylene diamine tetra-acetic acid, pH 8.0) containing 0.45% NP-40, 0.45% Tween-20, and 100 µg/ml proteinase K, and incubated for 18 h at $37\,°C$. Samples were extracted twice with phenol, once with phenol/chloroform (1:1), and twice with chloroform. The DNA was then precipitated with cold ethanol, air-dried, and resuspended in $50-100$ µl of TE buffer. CSF specimens were tested without any prior treatment. Urine specimens (approximately 15 ml) were centrifuged ($1500 \times g$, 15 min), and the sediments were suspended in an equal volume of PCR buffer and boiled for 10 min. Pellets with gross evidence of blood were washed once with TE to lyse red cells and pelleted in a microcentrifuge before the addition of PCR buffer and boiling. The paraffin-preserved needle biopsies were processed as described by Shibata et al. (1988). Briefly, several $5-10$ µm thin sections were placed in a microcentrifuge tube, deparaffinized with xylene, and centrifuged. The xylene was decanted, and the pellet washed twice with 95% ethanol and desiccated. Two types of negative controls were used. One blank tube for every five specimens was included in all manipulations involving DNA extraction or the processing of paraffin-embedded tissues. These blanks were tested with the samples. In addition, a tube containing water was inserted as every sixth specimen during PCR testing. All specimen processing and preparation of reaction mixtures were performed in laminar flow hoods used exclusively for those purposes.

Polymerase Chain Reaction

PCR was performed with minor modifications of reaction conditions for amplification and methods for analysis of amplification products (Arthur et al. 1989). Briefly, target sequences were amplified in a total reaction volume of 50 µl containing 10 µl of sample DNA, 200 µM of each dNTP, 0.5 µM of oligonucleotide primers PEP-1 (5′-AGTCTTTAGGGTCTTCTACC-3′) and PEP-2 (5′-GGTGCCAACCTATGGAACAG-3′), 1.25 U Taq polymerase (Perkin Elmer Cetus, Norwalk, Conn.), and 1 × reaction buffer (50 mM KCl, 10 mM TRIS-HCl, pH 8.3, 1.5 mM MgCl, and 0.01% wt/vol gelatin). For the testing of paraffin-embedded tissue, the 50-µl reaction mixture was added to the tissue fragments, after they were processed as described above. Reaction mixtures were overlaid with mineral oil and subjected to 40 cycles of amplification on a DNA thermal cycler (Perkin Elmer Cetus). Phase setting was 1 min at 94 °C, 30 s at 42 °C and 1 min at 72 °C per cycle. A 10-min denaturing step (94 °C) was included in the first cycle when testing the deparaffinized tissue specimens.

Amplified JCV and BKV DNA sequences in reaction products were identified by agarose gel electrophoresis and/or by slot-blot hybridization. The JCV and BKV amplified products are 173 and 176 bp, respectively. *Bam*HI digestion of the amplified JCV fragment produces 120- and 53-bp fragments, whereas the amplified region of BKV does not have a *Bam*HI site and is not cleaved. In the agarose gel method, 10 µl of the PCR reaction product (prior to and after treatment with *Bam*HI) was electrophoresed on 4% agarose gels (3% NuSieve + 1% SeaKem; FMC Bioproducts, Rockland, M.) containing ethidium bromide and examined for bands of the appropriate size.

Slot-blot hybridizations were performed as described previously using duplicate filters for JCV and BKV analysis (Arthur et al. 1989). JCV-specific (JEP-2: 5′-TGGGATCCTGTGTTTTCATC-3′) and BKV-specific (BEP-2: 5′-GAGAATCTGCTGTTGCTTCT-3′) oligoprobes of 20 nucleotides in length were substituted for the 40-mer probes described previously. Two ng/ml of oligonucleotide probe (either BEP-2 or JEP-2), end-labeled with [^{32}P]-ATP and T4 polynucleotide kinase, was added and incubated for 1 h at 42 °C. The filter was then washed for 5 min at room temperature in 3 × SSPE (20 × SSPE is 3.6 M NaCl, 200 mM NaH$_2$PO$_4$, pH 7.4 and 20 mM EDTA, pH 7.4), three times for 5 min in 1 × SSPE containing 0.1% sodium dodecyl sulfate (SDS), and once for 10 min at 55 °C in 5 × SSPE with 0.1% SDS.

Autoradiograms were prepared by exposure of filters to Kodak X-omat XAR-5 film with intensifying screens for 1–4 h.

Results and Discussion

Frozen and paraffin-preserved brain tissues from all cases of PML were positive for JCV by PCR. BKV was not detected in any tissue or CSF specimens from these cases. We and others have previously described the amplification of JCV DNA sequences in paraffin-preserved brain tissue from patients with PML (Arthur et al. 1989; Telenti et al. 1990). In addition to brain tissue, JCV was detected in the spleen, adrenal gland, kidney, bone marrow, and CSF from autopsy tissues from one case; the liver tissue was negative. Detection of JCV in multiple organs from PML cases has been described before by Grinnell et al. (1983), who used a less sensitive dot-blot procedure. Of the tissues examined in that study, the brain, kidney, lung, liver, and spleen were positive for JCV in 100%, 78%, 60%, 22%, and 22% of PML cases, respectively. The copy numbers of JCV were the lowest in the liver and spleen. Our detection of JCV in the bone marrow and spleen of the autopsy case is in agreement with the observation of Houff et al. (1988), who used an in situ hybridization technique to demonstrate the presence of JCV in the bone marrow from two PML patients. Spleen tissue from one of these cases was examined and was positive. To our knowledge, the present study represents the first description of JCV in the adrenal gland. Examination of additional tissues will be required to study the distribution of JCV in organs of patients with PML.

All six specimens obtained by CT-guided stereotactic biopsy of patients with hypodense lesions on CT scan were positive for JCV. Histological changes typical of PML were evident in one case, lesions atypical for PML were present in one case, lesions suggestive of PML were found in two, and the findings were nonspecific in two. Immunohistochemical staining for polyomavirus common antigen (Budka and Shah, 1983) confirmed the presence of a papovavirus in the cases with typical atypical and suggestive lesions but was negative in the two patients with nonspecific findings. PCR was instrumental in making the diagnosis in these two patients with clinical and CT scan findings consistent with PML, but whose specimens were collected from areas in which typical cytological changes were not evident. Furthermore, samples of limited size can be tested, which makes PCR particularly attractive for analyzing needle biopsy specimens. Whenever possible, CT-guided stereotactic biopsy is now being used instead of open brain biopsy to reduce the amount of tissue damage.

A potential limitation of examining paraffin-embedded tissue is the inability to amplify DNA in some specimens. The type and length of fixation and the length of storage of the specimens have been shown to degrade or make the DNA inaccessible for amplification (Greer et al. 1991). Using two different primer pairs, Telenti et al. (1990) detected JCV in 20 of 24 paraffin-embedded biopsy and autopsy specimens from PML cases. These authors noted the loss of a positive JCV signal in a specimen which was subjected to prolonged formalin fixation. They also observed that biopsy specimens produced stronger signals than autopsy specimens and attributed this phenomenon to shorter fixation times.

The examination of CSF from PML cases was conducted in an effort to make a definitive diagnosis without the need for the more invasive brain biopsy procedure. Of the eight specimens collected from AIDS patients during life, two (25%) were positive for JCV. The CSF collected from the autopsy case was JCV-positive, but there is some uncertainty about whether this specimen was contaminated with brain tissue during collection. Although the ability to detect JCV in the CSF represents an improvement over previous diagnostic techniques, the low frequency of detection limits the value of making the diagnosis of PML on this basis.

In studying polyomaviruria as measured by PCR, we have extended our initial studies of nonimmunosuppressed subjects and continue to find that approximately 10% of urine samples are positive for JCV and/or BKV. In the present study of sequentially collected specimens from healthy adult males, 8 of 30 (27%) subjects had one or more positive specimens over the 5–6 year period. Overall, 20 of 210 (9.5%) specimens were positive, with 18 and 1 specimens positive for JCV and BKV, respectively. One specimen was positive for both viruses. Assuming 60%–70% of adults are JCV-seropositive and therefore at risk for reactivation, the proportion of JCV-positive subjects observed above suggests that the reactivation rate in this group of individuals was approximately 40%. Five of the eight (62%) subjects had multiple positive specimens; in two of these subjects 4 and 6 of the 7 specimens collected from each individual were JCV-positive. None of the PCR-positive specimens were positive by a less sensitive filter, in situ hybridization (FISH) method (Arthur et al. 1985), indicating that viral replication in these subjects was at a low level. Our observation that JC viruria, and not BK viruria, occurs in a reasonably high proportion of nonimmunosuppressed individuals has also been noted by Kitamura et al. (1990). These authors made similar observations in urology and internal medicine clinic patients using hybridization with ^{32}P-labeled, molecularly cloned, genomic JCV and BKV probes. They reported that the frequency of JCV shedding increased as patients became older. In their study, virus was concentrated from 35-ml volumes of urine by ultracentrifugation and the DNA extracted, cleaved with restriction endonucleases, and analyzed for each virus by gel electrophoresis and Southern transfer hybridization. PCR can be performed in a less cumbersome and more timely manner and requires a smaller sample volume.

In contrast, a much higher proportion of specimens from immunocompromised patients were positive, mostly for BKV. Of 68 urine samples, 42 (62%) were positive, mostly for either JCV or BKV. A total of 26 (38%) was BKV-positive, 5 (7%) were JCV-positive, and 3 (4%) contained both viruses. The presence of BKV in the specimens was expected since the vast majority of samples was taken from bone marrow recipients with hemorrhagic cystitis (Arthur et al. 1986; Apperley et al. 1987). The increased frequency of JCV detection in these recipients, as compared with our previous studies using FISH (Arthur et al. 1988), reflects the increased sensitivity of PCR.

The implications for the use of PCR for the diagnosis of diseases associated with JCV and BKV infections are the following. PCR is a sensitive, specific, and rapid method for the diagnosis of PML when examining brain tissue. At present, there is no evidence that JCV persists in normal brain tissue. If this is true, JCV must be transmitted to the brain after reactivation; therefore, the presence of any detectable virus in brain tissue or CSF may have pathogenic implications. The presence of JCV in multiple organs in patients with PML probably reflects the distribution of hematogenous, reactivated virus. Certainly, detection of JCV in sites where the virus is known to persist, such as the kidney and possibly bone marrow (Houff et al. 1988) or urine has little diagnostic value.

The significance of detecting JCV or BKV in the urine or tissues of the urinary tract is less clear. High-level viruria associated with marrow transplantation and other immunologic impairment can be readily detected by PCR. In nonimmunosuppressed individuals, low-level JCV replication detectable by PCR may occur in approximately 25% of subjects. PCR is a useful diagnostic tool for confirming the presence of BKV in marrow transplant patients with hemorrhagic cystitis, but transplant patients can shed BKV without any evidence of disease (Arthur et al. 1986). Additional studies of immunosuppressed populations that consider the type (JCV or BKV) and level of viral replication will be required to define the pathologic significance of JCV and BKV in urine.

One additional application of PCR for studying JCV and BKV is worthy of mention. Not only does PCR have utility in detecting virus in clinical specimens, but the technique can also be used to characterize genomic variants. The strategy that has been used to study genomic variation is to amplify, directly from urine samples, the virus regulatory region. This region contains the highest diversity of sequences among different strains. This approach avoids the inherent selection and variation associated with virus propagation in cell culture. Sequences for primers are selected from conserved areas of the early and late genes of the virus that flank the noncoding regulatory region. The nucleotide sequence of the amplified product is determined directly by asymmetric PCR or after cloning. Variants of BKV from bone marrow transplant recipients, JCV from renal allograft recipients, and BKV and JCV from persons with AIDS and a variety of other conditions have been examined using this approach (Fleagstad et al. 1991; Negrini et al. 1991; Yogo et al. 1991). Several schemes for typing these viruses have resulted. These studies have led to the description of ancestral prototypic genomes (archetypes) for both JCV and BKV. Using sequence and matrix analysis to compare BKV and JCV archetypes, Negrini et al. (1991) have shown a relationship between the two viruses and have speculated on the functional role of elements in the noncoding region. PCR has significantly increased the ease with which it is possible to obtain data about various genetic variants that can be used to help ascertain their relationship with diseases in man.

Acknowledgments. I thank Keeti Shah for helpful discussions and Justin McArthur and Sheila Gillespie for providing specimens and data from PML patients.

References

Apperley JF, Rice SJ, Bishop JA, Chia YC, Krausz T, Gardner SD, Goldman JM (1987) Late-onset hemorrhagic cystitis associated with urinary excretion of polyomaviruses after bone marrow transplantation. Transplantation 43:108–112

Arthur RR, Shah KV (1988) Papovaviridae: the polyomaviruses. In: Halonen P, Lennette E Jr, Murphy F (eds) The laboratory diagnosis of infectious diseases: principles and practice, vol 2. Springer, Berlin Heidelberg New York, pp 317–332

Arthur RR, Shah KV (1989) Occurrence and significance of papoviruses BK and JC in the urine. In: Melnick J (ed) Progress in medical virology, vol 36. Karger, Basel, pp 42–61

Arthur RR, Beckmann AM, Li CC, Saral R, Shah KV (1985) Detection of the human papovavirus BK in urine of bone marrow transplant recipients: comparison of DNA hybridization with ELISA. J Med Virol 16:29–36

Arthur RR, Shah KV, Baust SJ, Santos GW, Saral R (1986) Association of BK viruria with hemorrhagic cystitis in recipients of bone marrow transplants. New Engl J Med 315:230–234

Arthur RR, Dagostin S, Shah KV (1989) Detection of BKV and JCV in urine and brain tissue by the polymerase chain reaction. J Clin Microbiol 27:1174–1179

Arthur RR, Shah KV, Charache P, Saral R (1988) BK and JC virus infections in recipients of bone marrow transplants. J Infect Dis 158:563–569

Berger JR, Kaszovitz B, Donovan Post MJ, Dickinson G (1987) Clincal review: progressive multifocal leukoencephalopathy associated with human immunodeficiency virus infection. Ann Intern Med 107:78–87

Brown P, Tsai T, Gajdusek DC (1975) Seroepidemiology of human papovaviruses: discovery of virgin populations and some unusual patterns of antibody prevalence among remote peoples of the world. Am J Epidemiol 102:331–340

Budka H, Shah KV (1983) Papovavirus antigens in paraffin sections of PML brains. Clin Biol Res 105:299–309

Chesters PM, Heritage J, McCance DJ (1983) Persistence of DNA sequences of BK virus and JC virus in normal human tissues and in diseased tissues. J Infect Dis 147:676–684

Flaegstad T, Sundsfjord A, Arthur R, Pedersen M, Traavik T, Subramani S (1991) Amplification and sequencing of the control regions of BK and JC virus from human urine by polymerase chain reaction. Virology 180:553–560

Gardner SD, MacKenzie EFD, Smith C, Porter AA (1984) Prospective study of the human polyomaviruses BK and JC and cytomegalovirus in renal transplant recipients. J Clin Pathol 37:578–586

Gillespie SM, Chang Y, Lemp G, Arthur R, Buchbinder S, Steimle A, Baumgartner J, Rando T, Neal D, Rutherford G, Schonberger L, Janssen R (1991) Progressive multifocal leukoencephalopathy in persons infected with Human Immunodeficiency virus, San Francisco, 1981–1989. Ann Neurol 30:597–604

Greer CE, Peterson SL, Kiviat NB, Manos MM (1991) PCR amplification from paraffin-embedded tissues. Effects of fixative and fixation time. Am J Clin Pathol 95:117–124

Grinnell BW, Padgett BL, Walker DL (1983) Distribution of nonintegrated DNA from JC papovavirus in organs of patients with progressive multifocal leukoencephalopathy. J Infect Dis 147:669–675

Heritage J, Chesters PM, McCance DJ (1981) The persistence of papovavirus BK DNA sequences in normal human renal tissue. J Med Virol 8:143–150

Hogan TF, Padgett BL, Walker DL (1984) Human polyomaviruses. In: Belshe RB (ed) Textbook of human virology. PSG Publishing, Littleton, pp 969–995

Houff SA, Major EO, Katz DA, Kufta CV, Sever JL, Pittaluga S, Roberts JR, Gitt J, Saini H, Lux W (1988) Involvement of JC virus-infected mononuclear cells from the bone marrow and spleen in the pathogenesis of progressive multifocal leukoencephalopathy. New Engl J Med 318:301–305

Kitamura T, Aso Y, Kuniyoshi N, Hara K, Yogo Y (1990) High incidence of urinary JC virus excretion in nonimmunosuppressed older patients. J Infect Dis 161:1128–1133

Krupp LB, Lipton RB, Swerdlow ML, Leeds NE, Llena J (1985) Progressive multifocal leukoencephalopathy: clinical and radiographic features. Ann Neurol 17:344–349

Negrini M, Sabbioni S, Arthur RR, Castagnoli A, Barbanti-Brodano G (1991) Prevalence of the archetypal regulatory region and sequence polymorphisms in nonpassaged BK virus variants. J Virol 65:5092–5095

Padgett BL, Walker DL (1973) Prevalence of antibodies in human sera against JC virus, an isolate from a case of progressive multifocal leukoencephalopathy. J Infect Dis 127:467–470

Padgett BL, Walker DL (1983) Virological and serologic studies of progressive multifocal leukoencephalopathy. Clin Biol Res 105:107–117

Shah KV, Daniel RW, Warszwaski R (1973) High prevalence of antibodies of BK virus, an SV40 related papovavirus, in residents of Maryland. J Infect Dis 128:784–787

Shibata DK, Arnheim N, Martin WJ (1988) Detection of human papilloma virus in paraffin-embedded tissue using the polymerase chain reaction. J Exp Med 167:225–230

Snider WD, Simpson DM, Nielsen S, Gold JW, Metroka CE, Posner JB (1983) Neurological complications of the acquired immune deficiency syndrome: analysis of 50 patients. Ann Neurol 14:403–418

Silver SA, Arthur RR, Erozan YS, Sherman ME, McArthur JC, Uematsu S (1991) Diagnosis of progressive multifocal leukoencephalopathy by computed tomography-guided stereotactic biopsy using cytopathologic and histopathologic examination in combination with immunostaining for SV-40 related antigen and the polymerase chain reaction for JC virus. Abtract, American Society of Cytology, Los Angeles

Telenti A, Aksamit AJ Jr, Proper J, Smith TF (1990) Detection of JC virus DNA by polymerase chain reaction in patients with progressive multifocal leukoencephalopathy. J Infect Dis 162:858–861

Walker D, Padgett BL (1983) Progressive multifocal leukoencephalopathy, In: Fraenkel-Conrat H, Wagner RR (eds) Comprehensive virology, vol 18. Plenum, New York, pp 161–193

Yogo Y, Kitamura T, Sugimoto C, Hara K, Iida T, Taguchi F, Tajima S, Kawabe K, Aso Y (1991) Sequence arrangement in JC virus DNAs molecularly cloned from immunosuppressed renal transplant patients. J Virol 65:2422–2428

Section V Airborne and Respiratory Viruses

Chapter 18 Diagnosis of Prenatal Rubella by Polymerase Chain Reaction

Linda Ho and George Terry

Summary

First trimester rubella carries a high risk of congenital abnormalities in the fetus. Evidence of an intrauterine infection is therefore of value for prenatal counselling. We present data to show that the polymerase chain reaction has advantages over other conventional techniques in confirming such a diagnosis, but problems remain in its application, and these are also discussed.

Introduction

Maternal viral infection during pregnancy is known frequently to involve the fetus. The incidence of fetal infection and the risk of congenital deformities are particularly high with rubella (Miller et al. 1982). When fetal infection is suspected, a direct intrauterine diagnosis can provide invaluable information for prenatal counselling. This is now possible because of the availability of fetal specimens which can be obtained with relatively low risk in suitably equipped and staffed obstetrics centres.

Several approaches have been adopted to diagnose fetal rubella infection. One involves the detection of specific immunoglobulin (IgM) in fetal blood by standard immunoassays (Daffos et al. 1984; Enders 1987; Ho-Terry et al. 1988). However, since a fetus does not usually develop IgM antibodies at a detectable level until the 18th–20th weeks of gestation, this method of diagnosis is not satisfactory in the care of women who are infected early in pregnancy when the risk of fetal deformities is highest (Miller et al. 1982). Viral infections can affect the development of the immunological system in utero and transplacental maternal IgG antibodies are also liable to modulate the fetal immunological response. False-negative results are sometimes obtained when fetal rubella is diagnosed in this way (Daffos et al. 1984; Enders 1987).

As an alternative, fetal rubella infection can be diagnosed by the detection of infectious virus in products of conception, e.g., chorionic villus samples

Department of Medical Microbiology and Department of Chemical Pathology, Faculty of Clinical Sciences, University College London, 46 Cleveland St., London W1P 6DB, UK

(CVS) and amniotic fluid during pregnancy or placental and fetal tissues after therapeutic abortion. This method of diagnosis has been carried out successfully in a limited number of laboratories, but in general the rate of rubella virus isolation after first trimester infection is variable (68%–90%) and dependent on viral load (Cooper et al. 1968; Rawls et al. 1968). A missed diagnosis of first trimester rubella presents a serious risk of delivery of an infant with congenital rubella defects.

We have examined aborted specimens from women with first trimester rubella by virus isolation. These specimens were cocultivated with VERO, African green monkey kidney cells for 4–5 days under standard laboratory conditions (Herrmann 1979). Any rubella virus, if present, is "biologically amplified". Virus released into the inoculated medium was detected in RK13, rabbit kidney cells by histochemical staining of infected cells with a mixture of rubella monoclonal antibodies (Turner 1986). In addition, RNA from the same culture was analysed by Northern blotting using a ^{32}P-labelled cloned rubella cDNA probe (Terry et al. 1986).

Results obtained from this study identified two categories of specimens with evidence of rubella virus infection. Neither of these categories show organ specificity. Category 1 (19/40) represents positive isolates which yield infectious rubella virus and rubella-specific RNA (40S genomic RNA and/or 24S mRNA) in inoculated cells (Fig. 1 a, lanes 1 and 5). Categorgy 2 (21/40) represents specimens from which no infectious virus can be isolated but which contain rubella-specific subgenomic and sub-24S RNA species demonstrable by Northern blotting (Fig. 1 a, lane 3). The precise mechanism of generation of subgenomic or sub-24S RNA species in rubella-infected human tissue is not known. However, virus particles with subgenomic RNA species frequently develop in rubella-infected tissue culture systems (Terry et al. 1986; Frey and Hemphill 1988). These defective interfering (DI) particles are not infectious but are known to interfere with the infectivity of intact particles.

To try to identify specimens which have been infected with rubella but which may contain very low levels of intact rubella RNA or which are not capable of being biologically amplified due to interference by DI particles, we have employed hybridisation of RNA extracted from specimens with a ^{32}P-labelled, cloned, rubella cDNA probe. Reconstruction experiments show that 1–2 pg of rubella-specific RNA can be detected in this way. This represents about 1000 infected cells. Assuming that between 1/1000 and 1/250 000 fetal cells are infected in intrauterine rubella infection (Rawls et al. 1968), this quantity of rubella RNA would require a specimen containing 10^6 to 2.5×10^8 fetal cells, equivalent to a wet weight of 3–750 mg of tissue. Although this poses no problem for retrospective studies using aborted specimens, it is a serious limitation for prenatal diagnosis since the average wet weight of a CVS is about 20–30 mg, so that a positive result will be obtained only in specimens with a relatively high rate of rubella infection, i.e., at least 1 in 10 000 fetal cells. About 30% of CVS made available for our study fall short of 20 mg wet weight. It is therefore evident that diagnosis of intrauterine rubella infection by detection of specific RNA would be more efficient and reliable if viral se-

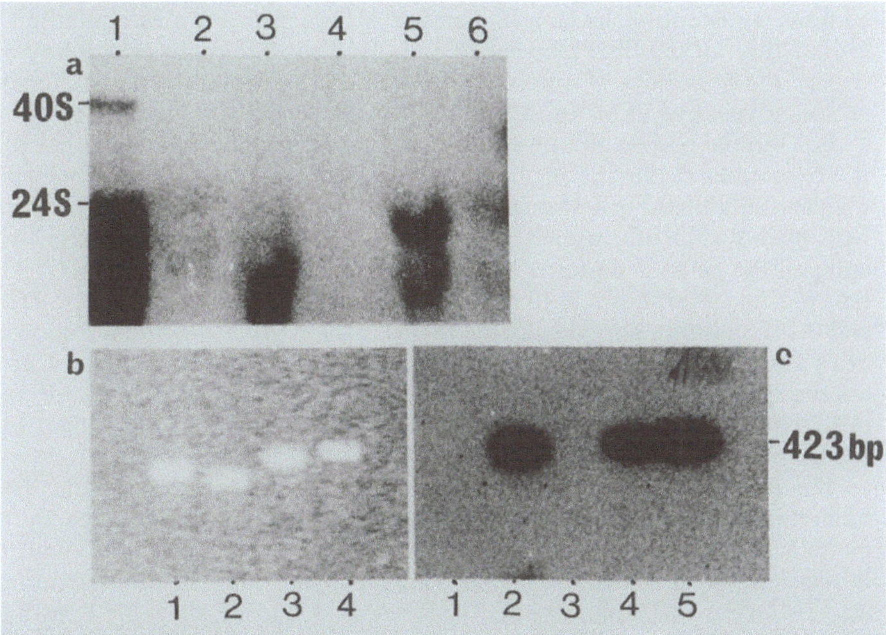

Fig. 1. a Northern blotting of RNA extracted from VERO cells cocultivated with tissues from rubella-infected fetuses. *Lanes 1, 3, 5* are the inoculated cultures, and *lanes 2, 4, 6* are the corresponding uninoculated controls. **b** Ethidium bromide-stained polymerase chain reaction (PCR) product generated by primers R1 and R2 (*lane 4*). *Lanes 1–3* represent PCR product restricted with *Bam*H1 alone, *Bam*H1 and *Xho*1 or *Xho*1 alone, respectively. **c** Southern blotting of PCR product generated by primers R1 and R2 with ^{32}P-labelled insert cDNA probe. *Lanes 1 and 2* represent an aborted specimen tested without and with reverse transcriptase. *Lanes 3 and 4* represent a chorionic villus sample tested without and with reverse transcriptase. *Lane 5* represents a PCR marker generated by primers R1 and R2 (423 bp)

quences in the clinical specimen could be amplified by polymerase chain reaction (PCR) prior to specific identification by hybridisation.

Materials and Methods

Sample Preparation

Whereas the sensitivity of PCR in the detection of DNA or RNA (after reverse transcription) is never in doubt, its application to the examination of clinical tissue specimens can present problems. In rubella, the infection is usually focal so that detection of rubella-specific RNA sequences in a placenta or an aborted fetus is susceptible to sampling errors. Although this can be minimised by examination of multiple biopsies (e.g. 5–10 from different parts of any one tissue), the possibility of a false-negative result cannot be ruled out. For this reason, a negative PCR result cannot exclude rubella virus infection.

For sample preparation, we use guanidinium isothiocyanate and hot phenol extraction (Maniatis et al. 1982). Briefly, the tissue to be examined is weighed, solubilised in 4 M guanidinium isothiocyanate at 60 °C and extracted with phenol-chloroform and chloroform at 60 °C. After precipitation with ethanol, the pellet is digested with proteinase K (200 µg/ml) in 0.1 M TRIS-HCl pH 7.4, 50 mM NaCl, 10 mM ethylene diamine tetra-acetic acid (EDTA) and 0.2% sodium dodecyl sulphate (SDS) at 37 °C for 30–60 min. Following extraction with phenol-chloroform and chloroform at 60 °C, the nucleic acids are precipitated with ethanol, washed twice with 75% ethanol to ensure complete removal of SDS and resuspended in 300 µl of 10 mM TRIS-HCl, 1 mM EDTA pH 8.0 (TE) containing 100 units/ml of RNasin.

The RNA is purified by selective elution from an anionic exchange resin immobilised in a micropipette tip using the protocol recommended by the manufacturer (Diagen). A Diagen tip 20 is equilibrated to binding conditions by washing with 400 mM NaCl, 50 mM 3(N-morpholino)propanesulfonic acid (MOPS), 1% ethanol, pH 7.0. RNA in 300 µl of TE with RNasin, a ribonuclease inhibitor (prepared as described above) is adjusted to 250 mM NaCl and 35 mM MOPS, pH 7.0, and applied to the resin. After extensive washing of the resin with equilibration buffer to remove proteins, nucleotides, and other impurities, RNA is eluted in 1.05 M NaCl, 50 mM MOPS, 2 M urea, 15% ethanol, pH 7.0. DNA remains bound to the resin. RNA is precipitated with isopropanol at -20 °C. After washing twice with 75% ethanol, the RNA is taken up in sterile water at a rate of 5 µl per 10 mg tissue. Recovery of RNA from Diagen tips is approximately 70%–80%. The final concentration of RNA obtained is 2–5 µg/ul.

Primers

Diagnostic Primers. Previously, we determined the complete nucleotide sequence of the region encoding the envelope protein E1 of rubella virus strain Judith and located the coding sequence for a major group of antigenic determinants between nucleotides 8984 and 9108 (Terry et al. 1988; Dominguez et al. 1990). The sequences flanking this region are highly conserved in various wild-type strains (Frey et al. 1986; Clarke et al. 1987; Vidgren et al. 1987; Nakhasi et al. 1989) and are likely to be present in most rubella virus-infected clinical specimens. Specific primers were designed to amplify this segment of the coding sequence by PCR. The sequences are located in the conserved regions close to useful restriction sites and are as follows:

R1: 5′-AACTTCAGCCCCAAGGGGCC-3′ (complementary to nucleotides 9210–9229, numbered according to Dominguez et al. 1990)

R2: 5'-CAACACGCCGCACGGACAAC-3' (identical to nucleotides 8807–8826

The size of the target sequence is 423 bp (Fig. 1 b, lane 4).

Confirmatory Primers. A positive PCR signal obtained by using R1 and R2 is confirmed by a second pair of primers, R3 and R4:

R3: 5'-GGGCCCTCCGGAGTCACGGC-3' (complementary to nucleotides 6664–6683)
R4: 5'-GTTTCGCCGCATCTCGTGGG-3' (identical to nucleotides 6446–6465)

The size of the target sequence is 238 bp.

Southern Blotting

All PCR products generated by primer pair R1 and R2 are electrophoresed on a 2% agarose gel and blotted onto Zeta-probe membrane (BioRad). PCR products with rubella specificity are identified by hybridisation with a cloned insert cDNA probe (comprising nucleotides 8827–9173), labelled by primer extension with [^{32}P-]dCTP (Feinberg and Vogelstein 1983). This probe recognises the target sequence but does not include the primer sequences. Hybridisation is at 42°C in 50% formamide for 16 h. The filter is washed at high stringency (68°C in 0.1 × SSC and 0.1% SDS) for 2 h and autoradiographed (Fig. 1 c) using an intensifying screen for 3–72 h. Digoxigenin-labelled probe (labelling carried out as recommended by the manufacturer, BCL) can also be used.

Reverse Transcription

Rubella virus cDNA is synthesised by reverse transcription (RT). A 2 µl sample of extracted RNA is denatured in 10 µl of 10 m*M* HEPES pH 6.9, 0.1 m*M* EDTA and 5 µ*M* hexanucleotide random primer (Pharmacia) at 90°C for 2 min. The sample is quenched in ice, and Moloney murine leukaemia virus (M-MuLV) reverse transcriptase (Pharmacia; 20 units in 10 µl of 100 m*M* TRIS-HCl, pH 7.5, 150 m*M* KCl, 6 m*M* MgCl$_2$, 20 m*M* dithiothreitol (DTT), 1 m*M* of each dNTP and 20 units of RNasin is added at room temperature, and further incubation is done at 37°C for 90 min. The reaction mixture is extracted with phenol-chloroform and chloroform and used directly as cDNA in the PCR without precipitation.

Every RT run includes (a) negative controls with no RNA or RNA extracted from normal placenta, (b) positive controls with RNA extracted from purified rubella virus or rubella-infected cells (Fig. 1 c, lane 5) or RNA transcribed from a vector (pSP18) carrying a cloned rubella E1 cDNA insert and

(c) identical reactions but with reverse transcriptase omitted, which are carried out for all RNA preparations (Fig. 1c, lanes 1 and 3).

Polymerase Chain Reaction

Our PCR protocol employs the heat-stable Taq polymerase (Saiki et al. 1988), and the final conditions are 2.5 µl cDNA, 10 mM TRIS-HCl, pH 8.3 (at 25 °C), 50 mM KCl, 1.5 mM MgCl$_2$, 0.01% gelatin, 25 pmol each of R1 and R2 primers (or R3 and R4 primers) and 1.25 units of Taq polymerase (Perkin-Elmer/Cetus) in a final volume of 50 µl. Initial denaturation is at 99 °C for 3.5 min. Enzyme is added at 70 °C. Amplification is usually 40 cycles of denaturation at 95 °C for 15 s followed by annealing/extension at 70 °C for 100 s. The final extension at 70 °C is for 8 min. Thermal cycling is carried out on an Intelligent Heating Block (Hybaid), and the times represent actual time at the given temperature.

Results

The detection of rubella virus infection in prenatal and aborted specimens is shown in Table 1. PCR was carried out using RNA extracted either directly from the specimens (i.e. indicative of rubella virus infection) or from VERO cells cocultivated with the specimens (i.e. indicative of rubella virus infection and the presence of infectious rubella virus). Rubella specificity was confirmed by Southern blotting. The combination of these techniques is capable of detecting 1/100th of the rubella RNA present in one infected baby hamster kidney (BHK) cell.

The results obtained from 20 patients (45 specimens) with serologically confirmed first trimester rubella are shown in Table 1a. Prenatal rubella infection (positive PCR in specimen) was found in 80% of pregnancies (patients 5–20). Some 75% of these (patients 5–16) were positive for infectious virus in the products of conception (positive PCR in cocultivated cells) and could have been identified by virus isolation. However, diagnosis by virus isolation would not have been possible in 25% (patients 17–20).

The products of conception from 14 patients (26 specimens) who either had other viral infections or rubella but not during the first trimester of pregnancy show no evidence of rubella infection (Table 1b).

Discussion

The protocol above describes a method (which is under constant review) of detection of rubella-specific RNA by both target amplification (PCR) and

Table 1. Detection of rubella-specific nucleic acid sequences by polymerase chain reaction (PCR). The specimens examined include chorionic villus samples, cells concentrated from amniotic fluid, aborted fetal organs and aborted placentas. **a** Patients with serologically confirmed first trimester rubella where testing for rubella RNA sequences by direct PCR and by PCR after cocultivation showed (A) concordant negative results; (B) concordant positive results; (C) concordant positive results with respect to patients but not in all specimens; (D) discordant results, positive by direct PCR but negative by PCR after cocultivation. **b** Patients with viral infections other than rebella (E) or with rubella infection but not during the first trimester of pregnancy (F, G, H)

Patient		Types of infection	No. of speci-mens exa-mined per patient	Direct PCR		PCR after co-cultivation	
Cate-gory	Nos.			Posi-tive	Nega-tive	Posi-tive	Nega-tive
a A	1	Rubella	3	0	3	0	3
	2		5	0	5	0	5
	3–4		1	0	1	0	1
B	5–12	Rubella	1	1	0	1	0
C	13	Rubella	3	1	2	3	0
	14		6	1	5	2	4
	15		2	2	0	1	1
	16		2	1	1	2	0
D	17	Rubella	11	1	10	0	11
	18–20		1	1	0	0	1
b E	1–4	Infection not related to rubella	1	0	1	0	1
F	5–6	Second trimester rubella infection	1	0	1	0	1
G	7–9	Reinfection	1	0	1	0	1
	10		8	0	8	0	8
H	11	Asymptomatic infection	3	0	3	0	3
	12	Contact	4	0	4	0	4
	13	Preconception	1	0	1	0	1
	14	Vaccination	1	0	1	0	1

signal amplification (Southern blotting). An advantage of the test is that it should be very robust against adverse handling of specimens. Rubella virus is thermolabile, and at 37 °C the half-life for infectivity is estimated to be 70 min. This property probably contributes to the number of clinical specimens seen with detectable rubella RNA species but no infectious rubella virus. A large measure of protection against non-ideal handling of specimens is provided by using the PCR since the RNA target size for loss of infectivity (e.g. RNase activity) is about 9800 bases (i.e. the intact genome), but for PCR is only 423 bases with the primers used here. This can probably be reduced to 100 bases by a different choice of primers.

To ensure an accurate diagnosis, the specificity of the PCR product is monitored by the following methods:

1. The size of the amplified fragment. In most cases the DNA band is directly visible by ethidium bromide staining (Fig. 1 b, lane 4). However, if RNA is used in vast excess or if the host DNA is not removed from the specimen before RT or high cycle numbers are employed to maximise sensitivity, non-specific amplified products accumulate, and further confirmatory tests are required.
2. Southern blot hybridisation with a rubella virus-specific probe which recognises the target sequences but does not include the primer sequences (Fig. 1 c, lanes 2, 4 and 5).
3. Testing for the presence and position of known restriction sites in the amplified target sequence. This assumes that the restriction sites will be present in the virus isolate under test. In the protocol described, digestion of the 423-bp target with *Bam*H1 alone, *Bam*H1 + *Xho*1 or *Xho*1 alone yields fragments of 367 bp, 347 bp and 403 bp which are distinguishable on a 2% agarose gel (Fig. 1 b, lanes 1, 2 and 3).
4. The inclusion of control reactions with reverse transcriptase omitted and reactions containing negative and positive templates in each PCR run in order to minimise incorrect results (Fig. 1 c, lanes 1 and 3).

Although rubella virus is antigenically stable, variants with different growth characteristics and with nucleic acid sequence variations have been observed. The primers we used have been found to be satisfactory for the detection of rubella sequences from two vaccine strains (RA27/3 and HPV77), three wild-type strains isolated in our laboratory and a laboratory-adapted strain (strain Judith) and would therefore be expected to detect most field viruses.

The results presented in Table 1 show that PCR can provide a diagnosis of prenatal rubella. However, due consideration should be given before the test is carried out because (a) fetal sampling involves invasive techniques with a risk of fetal loss (Medical Research Council 1991) and (b) diagnosis once established carries grave consequences for the fetus.

It is evident that a positive detection of rubella nucleic acid sequences is useful for immediate prenatal counselling. However, a negative diagnosis should entail further clinical and virological follow-up. In the laboratory, amniotic fluid can be obtained at 20–22 weeks of gestation and the cells tested for rubella-specific nucleic acid sequences, with and without cocultivation, by PCR. A sample of fetal blood can also be obtained simultaneously for estimation of total fetal IgM and detection of fetal rubella-specific IgM.

Using these necessarily elaborate procedures, PCR has been found to be a valuable tool in the investigation of prenatal rubella under the following circumstances.

1. Patients with inconclusive IgM detection, i.e. virus-specific IgM at equivocal levels or non-specific reactions of IgM with positive and negative antigens

2. Patients with preconception rubella
3. Individuals who become pregnant within 1 month of vaccination
4. Patients who are not vaccinated (quite frequently women between the age of 35–45 years with unknown immune status) and develop first trimester rubella
5. Patients with history of vaccination who develop virus-specific IgM on reinfection
6. Patients who have previously been found to be naturally immune and develop virus-specific IgM on reinfection

From the feedback we have received, our positive diagnoses have provided some assistance to patients and clinicians in deciding on the course of the pregnancy. Our negative diagnoses have also provided a small measure of relief from uncertainty in those women who subsequently go on to deliver healthy infants. No amount of effort in the laboratory can match the courage of these mothers.

References

Clarke DM, Loo TW, Hui I, Chong P, Gillam S (1987) Nucleotide sequence and in vivo expression of rubella virus 24S subgenomic messenger RNA encoding the structural proteins E1, E2 and C. Nucleic Acids Res 15:3041–3057

Cooper LZ, Fedun BA, Matters BA, Krugman S (1968) Maternal rubella and the risk to the fetus. International symposium on rubella vaccines. Symp Series Immunobiol Standard 11:73–78

Daffos F, Granteot-Keros L, Lebon P, Forestier F, Pavlovsky MC, Chartier M, Pillot J (1984) Prenatal diagnosis of congenital rubella. Lancet ii:1–3

Dominguez G, Wang C-Y, Frey TK (1990) Sequence of the genome RNA of rubella virus: evidence for genetic rearrangement during togavirus evolution. Virology 177:225–238

Enders G (1987) Prenatal diagnosis of intrauterine rubella. European Journal for the Clinical Study and Treatment of Infection 15:162–164

Enders G, Knotek F (1989) Rubella IgG total antibody avidity and IgG subclass-specific antibody avidity assay and their role in the differentiation between primary rubella and rubella reinfection. Infection 17:218–226

Feinberg AP, Vogelstein B (1983) A technique for radiolabelling DNA restriction endonuclease fragments to high specific activity. Anal Biochem 132:6–13

Frey TK, Hemphill ML (1988) Generation of defective-interfering particles by rubella virus in VERO cells. Virology 164:22–29

Frey TK, Marr LD, Hemphill ML, Dominguez G (1986) Molecular cloning and sequencing of the region of the rubella virus genome coding for glycoprotein E1. Virology 154:228–232

Herrmann KL (1979) Rubella virus. In: Lennette EH, Schmidt NJ (eds) Diagnostic procedures for viral, rickettsial and chlamydial infection, 5th edn. American Public Health Association, Washington DC, pp 725–766

Ho-Terry L, Terry GM, Londesborough P, Rees KR, Wielaard F, Denissen A (1988) Diagnosis of fetal rubella infection by nucleic acid hybridization. J Med Virol 24:75–182

Maniatis T, Frisch EF, Sambrook J (1982) Molecular cloning. A laboratory manual. Cold Spring Harbor Laboratory, Cold Spring Harbor

Medical Research Council (1991) European trial of chorion villus sampling. MCR working party on the evaluation of chorion villus sampling. Lancet 337:1491–1499

Miller E, Cradock-Watson JE, Pollock TM (1982) Consequences of confirmed maternal rubella at successive stages of pregnancy. Lancet ii:781–784

Nakhasi HL, Thomas D, Zheng D, Liu TY (1989) Nucleotide sequence of capsid, E2 and E1 protein genes of rubella virus vaccine strain RA27/3. Nucleic Acids Res 17:4393–4394

Rawls WE, Desmyter J, Melnick JL (1968) Serological diagnosis and fetal involvement in maternal rubella. JAMA 203:627–631

Saiki RK, Gelfand DH, Stoffel S, Scharf SJ, Higuchi R, Horn GT, Mullis KB, Ehrlich HA (1988) Primer-directed enzymatic amplification of DNA with a thermostable DNA polymerase. Science 239:487–491

Terry GM, Ho-Terry L, Cohen A, Londesborough P (1985) Rubella virus RNA: effect of high multiplicity passage. Arch Virol 86:29–36

Terry GM, Ho-Terry L, Warren RC, Rodeck CH, Cohen A, Rees KR (1986) First trimester prenatal diagnosis of congenital rubella: a laboratory investigation. BMJ 292:930–933

Terry GM, Ho-Terry L, Londesborough P, Rees KR (1988) Localization of the rubella E1 epitopes. Arch Virol 98:189–197

Turner BM (1986) Use of alkaline phosphatase-conjugated antibodies for detection of protein antigens on nitrocellulose filters. Methods Enzymol 121:848–855

Vidgren G, Takkinen K, Kalkkinen N, Kaariainen L, Pettersson RF (1987) Nucleotide sequence of the genes coding for the membrane glycoproteins E1 and E2 of rubella virus. J Gen Virol 68:2347–2357

Chapter 19 Detection of Measles Virus in Subacute Sclerosing Panencephalitis Brain Tissue

Mark S. Godec

Summary

Subacute sclerosing panencephalitis (SSPE) is a progressive human neurologic disorder caused by reactivation of latent measles virus in the central nervous system (CNS). Prevalence of the disorder is linked to the age, sex, race and measles vaccination status of affected individuals. Onset is gradual, with subtle clinical symptoms initially and diffuse neurologic signs later. Death ensues within months to years. Pathologically, the brain shows widespread but patchy inflammation, necrosis and gliosis with inclusion bodies in the gray and white matter. Diagnosis of SSPE is made through measurement of elevated titers of anti-measles antibodies in the serum and cerebrospinal fluid. Adaptation of polymerase chain reaction (PCR) to RNA genomic systems by the application of reverse transcriptase as a preliminary step (RT/PCR) provides a novel method of detecting measles genome in SSPE tissue with a high degree of sensitivity and specificity. Using primer pairs designed to amplify segments of all 5 major structural protein genes of measles virus, RT/PCR was used to amplify these genes in RNA extracted from frozen and formalin-fixed, paraffin-embedded SSPE brain tissue but not in RNA from control brain tissue. The sensitivity of this technique is enhanced by using internal (nesting) primers and a second round of amplification. The products generated by RT/PCR may be sequenced directly after minimal processing to increase the information obtained using this method.

Introduction

Subacute sclerosing panencephalitis (SSPE) is a slowly progressive neurological disorder caused by measles virus persisting and replicating in the CNS (Sever 1983). A viral etiology for SSPE was first proposed in 1933 by Dawson who observed intraneuronal inclusions in patients suffering from the disorder (Dawson 1933). This hypothesis was confirmed in 1969 when a measles-like

Laboratory of Central Nervous System Studies, National Institute of Neurological Disorders and Stroke, National Institutes of Health, Rm. 5B21, Bldg. 36, Bethesda, MD 20892, USA

agent was isolated by cocultivation of neural tissue from SSPE patients with continuous cell lines (Chen et al. 1969; Horta-Barbosa et al. 1969; Payne and Baublis 1969). Subsequently, many studies using a variety of techniques have explored the role of measles virus, an RNA virus, in the pathogenesis of SSPE (Wechsler and Meissner 1982). This disorder is a paradigm of a chronic human neurological disorder caused by the reactivation or persistence of an RNA virus.

SSPE is worldwide in distribution and affects primarily children and adolescents, although a case has been described in a 32-year-old individual (Cape et al. 1973). The average age of onset is approximately 14 years (Dyken 1985). The risk of SSPE is increased by measles infection under the age of 2 years, by living in rural areas, by sex (boys are more than twice as likely as girls to develop the disease), and by race (whites are 4 times as likely as blacks to develop the disease (Halsey et al. 1980; Modlin et al. 1979). Immunization with attenuated measles vaccine reduces the risk of developing SSPE tenfold (Centers for Disease Control 1982). Survival ranges from a few months to a few years depending on the type of SSPE. Patients with classic SSPE survive 1–3 years, while 10% of patients with SSPE have a rapidly progressive course averaging 3 months' duration, and another 10% have a slower course of disease lasting 4–8 years (Jabbour et al. 1975).

Clinically (Jabbour et al. 1969), SSPE presents in a subtle fashion with the affected individual initially demonstrating mild alterations in behavior and cognition which become progressively worse until the individual is completely demented. The course of the disease ranges from acute to chronic as noted above, but the typical signs of encephalitis are absent. Neurologic signs are diffuse and may include bilateral myoclonic jerks, dystonias, ataxia, generalized convulsions, and visual disturbances. Death usually occurs from diseases such as pneumonia which accompany chronically debilitating illnesses.

SSPE is diagnosed by the measurement of elevated blood and CSF titers of antibodies against measles virus (Connolly et al. 1967) and by EEG abnormalities (Cobb 1966). The differential diagnosis includes diseases of childhood which present with an alteration of mental status such as progressive rubella panencephalitis, cerebral storage diseases, Alper's disease, leukodystrophies, and demyelinating disorders of childhood (Asher 1991). No treatment reverses the course of the disease, though inosiplex, in conjunction with supportive therapy, prolongs its course (DuRant and Dyken 1983; Jones et al. 1982).

Pathologically, the brain affected by SSPE shows diffuse areas of inflammation in both the white and gray matter, hence the term panencephalitis. Distribution of the virus and inflammation is not uniform; thus, it is possible that a single brain biopsy may not demonstrate the disease (Budka et al. 1982). Perivascular cuffing, neuronal loss, and demyelination with remyelination become prominent as the disease progresses (Jabbour et al. 1969). Measles antigen can be found by immunocytochemistry analysis in neurons and oligodendroglial cells (Budka et al. 1982). Electron microscopy reveals intranuclear tubular structures which resemble paramyxovirus nucleocapsids (Bouteille et al. 1965).

The development of the polymerase chain reaction (PCR) by Kary Mullis at Cetus Corporation has provided a unique and powerful tool for the detection of viral genomes in a variety of human disorders (Mullis et al. 1986; Saiki et al. 1988). Adaptation of this technique to RNA by the addition of reverse transcriptase and an RNAse inhibitor to the PCR reaction mixture allows the detection of RNA viruses by conversion of the RNA genome to cDNA on which Taq polymerase can operate (Godec et al. 1990). In vitro amplification of selected cDNA sequences followed by confirmation of the PCR product identity by restriction enzyme analysis, hybridization with a labelled probe, or sequencing of the PCR product allows a specific and sensitive system of viral detection. The use of internal or nested primers operating on the first PCR product in a second round of amplification improves PCR sensitivity by approximately 3 orders of magnitude (logarithmic) (Godec et al. 1992).

Materials and Methods

Control measles RNA was obtained by culturing the Edmonston strain of measles virus (American Type Culture Collection) in Vero E6 cells. Frozen SSPE and control brain tissues (not age-matched) were obtained from the Laboratory of Central Nervous System Studies (LCNSS) specimen collection and the National Neurological Research Bank. Formalin-fixed and paraffin-embedded SSPE brain tissues were provided by the Psychoneurological Institute of Warsaw, Poland. The SSPE cases were confirmed by clinical course and neuropathological examination, and most had markedly elevated titers of antibodies to measles in the serum and CSF. Table 1 gives the ages and sexes of the patients as well as descriptions of the specimens used in this study.

Thin sections were cut from formalin-fixed, paraffin-embedded brain tissues, deparaffinized, and labelled immunocytochemically for measles virus antigen using rabbit anti-measles serum (1:2000 dilution) followed by biotinylated antirabbit IgG, peroxidase-conjugated avidin-biotin complex (ABC, Vector Labs), diaminobenzidine (DAB), and hydrogen peroxide. This procedure was not performed on the frozen brain tissue because most of those specimens were liquified or fragmented.

Formalin-fixed, paraffin-embedded brain tissues were processed by 2 methods to extract the nucleic acid: boiling of thin sections (Shibata et al. 1988) and digestion of whole paraffin blocks (Rupp and Locker 1988). Thin sections (approximately 1 cm^2 by 10 μm thick) were cut with disposable blades on a microtome and placed in Eppendorf tubes. Each section was deparaffinized with 1 ml of xylene followed by 2 washes of absolute ethanol (1 ml per wash). The residual tissue sections were dried by vacuum centrifugation and diluted with 50 μl of 10 mM TRIS (pH 7.4) and 0.1 mM EDTA or distilled water. The tissue suspension was boiled for 5 min at 100 °C and the entire volume used for PCR. Alternatively, whole blocks of formalin-fixed, paraffin-embedded brain tissues were extracted after thin sections had been

Table 1. SSPE and control brain tissues used for nucleic acid extraction and RT/PCR

Specimen number	Age/sex of patient	Diag- nosis	Type of specimen	Time of storage (years)	Wt/vol of specimen
Frozen brain tissue					
1	10/M	SSPE	NA[a]	20	0.71 g
2	16/M	SSPE	NA[b]	26	0.63 ml
3	15/M	SSPE	NA[b]	27	0.32 ml
4	4/M	SSPE	Frontal cortex	NA	0.75 g
5	14/M	SSPE	Frontal cortex	21	0.78 g
6	6/M	SSPE	Frontal cortex	14	0.10 g
7	21/F	NNC	NA	15	0.95 g
8	63/M	NNC	Frontal cortex	3	0.85 g
9	44/M	NNC	Frontal cortex	2	0.56 g
10	55/F	HD	Frontal cortex	NA	1.09 g
11	7/F	SSPE	Frontal cortex[c]	23	0.2 ml
12	6/M	SSPE	NA[c]	22	0.2 ml
13	8/M	SSPE	NA[c]	22	0.5 ml
14	16/M	SSPE	Occipital cortex[c]	20	0.4 ml
Formalin-fixed, paraffin-embedded brain tissue					
15a	19/F	SSPE	Cortex	4	0.27 g
15b	19/F	SSPE	Cortex	4	10-μm thin section
16	15/M	SSPE	Cortex	7	0.21 g
17a	19/F	SSPE	Cortex	7	0.34 g
17b	19/F	SSPE	Cortex	7	0.20 g
17c	19/F	SSPE	Cortex	7	10-μm thin section
18	9/F	SSPE	Cerebellum	5	0.72 g
19	NA/NA	NNC	Cortex	2/3	0.43 g
20a	7/NA	SSPE	Cortex	9	0.35 g
20b	7/NA	SSPE	Cortex	9	10-μm thin section
21	10/F	SSPE	Cortex	8	10-μm thin section

[a]   Information not available
[b]   Specimen found liquified on thawing
[c]   Brain suspension
NNC, Nonneurological control; HD, Huntington's disease; SSPE, subacute sclerosing panencephalitis; RT/PCR, reverse transcriptase polymerase chain reaction
More than a single piece of brain tissue was available for specimens 15, 17, and 20, each represented by an entry in the table

cut for immunocytochemistry, H&E staining, or thin-section PCR. Excess paraffin was trimmed from the paraffin blocks, and the remaining formalin-fixed tissue was minced with a disposable blade and then placed in 10 ml of the following solution: 10 mM NaCl, 500 mM TRIS-HCl (pH 7.6), 20 mM EDTA, 10 mg/ml aurintricarboxylic acid (Sigma), 1% SDS, and 0.5 mg/ml proteinase K. This mixture was incubated at 48 °C for 16 h and extracted with phenol and the RNA precipitated with 2 M lithium chloride over 16 h at 0 °C. The RNA was collected by centrifugation and redissolved in 10 ml of 10 mM TRIS (pH 7.6) with 1 mM EDTA. The RNA was reprecipitated with 2 M

lithium chloride over 16 h at 0 °C and again collected by centrifugation. The specimens were dried by vacuum centrifugation and then reconstituted with 300 µl of sterile distilled water and frozen at −70 °C.

Vero cells and frozen brain tissues were homogenized individually with autoclaved disposable tissue grinders in 5.0 ml of 4 *M* guanidine isothio-cyanate (GIT, Fluka) and 0.1 *M* 2-mercaptoethanol (Chirgwin et al. 1979). The homogenates were layered on 4.0 ml of 5.7 *M* cesium chloride (CsCl) buffered with sodium acetate and placed in an ultracentrifuge (Beckman SW 41 rotor) at 32 000 rpm for 21 h at 20 °C. After centrifugation, the upper GIT layer was decanted and discarded, while the lower CsCl layer containing the RNA and DNA was saved. The RNA pellet recovered from the bottom of the CsCl layer was washed with 300 µl of 0.3 *M* sodium acetate (pH 6.0) and the RNA precipitated by adding 750 µl of absolute ethanol, chilling to −70 °C, and centrifuging. The RNA was washed twice with 80% ethanol, centrifuged, the ethanol decanted, and the sample dried overnight at room temperature in a covered, open-ended Eppendorf tube. The dried RNA was diluted with 100 µl of sterile distilled water and frozen at −70 °C. The upper CsCl layer was treated with 40 ml of 80% ethanol at −20 °C to precipitate the DNA. The DNA was centrifuged and the ethanol and CsCl decanted. The DNA was washed with 10 ml of 80% ethanol, centrifuged, and the ethanol discarded. It was left at room temperature for a few hours until just damp, diluted with 1 or 2 ml of 10 m*M* TRIS (pH 7.4) and 0.1 m*M* EDTA and then frozen at −70 °C.

To determine the amount and purity of the nucleic acid extracted from the frozen or paraffin-embedded brain specimens, each RNA or DNA sample was measured spectrophotometrically at 260 and 280 nm.

The sequences of the 5 major structural protein genes of measles virus (nucleocapsid protein, phosphoprotein, matrix protein, fusion protein, and hemagglutinin protein) were obtained using the Sequence Analysis Software Package produced by the University of Wisconsin Genetics Computer Group (Devereux et al. 1984). Oligonucleotide primer pairs to each of these 5 gene regions were selected to amplify a region of the target sequence 400–500 bases long with restriction sites for either *Alu*I (recognizing AG′CT) or *Ava*II (recog-nizing G′GwC C) located off-center in the target sequence. The primers were synthesized on an Applied Biosystems DNA synthesizer (model 391-03 or 380B). The primer sequences and related information are shown in Table 2.

Transcription of RNA specimens to cDNA was done by adding reverse transcriptase (RT) and an RNase inhibitor to the PCR reaction mixture. Each PCR reaction was prepared in a 0.5-ml Eppendorf tube by combining the following reagents to a final volume of 100 µl: 5.6 units of avian myeloblastosis RT (AMV RT, Promega); 40 units of RNasin (Promega); 5 units of Taq polymerase (Perkin-Elmer-Cetus); 200 µ*M* each of dATP, dCTP, dGTP, dTTP; 1.0 µ*M* of each primer; 50 m*M* KCl; 10 m*M* TRIS-HCl (pH 8.3); 1.5 m*M* MgCl$_2$; 0.01% (w/v) gelatin, and sterile distilled water (q.s.). One drop of mineral oil was added to cover the reaction mixture, and, as the final step, 0.5–10 µg of extracted nucleic acid in 10 µl of sterile distilled water was added.

Table 2. Oligonucleotide primer pairs for measles virus RT/PCR

Gene region	Primer symbol	Primer sequences	Product size (base pairs)	Location in gene region (bases)	Restriction enzyme and fragment sizes (base pairs)
Nucleocapsid protein	MNP 1 MNP 2	5′ GGTTCGGATGGTTCGAGAACA 3′ 5′ GGTTCATCAAGGACTCAAGTG 3′	477	428– 904 [a]	AluI: 138/11/328
Phosphoprotein	MPP 1 MPP 2	5′ CTGTTCTCTGATGTCCAAGAT 3′ 5′ AACCCGACGGCTGAGCTCATC 3′	425	993–1417 [b]	AluI: 75/291/18/41
Matrix protein	MMP 1 MMP 2	5′ CCTGGTCTAGGCGACAGGAAG 3′ 5′ CACCAGCAGGTTGAAGGCCAC 3′	459	163– 621 [c]	AluI: 171/288
Fusion protein	MFP 1 MFP 2	5′ GGCAATTGAGGCAATCAGACA 3′ 5′ CTTGAGAGCCTATGTTGTACG 3′	452	1066–1517 [d]	AvaII: 190/262
Hemagglutinin	MHA 1 MHA 2	5′ CCAACCGACATGCAATCCTGG 3′ 5′ AGGCTAGGTTCTTCATTGGCG 3′	403	1008–1410 [e]	AluI: 279/124

[a] Rozenblatt et al. 1985
[b] Bellini et al. 1985
[c] Bellini et al. 1986
[d] Buckland et al. 1987
[e] Alkhatib and Briedis 1986

The Eppendorf tube containing the reaction mixture was briefly centrifuged and placed in a thermal cycler (Perkin-Elmer-Cetus) programmed for the following thermal profile: 42 °C for 1 h, 94 °C for 1 min, 55 °C for 1 min, 72 °C for 3 min. The final 3 steps were repeated for 40 cycles. When testing DNA specimens, the RNasin and AMV RT were replaced with sterile distilled water, and the 42 °C preliminary incubation was omitted from the thermal cycler program.

PCR products were detected under ultraviolet light in 1.2% or 4.0% agarose gels after electrophoresis in TBE buffer (100 mM TRIS, pH 7.8, 100 mM boric acid, and 2 mM EDTA) and ethidium bromide staining. Product identity was confirmed by enzymatic cleavage by *Alu*I or *Ava*II to fragments of known size. Twenty units of restriction enzyme in the appropriate buffer were mixed with 20 µl of the PCR product and incubated for 1 h at 37 °C. Restriction fragments were also visualized in 1.2% or 4.0% agarose gels by ethidium bromide staining after electrophoresis in TBE buffer. To minimize the possibility of contamination with PCR products or copies of measles RNA or cDNA during this study, the PCR reactions and gel electrophoresis were performed in a separate room from the one in which the brain specimens and PCR reagents were handled. Disposable gloves were worn and changed before handling each specimen. Templates, primers, and reagents were measured with positive displacement pipettes which were used once and then discarded. The DNA or RNA template was always added last to a reaction mixture. PCR reagents, primers, and templates were prepared in aliquots to reduce the risk of contamination (Kwok and Higuchi 1989).

Results

Following amplification by RT/PCR, products of expected size were generated by all 5 primer pairs using RNA extracted from measles-infected Vero cells but not with RNA from uninfected Vero cells. Product identity was confirmed by the cleavage of each product with *Alu*I or *Ava*II to yield fragments of predicted size (Fig. 1).

RNA extracted from frozen SSPE and control brain tissues (specimens 1–10) were then tested by RT/PCR with all 5 primer pairs. These experiments are summarized in Table 3, which shows that at least one segment of the measles genome was detected in RNA from each SSPE case and that all 5 major structural protein genes of the measles virus were detected in 5 of the 6 SSPE cases. PCR product identity was confirmed by cleavage with *Alu*I or *Ava*II in every positive case except for specimen 2, for which the MMP product did not cleave with *Alu*I and the MPP products gave fragments of the incorrect size after treatment with *Alu*I. Measles PCR products were not detected after RT/PCR with RNA from any control brain using all 5 primer pairs. Figure 2 compares the PCR products generated with RNA from specimen 1, an SSPE case, with specimen 7, a control case. The restriction fragments of the PCR

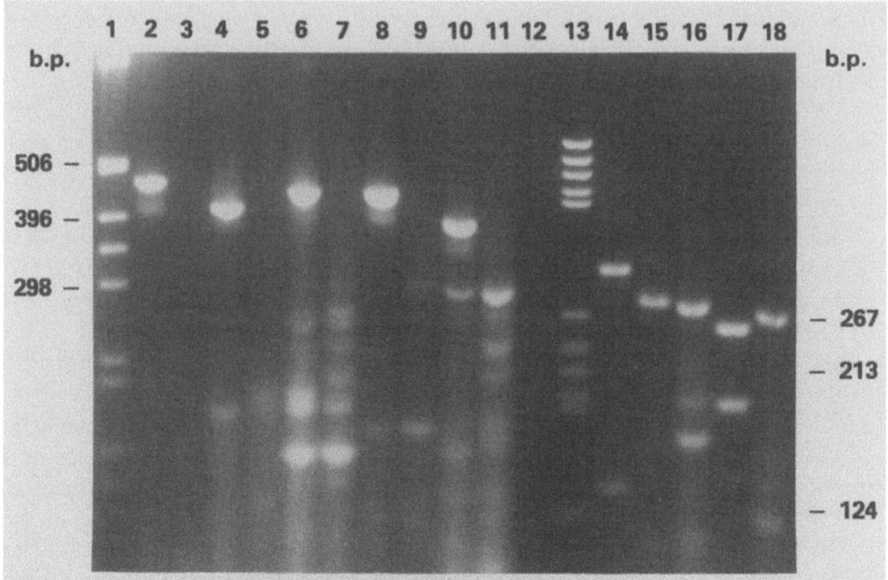

Fig. 1. Ethidium-stained agarose gel showing PCR products amplified by RT/PCR from the RNA of measles-infected (lanes 2, 4, 6, 8, 10) and uninfected (lanes 3, 5, 7, 9, 11) Vero cells. Lane *1*, 1-kb DNA ladder; lanes *2* and *3*, MNP products; lanes *4* and *5*, MPP products; lanes *6* and *7*, MMP products; lanes *8* and *9*, MFP products; lanes *10* and *11*, MHA products; lane *12*, blank; lane *13*, 500-bp DNA ladder; lanes *14–18* show the respective *Alu*I or *Ava*II digests of the products in lanes 2, 4, 6, 8, 10. Some smaller restriction fragments are not well visualized in this photograph

Table 3. Results of RT/PCR on frozen SSPE and control brain specimens

Specimen number	Diagnosis	MNP	MPP	MMP	MFP	MHA
1	SSPE	+	+	+	+	+
2	SSPE	+	+	+	+	+
3	SSPE	+	±[a]	+	+	+
4	SSPE	+	+	+	+	+
5	SSPE	−	−	±	+	−
6	SSPE	+	+	+	+	+
7	NNC	−	−	−	−	−
8	NNC	−	−	−	−	−
9	NNC	−	−	−	−	−
10	HD	−	−	−	−	−

[a] Trace positive result

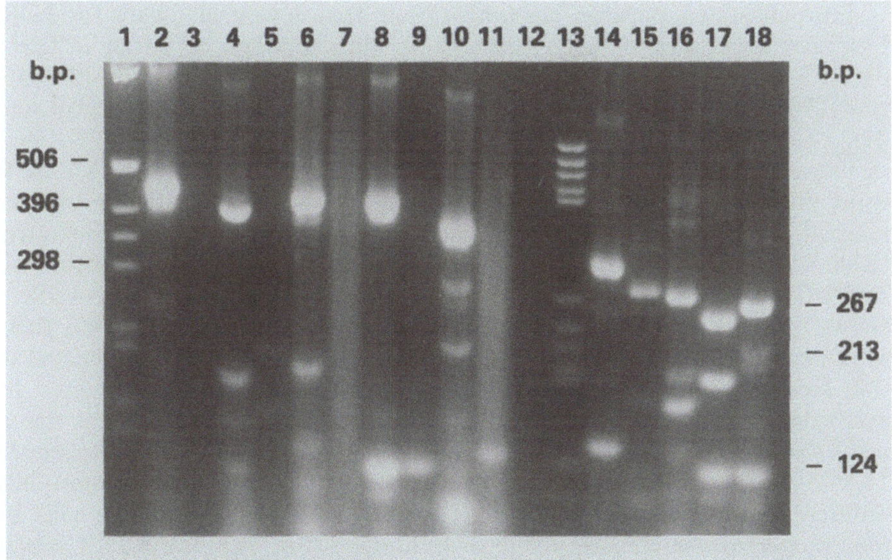

Fig. 2. Ethidium-stained agarose gel showing PCR products amplified by RT/PCR from RNA of specimen 1 (SSPE patient, lanes 2, 4, 6, 8, 10) and specimen 7 (control patient, lanes 3, 5, 7, 9, 11) brain tissues. Lane *1*, 1-kb DNA ladder; lanes *2* and *3*, MNP products; lanes *4* and *5*, MPP products; lanes *6* and *7*, MMP products; lanes *8* and *9*, MFP products; lanes *10* and *11*, MHA products; lane *12*, blank; lane *13*, 500-bp DNA ladder; lanes *14–18* show the respective *Alu*I or *Ava*II digests of the products in lanes 2, 4, 6, 8, 10. Some smaller restriction fragments are not well visualized in this photograph

products from specimen 1 are also shown in Fig. 2. The pattern of PCR products and restriction fragments is very similar to that observed in Fig. 1. Note that the MPP primers generated the least amount of PCR product and the MFP primers, the most. This difference in the amounts of PCR product produced by these 2 primer pairs was observed in most of the experiments.

To determine whether DNA transcripts of measles virus RNA might be present in SSPE brain tissue, PCR was performed with DNA extracted from specimens 1–6 using all 5 primer pairs while omitting the RT and RNasin from the reaction mixtures. DNA transcripts of measles virus genome were not detected in any of these 6 SSPE brain specimens (data not shown).

Four frozen 10% suspensions of SSPE brain in PBS were also examined by PCR (specimens 11–14). These 4 SSPE cases had been donated to the LCNSS in the late 1960s and inoculated intracerebrally into primates as brain homogenates. Less than 50 mg of total brain tissue was present in each specimen. After GIT-CsCl extraction, no spectrophotometrically measurable nucleic acid was obtained. RT/PCR with all 5 primer pairs using the putative RNA extracts as well as 10-µl aliquots of the original unextracted brain suspensions itself yielded no PCR products (data not shown).

Formalin-fixed, paraffin-embedded brain tissues were also used for PCR. Genomic sequences of DNA viruses in thin sections of formalin-fixed, paraffin-embedded tissue have been successfully detected by PCR (Shibata et al. 1988), and this was attempted with 4 SSPE specimens (15b, 17c, 20b, and 23).

Deparaffinized, boiled thin sections of SSPE brains were tested by RT/PCR using the MNP primers, but no PCR product was detected after several attempts (Table 4). As an alternative, whole paraffin blocks of SSPE and control tissue were used for nucleic acid extraction after each specimen was examined by immunocytochemical labelling for evidence of measles antigen. Table 4 shows the results of RT/PCR and immunocytochemical labelling on these specimens. Although measles virus genomic sequences were not detected in thin sections from specimens 15b, 17c, and 20b, the RNAs extracted by aurintricarboxylic acid, detergent, and protease from blocks of these specimens were strongly positive by RT/PCR using all 5 primer pairs. In specimen 17, two different areas of the cortex were tested by RT/PCR with substantially different results. Specimen 17a showed only rare antigen-positive cells by immunoperoxidase staining for measles antigen (Fig. 3a), and RT/PCR was successful only with the MMP and MFP primers which gave much smaller amounts of PCR products. Specimen 17b was strongly positive by immunoperoxidase staining (Fig. 3b), and RT/PCR yielded large quantities of product using all 5 primer pairs. In specimen 18, cerebellar tissue from a neuropathologically confirmed case of SSPE, the extracted RNA yielded a small amount of PCR product with the MFP primers, although this specimen was completely negative for measles antigen by immunocytochemistry.

Table 4. Results of immunocytochemistry and RT/PCR on formalin-fixed, paraffin-embedded SSPE and control brain specimens

Specimen number	Diagnosis	Measles antigen	MNP	MPP	MMP	MFP	MHA
15b	SSPE-TS[a]	+	−	NP[b]	NP	NP	NP
17c	SSPE-TS	+	−	NP	NP	NP	NP
20b	SSPE-TS	+	−	NP	NP	NP	NP
21	SSPE-TS	+	−	NP	NP	NP	NP
15a	SSPE	+	+	+	+	+	+
16	SSPE	+	+	+	+	+	+
17a	SSPE	±	−	−	+	+	−
17b	SSPE	+	+	+	+	+	+
18	SSPE	−	−	−	−	±	−
19	NNC	−	−	−	−	−	−
20a	SSPE	+	+	+	+	+	+

[a] Thin section from a paraffin block
[b] Test not performed

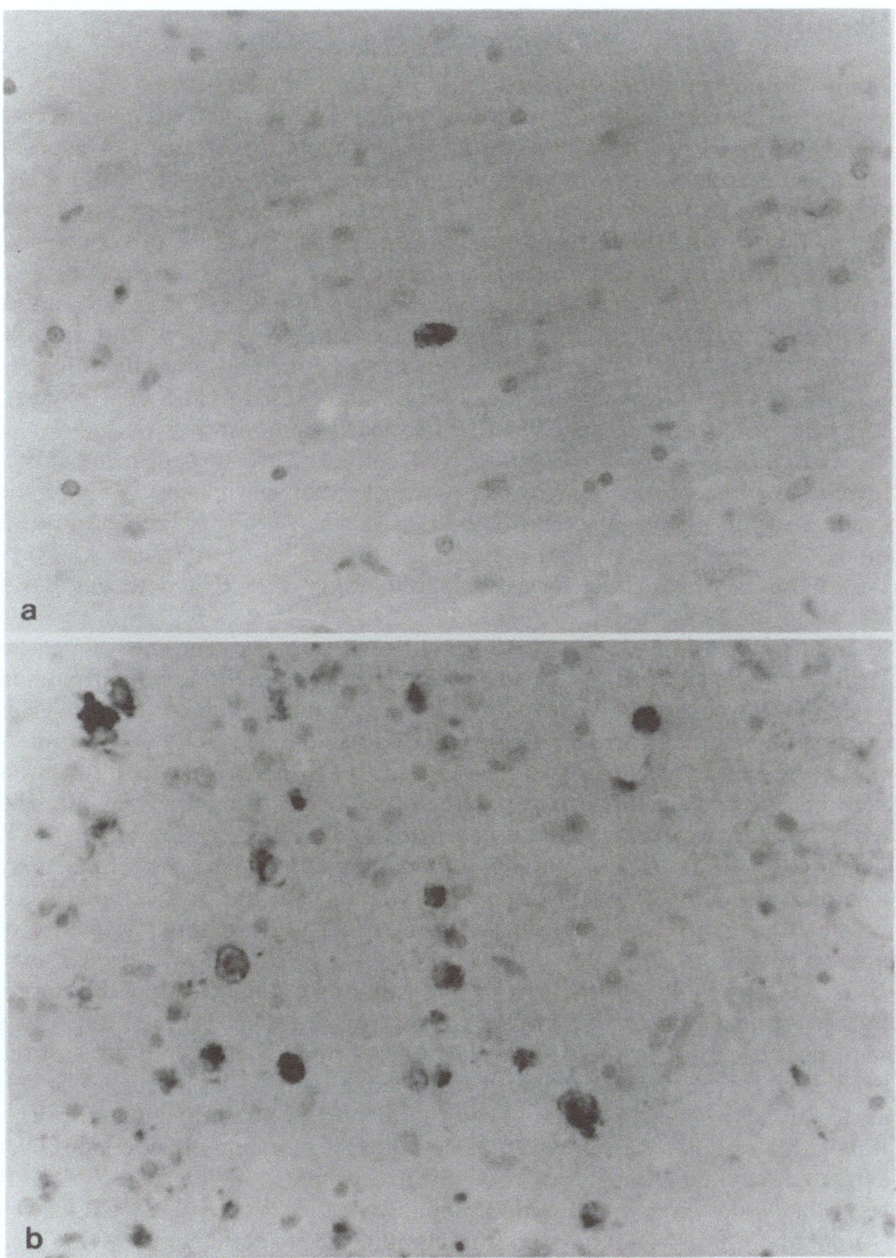

Fig. 3a, b. Subcortical white matter from specimen 17 (SSPE patient), labelled by immunocytochemistry for measles antigen. **a** Minimally infected area showing a rare measles-antigen-positive cell in the center of the field; **b** heavily infected area with many antigen-positive oligodendrocytes. × 125

Discussion

Many chronic neurological disorders of man are of undetermined etiology. A viral cause has been suggested for some of these disorders, and the possible role of RNA viruses has been investigated in connection with diseases such as multiple sclerosis (Gonzalez-Scarano and Nathanson 1985), amyotrophic lateral sclerosis (Kascsak et al. 1982), Parkinson's disease (Elizan and Casal 1983), and schizophrenia (Kaufmann et al. 1983; Torrey 1988). No conclusive etiologic link between RNA viruses and these disorders has been made despite years of study using various techniques. The development of the PCR and its adaptation to RNA (RT/PCR) provides a uniquely sensitive and specific method of detecting viral genomic sequences in these and other disorders.

SSPE is a chronic neurologic disorder of man caused by the measles virus, which has an RNA genome. The disease clearly demonstrates that an RNA virus can remain latent in the human CNS for years, even decades, and subsequently emerge, producing slowly progressive damage in the CNS and the eventual death of the affected individual. As such, SSPE is the best known model for other disorders that may also be caused by RNA viruses and presents a unique opportunity for assessing the utility of PCR in detecting RNA viruses in these disorders.

In this study, primer pairs for all 5 major structural protein genes of measles virus were used successfully for RT/PCR. The use of multiple primer pairs in detecting a given viral genome by PCR allows greater confidence that the virus in question is actually present. Some variability in the sensitivity of the primers was noted, with the MFP primers being most sensitive in detecting measles viral cDNA and the MPP being the least sensitive. The MMP, MNP, and MHA primers were intermediate in sensitivity. Primer selection affects PCR sensitivity, but the success rate in amplification with a given primer pair may also be affected by the viral genome. If a virus is defective or incomplete, it is possible that a given primer pair may not amplify the genomic sequence of interest (Brown et al. 1971).

To enhance PCR sensitivity, a second round of amplification with internal (nested) primers may be performed (Mullis and Faloona 1987). Table 5 lists sequences of nested primers for measles virus which have been used with the primers in Table 2. In experiments performed in the LCNSS, nested PCR was as sensitive as a single PCR amplification followed by Southern blot hybridization with a radiolabelled probe in detecting HIV genomic sequences in DNA extracted from lymphocytes of HIV-infected primates (Godec et al. 1992). It seems likely that nested PCR will often increase the specificity of PCR as well (Mullis and Faloona 1987). In another unpublished experiment involving serial tenfold dilutions of measles-infected Vero cell RNA in distilled water amplified by nested RT/PCR with the MFP 1,2,3,4 primers shown in Tables 2 and 5, PCR product was detected in dilutions down to 1×10^7. Knowing that the total amount of RNA in the starting material was 1.5 µg, that a single strand of measles viral RNA weighs approximately 8.4×10^{-12} µg (Crowley et al. 1987), and assuming that no more than 1% of the total RNA extracted from

Table 5. Internal oligonucleotide primer pairs for measles virus RT/PCR

Gene region	Primer symbol	Primer sequences	Product size (base pairs)	Location in gene region (bases)	Restriction enzyme and fragment sizes (base pairs)
Nucleocapsid protein	MNP 3 MNP 4	5′ TGAAGTGCAAGACCCTGAGGG 3′ 5′ TTCATGCAGTCCAAGAGCAGG 3′	400	465– 864[a]	*Alu*I: 101/11/238
Phosphoprotein	MPP 3 MPP 4	5′ CCTTGGCCAAAATACACGAGG 3′ 5′ AAATTCCTTCAGCAGCTGTCC 3′	350	1024–1373[b]	*Alu*I: 44/291/15
Matrix protein	MMP 3 MMP 4	5′ TGTACATGTTTCTGCTGGGGG 3′ 5′ CCAGCATTCTTCTAGGAACGG 3′	384	197– 580[c]	*Alu*I: 137/247
Fusion protein	MFP 3 MFP 4	5′ GATGATATTGGCTGTTCAGGG 3′ 5′ CTTAATCTCGGACAGCGTCGG 3′	367	1099–1465[d]	*Ava*II: 157/210
Hemagglutinin	MHA 3 MHA 4	5′ GATGATCCAGTGATAGACAGG 3′ 5′ AATACACATTGTTGTGGTTGG 3′	334	1044–1377[e]	*Alu*I: 243/91

[a] Rozenblatt et al. 1985
[b] Bellini et al. 1985
[c] Bellini et al. 1986
[d] Buckland et al. 1987
[e] Alkhatib and Briedis 1986

the measles-infected Vero cells was from the viral genome, it can be calculated that no more than 180 copies of measles viral RNA were present in the highest dilution positive by nested RT/PCR. Given this calculation, the abrupt disappearance of PCR product after the highest positive dilution, and the probability that the viral genome copy number in the highest positive dilution is much lower due to the aggregation of the RNA in solution, it is apparent that a very small number of genomic copies may be detected in a single nested RT/PCR reaction and that this number may approach a single genomic copy (Li et al. 1988).

The advantage in sensitivity afforded by PCR for the detection of viruses is partially offset by the potential for contamination (Kwok and Higuchi 1989). This is particularly important when using nesting primer pairs where a major potential source of "carry-over" contamination occurs when products from the first PCR reaction mixture are transferred to new reaction tubes for the second round of amplification. Known positive samples should be handled last, gloves changed before each specimen is handled, and only positive-displacement pipettes used. Clean disposable absorbent pads should be placed over the work area and changed when the PCR is complete. Any positive specimen would be treated with suspicion and retested. Amplification of a product from an RNA virus in the absence of RT in the first round of amplification suggests that the product arose from contamination with PCR product or cDNA from a previous reaction.

An important example of contamination detected by nested PCR occurred during ongoing studies of various chronic neurologic disorders by nested RT/PCR. Several frozen brain specimens were obtained from a source outside the LCNSS including SSPE brain specimens. Unfortunately, while processing the brain tissue for shipment, the SSPE brain specimens were handled and cut first, followed by multiple sclerosis (MS), Alzheimer's disease, Parkinson's disease, and control brain tissue. Gloves were not changed between handling specimens, nor was the device used to cut pieces from the brain specimens (a dental drill with a grinding tip). Subsequently, most MS and control brain samples tested positive by nested RT/PCR for measles virus genome by all 5 primer pairs. Omitting RT abolished the signal, suggesting that the measles genome had been transferred from the SSPE cases to the other brain specimens during handling. Only traces of the measles genome could have been present on the other brain specimens, demonstrating the power of this technique to detect very small quantities of viral genome.

The identity of a PCR product may be confirmed by restriction enzyme analysis (as was done in this study), hybridization with a labelled cDNA probe, or direct sequencing (Saiki et al. 1988). Restriction enzyme analysis of PCR products has the advantage of being easy, rapid, and inexpensive. If the products and fragments are compared with a known positive control, specificity is assured. This method suffers from the disadvantage of being unable to cleave a PCR product if a mutation is present at the restriction site. SSPE strains of measles virus have multiple point mutations in their genomes, especially in the matrix protein gene (Cattaneo et al. 1988). This problem was encountered with the MMP PCR product from specimen 2 which was not cleaved with *Alu*I and the MPP product from the same specimen which was cleaved to products of the wrong sizes. It is now possible to subject a PCR product directly to automated DNA sequencing after minimal effort in preparing the sample (L. McBride et al., personal communication). It is likely that this method will supplant all others in confirming the identity of PCR products because of its simplicity and because of the importance of the information it provides. By selecting a variable region of a viral genome for amplification, it is possible to compare variations in isolates from one individual to another and even within one individual at different times. Also, direct sequencing of PCR products makes the detection of any contamination easier; PCR products contaminating primers, equipment, or reagents will have identical sequences (Bangham et al. 1989).

The mechanism by which measles virus remains latent in the human CNS remains unknown. It has been postulated that a helper retrovirus reverse transcribes measles RNA to cDNA and inserts this sequence into the human genome, allowing conservation of the measles genome in vivo for prolonged periods (Zhdanov 1975). To verify this hypothesis, DNA from specimens 1–6 were tested by PCR with the RT and RNase inhibitor omitted from the reaction mixture. In no instance was a PCR product detected with any of the 5 primer pairs, suggesting that measles cDNA was not present in the samples tested, only measles RNA. It is possible, however, that such measles cDNA transcripts were present in other areas of the subjects' brains. For any subject, only a tiny fraction of the brain was used for nucleic acid extraction, and only a fraction of the extracted nucleic acid was used for PCR. Thus, the existence of cDNA copies of the measles genome cannot be excluded with certainty in SSPE.

In this study, RNA for RT/PCR was obtained by extraction of both frozen and formalin-fixed, paraffin-embedded SSPE brain tissue. Although extracts from frozen brain suspensions and thin sections from paraffin blocks failed to yield detectable amounts of PCR products, substantial amounts of PCR products were generated from almost all other specimens, regardless of the specimen age or condition. The two oldest frozen specimens in the study (specimens 2 and 3, stored for 26 and 27 years, respectively), which were completely liquified on thawing, gave large amounts of PCR products after extraction by GIT-CsCl centrifugation, reverse transcription, and enzymatic amplification. Although formalin fixation degrades nucleic acid, especially if the formalin becomes acidic or the fixation time is prolonged (Ben-Ezra et al. 1991; Greer et al. 1991), the formalin-fixed, paraffin-embedded SSPE brain tissues extracted by digestion and lithium chloride precipitation gave excellent results by RT/PCR, although this method requires destruction of the paraffin block. Success in RT/PCR correlated better with evidence of viral antigen in a given area of the brain than with specimen age or condition. Immunoperoxidase labelling of measles antigen in specimen 17 demonstrated that certain areas of the brain contained more measles virus than others (Fig. 3). RT/PCR detected viral genome with greater success in portions of the brain that stained heavily for viral antigen by immunoperoxidase labelling than in areas where less viral antigen was apparent (Table 4). A similar situation arose with specimen 18, cerebellar tissue from an SSPE patient, which showed no viral antigen in neurons or glial cells by immunoperoxidase labelling but which was weakly positive for viral genome using the MFP primer pair. In this specimen, and in any other specimen used in this study, it is possible that the PCR signal arose entirely or in part from measles-infected lymphocytes in the circulatory system rather than from infected CNS cells. Peripheral lymphocytes from patients with SSPE have been shown to contain measles virus genome by in situ hybridization (Fournier et al. 1985), and these may serve as a source of genome for amplification. This fact demonstrates a limitation of PCR in detecting viral genomes in chronic neurologic diseases: although PCR is an exquisitely sensitive and specific method for detecting viral genomes, it provides no information on the location of the genome within the tissue being tested. Any cell in the tissue used for nucleic acid extraction may serve as a source for the viral genome and give detectable product following PCR.

The failure to amplify measles virus genomic sequences in SSPE brain specimens where antigen was not detected suggests that a proper selection of tissue for study is important for this neurological disorder and probably other neurological disorders as well. A substantial increase in sensitivity can be achieved by using nesting primer pairs in sequential rounds of amplification; however, if there is no viral genome in the tissue selected for nucleic acid extraction, PCR will not be successful. In MS, areas showing active demyelination might logically be investigated first by PCR, while in amyotrophic lateral sclerosis, degenerating motor neurons in the motor cortex and anterior horns of the spinal cord would be of prime interest. In other diseases such as Alzheimer's disease (basal forebrain nuclei and cortex) and Parkinson's disease (substantia nigra), there are anatomic areas with the greatest histopathological changes that might be most suitable for investigation by PCR. For any particular disease, it is impossible to know with certainty that a specific area of the brain will contain virus, but it is clear that in SSPE, testing random areas of the brain in the hope of detecting a virus by PCR is not adequate. A directed search sampling areas with the most severe neuropathology should be conducted for any of these diseases.

The success of PCR in detecting measles virus in SSPE brain tissue demonstrates the utility of PCR in investigating potential etiologic relationships between viruses and chronic human neurologic disorders. The amplification from brain or other tissue of specific genomic sequences from any virus is limited only by the availability of the viral sequence information necessary for designing the primer pairs. In the case of SSPE, measles virus may be detected in peripheral lymphocytes as well as in brain tissue, which presents a simple method of diagnosis by phlebotomy, extraction of RNA from the white cell fraction, and RT/PCR. It is likely that PCR, enhanced by the use of nesting primers and coupled with direct sequencing of the PCR products, will become the method of choice for detecting and analyzing viruses in chronic and acute diseases affecting man.

Acknowledgments. Mark S. Godec is a Senior Staff Fellow in the Laboratory of Central Nervous System Studies at the National Institutes of Health. Dr. Peggy T. Swoveland supplied the photographs of immunoperoxidase-labelled SSPE brain tissue. Drs. J. Dymecki and I. Dzidusko of the Psychoneurological Institute, Warsaw, Poland, and Dr. W. Tourtellotte of the National Neurological Research Bank generously provided SSPE and control brain tissue for this study. Drs. D. Camenga and J. Woyciehowska translated the clinical records of patients from Poland. Dr. David M. Asher is thanked for providing editorial assistance.

References

Alkhatib G, Briedis DJ (1986) The predicted primary structure of the measles virus hemagglutinin. Virology 150:479–490

Asher DM (1991) Slow viral infections of the human nervous system. In: Scheld WM, Whitley RJ, Durack DT (eds) Infections of the central nervous system. Raven, New York, pp 145–166

Bangham CRM, Nightingale S, Cruickshank JK, Daenke S (1989) PCR analysis of DNA from multiple sclerosis patients for the presence of HTLV-1. Science 246:821–824

Bellini WJ, Englund G, Rozenblatt S, Arnheiter H et al. (1985) Measles virus P gene codes for two proteins. J Virol 53:908–919

Bellini WJ, Englund G, Richardson CD, Rozenblatt S et al. (1986) Matrix genes of measles virus and canine distemper virus: cloning, nucleotide sequences, and deduced amino acid sequences. J Virol 58:408–416

Ben-Ezra J, Johnson DA, Rossi J, Cook N et al. (1991) Effect of fixation on the amplification of nucleic acids from paraffin-embedded material by the polymerase chain reaction. J Histochem Cytochem 39:351–354

Bouteille M, Fontaine C, Vedrenne C, Delarue J (1965) Sur un cas d'encéphalite subaiguë à inclusions. Etude anatomo-clinique et ultrastructurale. Rev Neurol (Paris) 113:454–458

Brown P, Cathala F, Gajdusek DC, Gibbs CJ Jr (1971) Measles antibodies in the cerebrospinal fluid of patients with multiple sclerosis. Proc Soc Exp Biol Med 137:956–961

Buckland R, Gerald C, Barker R, Wild TF (1987) Fusion glycoprotein of measles virus: nucleotide sequence of the gene and comparison with other paramyxoviruses. J Gen Virol 68:1695–1703

Budka H, Lassmann H, Popow-Kraupp T (1982) Measles virus antigen in panencephalitis: an immunomorphological study stressing dendritic involvement in SSPE. Acta Neuropathol (Berl) 56:52–62

Cape CA, Martinez AJ, Robertson JT, Hamilton R et al. (1973) Adult onset of subacute sclerosing panencephalitis. Arch Neurol 28:124–127

Cattaneo R, Schmid A, Billeter MA, Sheppard RD et al. (1988) Multiple viral mutations rather than host factors cause defective measles virus gene expression in a subacute sclerosing panencephalitis cell line. J Virol 62:1388–1397

Centers for Disease Control (1982) Subacute sclerosing panencephalitis surveillance – United States. MMWR 31:585–588

Chen TT, Watanabe I, Zeman W, Mealey J (1969) Subacute sclerosing panencephalitis: propagation of measles virus from brain biopsy in tissue culture. Science 163:1193–1194

Chirgwin JM, Przybyla AE, MacDonald RJ, Rutter WJ (1979) Isolation of biologically active ribonucleic acid from sources enriched in ribonuclease. Biochemistry 18:5294–5299

Cobb W (1966) The periodic events of subacute sclerosing leucoencephalitis. Electroencephalogr Clin Neurophysiol 21:278–294

Connolly JH, Allen IV, Hurwitz LJ, Millar JHD (1967) Measles-virus antibody and antigen in subacute sclerosing panencephalitis. Lancet 1:542–544

Crowley J, Dowling P, Menonna J, Schanzer B et al. (1987) Molecular cloning of 99% of measles virus genome, positive identification of 5′ end clones, and mapping of the L gene region. Intervirology 28:65–77

Dawson JR (1933) Cellular inclusions in cerebral lesions of lethargic encephalitis. Am J Pathol 9:7–16

Devereux J, Haeberli P, Smithies O (1984) A comprehensive set of sequence analysis programs for the VAX. Nucleic Acids Res 12:387–395

DuRant RH, Dyken PR (1983) The effect of inosiplex on the survival of subacute sclerosing panencephalitis. Neurology 33:1053–1055

Dyken PR (1985) Subacute sclerosing panencephalitis: current status. Neurol Clin 3:179–196

Elizan TS, Casal J (1983) The viral hypothesis in Parkinsonism. J Neural Transm Suppl 19:75–88

Fournier JG, Tardieu M, Lebon P, Robain O et al. (1985) Detection of measles virus RNA in lymphocytes from peripheral-blood and brain perivascular infiltrates of patients with subacute sclerosing panencephalitis. N Engl J Med 313:910–915

Godec MS, Asher DM, Swoveland PT, Eldadah ZA et al. (1990) Detection of measles virus genomic sequences in SSPE brain tissue by the polymerase chain reaction. J Med Virol 30:237–244

Godec MS, Asher DM, Murray RS, Shin ML et al. (1992) Absence of measles, mumps and rubella viral genomic sequences from multiple sclerosis brain tissue by polymerase chain reaction. Ann Neurol (in press)

Gonzalez-Scarano F, Nathanson N (1985) Viral etiology of multiple sclerosis: a critique of the evidence. In: Maramorosch K, McKelvey JJ Jr (eds) Subviral pathogens of plants and animals: viroids and prions. Academic, Orlando, pp 465–481

Greer CE, Peterson SL, Kiviat NB, Manos MM (1991) PCR amplification from paraffin-embedded tissues. Am J Clin Pathol 95:117–124

Halsey NA, Modlin JF, Jabbour JT, Dubey L et al. (1980) Risk factors in subacute sclerosing panencephalitis: a case-control study. Am J Epidemiol 111:415–424

Horta-Barbosa L, Fuccillo DA, Sever JL, Zeman W (1969) Subacute sclerosing panencephalitis: isolation of measles virus from a brain biopsy. Nature 221:974

Jabbour JT, Garcia JH, Lemmi H, Ragland J et al. (1969) Subacute sclerosing panencephalitis: a multidisciplinary study of eight cases. JAMA 207:2248–2254

Jabbour JT, Duenas DA, Modlin J (1975) SSPE: clinical staging, course, and frequency, 1975. Arch Neurol 32:493–494

Jones CE, Dyken PR, Huttenlocher PR, Jabbour JT et al. (1982) Inosiplex therapy in subacute sclerosing panencephalitis: a multicentre, non-randomized study in 98 patients. Lancet 1:1034–1037

Kascsak RJ, Carp RI, Vilcek JT, Donnenfeld H et al. (1982) Virological studies in amyotrophic lateral sclerosis. Muscle Nerve 5:93–101

Kaufmann CA, Weinberger DR, Yolken RH, Torrey EF et al. (1983) Viruses and schizophrenia. Lancet 2:1136–1137

Kwok S, Higuchi R (1989) Avoiding false positives with PCR. Nature 339:237–238

Li H, Gyllensten UB, Cui X, Saiki RK et al. (1988) Amplification and analysis of DNA sequences in single human sperm and diploid cells. Nature 335:414–417

Modlin JF, Halsey NA, Eddins DL, Conrad JL et al. (1979) Epidemiology of subacute sclerosing panencephalitis. J Pediatr 94:231–236

Mullis KB, Faloona FA (1987) Specific synthesis of DNA in vitro via a polymerase-catalyzed chain reaction. Methods Enzymol 155:335–350

Mullis K, Faloona F, Scharf S, Saiki R et al. (1986) Specific enzymatic amplification of DNA in vitro: the polymerase chain reaction. Cold Spring Harbor Symp Quant Biol 51:263–273

Payne FE, Baublis JV, Itabashi HH (1969) Isolation of measles virus from cell cultures of brain from a patient with subacute sclerosing panencephalitis. NEJM 281:585–589

Rozenblatt S, Eizenberg O, Ben-Levy R, Lavie V et al. (1985) Sequence homology within the morbilliviruses. J Virol 53:684–690

Rupp GM, Locker J (1988) Purification and analysis of RNA from paraffin-embedded tissues. BioTechniques 6:56–60

Saiki RK, Gelfand DH, Stoffel S, Scharf SJ et al. (1988) Primer-directed enzymatic amplification of DNA with a thermostable DNA polymerase. Science 239:487–491

Sever JL (1983) Persistent measles infection of the central nervous system: subacute sclerosing panencephalitis. Rev Infect Dis 5:467–473

Shibata DK, Arnheim N, Martin WJ (1988) Detection of human papilloma virus in paraffin-embedded tissue using the polymerase chain reaction. J Exp Med 167:225–230

Torrey EF (1988) Stalking the schizovirus. Schizophr Bull 14:223–229

Wechsler SL, Meissner HC (1982) Measles and SSPE viruses: similarities and differences. Prog Med Virol 28:65–95

Zhdanov VM (1975) Integration of viral genomes. Nature 256:471–473

Chapter 20 The Use of the Polymerase Chain Reaction in Influenza Virus Detection and Characterization*

Augustine Rajakumar[1], Michelle Inkster[1], Ian D. Manger[1],
Ella M. Swierkosz[2], and Irene T. Schulze[1]

Summary

We have shown that the sequence of the hemagglutinin gene of influenza virus can be determined from virus recovered from nasopharyngeal swabs collected for diagnostic purposes by using the polymerase chain reaction (PCR) to amplify cDNA made by reverse transcriptase (Rajakumar et al. 1990). These studies suggest that PCR might be used in the routine diagnosis of influenza. Our experience indicates that the success of such a procedure would depend on (1) obtaining adequate cDNA synthesis from all virus strains encountered in clinical samples, (2) eliminating problems of DNA contamination, (3) ensuring production of amplified DNA from the appropriate influenza virus genes, and (4) developing appropriate methods to identify virus strains. The feasibility of such an approach to diagnosis is discussed with respect to sensitivity, reliability, virus strain identification, speed, and cost.

Introduction

Influenza is an acute respiratory disease with a short incubation period and a high attack rate. It occurs annually and produces epidemics of varying severity which often reach worldwide proportion, causing excess mortality, especially in the elderly and in individuals with chronic, debilitating disease.

The causative agent of this disease is an enveloped virus (orthomyxovirus). It consists of five nonglycosylated proteins (three polymerase proteins, a nucleoprotein, and a matrix protein), which are enclosed in a lipid bilayer, and two integral membrane glycoproteins, hemagglutinin (HA) and neuraminidase (NA). (For a detailed review of the structure and epidemiology of these viruses, see Murphy and Webster 1990; Wiley and Skehel 1990).

[1] Department of Microbiology, St. Louis University Medical Center, 1402 South Grand Blvd., St. Louis, M. 63104, USA
[2] Diagnostic Virology Laboratory, Cardinal Glennon Children's Hospital and Department of Pediatrics/Adolescent Medicine and Pathology, St. Louis University Medical Center, 1465 South Grand Blvd., St. Louis, M. 63104, USA
* This work was supported in part by grants AI10097 and AI23520 from the National Institutes of Health.

Epidemic strains are classified as A or B depending on the antigenic properties of the internal nonglycosylated proteins. A third type of orthomyxovirus, influenza C, has internal antigens which are distinct from those of the A and B strains. Influenza C does not cause serious epidemic disease. Differences between the surface glycoproteins of individual influenza A isolates permit their further division into antigenically distinct subtypes. Based on the antigenic properties of the HA, 14 distinct (noncrossreacting) influenza A subtypes have been identified, three of which have been isolated from humans and the remainder, from lower animals and birds (Murphy and Webster 1990; Kawaoka et al. 1990).

Ongoing surveillance in both human and animal populations indicates that small sequential antigenic changes induced by point mutations in the viral surface antigens (antigenic drift) occur within the various influenza A subtypes. Thus, the accumulation of amino acid substitutions in the HA over a number of years can result in the production of virus strains which are only distantly related to earlier isolates of the same subtype.

Investigation has also indicated that movement of influenza A virus strains from one animal species to another or to man can occur. Lower animals are therefore considered to be natural reservoirs from which new subtypes of influenza A can be introduced into the human population. This is thought to be the mechanism by which drastic changes in the properties of the surface antigens (antigenic shift) suddenly occur. Influenza B strains have been isolated only from man, and the surface antigens of all B strains are related. The basis for this difference between influenza A and B viruses is not known.

Infection with either influenza A or B strains constitutes a serious public health problem worldwide. Influenza vaccines are only of limited value because of changes in the antigenic properties of the virion surface antigens. Since it is the HA which stimulates the formation of protective antibodies, antigenic changes in this glycoprotein are the most significant ones in overcoming the protective effects of naturally acquired immunity or vaccination. The unpredictable nature of these changes, along with the ever-present possibility of introducing an influenza A virus into the human population from an animal reservoir, requires ongoing virus isolation and characterization in preparation for the annual formulation of influenza vaccines.

The HA is responsible for attachment of the virus to cellular receptors and for fusion of the viral envelope with membranes of endocytotic vesicles (reviewed in Wiley and Skehel 1990). Its active form is a trimer composed of identical subunits, each of which consists of a distal HA1 region and an HA2 region by which the trimer is inserted into the viral envelope. The 3-dimensional structure of the trimer has been determined by X-ray crystallography, and the amino acid sequence of numerous strains has been so the position of individual amino acids within the structure is known.

The extent of variation in the antigenic properties of the HA is a consequence of at least three factors: the genomic structure, the properties of the viral RNA-dependent RNA polymerase, and the ability of the HA to accommodate variation in its amino acid sequence. The genome consists of eight segments of negative-strand RNA which can reassort upon double infection. The RNA-dependent RNA polymerase is responsible for the synthesis of positive (messenger and genome-template) and negative (genomic) polarity RNA species during infection and replication. It exhibits a comparatively high frequency of nucleotide misincorporation, so that all influenza virus gene products constantly undergo variation in amino acid sequence which is restricted by the need to conserve function. HA and NA are under additional selective pressures from the host immune system. In the case of HA, amino acid substitutions which do not disrupt its 3-dimensional structure or change the amino acids in the receptor binding pocket appear to be well tolerated, resulting in antigenic variation and in changes in receptor binding properties and fusion activity.

The high frequency of amino acid substitution in HA has also been implicated in host-dependent selection during the growth of the influenza viruses in various laboratory animals and in cell cultures (reviewed in Schulze 1987; Aytay and Schulze 1991; Katz et al. 1987). Changes in the amino acid sequence of HA can alter virus host range at the cellular level, enabling minority forms within a virus population to predominate if they have a growth advantage during cultivation in the laboratory. The concern has therefore arisen that minor heterogeneity within clinical samples or mutations which occur during isolation and growth in the laboratory could lead to virus populations which are sufficiently different from those in clinical isolates that they are no longer suitable for use as vaccines.

With the advent of the polymerase chain reaction (PCR; Saiki et al. 1988), it became possible to design a method for determining the sequence of the HA gene without growing the virus in eggs or in cultured cells. Since all eight segments of the influenza virus genome have the same terminal sequences which are conserved in all virus strains (Desselberger et al. 1980), we reasoned that it should be possible to use PCR to amplify cDNA copies of all of the segments and that the sequence of any of the viral genes could be determined from very small amounts of starting virus.

The studies presented here were originally undertaken for the purpose of obtaining the amino acid sequence of HA of virus obtained directly from the nasopharynx so that any changes which might be associated with the subsequent growth of the same virus in embryonated eggs or in cell culture could be detected and evaluated (Rajakumar et al. 1990). In the process of carrying out those experiments we have determined some of the factors which need to be addressed if PCR is to be used for the diagnosis of influenza. Our findings are reported here.

Materials and Methods

Except as noted, the materials and methods used were those previously described (Rajakumar et al. 1990).

Virus Isolation and Growth. Nasopharyngeal swabs collected for the diagnosis of febrile respiratory illness were used in this study. Swabs which contained influenza virus (as determined by isolation and identification following growth in primary rhesus monkey kidney cells) were stored at $-70\,^{\circ}$C in transport medium. Virus used in this study was reisolated directly from the transport medium by inoculation into the allantoic cavity of 10-day-old embryonated chicken eggs. Virus was harvested after 48 h at $35\,^{\circ}$C and stored at $-70\,^{\circ}$C.

Avian influenza virus of the H1 subtype (A/duck/Wisconsin/1938/80) was obtained from duck intestinal tract tissue (kindly supplied by Dr. Virginia Hinshaw, University of Wisconsin-Madison) and was grown in Madin Darby canine kidney (MDCK) cells for 10 passages prior to use in these experiments. Plaque-purified influenza A (H1N1), strain WSN, which had been grown in Madin Darby bovine kidney (MDBK) cells was also used in these studies.

RNA Extraction. When extracting RNA from nasopharyngeal virus, both virus and cellular material in the transport medium were concentrated by high-speed centrifugation. RNA was extracted from an aliquot containing 50–100 hemagglutinating units (HAU) of virus, using phenol-chloroform following digestion with proteinase K. Since no attempt was made to remove cellular debris prior to extraction of the RNA, the extracted RNA was treated with RNase-free DNase prior to cDNA synthesis.

We have also used a guanidinium thiocyanate procedure for RNA extraction (Chomczynski and Sacchi 1987). In our hands, RNA prepared in this way proved to be less useful for production of full-length copies of amplified HA DNA. It has, however, been used for the synthesis and amplification of short stretches of DNA representing internal HA sequences (Bressoud et al. 1990).

Synthesis of cDNA and Hemagglutinin Gene Amplification Using PCR. cDNA copies of the virion RNAs were made using avian myeloblastosis virus (AMV) reverse transcriptase (RT) and a primer consisting of the sequence d(AGCAAAAGCAGG). Since this 12-nucleotide sequence is complementary to the 3′-prime-ends of all of the virion segments, the RT product contained DNA transcripts of all of the viral genes.

HA1 and part of the HA2 region were then amplified using PCR. The reaction mixtures (100 µl) consisted of the cDNA:RNA complex obtained by reverse transcription, 10 mM TRIS-HCl, pH 8.3, 50 mM KCl, 1.5 mM MgCl$_2$, 0.001% gelatin, 2.5 U Taq polymerase, 200 µM of each dNTPs, and 20 pmol of each of two primers (see Results). The reaction mixtures were overlaid with light mineral oil, and amplification was carried out for 25 cycles each consisting of 1 min of denaturation at $94\,^{\circ}$C, 2 min at $35\,^{\circ}$C or $50\,^{\circ}$C for annealing, and 2 min at $72\,^{\circ}$C for extension. Following electrophoresis in agarose gels, DNA of appropriate size for the primer pair was recovered and reamplified to ensure adequate amounts of DNA for multiple sequencing reactions.

We have found that pretreatment of the RNA with methyl mercuric hydroxide as described (Davis et al. 1986) facilitates the synthesis of full-length

HA cDNAs by RT. Methyl mercuric hydroxide reduces the secondary structure in the RNA.

Nucleotide Sequence Analysis. DNA copies of the HA gene were sequenced directly from the amplified pool by the dideoxy termination method using Sequenase kits. Approximately 300–400 ng of amplified DNA was used for each reaction.

Results

Sequence of an H1 HA Determined from PCR-Amplified DNA

In our published work using PCR (Rajakumar et al. 1990), we studied the part of the HA gene which encodes the HA1 region and the HA1/HA2 cleavage site (nucleotides 1–1095 of the plus-strand RNA). These regions were studied because they have been implicated in antigenic variation, receptor binding, and host-dependent selective processes.

The primers used to amplify HA DNA from cDNA are shown in Fig. 1. The regions of the HA gene represented in the DNA synthesized using various primer pairs are also shown. PCR was used to amplify cDNA made from virus taken directly from the nasopharynx (strain A/SL/2/87) and from virus which had been grown for one passage in embryonated chicken eggs. The two determinations yielded identical amino acid sequences (Rajakumar et al. 1990), indicating that growth in embryonated chicken eggs did not result in selection

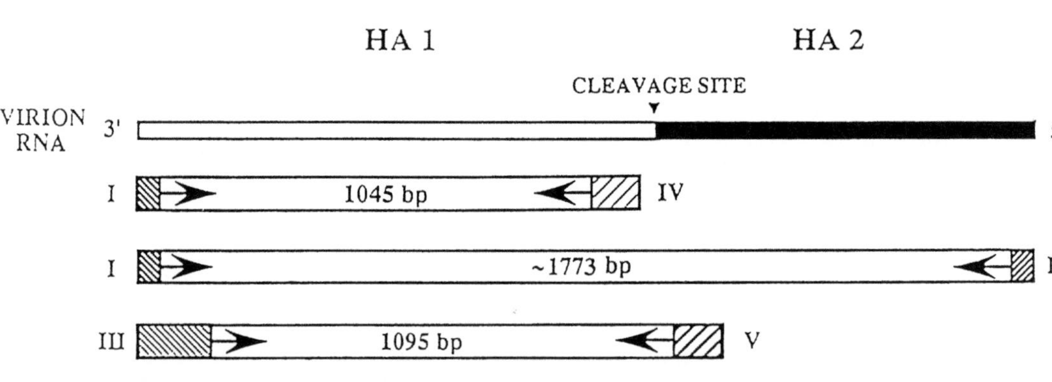

Fig. 1. Primers used to amplify hemagglutinin (HA) cDNA. Primer *I*, d(AG-CAAAAGCAGG), is complementary to the first 12 nucleotides at the 3'-end of all of the virion RNAs, primer *II*, d(AGTAGAAACAAGG), to the first 13 nucleotides at the 3'-end of all cDNAs, primer *III*, d(AGCAAAAGCAGGGGAAAATAAAAACAACCAAAATG), is complementary to the first 35 bases at the 3'-end of A/SL/2/87 HA virion RNA, primer *IV*, d(ATGTTCCTTAGTCCTGTAACCAT), to nucleotides 1045–1023 of the HA cDNA, and primer V, d(CAATGAAACCGGCAATGGCTCC), to nucleotides 1095–1074 of A/USSR/77 HA cDNA. (Reproduced from Rajakumar et al. 1990).

of virus with an altered HA sequence. Since both of these sequences were determined directly from the PCR-generated pool of DNA, it represents the majority sequence of both the nasopharyngeal and the egg-grown virus populations. In subsequent experiments, we have investigated the heterogeneity of the HA gene within the nasopharyngeal virus population. Our data indicate that sequence differences in individual HA genes are not detected by the technique used here and that sequencing the DNA directly from the amplified pool does, indeed, generate the same majority sequence as that obtained by sequencing a large number of cloned HA genes (A. Rajakumar et al., manuscript in preparation).

Effect of Primer Sequence on the Amplified DNA

In the process of preparing amplified DNA with different primer sets, we observed significant differences in the properties of the PCR product. Amplified DNA about 1000 bp long was obtained from two clinical samples, A/SL/1/87 and A/SL/2/87, using a primer which is complementary to the conserved 3'-terminus of all genome segments (primer I) and a HA-specific primer (primer IV). The amplified DNA obtained from the two clinical samples is shown in Fig. 2 A. This DNA could be resolved into three bands, one about 1050 bp in length, and two which were slightly smaller (Fig. 2 B, lane 1). The individual bands appeared to represent discrete populations of DNA molecules since each retained its original mobility when excised from the gel and subjected to a second round of electrophoresis (Fig. 2B, lanes 2–4). The origin and significance of these bands were investigated to determine how, if at all, they were related to the HA genome segment. We found that the amounts of the two smaller bands obtained using the primers described above could be reduced by increasing the annealing temperature to 50 °C and could be totally eliminated by using primers III and IV. In addition, Southern blots using HA-specific probes indicated that only the largest of the three DNA bands contained HA gene sequences. Lastly, limited sequence analysis of the smaller DNAs indicated that they were unrelated to any influenza virus gene, whereas the largest DNA gave the sequence from which the published one was derived. Thus, we concluded that the primer set I and IV was capable of amplifying DNA which was unrelated to the influenza virus genome, although the larger of the two (primer IV) was designed to be specific for the HA gene. The possibility suggested earlier (Rajakumar et al. 1990) that these DNAs represented defective HA genes with internal deletions was therefore eliminated.

Further evidence that primers which are complementary to the conserved regions at the ends of the genome segments are too short to confer adequate specificity is presented in Fig. 3. Agarose gel patterns of PCR-amplified DNA obtained from an avian strain of influenza A using two sets of primers are presented. When these short primers (primers I and II, consisting of 12 and 13 nucleotides, respectively) were used and annealing was carried out at 35 °C (lane 2), an array of bands of different sizes was observed, none of which

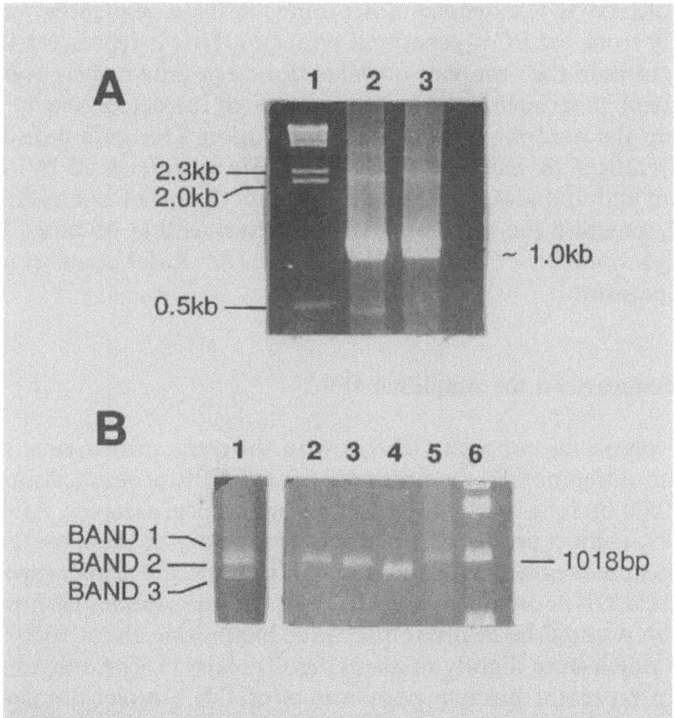

Fig. 2A, B. Agarose gel electrophoresis of the amplified DNA. **A** λ-*Hin*dIII marker DNA (*lane 1*), amplified DNA copies of HA RNA from two clinical samples, A/SL/1/87 (*lane 2*) and A/SL/2/87 (*lane 3*) using primers I and IV. **B** Amplified A/SL/2/87 HA DNA separated into three bands (*lane 1*), purified bands 1, 2, and 3 (*lanes 2, 3, and 4*, respectively), WSN (W. Smith Neurotropic, variant of original 1933 human isolate) HA DNA amplified as in **A** (*lane 5*), marker DNA (*lane 6*). (Modified from Rajakumar et al. 1990)

represented the entire HA gene (1.78 kb). The region of the gel which should have contained DNAs from 0.8 to 2.2 kb in length was then excised, and the DNA was purified and reamplified using the same primers at an annealing temperature of 35 °C. Again, only short DNA copies were found (lane 3). The T_m of each of these primers is approximately 36 °C, requiring that relatively low annealing temperatures be used. When longer primers designed to amplify the HA DNA (primers VI and VII; see legend to Fig. 3) were used and annealing was carried out at 35 °C, the PCR product contained DNA the size of the complete HA gene (compare lanes 4 and 6). Because of the higher T_m of these longer primers, the annealing temperature could be increased, and (as shown in lane 5) slightly more HA DNA was obtained when annealing was carried out at 50 °C. Lanes 4 and 5 show, however, that large amounts of a small DNA was still made at both temperatures when the more specific primers were used. Sequence analysis of this DNA has indicated that it is a copy of the influenza virus nonstructural protein (NS) gene. It was amplified along with the HA

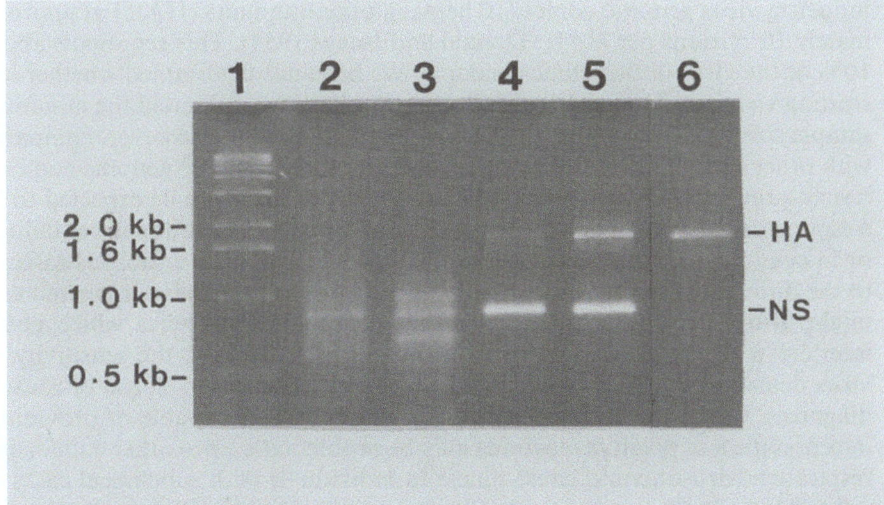

Fig. 3. Agarose gel electrophoresis of amplified DNA obtained from an avian H1 strain using different primers and annealing temperatures. *Lane 1*, marker DNA; *lane 2*, cDNA amplified using primers I and II at 35 °C. DNA eluted from lane 2 was reamplified at 35 °C (see text) using primers I and II (*lane 3*) or primers VI, d(GCGTGGATCCAG-**CAAAAGCAGGGGA**), and VII, d(CGCGAAGCTT**AGTAGAAACAAGGGTGTT**) (*lane 4*). *Lane 5*, same as lane 4 except that the annealing temperature was 50 °C. *Lane 6*, WSN DNA amplified using conditions as in lane 5. Sequences complementary to viral DNA are shown in *bold*. The 5′-extensions were used to add restriction sites for use in other experiments. *HA*, hemagglutinin gene; *NS*, nonstructural protein gene

DNA because the terminal sequences adjacent to the conserved regions are highly similar in these two genes. Thus, increasing the length of the primer and using a higher annealing temperature reduced nonspecific binding (binding at sites other than the target site) and increased the synthesis of the two specifically primed influenza DNAs.

Discussion

Our success and that of others in using PCR to investigate HA sequences directly from the upper respiratory tract (Rajakumar et al. 1990; Bressoud et al. 1990; Katz et al. 1990; Robertson et al. 1990) suggest that this procedure may be useful for the diagnosis of influenza and for the characterization of new virus isolates. In the development of these procedures, the following factors must be kept in mind.

1. New diagnostic procedures must be at least as sensitive for detecting virus as those already in use. When we sequenced the HA genes directly from the nasopharynx, we started PCR reactions with cDNA obtained from about 10^8

influenza virus genome copies (10 hemagglutinating units (HAU) at approximately 10^7 virions per HAU; Donald and Isaacs 1954). This represents about 10% of the virus in this clinical sample. We have not determined whether less starting virus would have been sufficient, nor have we estimated the minimum sample size needed to infect rhesus monkey kidney cells. However, compared with other uses of PCR following reverse transcription, 10^8 genome copies is a very large number with which to start. PCR can therefore be expected to be a more sensitive way to detect the presence of virus than isolation in cell culture or in eggs. In addition, detection by PCR would eliminate problems relating to the differences in the susceptibility of cell lines to influenza virus strains. It might also detect virus in samples that contained antibodies which could interfere with the growth of virus in cell cultures. Clearly, the sensitivity of virus detection by PCR would need to be evaluated prior to use in routine diagnosis. Given the extreme sensitivity which PCR is capable of providing, interpretation of positive reactions may be problematic since other coinfecting respiratory viruses could cause illness in individuals with subclinical cases of influenza.

Since influenza is an RNA virus, the necessity for reverse transcription prior to amplification places additional constraints on the utility of the system. In fact, the success of the procedure described here depends largely on the quality of the RNA used to generate the cDNA which is to be amplified. In a diagnostic laboratory setting, the established techniques would therefore have to ensure adequate cDNA synthesis from all influenza viruses including previously uncharacterized strains, such as those which might be in clinical samples at the beginning of an influenza virus season. For example, we have encountered problems due to secondary structure in the RNA with some virus strains and not with others, even within the same influenza A subtype. In our hands methyl mercuric hydroxide has proven useful for eliminating these effects, but the hazardous nature of this compound calls into question its use in a routine diagnostic procedure. Other denaturing agents such as heat or formaldehyde may be useful. The use of a single enzyme from *Thermus thermophilus* for reverse transcription at high temperature (thereby eliminating problems associated with RNA secondary structure) and PCR amplification (Myers and Gelfand 1991) may provide a simple solution to this problem.

Lastly, extreme care must be taken to avoid DNA contamination, especially since both negative and positive samples would be processed at the same time throughout the influenza season. Cross-contamination between patient samples as well as contamination with DNA produced in previous amplifications are major problems. Such contamination could constitute a serious impediment to the routine use of PCR in the diagnosis of a disease such as influenza. Laboratory practices for reducing potential contamination and protocols for "PCR sterilization" have been reviewed elsewhere (Erlich et al. 1991; Persing 1991).

2. New diagnostic procedures must provide as much information about the virus isolates as do existing procedures. In the case of influenza diagnosis, this requires identification of new isolates as type A or B, and when type A is

found, the subtype must be determined. This can be accomplished only if the authenticity of the PCR-amplified DNA is established either by hybridization with appropriate probes or by sequence analysis. Routine use of PCR for diagnostic purposes therefore requires (a) that the techniques used ensure production of amplified DNA from the appropriate influenza virus genes and (b) that probes are available for identifying the virus strains.

With respect to obtaining DNA copies of the appropriate influenza virus genes, we show here the effects of insufficient primer specificity on the amplified DNA. DNA of the expected size may be made in such small amounts that it cannot be detected visually as a band on an agarose gel after a single amplification. Nonspecific priming due to inadequate primer length can generate DNAs of various sizes which may or may not be related to the influenza virus genome. We also show that the PCR product may contain specifically primed DNAs of more than one kind if, as with the HA and NS genes, a single primer set amplifies both DNAs with comparable efficiency. These observations emphasize the need for primers of adequate specificity for the production of the amplified DNA.

Based on our observations and the use of PCR in other systems, one can envision procedures which would detect influenza viruses and identify their type by using gene-specific primer pairs to amplify the DNA of a number of the influenza virus genes within one PCR reaction. Hybridization probes could then be used to identify the virus as type A or B. These probes would consist of sequences found in the matrix and/or the nucleoprotein genes of one of these types but not the other. For example, the matrix protein gene of influenza type B strains is about 160 nucleotides longer than that of type A, and various regions of the influenza B gene are unrelated in sequence to the comparable regions in influenza A (Briedis et al. 1982). This rapid identification by type would permit early administration of amantadine therapy in the case of type A but not B infections (Murphy and Webster 1990).

Identification of influenza A strains by subtype could also be done from the same amplified DNA using probes against regions of the HA gene which show little variation over time. For example, few amino acid substitutions are seen along the stalk (Wiley and Skehel 1990), although H1 strains differ from H3 strains in this region. Probes consisting of the noncoding region at the 3'-end of the HA gene of these two subtypes could also be used for subtype identification (Nobusawa et al. 1991). The use of probes for conserved regions within an influenza A subtype would enable diagnostic laboratories to identify strains by subtype despite the time-dependent strain variation (antigenic drift) which presently necessitates frequent introduction of new reagents.

The introduction of stable reagents for subtype identification would, however, increase the need for surveillance. Amino acid substitutions on the tip of the HA which induce antigenic changes within a subtype would go undetected by probes for conserved sequences. Complete strain characterization would, therefore, still be needed as early as possible within an influenza season so that judgements could be made about the potential efficacy of the vaccines in current use. This would be best accomplished by using PCR-amplified HA

DNA to obtain the sequence of virus strains isolated throughout the influenza season.

In carrying out these sequence determinations, care should be taken to avoid errors due to nucleotide misincorporation by Taq polymerase so that the sequence obtained will be as close as possible to that in the clinical samples. Misincorporations by Taq polymerase are rare (Erlich et al. 1991) and essentially random, and only those which occur during the first few cycles of amplification will be present in the PCR product in sufficient numbers to be observed by direct sequencing. Starting the first cycle with a large number of templates as we have done also reduces the probability that specific misincorporations will be detected, unless cloned DNA is used to determine the sequence. It is therefore advantageous to sequence directly from the amplified pool or from a mixed population of DNA clones (Padgett et al. 1991) so that the sequence of the majority of the HA genes in the virus population is obtained.

3. Other factors being comparable, new diagnostic procedures must be as rapid and cost effective as those already in use. Isolation and identification of influenza virus isolates, including determination of influenza A strains, requires from 2 to 5 days, depending on the amount of virus shed at the time of sampling. Some 80% of positive samples can be identified and characterized within 3 days (Minnich and Ray 1987). A procedure requiring virus concentration, RNA extraction, reverse transcription, DNA amplification, and strain identification by hybridization techniques would require a minimum of 2–3 days. It would, however, require more personnel time than the present method of strain identification and would be more costly to carry out. With respect to time and effort, the use of PCR presently has little, if any, advantage over virus isolation and identification.

In conclusion, the use of PCR for the identification and characterization of influenza infection, while presently feasible, requires refinement in order for the process to be sensitive, reliable, rapid, and cost effective. Given the progress in the field, improvement in both reagents and equipment should make this possible within the near future. The technique is clearly already well enough developed so that it is very valuable for epidemiological studies and vaccine design.

Acknowledgements. We wish to thank Mrs. Tana M. McKerrow for skilled technical assistance.

References

Aytay S, Schulze IT (1991) Single amino acid substitutions in the hemagglutinin can alter the host range and receptor binding properties of H1 strains of influenza A virus. J Virol 65:3022–3028

Bressoud A, Whitcomb J, Pourzand C, Haller O, Cerutti P (1990) Rapid detection of influenza virus H1 by the polymerase chain reaction. Biochem Biophys Res Comm 167:425–430

Briedis DJ, Lamb RA, Choppin PW (1982) Sequence of RNA segment 7 of the influenza B virus genome: partial amino acid homology between the membrane proteins (M1) of influenza A and B viruses and conservation of a second open reading frame. Virology 116:581–588

Chomczynski P, Sacchi N (1987) Single step method of RNA isolation by acid guanidinium thiocyanate-phenol-chloroform extraction. Anal Biochem 162:156–159

Davis LG, Dibner MD, Battey JF (1986) Basic methods in molecular biology. Elsevier, New York, pp 199–202

Desselberger U, Racaniello VR, Zazra JJ, Palese P (1980) The 3' and 5' terminal sequence of influenza A, B and C virus RNA segments are highly conserved and show partial inverted complementarity. Gene 8:315–328

Donald HB, Isaacs A (1954) Counts of influenza virus particles. J Gen Microbiol 10:457–464

Erlich HA, Gelfand D, Sninsky JJ (1991) Recent advances in the polymerase chain reaction. Science 252:1643–1651

Katz JM, Naeve CW, Webster RG (1987) Host cell-mediated variation in H3N2 influenza viruses. Virology 156:386–395

Katz JM, Wang M, Webster RG (1990) Direct sequencing of the HA gene of influenza (H3N2) virus in original clinical samples reveals sequence identity with mammalian cell-grown virus. J Virol 64:1808–1811

Kawaoka Y, Yamnikova S, Chambers TM, Lvov DK, Webster RG (1990) Molecular characterization of a new hemagglutinin, subtype H14, of influenza A virus. Virology 179:759–767

Minnich LL, Ray CG (1987) Early testing of cell cultures for detection of hemadsorbing viruses. J Clin Microbiol 25:421–422

Murphy BR, Webster RG (1990) Orthomyxoviruses. In: Fields BN, Knipe DM (eds) Virology, vol 1, 2nd edn. Raven, New York, pp 1091–1152

Myers TW, Gelfand DH (1991) Reverse transcription and DNA amplification by a *Thermus thermophilus* DNA polymerase. Biochemistry 30:7661–7666

Nobusawa E, Aoyama T, Kato H, Suzuki Y, Tateno Y, Nakajima K (1991) Comparison of complete amino acid sequences and receptor-binding properties among 13 serotypes of hemagglutinins of influenza A viruses. Virology 182:475–485

Padgett KA, Mullersman JE, Saha BK (1991) A novel approach to sequencing DNA amplified by the polymerase chain reaction. Biotechniques 11:56–58

Persing DH (1991) Polymerase chain reaction: trenches to benches. J Clin Microbiol 29:1281–1285

Rajakumar A, Swierkosz EM, Schulze IT (1990) Sequence of an influenza virus hemagglutinin determined directly from a clinical sample. Proc Natl Acad Sci USA 87:4154–4158

Robertson JS, Bootman JS, Nicolson C, Major D, Robertson EW, Wood JM (1990) The hemagglutinin of influenza B virus present in clinical material is a single species identical to that of mammalian cell-grown virus. Virology 179:35–40

Saiki RK, Gelfand DH, Stoffel S, Scharf SJ, Higuchi R, Horn GT, Mullis KB, Erlich HA (1988) Primer-directed enzymatic amplification of DNA with a thermostable DNA polymerase. Science 239:487–491

Schulze IT (1987) Genetic changes in the influenza viruses during growth in cultured cells. In: Maramorosch K (ed) Advances in cell cultures, vol 5. Academic, Orlando, pp 59–96

Wiley DC, Skehel JJ (1990) Viral membranes. In: Fields BN, Knipe DM (eds) Virology, vol 1, 2nd edn. Raven, New York, pp 63–85

Chapter 21 Polymerase Chain-Reaction (PCR) Detection of Rhinoviruses

Tim Skern[1], Heinrich Kovar[2], Gunhild Jug[2], Herbert Auer[3],
Helge Torgensen[1], Klaus Hartmuth[1], and Dieter Blaas[1]

Summary

The polymerase chain reaction (PCR) was applied to the problem of rhinoviral serotyping and the detection of mutations in large DNA fragments. For serotyping, a region of about 380 nucleotides from the 5'-untranslated region of the genomes of six human rhinovirus serotypes was amplified by the PCR using one pair of primers. The DNAs were cleaved with selected restriction enzymes, and the fragments were analyzed on polyacrylamide gels. The cleavage patterns enabled the characterization of the different serotypes.

For the detection of mutations, cDNAs corresponding to the whole capsid region of rhinovirus variants stable at low pH were amplified. Restriction fragments of the amplified DNAs were analyzed on two-dimensional gels, separating in the first dimension according to size and in the second dimension according to the secondary structure of the single strands. When wild type and mutants were compared, a different migration of the single strands from two fragments was observed. The fragments of interest were eluted from the gel and directly sequenced. The method presented thus allows rapid identification of the location and of the exact nature of point mutations.

Introduction

Human rhinoviruses, the main causative agent of the common cold, constitute a serologically heterogeneous group belonging to the family *Picornaviridae* (Stott and Killington 1972; Hamparian et al. 1987). They contain an RNA genome of positive polarity encoding a polyprotein of about 250 kDa which is autocatalytically processed to the mature viral proteins (Stanway 1990; Rueckert 1990). The first processing event is the severing of the P1 from the P2–P3 region (for nomenclature, see Rueckert and Wimmer 1984) catalysed by the

---

[1]   Institute of Biochemistry, University of Vienna, Währingerstrasse 17, A-1090 Vienna, Austria
[2]   Children's Cancer Research Institute, St. Anna Kinderspital, Kinderspitalgasse 6, A-1090 Vienna, Austria
[3]   Institute of Molecular Pathology, Dr. Bohr-Gasse 7, A-1030 Vienna, Austria

proteinase 2A (Sommergruber et al. 1989). P1 is subsequently cleaved by the second viral proteinase 3C, giving rise to the capsid proteins VP0, VP1, and VP3; cleavages in P2 and P3 lead to the nonstructural proteins. VP0 is processed into VP2 and VP4 by an unknown proteinase during the assembly of the capsid with the genomic RNA (Palmenberg 1987; Harber et al. 1991).

Within the picornavirus family, rhinoviruses are unique in possessing at least 102 antigenically distinct serotypes (Hamparian et al. 1987; Uncapher et al. 1991). Thus, hyperimmune sera prepared against each rhinovirus serotype show only low crossreactivity (Halfpap and Cooney 1983); such sera are used for the assignment of unknown isolates by the neutralization of viral infectivity (Cooney et al. 1982; Hamparian et al. 1987; Kellner et al. 1988). The molecular basis of this serotypic diversity has been elucidated in the last few years, using techniques of molecular cloning and sequencing and X-ray crystallography and by examining viral mutants which can no longer be neutralized by monoclonal antibodies.

At present, the complete sequences of the genomes of four serotypes have been published (HRV1B, HRV2, HRV14, HRV89; Hughes et al. 1988; Stanway et al. 1984; Skern et al. 1985; Callahan et al. 1985; Duechler et al. 1987); the sequences of two additional serotypes have also been determined (HRV9, Leckie and Almond, pers. comm.; HRV85, Stanway, pers. comm.). Furthermore, the three-dimensional structure of the icosahedral capsid of two representatives (HRV14 and HRV1A) have been solved by X-ray crystallography (Rossmann et al. 1985; Kim et al. 1989). These data reveal a close similarity of the derived amino acid sequences of the diverse serotypes, with the nonstructural proteins less diverged than the capsid proteins, which are subject to immunological pressure.

The greatest divergence between the capsid proteins VP1, VP2, and VP3 was found to be at the surface loops which connect the characteristic eight stranded β-barrel structures. It is these loops which are changed in viral mutants capable of escaping neutralization by monoclonal antibodies (Sherry et al. 1986; Rossmann et al. 1985; Skern et al. 1987; Duechler et al. 1987). Therefore, the differences in these four loops are responsible for the production of antibodies which are specific for each serotype and allow them to be distinguished from one another.

Application of PCR Technology to Problems in Rhinovirus Research

We have applied PCR technology to the identification of a given rhinovirus serotype and the localization of point mutations in the viral capsid region (Torgersen et al. 1989b; Kovar et al. 1991). The diagnosis of a rhinovirus infection itself is a difficult procedure, as it is based on the physical characteristics of viral isolates from nasal secretions. Furthermore, the identification of a given serotype is dependent on laborious neutralization assays of viral infectivity by reference antisera. The first part of this chapter describes the application of PCR to this problem.

The second part deals with the identification of point mutations in the viral capsid region. Viral mutants raised, for example, by selection against changes in the environment may contain just one or two alterations in the capsid sequence; as this region encompasses about 2.5 kb, detection of mutations by cDNA cloning and DNA sequencing is a tedious task. We describe a method which allows the rapid localization of the mutation(s) and also the determination of their sequence.

Materials and Methods

Viruses

HRV2 was originally obtained from Dr. Tyrrell, Common Cold Centre, Salisbury, UK; HRV14 (kindly provided by Dr. A. Popow, Institute for Virology, Vienna) and all other strains were from the American Type Culture Collection (ATCC). All viruses were plaque purified before use and were prepared as described (Skern et al. 1985, 1991).

cDNA Synthesis from the 5′ Untranslated Region and PCR Amplification

RNA was prepared by treatment of polyethylene glycol 6000 (PEG)-concentrated viral suspension or of cleared cell lysate with 1% sodium dodecyl sulfate (SDS) and 10 mM ethylene diamine tetra-acetic acid (EDTA). After extraction with phenol/chloroform, carrier tRNA was added, and the RNA was precipitated with ethanol. cDNA was synthesized using the entire RNA preparation, 10 pmoles of primer 2 (complement to nucleotides 531–544 of the HRV2 sequence), and 10 U of reverse transcriptase (Super RT, Anglian Biotechnology) in 20 µl final volume following the supplier's instructions.

PCR was carried out in 50 µl total volume with 10 µl of the cDNA preparation, 100 pmoles each of primer 1 (corresponding to nucleotides 161–181 of the HRV2 sequence) and primer 2, 0.4 mM of all four dNTPs, 2 U of Taq DNA polymerase (Cetus) in the supplied buffer as described (Torgersen et al. 1989a) for 30 cycles at settings of 92 °C for 2 min, 40 °C for 3 min, and 70 °C for 3 min. The reaction mixtures (with or without restriction digestion) were analyzed directly on 6% polyacrylamide gels (Maniatis et al. 1982).

cDNA Synthesis and Amplification of the Entire Viral Capsid Region

First strand cDNA from the 5′-end of viral RNA spanning the capsid region was synthesized by reverse transcription using 90 pmol of primer 4 (complementary nucleotides 3262–3243 of the HRV14 sequence) and was desalted by spun-column chromatography through Sephadex G50 (Maniatis et al. 1982).

Half of the cDNA was added to 40 pmol of oligonucleotide 3 (corresponding to nucleotides 546–562 of the HRV14 sequence), and the mixture was lyophilized. The DNA was dissolved in 10 µl of PCR reaction mix containing 67 mM TRIS-HCl (pH 8.8), 16 mM $(NH_4)_2SO_4$, 6.7 mM $MgCl_2$, 6.7 µM EDTA, 10 mM 2-mercaptoethanol, dNTPs at 1.5 mM each, 170 µg/ml bovine serum albumin (DNase free, Pharmacia), 2.5 U Taq-polymerase (AmpliTaq, USB) and overlaid with paraffin oil (Jeffrey et al. 1988). For direct labelling of PCR products, 10–30 µCi α-[^{32}P]dCTP was included. The DNA was denatured at 95 °C for 10 min, and PCR was performed at 95 °C for 1 min, 56 °C for 1 min, and 70 °C for 15 min for 15 cycles followed by one annealing step at 56 °C for 1 min and one extension step at 72 °C for 15 min.

Two-Dimensional, Single-Strand Polymorphism Analysis

Restriction enzymes cleaving the DNA into fragments of a size easily resolved on polyacrylamide gels were determined with the program "SMALLS". This program, written in Fortran, searches for all combinations of up to three enzymes yielding fragments within given size limits.

Restriction digestions of PCR products were performed according to the manufacturer's recommendations followed by phenol/$CHCl_3$/isoamylalcohol extraction. Nucleic acids were precipitated in the presence of 1 µg of tRNA and dissolved in 5 µl water, lyophilized, and taken up in 2 µl 50% formamide, 10 mM EDTA, 0.025% bromophenol blue, 0.025% xylene cyanol. Samples were heated to 80 °C for 5 min and immediately applied to denaturing tube gels. First dimension electrophoresis was carried out in siliconized glass capillaries (60 mm, inner diameter 0.6–0.8 mm) which had been filled with a denaturing gel mix consisting of 6% polyacrylamide and 7.6 M urea in TBE (0.089 M TRIS-borate, 0.089 M boric acid, 2 mM EDTA) using a syringe. Electrophoresis was carried out at 65 °C for 10–30 min at 400 V in preheated TBE buffer in a tube cell with a water-thermostable casing. A thorough degassing of the gel mix and the hot electrophoresis buffer prior to use was found to be essential.

Tube gels were expelled from the capillaries using compressed air and placed horizontally on top of a 0.7 mm slab gel (33 × 41 cm) made of 6% polyacrylamide, 90 mM TRIS-borate, pH 8.3, and 4 mM EDTA and were overlaid with the same buffer and a few microliters of tracking dye (0.1% bromophenol blue in 4 M urea, 50% sucrose, 50 mM EDTA). Electrophoresis was performed in a nucleic acid sequencing cell at 4 °C using precooled buffer at 30 W for 5–15 min and then overnight at 3–4 W. Gels were dried on filter paper and autoradiographed.

Sequencing of Products. PCR products were electroeluted from the polyacrylamide gel and sequenced (Sanger et al. 1977) using T7 DNA polymerase (Pharmacia).

Sequencing of Generated Fragments. Unlabelled cDNA was amplified and restricted with *Hae*III. An aliquot of the fragments was dephosphorylated by calf intestine phosphatase, 5'-end-labelled with 50 µCi γ-[^{32}P]ATP by T4 polynucleotide kinase, and subjected to 2-dimensional, single-strand, conformational polymorphism (2DSSCP) analysis. Spots corresponding to polymorphic single strands were cut from the dried gel and submerged in 50 µl of 10 mM TRIS-HCl (pH 7.6) and 1 mM EDTA. The filter paper was removed from the swollen gel and DNA was eluted at 37 °C with one change of buffer in Eppendorf vials overnight. All vials were preincubated with denatured salmon sperm DNA for 1 h at 37 °C to minimize loss of labelled single-stranded DNA due to adsorption. Eluted DNA was recovered by centrifugation through siliconized glass wool. The remainder of unlabelled restricted PCR product was added, the mixture heated to 95 °C for 5 min and held at 37 °C for 30 min in order to convert the labelled ssDNA to dsDNA. Salmon sperm DNA (5 µg) was added, and the DNA was precipitated, washed with 70% ethanol, and subjected to Maxam-Gilbert (1980) sequencing.

Results

Use of Sequence Variation in the 5'-Untranslated Region of the HRV Genome for Typing

The 5'-untranslated region (5'-UTR) of the RNA genome of human rhinoviruses comprises about 600 bases. When the six known sequences are aligned, it becomes apparent that this region contains two blocks of identical sequence stretches flanking variable and for each serotype characteristic nucleotide sequences. As these blocks of 21 and 23 nucleotides (beginning at positions 161 and 531, respectively; numbering according to HRV2, Skern et al. 1985) are at a distance of some 370 bases, they are well suited to amplification using the PCR (Saiki et al. 1988).

Amplification experiments were carried out using serotypes HRV1A, HRV1B, HRV2, HRV14, HRV49, HRV70, and HRV89. In all cases, a DNA fragment of about 380 bp was obtained (data not shown). In the absence of sequence information from serotypes HRV1A, and HRV49, the amplified DNA was eluted from the gel, and 219 and 241 nucleotides were determined by the chain termination method for HRV1A and HRV49, respectively. Based on these sequences, restriction enzymes were selected which would allow unambiguous identification of serotypes (Table 1); in case of HRV1A and HRV49 it was assumed that no further sites for the selected enzymes were present in the regions which had not been sequenced.

The rationale behind the approach for serotype identification using enzymatic digestion of the fragments is shown in Fig. 1 for six HRVs; the scheme can easily be extended to accommodate further serotypes. As the enzyme *Ban*II cuts the amplified fragment of HRV2, HRV49, and HRV89 but not

Table 1. Restriction sites present in the 380-bp amplified fragment

	Serotype					
	HRV1A	HRV1B	HRV2	HRV14	HRV49	HRV89
*Ban*II	–	–	255,129	–	255,129	298,90
*Bgl*II	281,106	281,106	–	–	–	–
*Dra*III	–	–	–	–	212,172	–
*Eco*RI	–	–	–	215,165	–	–
*Hin*PI	208,179	–	NU	–	NU	–
*Hin*dIII	–	–	297,87	–	–	–
*Pvu*II	NU	260,127	–	–	–	–
*Rsa*I	NU	–	–	NU	–	294,94

Sizes are given in base pairs. A dash indicates that no site is present for a particular enzyme. NU indicates a site is present but that it was not used in this study

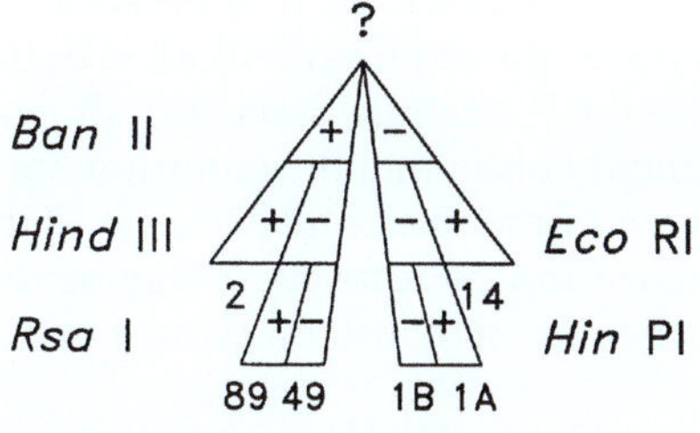

Fig. 1. Rationale of hierarchical polymerase chain reaction (PCR)/restriction enzyme based rhinovirus typing. The 5′-untranslated region of the viral RNA is amplified, and the DNA is restricted by selected endonucleases. Polyacrylamide gel electrophoretic analysis for the presence or absence of a restriction site enables the identification of a particular rhinovirus serotype

those of HRV1A, HRV1B and HRV14, the six serotypes can be divided into two groups. Similarly, *Hin*dIII and *Eco*RI were used to subdivide the *Ban*II positive serotypes (*Hin*dIII) and the *Ban*II negative serotypes (*Eco*RI). The gel analysis of the fragments thus obtained allows the identification of HRV2 and HRV14. The remaining serotypes of each branch can be identified by restriction with the additional enzymes, *Rsa*I (*Ban*II positive) and *Hin*PI (*Ban*II negative). It should be pointed out that the fragment lengths contain additional information and were also used for discrimination of the various serotypes. The results of the restriction analysis are shown in Fig. 2. Fragments from each serotype were restricted with *Ban*II and the products analysed on polyacrylamide gels. The amplified fragments of HRV2, HRV49, and HRV89 are cleaved by this enzyme (Fig. 2a), whereas the fragments from HRV1A, HRV1B, and HRV14 are not (Fig. 2b). Furthermore, the different lengths obtained of the fragments from HRV2 and HRV89 allow these serotypes to be distinguished from one another (see Table 1). The characteristic *Hin*dIII site of HRV2 and the *Rsa*I site of HRV89 confirmed this identification (the 96-bp *Rsa*I fragment is too faint to be seen on this gel). HRV2 and HRV49 could be differentiated using the absence of the *Hin*dIII site and the presence of the *Dra*III site in HRV49.

For the serotypes not possessing a *Ban*II site (Fig. 2b), HRV14 was identified with the *Eco*RI site. Since HRV1B and HRV1A both have a *Bgl*II site at the same position, the *Hin*PI site in HRV1A allowed discrimination between the two serotypes. The presence of only two fragments with *Hin*PI indicates that this site is unique in the amplified fragment.

Identification of Mutations in the Capsid Region of HRV14 Acid-Resistant Mutants

Rhinovirus mutants have been used in the determination of antigenic sites by analyzing the sequence of variants escaping neutralization by monoclonal antibodies (Sherry and Rueckert 1985; Sherry et al. 1986; Skern et al. 1987; Appleyard et al. 1990), for the localization of amino acid changes involved in low pH-resistant HRV14 mutants (Skern et al. 1991), and for the identification of the viral proteins involved in certain steps of the viral life-cycle. Although the approximate region of the genome to be examined is known in these cases (e.g. the capsid region), cloning and sequencing of the required 2–3 kb is both labor-intensive and time-consuming. The 2DSSCP analysis described allows first the identification of a small cDNA fragment containing the mutation and second permits the determination of the nature of the nucleotide change. The methodology is based on 2-dimensional separation of restriction fragments obtained from PCR products encompassing the region of interest (Kovar et al. 1991).

HRV14 variants resistant to treatment at pH 4.5 (HRV14-as) have been characterized previously (Skern et al. 1991). The genomic regions between nucleotides 546 and 3262 of the various mutants were amplified; *Hae*III was selected as the most suitable restriction enzyme by the program "SMALLS." This was in spite of the fact that out of the eight predicted fragments [300 (A), 98 (B), 378 (C), 902 (D), 155 (E), 66 (F), 273 (G), and 544 (H) bp], three (C, D, and H) were expected not to be resolved in the nondenaturing gel. The labelled restriction fragments were analyzed by 2DSSCP. As shown in Fig. 3, all *Hae*III fragments of wild-type virus and of the isolate HRV14-as3 resolved in the first dimension according to their size gave rise to only two spots in the 2DSSCP analysis corresponding to the different conformations of the complementary single strands. When the autoradiographs were superimposed it became clear that fragments G and B gave rise to different spots in HRV14 wild-type and HRV14-as3. Control experiments with *Mbo*II- or *Rsa*I-digested DNA showed that the regions corresponding to the *Hae*III fragments C, D, and H were not changed when compared with wild type and were therefore not considered further (data not shown). The patterns of the G fragments from wild type and the isolates HRV14-as3 and -as5 are presented in the inset to Fig. 3; since the G fragments from this isolate were clearly different from the ones derived from all other mutants analyzed, it was inferred that a different mutation was present. DNA sequencing was therefore carried out on 5'-end-labelled single strands eluted from the 2DSSCP gel. Two single-base substitu-

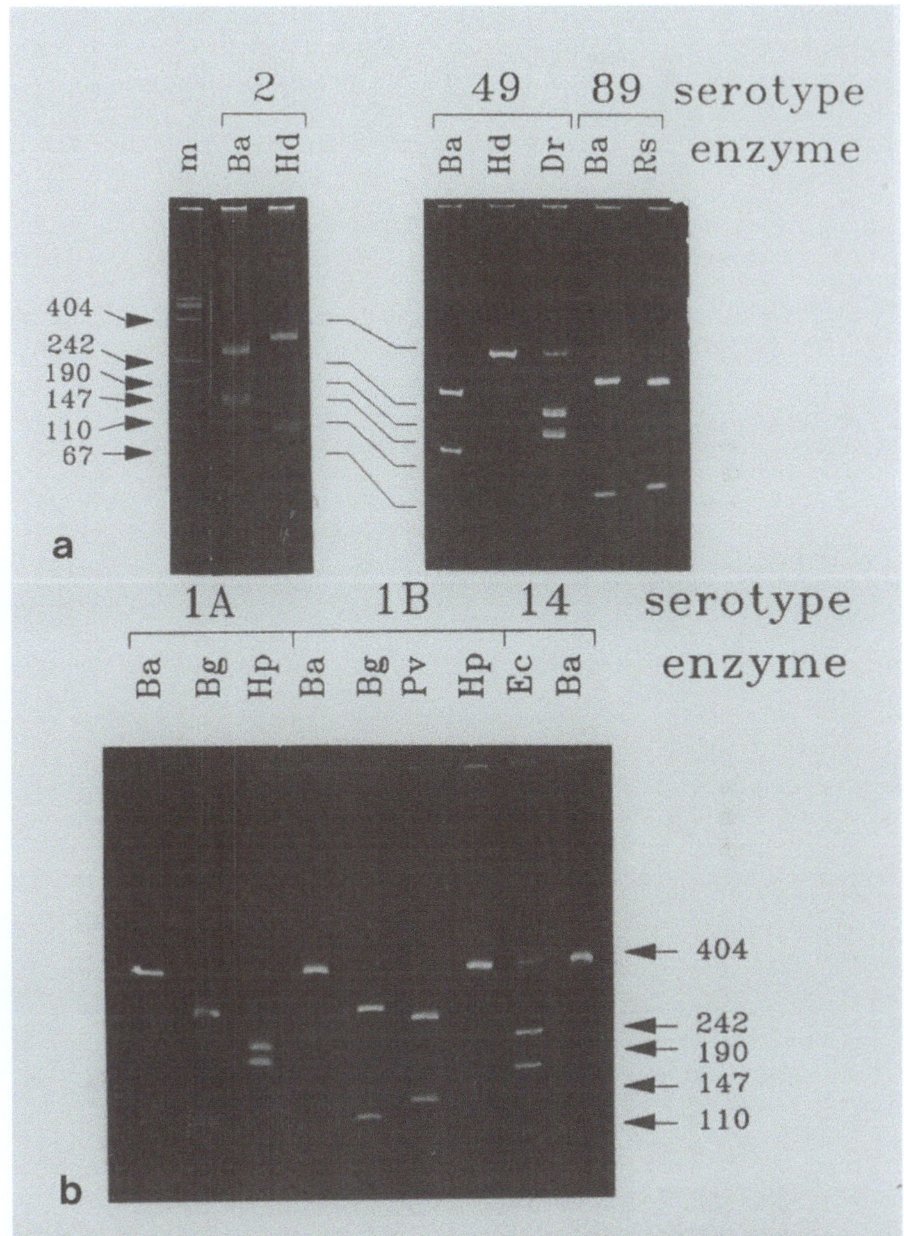

Fig. 2a, b. Polyacrylamide gel analysis of restriction fragments obtained from the 380-bp polymerase chain reaction (PCR) product of cDNA from various rhinovirus serotypes. Amplified DNA fragments were digested with the enzymes shown in Table 1. The serotypes and enzymes are given above the individual lanes. Ba, *Ban*II; Bg, *Bgl*II; Dr, *Dra*III; Ec, *Eco*RI; Hn, *Hin*PI; Hd, *Hin*dIII; Pv, *Pvu*II; Rs, *Rsa*I; m, markers. The sizes of the markers are indicated in base pairs. **a** *Ban*II-positive serotypes; **b** *Ban*II-negative serotypes

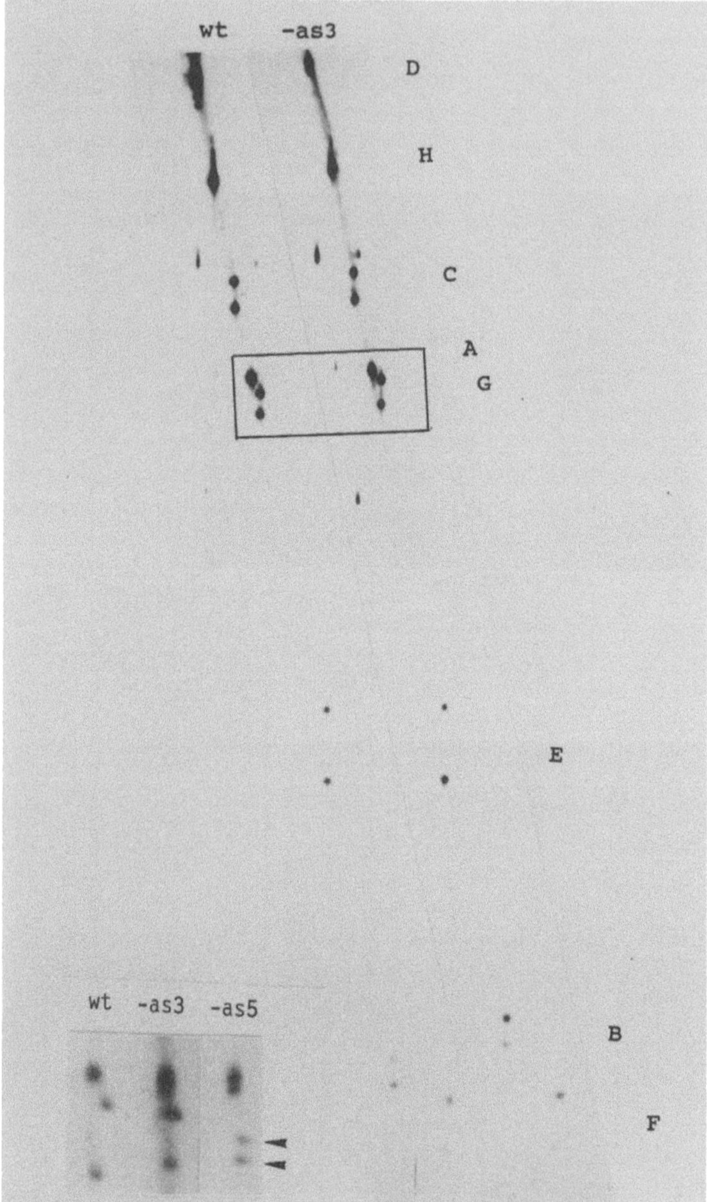

Fig. 3. Two-dimensional, single-strand, conformational polymorphism analysis of HRV14-as3 and wild type. Fragments G (*boxed*) and B give rise to polymorphic single strands. *Inset,* Close-up of the enclosed region (a different gel is shown). In addition to wild type and as3, the polymorphism of fragment G is also shown for the isolate as5. The second dimension gels were run for a prolonged period of time in order to obtain better resolution. *Arrowheads* indicate the single-stranded fragments eluted from the gel and sequenced by the Maxam-Gilbert method

tions were detected, one being identical to the alteration identified in VP1 of HRV14-as1, -as2, -as4, and -as6; the second, an A→G transition at position 2609, results in a Glu95:Gly amino acid change in VP1 (Kovar et al. 1991).

Discussion

Rhinovirus Typing

For the application of PCR to rhinovirus typing, two primers derived from regions of sequence identity in the 5′-UTR of the genome of four previously sequenced HRVs were used to amplify a 380-bp DNA fragment from cDNA of 7 HRVs. Amplification was thus possible for the three serotypes (HRV1A, HRV49, and HRV70) whose sequences were unknown; this suggests that the conserved sequences will be found in all HRVs. Sequence data were then determined to enable appropriate restriction enzymes to be selected. Restriction analysis of DNA using a combination of enzymes for each serotype allowed an unambiguous assignment and provides the basis for a simple and rapid method for rhinovirus typing.

The significance of the segmentation of the 5′-UTR of rhinoviruses into conserved and variable regions has been discussed extensively (Rivera et al. 1988). For this property to be used as a basis for typing rhinoviruses, two conditions must be met. The conserved regions must be present in all serotypes to allow amplification to occur; however, the sequences between them must be variable enough to ensure sufficient differences in restriction sites. The first criterion is addressed by the presence of the blocks of 23 (sequence 531–553 in HRV2) and 21 (sequence 161–181 in HRV2) identical nucleotides in the serotypes from which the sequence is available. As DNA amplification was also possible in the remaining three serotypes, it can be assumed that the structure of their 5′-UTR is extremely similar. The two sets of closely related HRV serotypes (pairs 1A and 1B and 2 and 49) were chosen to test whether characteristic sites could be found which would allow discrimination between them. When sequenced regions of the amplified fragment for each set were compared, useful restriction sites were found, allowing both positive and negative identification. This suggests that sufficient differences will be present between other rhinovirus serotypes to enable a catalogue of restriction sites to be established. Furthermore, defining groups of serotypes based on the presence or absence of a selected site enables a hierarchical method of typing to be devised (see Fig. 1). Thus, the serotypes could be divided into two groups depending on the presence or absence of a *Ban*II site, followed by a combination of enzymes to identify the individual serotypes. This approach lends itself well to the typing of unknown viruses. For instance, had the six serotypes been presented in a blind trial, digestion with *Ban*II followed by *Hin*dIII would have rapidly identifed three serotypes (HRV14, HRV1A, and HRV1B).

It is desirable that the method should be as specific for rhinoviruses as possible; examination of sequence data banks showed that these primers would also result in amplification of RNA from all strains of poliovirus types 1, 2, and the strain 23127 of type 3 (Cameron 1988). It is known that poliovirus is secreted from the nasal membranes in individuals having received Sabin vaccine; cDNA from types 1 and 2 will thus be amplified. However, this problem can be overcome by the use of the characteristic restriction sites which are available for poliovirus. No amplification of coxsackieviruses is expected with these primers under stringent amplification conditions.

The amount of material for an amplification as described here is similar to that obtained after a single passage of viruses in cell culture; however, it should be feasible to work directly with nasal washings as described by Gama et al. (1988) since the 380-bp fragment was generated using dilutions of supernatant estimated to contain one infectious virus particle (data not shown). It is thus probable that the method described here will be more sensitive and informative than the detection of rhinovirus serotypes by hybridization with oligonucleotides derived from the 5′-UTR (Bruce et al. 1990; Hyypiä et al. 1989).

Restriction enzyme analysis in conjunction with DNA amplification technology can thus be used to identify rhinovirus serotypes, providing a useful method for rapid typing. Moreover, it can be envisaged that this procedure might be applicable to other viruses exhibiting a similar arrangement of conserved blocks of sequence between different serotypes.

Determination of Mutations

The method of detecting mutations in long DNA fragments by amplification of cDNA derived from human rhinoviruses by 2DSSCP described here offers a number of advantages. First, all steps including reverse transcription, amplification, mutation detection, and sequencing of the fragment containing the alteration can be performed using a single pair of oligonucleotides with standard equipment. Second, several analyses can be carried out simultaneously. As the restriction fragments are separated under denaturing conditions in the first dimension, reassociation is prevented, and there are no difficulties arising from the presence of double-stranded DNA in the second dimension.

The detection of four different mutations in the six viral isolates examined clearly showed that sequences differing in only one nucleotide possessed different migration charateristics (Kovar et al. 1991). The identification of the one previously unknown mutation was shown to be possible by chemical sequencing of the labelled DNA eluted from the dried gel, demonstrating the sensitivity of the technique. It cannot, however, be excluded that mutations localised close to one end of a fragment would not sufficiently influence the conformation of the DNA to change the migration and would therefore not be detected by 2DSSCP analysis. Here, digestions with *Rsa*I and *Mbo*II were performed to ensure that the regions represented by larger *Hae*III fragments did not contain mutations. Comparison of the time necessary for the detection of the mutations in HRV-as1 to -as3 by cDNA cloning and DNA sequencing (several weeks) and those in HRV14-as4 to -as6 by 2DSSCP (a few days) illustrates the rapidity of the latter technique. The mutation in HRV14-as5 (A→G at nucleotide 2609; Glu95:Gly) was determined within 48 h and characterized within less than a week. Thus, 2DSSCP represents a significant improvement in detecting mutations in large PCR fragments, due to its speed, its simplicity, and its ability to detect practically all mutations.

References

Appleyard G, Russell SM, Clarke BE, Speller SA, Trowbridge M, Vadolas J (1990) Neutralization epitopes of human rhinovirus type-2. J Gen Virol 71:1275–1282

Bruce CB, Gama RE, Hughes PJ, Stanway G (1990) A novel method of typing rhinoviruses using the product of a polymerase chain reaction. Arch Virol 113:83–87

Callahan PL, Mizutani S, Colonno RJ (1985) Molecular cloning and complete sequence determination of RNA sequence of human rhinovirus type 14. Proc Natl Acad Sci USA 82:732–736

Cameron GN (1988) The EMBL data library. Nucleic Acids Res 16:1865–1867

Cooney MK, Fox JP, Kenney GE (1982) Antigenic groupings of 90 rhinovirus serotypes. Infect Immun 37:642–647

Duechler M, Skern T, Sommergruber W, Neubauer C, Gruendler P, Fogy I, Blaas D, Kuechler E (1987) Evolutionary relationships within the human rhinovirus genus; comparison of serotypes 89, 2 and 14. Proc Natl Acad Sci USA 84:2605–2609

Gama RE, Hughes PJ, Bruce CB, Stanway G (1988) Polymerase chain reaction amplification of rhinovirus nucleic acids from clinical material. Nucleic Acids Res 16:9346

Halfpap LM, Cooney MK (1983) Isolation of rhinovirus intertypes related to either rhinovirus 12 and 78 or 36 and 58. Infect Immun 40:213–218

Hamparian VV, Colonno RJ, Cooney MK, Dick EC, Gwaltney JM Jr, Hughes JH, Jordan WS, Kapikian AZ, Mogabgab WJ, Monto A, Philips CA, Rueckert RR, Schieble JH, Stott EJ, Tyrrell DAJ (1987) A collaborative report: rhinoviruses – extension of the numbering system from 89 to 100. Virology 159:191–192

Harber JJ, Bradley J, Anderson CW, Wimmer E (1991) Catalysis of poliovirus VP0 maturation cleavage is not mediated by serine-10 of VP2. J Virol 65:326–334

Hughes PJ, North C, Jellis CH, Minor PD, Stanway G (1988) The nucleotide sequence of human rhinovirus 1B: molecular relationships within the rhinovirus genus. J Gen Virol 69:49–58

Hyypiä T, Auvinen P, Maaronen M (1989) Polymerase chain reaction for human picornaviruses. J Gen Virol 70:3261–3268

Jeffrey AJ, Wilson V, Neumann R, Keyte J (1988) Amplification of human minisatellites by the polymerase chain reaction: towards DNA fingerprinting of single cells. Nucleic Acids Res 16:10953–10971

Kellner G, Popow-Kraupp T, Kundi M, Binder C, Wallner H, Kunz C (1988) Contribution of rhinoviruses respiratory infections in childhood: a prospective study in a mainly hospitalized infant population. J Med Virol 25:455–469

Kim S, Smith TJ, Chapman MS, Rossmann MG, Pevear DC, Dutko FJ, Felock PJ, Diana GD, McKinlay MA (1989) Crystal structure of human rhinovirus serotype-1A (Hrv1A). J Mol Biol 210:91–111

Kovar H, Jug G, Auer H, Skern T, Blaas D (1991) Two dimensional single-strand conformational polymorphism analysis: a useful tool for the detection of mutations in long DNA fragments. Nucleic Acids Res 19:3507–3510

Maniatis T, Fritsch F, Sambrook J (1982) Molecular cloning, a laboratory manual. Cold Spring Harbor Laboratory, Cold Spring Harbor

Maxam AM, Gilbert W (1980) Sequencing end-labelled DNA with base-specific chemical cleavages. Methods Enzymol 65:499–560

Palmenberg A (1987) Picornaviral processing: some new ideas. J Cell Biochem 33:191–198

Rivera VM, Welsh JD, Maizel JV (1988) Comparative sequence analysis of the 5'-non-coding region of enteroviruses and rhinoviruses. Virology 165:42–50

Rossmann MG, Arnold E, Erickson JW, Frankenberger EA, Griffith JP, Hecht HJ, Johnson JE, Kamer G, Luo M, Mosser AG, Rueckert RR, Sherry B, Vriend G (1985) Structure of a human common cold virus and functional relationship to other picornaviruses. Nature 317:145–153

Rueckert RR (1990) Picornaviridae and their replication. In: Fields BN (ed) Virology, 2nd edn, vols 1–2. Raven, New York, pp 507–548

Rueckert RR, Wimmer E (1984) Systematic nomenclature of picornavirus proteins. J Virol 50:957–959

Saiki RK, Gelfand DH, Stoffel S, Scharf SJ, Higuchi R, Horn GT, Mullis KB, Erlich HA (1988) Primer directed enzymatic amplification of DNA with a thermostable DNA polymerase. Science 239:487–491

Sanger F, Nicklen S, Coulson AR (1977) DNA sequencing with chain-terminating inhibitors. Proc Natl Acad Sci USA 74:5463–5467

Sherry B, Rueckert RR (1985) Evidence for at least two dominant neutralization antigens on human rhinovirus 14. J Virol 53:137–143

Sherry B, Mosser AG, Collono RJ, Rueckert RR (1986) Use of monoclonal antibodies to identify four neutralization immunogens on a common cold picornavirus, human rhinovirus 14. J Virol 57:246–257

Skern T, Sommergruber W, Blaas D, Gruendler P, Fraundorfer F, Pieler C, Fogy I, Kuechler E (1985) Human rhinovirus 2: complete nucleotide sequence and proteolytic processing signals in the capsid protein region. Nucleic Acids Res 13:2111–2126

Skern T, Neubauer C, Frasel L, Gruendler P, Sommergruber W, Zorn M, Kuechler E, Blaas D (1987) A neutralizing epitope on human rhinovirus type 2 includes amino acid residues between 153 and 164 of virus capsid protein VP2. J Gen Virol 68:315–323

Skern T, Torgersen H, Auer H, Kuechler E, Blaas D (1991) Human rhinovirus mutants stable at low pH. Virology 183:757–763

Sommergruber W, Zorn M, Blaas D, Fessl F, Volkmann P, Maurer-Fogy I, Pallai P, Merluzzi V, Matteo M, Skern T, Kuechler E (1989) Polypeptide 2A of human rhinovirus type 2: identification as a protease and characterization by mutational analysis. Virology 169:68–77

Stanway G (1990) Structure, function and evolution of picornaviruses. J Gen Virol 71:2483–2501

Stanway G, Hughes PJ, Mountford RC, Minor PD, Almond JW (1984) The complete nucleotide sequence of a common cold virus: human rhinovirus 14. Nucleic Acids Res 12:7859–7875

Stott EJ, Killington RA (1972) Rhinoviruses. Annu Rev Microbiol 26:503–525

Torgersen H, Blaas D, Skern T (1989a) Low cost apparatus for primer-directed DNA amplification. Anal Biochem 176:33–35

Torgersen H, Skern T, Blaas D (1989b) Typing human rhinoviruses based on sequence variations in the 5'-non-coding region. J Gen Virol 70:3111–3116

Uncapher CR, Dewitt CM, Colonno RJ (1991) The major and minor group receptor families contain all but one human rhinovirus serotype. Virology 180:814–817

Chapter 22 Polymerase Chain Reaction Diagnosis of Human Parvovirus B19

Jonathan P. Clewley

Summary

Polymerase chain reaction (PCR) is being widely applied in virological laboratories, although it is not yet a routine technique. It has been used for the detection of parvovirus B19 DNA in serum from acute and chronic cases of infection and also for the detection of B19 DNA in blood products. Although for the diagnosis of acute infections PCR is of limited value to laboratories that already have available specific anti-B19 IgM, IgG and DNA hybridization tests, it can be established by any other laboratory that does not have access to the necessary reagents for these tests. Moreover, chronic B19 infections and their treatment can be monitored by PCR. As in most other areas of molecular biology, PCR has considerable potential as a research method for studying parvoviruses.

Introduction

Human parvovirus B19 is a small icosahedral virus which has a single-stranded (ss) DNA genome of approximately 5.6 kb encapsidated by non-glycosylated proteins (Clewley 1984; Cotmore et al. 1986; Shade et al. 1986; Astell and Blundell 1989; Deiss et al. 1990). Strands of either positive or negative polarity may be packaged into virions, so that double-stranded molecules are formed on chemical extraction (Summers et al. 1983; Clewley 1984; Cotmore and Tattersall 1984). There is very little sequence divergence between B19 isolates (Morinet et al. 1986; Mori et al. 1987; Turton et al. 1990; Umene and Nunoue 1990), and no convincing evidence for any correlation between the B19 genome sequence (the genotype) and disease pathology (the phenotype) (Mori et al. 1987).

B19 is the only known pathogenic human parvovirus, although other parvoviruses, the adeno-associated viruses, and the zoonotic virus H1 may also infect man (Toolan 1990). Virus is usually spread by the respiratory route, and it replicates in erythroid progenitor cells (Anderson 1985a; Pattison 1987). The

PHLS Virus Reference Laboratory, 61 Colindale Avenue, London NW9 5HT, UK

Table 1. Parvovirus B19 disease

Minor febrile illness with a rash (erythema infectiosum)
Acute onset arthropathy
Transient aplastic crisis
Chronic anaemia in immunocompromised patients
Spontaneous abortion/hydrops fetalis

Pattison 1987; L. J. Anderson 1989; Frickhofen and Young 1989

most common manifestation of B19 infection is an acute exanthematous rash illness of childhood called fifth disease or erythema infectiosum (M. J. Anderson 1987; L. J. Anderson 1989). This disease is frequently mild and self-limiting; however, there can be several complications (Frickhofen and Young 1989; Table 1). Transient aplastic crisis may result from infection of patients with decreased red cell lifespan, e.g. in sickle cell anaemia or hereditary spherocytosis. Adults may develop acute arthropathy which, although it usually resolves in several weeks, may in some cases last for months or years. This resembles rheumatoid arthritis, but persistence of virus has not been demonstrated (White et al. 1985; Cohen et al. 1986; Frickhofen and Young 1989). An uncommon consequence of infection during pregnancy is fetal death (Levy and Read 1990; PHLS Working Party on Fifth Disease 1990), which is thought to be due to a persistent infection of the developing fetus not neutralised by maternal antibodies. Persistent parvovirus infection also occurs in patients with congenital or acquired immunodeficiencies. In these cases an inadequate antibody response allows the virus to continue replicating at a low level. This may manifest itself as sudden, sustained bone marrow failure (Young 1988).

Diagnosis of B19 Infection

B19 infection is diagnosed either by the demonstration of the virus itself or by the presence of a specific host antibody response, either IgM or IgG (Cohen 1986). The virus was discovered by use of counterimmunoelectrophoresis (CIE) and electron microscopy (EM) during routine screening of blood donor sera for hepatitis B virus surface antigen (Cossart et al. 1975). Although CIE is not much used today, EM remains a useful method for confirmation of the presence of the virus (Field et al. 1991). For many years the only source of B19 virus was from viraemic blood donors or patients. This source provided antigen for serological tests (Cohen 1986) and also DNA for cloning the viral genome (Cotmore and Tattesall 1984; Morinet et al. 1986; Shade et al. 1986; Mori et al. 1987; C. S. Brown et al. 1990; Clewley et al. 1990; Umene and Nunoue 1990). Limited supplies of the native B19 antigen have restricted development of serological tests to a few centres. However, tests based on recombinant expressed viral proteins and synthetic peptides are becoming available (C. S. Brown et al. 1990; Fridell et al. 1991). The virus can be grown with difficulty in primary bone marrow cultures (Ozawa et al. 1986) or in fetal liver cells (Yaegashi et al. 1989; K. E. Brown et al. 1991), neither of which are very practical culture systems.

About a week after a susceptible individual is infected with B19, an intense viraemia arises, lasting 5–7 days (Fig. 1). The viraemia is often accompanied by non-specific feverish symptoms; rash and arthralgia, if they occur, develop about 2 weeks after infection. Specific IgM and IgG is detectable towards the end of the viraemia; specific IgG is much more long lasting and may indicate protective immunity for life, although there is evidence that some individuals can be re-infected (M. J. Anderson et al. 1985a; Pattison 1987). Thus, a positive laboratory test for the presence of B19-specific IgM is evidence of a recent infection, and the presence of B19-specific IgG in the absence of IgM is evidence of a past infection. Serological tests for virus antigen have been developed in some laboratories. These give positive results in the early stage of infection but may not be sensitive enough to detect declining levels of virus (Cohen 1986).

Soon after the fragments of the virus genome were cloned, dot-blot hybridisation tests using ^{32}-P-labelled probes were described for the detection of B19 DNA (M. J. Anderson et al. 1985b; Clewley 1985). These tests have been progressively refined and non-radioactive formats developed for use in routine

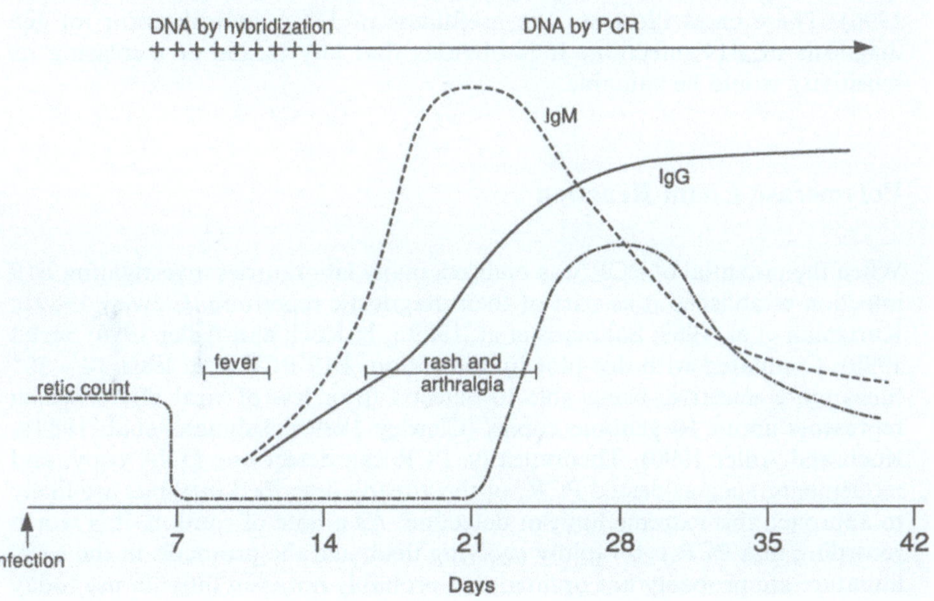

Fig. 1. The course of infection with parvovirus B19. Viraemia is indicated by + + +: retic, reticulocyte

laboratories (Cunningham et al. 1988; Mori et al. 1989; Azzi et al. 1990; Clewley et al. 1990; Zerbini et al. 1990; Musiani et al. 1991). A laboratory diagnosing B19 infections is therefore likely to use regularly IgM, IgG and dot-blot tests (Clewley 1989 a). When available, EM provides additional support for the diagnosis of B19 infection.

DNA hybridisation is both a routine test for B19 virus and a means of screening for virus antigen in blood donations (Cohen et al. 1990). It has been used to define the incubation period and course of infection of B19 and to show that transmission occurs via the respiratory tract (Fig. 1; Plummer et al. 1985; M. J. Anderson et al. 1985 a). Hybridisation has also been used to detect B19 DNA in the synovial fluid of a patient with acute arthralgia who had a serologically proven parvovirus infection (Dijkmans et al. 1988). The investigation of three of the most life-threatening consequences of B19 infection has also benefited from the use of DNA hybridisation. First, the presence of virus in cases of hydrops fetalis associated with B19 infection has been shown by hybridisation (T. Brown et al. 1984; Gray et al. 1986; Bond et al. 1986; Anand et al. 1978; Clewley et al. 1987; Weiland et al. 1987; Porter et al. 1988; Weiland et al. 1989; van Elsacker-Neile et al. 1989; Nascimento et al. 1991). Second, hybridisation has been used to show B19 DNA in cases of transient aplastic crisis in patients with underlying hereditary haemolytic disorders (Saarianen et al. 1986; Kurtzman et al. 1987, 1988, 1989; Malarme et al. 1989; Rechavi et al. 1989). Third, patients with immunodeficiency and persistent B19 infection have been treated with immunoglobulin containing B19-specific antibodies after diagnosis was established by dot-blot hybridisation (Frickhofen et al. 1990). These cases illustrate the usefulness of DNA hybridisation for the diagnosis of B19 infection. It is obvious that any means of increasing its sensitivity would be valuable.

Polymerase Chain Reaction

When the potential of PCR was realised, many laboratories investigating B19 infection established it as part of their diagnostic repertoire (Clewley 1989 b; Kurtzman et al. 1989; Salimans et al. 1989 a, b; Koch and Adler 1990; Sevall 1990). Compared with dot-blot hybridization, B19 PCR is at least $10^3 - 10^4$ times more sensitive, being able to detect 1 fg or less of viral DNA, which represents about 10 genome copies (Clewley 1989 b; Salimans et al. 1989 a; Koch and Adler 1990). Theoretically, PCR can detect one DNA copy, and refinements, such as nested PCR, of the already described methods are likely to approach this extreme limit of detection. As a note of caution, it is worth recording that PCR is a rapidly evolving field, and the protocols in the early literature are probably not optimal and probably not even those in use today in the laboratories that initially described them. As B19 DNA shows only minimal sequence variation between isolates, most of the coding regions, both non-structural and structural, can be used as a suitable amplification target.

The only constraints are those that apply to the design of any PCR primers (Kwok et al. 1990). The terminal repeats of the genome are probably not good PCR targets, except for special purposes (see below). Conditions for amplification (temperature, time, buffers, etc.) are not specifically recommended, and standard PCR manuals should be consulted (Innis et al. 1990; Clewley 1991). Most manufacturers of thermostable polymerases supply buffers and detail conditions suitable for successful amplification. It is usually necessary to extract DNA from serum or another specimen to achieve a reproducible amplification of the target by PCR. This can be achieved by proteinase K and phenol-chloroform methods, or by chaotropic disruption and binding of the DNA to silica (using commercial kits, e.g. Geneclean, Bio 101; Isogene, Perkin Elmer Cetus; or by the method of Boom et al. 1990). To confirm the specificity of the amplification, the PCR product can be detected by gel electrophoresis and solution hybridisation with an oligonucleotide probe (Gibson et al. 1991). Possible applications of PCR are listed in Table 2.

Applications for B19 investigation

Detection of DNA in B19 viraemia. For most routine purposes a dot-blot hybridisation test is sufficient for examining specimens for the presence of B19 DNA. Virus titres are very high during acute viraemia (perhaps 10^{12} particles per ml of serum; M. J. Anderson et al. 1985 b; Saarinen et al. 1986; Mori et al. 1987). Higher and/or more persistent levels of viraemia occur in patients with transient aplastic crisis than in other viraemic patients; the lowest levels are observed in immunocompromised patients (Saarinen et al. 1986; Frickhofen et al. 1990). This makes detection of viral DNA in serum relatively straightforward. The greater sensitivity of PCR allows B19 DNA to be detected both before and after it can be detected by hybridisation (Fig. 1; Clewley 1989 b; Koch and Adler 1990). Since the start of viraemia precedes the onset of clinical symptoms, it is unlikely that there will be many specimens submitted to a

Table 2. Uses of polymerase chain reaction (PCR) for B19 DNA detection

---

Extending the diagnostic window through which DNA can be detected in self-limiting infections
Detecting DNA in persistent B19 infections
Detecting B19 DNA in fetal tissues
Possible discrimination between virus in maternal blood from that which has crossed the placenta and replicated in the fetus
Detecting B19 DNA in blood products, e.g. clotting factor concentrates
Cloning B19 DNA from tissues where there is insufficient virus for conventional methods to be successful
Searching for B19 DNA integrated into chromosomal DNA
Searching for uncharacterised parvoviruses
Studying B19 replication in culture, for instance to detect spliced mRNAs

---

laboratory, at least from otherwise healthy individuals, that are hybridisation, B19-specific IgM and IgG negative, but PCR positive. Also, since PCR is able to detect B19 DNA in acute infections for much longer than hybridisation (Clewley 1989 b), it is not likely to be very useful to a laboratory that has IgM and IgG tests available. In this case, a PCR-positive result for serum from a healthy individual contributes little to a diagnosis based on an B19-specific IgM and IgG-positive result. However, patients with acute arthritis may only be examined for markers of B19 infection if their symptoms persist for longer than 2 months (B. J. Cohen, personal communication). At this time B19-specific IgM is likely to have declined below detectable levels, but PCR may be positive for B19 DNA if that virus is involved in the illness.

At the time of writing (mid-1991) B19 serological tests are not widely available, and a laboratory may not want to or be able to obtain cloned B19 DNA for establishing hybridisation tests. Given access to DNA synthesisers and molecular biology reagents, all that is required for setting up a PCR assay is sequence information freely available in the literature (Shade et al. 1986). The main disadvantage of relying on PCR as a sole diagnostic method is that false-positive results due to carry-over or contamination and false-negative results due to inhibition of *Taq* polymerase are both possible. Thus, a PCR test on its own may not be a practical diagnostic tool.

Detecting DNA in B19 Persistent Infections. Immunocompromised patients who are infected with B19 fail to neutralise and clear the infection and often maintain a low level of viraemia (Kurtzman et al. 1987, 1988, 1989; Frickhofen and Young 1989; Frickhofen et al. 1990). Thus, the virus continues to replicate in the bone marrow and may cause severe and prolonged disease. These patients can be treated with normal immunoglobulin and the fate of the virus can be monitored by PCR (Kurtzman et al. 1989; Frickhofen et al. 1990). A simple and unambiguous PCR and a means for quantifying the amount of B19 DNA present will be necessary if this type of investigation and treatment is to spread beyond a few specialist centres.

Detecting B19 DNA in Fetal Tissues. The involvement of B19 in fetal death was first shown by retrospective diagnostic testing of aborted fetal tissues (T. Brown et al. 1984; Anand et al. 1987; Clewley et al. 1987). As B19 DNA was detectable by hybridisation in several organs and viral particles were observed by EM, the virus was obviously present at a high titre. It was thus considered to be the causal agent of the fetal death. However, these initial reports did not establish the degree of risk to the fetus when the mother is infected. A prospective study has since shown that the risk of fetal death is low and that a favourable outcome of the pregnancy is very likely (PHLS Working Party on Fifth Disease 1990). This study involved 190 women with a serologically confirmed B19 infection. The transplacental transmission rate was estimated to be 33%. Thirty women experienced fetal loss, and fetal tissue or products of conception were available for testing from 14 of these. DNA was found by hybridisation to be present in 8 cases. Based on this finding, the risk of fetal death due to B19 in pregnancy was estimated to be 9%. Thus, the virus does not always cross the placenta, and when it does, the fetus often survives.

Placental tissues from women in the PHLS study who gave birth to healthy infants were tested for B19 DNA by PCR, and 74 of 89 (83%) were found to be positive (Clewley 1989b). These tissues were negative for B19 DNA by dot-blot hybridisation, suggesting that the virus was present only at low titre. As the placental specimens contained maternal blood, the positive PCR result does not necessarily indicate that the infant was exposed to the virus. Also, the presence of maternal blood meant that the rate of transplacental transmission could not be estimated from this experiment.

Török and colleagues (1992) have suggested that PCR, applied to amniotic fluid, could be used to diagnose intrauterine B19 infection and to identify pregnancies that might benefit from therapeutic intervention. They studied 56 pregnant woman for evidence of B19 infection and found DNA by PCR in 23. In the majority of the 23 cases B19 DNA was found by PCR in the amniotic fluid as well as in maternal and fetal blood. However, the serological findings were not entirely consistent with the PCR result in 8 of the cases. All of these 8 were negative for maternal and fetal B19-specific IgM, and 4 were also negative for maternal and fetal B19-specific IgG. It is apparent that more studies are required before PCR can be used with confidence to diagnose intrauterine infection.

Detection of B19 DNA in Blood Products. It has been estimated that about 1 in 20000 blood donors is, by chance, viraemic for B19 at the time of giving blood (Cohen et al. 1990). Clotting factor concentrates are prepared from very large numbers of donations, and it would therefore be expected that some preparations contain B19 virus. Moreover, as parvoviruses are very resistant to most inactivation methods, these preparations could transmit B19. This has been observed to be the case (Corsi et al. 1988), and B19 DNA has been demonstrated by PCR in a preparation of heat-treated Factor IX concentrate implicated in the transmission of symptomatic B19 illness (Lyon et al. 1989). However, the demonstration of B19 DNA by PCR in a blood product does not necessarily establish the presence of infectious virus in that preparation. For instance, the DNA could be degraded, but still amplifiable. PCR sequencing of virus DNA from patients and the blood products they have received will be useful in further clarifying such cases.

Cloning of B19 DNA. From viraemic serum B19 DNA can be purified and relatively easily cloned (Mori et al. 1987; Clewley et al. 1990). However, it is not as simple to prepare intact viral DNA from other sources; for instance, the viral DNA in aborted fetal tissues is mostly degraded (Clewley et al. 1987). Also, blood products are likely to contain very low levels of B19 DNA. By PCR, B19 DNA from these materials can be amplified and further character-ised, and using primers located in the termini of the B19 genome, it is possible to amplify almost the whole genome (J. P. Clewley, unpublished observations). It should be possible to clone this PCR product for sequencing studies.

Searching for Integrated B19 DNA. The DNA of the human adeno-associated virus type 2 is able to integrate into chromosomal DNA (Carter et al. 1990). It is possible that B19 DNA could also integrate into the chromosome of a suitable host cell and that these infected cells could be involved in some chronic and persistent B19 conditions. It can be argued that this is unlikely, but inverse PCR does provide a means of investigating the possibility.

Searching for Uncharacterised Parvoviruses. Besides the adeno-associated viruses and B19, the presence of other human parvoviruses is suspected (Paver and Clarke 1976; Simpson et al. 1984). There are also diseases with as yet unknown etiologies that could be caused by viruses, perhaps parvoviruses (Bissenden and Hall 1990). It is possible to design generic primers capable of amplifying conserved sequences within virus groups (Mack and Sninsky 1988), and this approach could be attempted for the parvoviruses. A region of conservation within the NS1 protein of parvoviruses has been identified by Shade and colleagues (1986), and consensus sequence or degenerate primers can amplify this region (J. P. Clewley, unpublished observations).

Replication of B19 in Culture and In Vivo. The transcription of mRNA of B19 that is replicating in bone marrow cells has been studied by S1 nuclease mapping (Ozawa et al. 1986, 1987). B19 has also been propagated in fetal liver cells (Yaegashi et al. 1989; K. E. Brown et al. 1991). As well as S1 mapping, PCR using primers spanning introns could be used to detect spliced mRNAs in experimentally infected cells.

Conclusions

The stability of the parvovirus genome and the availability of published primer sequences of proven efficiency make B19 a good candidate for PCR-based studies, and these can be expected to elucidate outstanding questions such as the role of the virus in fetal illness and death and the presence of the virus in blood products. Amplification of conserved regions may be a tool for detecting other human or animal parvoviruses.

Acknowledgements. I thank Mark Broughton for technical assistance and Kevin Brown, Bernard Cohen, Julie Mori and Philip Mortimer for helpful discussions.

References

Anand A, Gray ES, Brown T, Clewley JP, Cohen BJ (1987) Human parvovirus infection in pregnancy and hydrops fetalis. N Engl J Med 316:183–186

Anderson LJ (1989) Human parvoviruses. J Infect Dis 161:603–608

Anderson MJ (1987) Parvoviruses as agents of human disease. Prog Med Virol 34:55–69

Anderson MJ, Higgins PG, Davis LR, Willman JS, Jones SE, Kidd IM, Pattison JR, Tyrell DAJ (1985a) Experimental parvovirus infection in humans. J Infect Dis 152:257–265

Anderson MJ, Jones SE, Minson AC (1985b) Diagnosis of human parvovirus infection by dot-blot hybridzation using cloned viral DNA. J Med Virol 15:163–172

Astell CR, Blundell MC (1989) Sequence of the right hand terminal palindrome of the human B19 parvovirus genome has the potential to form a 'stem plus arms' structure. Nucleic Acid Res 14:5857

Azzi A, Zakrewska K, Gentilomi A, Musiani M, Zerbini M (1990) Detection of B19 parvovirus infections by a dot-blot hybridization assay using a digoxigenin labelled probe. J Virol Methods 27:125–133

Bissenden JG, Hall S (1990) Kawasaki syndrome: lessons for Britain. Br Med J 300:1025–1026

Bond PR, Caul EO, Usher J, Cohen BJ, Clewley JP, Field AM (1986) Intrauterine infection with human parvovirus. Lancet i:448–449

Boom R, Sol CJA, Salimans MMM, Larsen CL, Wertheim-van Dillen PME, van der Noorda J (1990) Rapid and simple method for purification of nucleic acids. J Clin Microbiol 28:495–503

Brown CS, Salimans MMM, Noteborn MHM, Weiland HT (1990) Antigenic parvovirus B19 coat proteins VP1 and VP2 produced in large quantities in a baculovirus expression system. Virus Res 15:197–211

Brown KE, Mori J, Cohen BJ, Field AM (1991) In vitro propagation of parvovirus B19 in primary foetal liver culture. J Gen Virol 72:741–745

Brown T, Anand A, Richie LD, Clewley JP, Reid TMS (1984) Intrauterine parvovirus infection associated with hydrops fetalis. Lancet ii:1033–1034

Carter BJ, Mendelsen E, Trempe JP (1990) AAV DNA replication, integration and genetics. In: Tijssen P (ed) Handbook of parvoviruses, vol I. CRC, Boca Raton, pp 166–226

Clewley JP (1984) Biochemical characterization of a human parvovirus. J Gen Virol 65:241–245

Clewley JP (1985) Detection of human parvovirus using a molecularly cloned probe. J Med Virol 15:173–181

Clewley JP (1989a) The application of DNA hybridization to the understanding of human parvovirus disease. In: Tenover FC (ed) DNA probes for infectious diseases. CRC, Boca Raton, pp 211–220

Clewley JP (1989b) Polymerase chain reaction assay of parvovirus B19 DNA in clinical specimens. J Clin Microbiol 27:2647–2651

Clewley JP (1991) The polymerase chain reaction: basic procedures and use for virus detection. In: Dale JW. Sanders PG (ed) Methods in gene technology. JAI, London, pp 219–238

Clewley JP, Cohen BJ, Field AM (1987) Detection of parvovirus B19 DNA, antigen and particles in the human fetus. J Med Virol 23:267–276

Clewley JP, Mori J, Turton J (1990) Molecular approaches for production of B19 antigen. Behring Inst Mitt 85:14–27

Cohen BJ (1986) Laboratory tests for the diagnosis of infection with B19 virus. In: Pattison JR (ed) Parvoviruses and human disease. CRC, Boca Raton, pp 69–83

Cohen BJ, Buckley MM, Clewley JP, Jones VE, Puttick AH, Jacoby RK (1986) Human parvovirus infection in early rheumatoid and inflammatory arthritis. Ann Rheum Dis 45:832–838

Cohen BJ, Field AM, Gudnadottir S, Beard S, Barbara JAJ (1990) Blood donor screening for parvovirus B19. J Virol Methods 30:233–238

Corsi OB, Azzi A, Morfini M, Fanci R, Ferrani PR (1988) Human parvovirus infection in haemophiliacs first infused with treated clotting factor concentrates. J Med Virol 25:165–170

Cossart YE, Field AM, Cant B, Widdows D (1975) Parvovirus-like particles in human sera. Lancet i:72–73

Cotmore SF, Tattersall P (1984) Characterization and molecular cloning of a human parvovirus genome. Science 226:1161–1165

Cotmore SF, McKie VC, Anderson LJ, Astell CR, Tattersall P (1986) Identification of the major structural and nonstructural proteins encoded by human parvovirus B19 and mapping of their genes by procaryotic expression of isolated genomic fragments. J Virol 60:548–557

Cunningham DA, Pattison JR, Craig RK (1988) Detection of parvovirus DNA in human serum using biotinylated RNA hybridisation probes. J Virol Methods 19:279–288

Deiss V, Tratschin J-D, Weitz M, Siegl G (1990) Cloning of the human parvovirus B19 genome and structural analysis of its palindromic termini. Virology 175:247–254

Dijkmans BAC, van Elsacker-Niele AMW, Salimans MMM, van Albada-Kuipers GA, de Vries E, Weiland HT (1988) Human parvovirus B19 DNA in synovial fluid. Arthritis Rheum 31:279–281

Field AM, Cohen BJ, Brown KE, Mori J, Clewley JP, Nascimento JP, Hallam NF (1991) Detection of B19 parvovirus in human fetal tissues by electron microscopy. J Med Virol 35:85–95

Frickhofen N, Young NS (1989) Persistent parvovirus B19 infections in humans. Microbiol Pathol 7:319–327

Frickhofen N, Abkowitz JL, Safford M, Berry JM, Antunez-de-Mayolo J, Astrow A, Cohen R, Halperin I, King L, Mintzer D, Cohen B, Young NS (1990) Persistent B19 parvovirus infection in patients infected with human immunodeficiency virus type I (HIV-1): a treatable cause of anemia in AIDS. Ann Intern Med 113:926–933

Fridell E, Trojar J, Mehlin H, Wahren B (1991) A cyclized peptide for studies of human parvovirus B19 infection. J Immunol Methods 138:125–128

Gibson KM, McLean KA, Clewley JP (1991) A simple and rapid method for detecting human immunodeficiency virus by PCR. J Virol Methods 32:277–286

Gray ES, Anand A, Brown T (1986) Parvovirus infections in pregnancy. Lancet i:208

Innes MA, Gelfand DH, Sninsky JJ, White TJ (eds) (1990) PCR protocols: a guide to methods and applications. Academic, San Diego

Koch WC, Adler SP (1990) Detection of human parvovirus B19 by using the polymerase chain reaction. J Clin Microbiol 28:65–69

Kurtzman GJ, Ozawa K, Cohen B, Hanson G, Oseas L, Young NS (1987) Chronic bone marrow failure due to persistent B19 parvovirus infection. N Engl J Med 317:287–294

Kurtzman GJ, Cohen BJ, Myers P, Amunullah A, Young NS (1988) Persistent B19 infection as a cause of severe chronic anaemia in children with acute lymphocytic leukemia. Lancet ii:1159–1160

Kurtzman G, Frickhofen N, Kimball RW, Jenkins DW, Nienhuis AW, Young NS (1989) Pure red-cell aplasia of 10 years' duration due to persistent parvovirus B19 infection and its cure with immunoglobulin therapy. N Engl J Med 321:519–523

Kwok S, Kellogg DE, McKinney N, Spasic D, Goda L, Levenson, Sninsky JJ (1990) Effects of primer-template mismatches on the polymerase chain reaction: human immunodeficiency type 1 model studies. Nucleic Acid Res 18:999–1005

Levy M, Read SE (1990) Erythema infectiosum and pregnancy related complications. Can Med Assoc J 143:849–858

Lyon DJ, Chapman CS, Martin C, Brown KE, Clewley JP, Flower AJE, Mitchell VE (1989) Symptomatic parvovirus B19 infection and heat treated factor IX concentrate. Lancet i:1085

Mack DH, Sninsky JJ (1988) A sensitive method for the identification of uncharacterised viruses related to known virus groups: hepadnaviurs model systems. Proc Natl Acad Sci USA 85:6977–6981

Malarme M, Vandervelde D, Brasseur M (1989) Parvovirus infection, leukemia, and immunodeficiency. Lancet i:1457

Mori J, Beattie P, Melton DW, Cohen BJ, Clewley JP (1987) Structure and mapping of the DNA of human parvovirus B19. J Gen Virol 68:2797–2806

Mori J, Field AM, Clewley JP, Cohen BJ (1989) Dot-blot hybridization assay of B19 virus DNA in clinical specimens. J Clin Microbiol 27:459–464

Morinet F, Tratschin J-D, Perol Y, Siegl G (1986) Comparison of 17 isolates of the human parvovirus B19 by restriction enzyme analysis. Arch Virol 90:165–172

Musiani M, Zerbini M, Gillellini D, Gentilomi G, La Placa M, Ferri E, Girotti S (1991) Chemiluminescent assay for the detection of viral and plasmid DNA using digoxigenin labelled probes. Anal Biochem 194 (in press)

Nascimento JP, Hallam NF, Mori J, Field AM, Clewley JP, Brown KE, Cohen BJ (1991) Detection of B19 parvovirus in human fetal tissues by in situ hybridisation. J Med Virol 33:77–82

Ozawa K, Kurtzman G, Young N (1986) Replication of the B19 parvovirus in human bone marrow cultures. Science 233:883–886

Ozawa K, Ayub J, Yu-Shu H, Kurtzman G, Shimada T, Young N (1987) Novel transcription map for the B19 (human) pathogenic parvovirus. J Virol 61:2395–2406

Pattison JR (1987) B19 virus – a pathogenic human parvovirus. Blood Rev 1:58–64

Paver WK, Clarke SKR (1976) Comparison of human fetal and serum parvo-like viruses. J Clin Microbiol 4:67–70

PHLS Working Party on Fifth Disease (1990) Prospective study of human parvovirus (B19) infection in pregnancy. Br Med J 300:1166–1170

Plummer FA, Hammond GW, Forward K, Sekla L, Thomspon LM, Jones SE, Kidd IM, Anderson MJ (1985) An erythema infectiosum-like illness caused by human parvovirus infection. N Engl J Med 313:74–79

Porter HJ, Khong TY, Evans MF, Chan VTW, Fleming KA (1988) Parvovirus as a cause of hydrops fetalis: detection by in situ DNA hybridisation. J Clin Pathol 41:381–383

Rechavi G, Vonsover A, Manor Y, Mileguir F, Shpilberg O, Kende G, Brok-Simoni F, Mandel M, Gotlieb-Stematski T, Ben-Bassat I, Ramot B (1989) Aplastic crisis due to human B19 parvovirus infection in red cell pyrimidine-5′-nucleotidase deficiency. Acta Haematol (Basel) 82:46–49

Saarinen UM, Chorba TL, Tattersall P, Young NS, Anderson LJ, Palmer E, Coccia PF (1986) Human parvovirus B19-induced epidemic acute red cell aplasia in patients with hereditary hemolytic anemia. Blood 67:1411–1417

Salimans MMM, Holsappel S, van de Rijke FM, Jiwa NM, Raap AK, Weiland HT (1989a) Rapid detection of human parvovirus B19 DNA by dot-hybridization and the polymerase chain reaction. J Virol Methods 23:19–28

Salimans MMM, van de Rijke FM, Raap AK, van Elsacker-Niele AMW (1989b) Detection of parvovirus B19 DNA in fetal tissues by in situ hybridization and polymerase chain reaction. J Clin Pathol 42:525–530

Sevall JS (1990) Detection of parvovirus B19 by dot-blot and polymerase chain reaction. Mol Cell Probes 4:237–246

Shade RO, Blundell MC, Cotmore SF, Tattersall P, Astell CR (1986) Nucleotide sequence and genome organization of human parvovirus B19 isolated from the serum of a child during aplastic crisis. J Virol 58:921–936

Simpson RW, McGinty L, Simon L, Smith CA, Godzeski CW, Boyd RJ (1984) Association of parvovirus with rheumatoid arthritis of humans. Science 224:1425–1428

Summers J, Jones SE, Anderson MJ (1983) Characterization of the genome of the agent of erythrocyte aplasia permits its classification as a human parvovirus. J Gen Virol 64:2527–2532

Toolan HW (1990) The rodent parvoviruses. In: Tijssen P (ed) Handbook of parvoviruses, vol II. CRC, Boca Raton, pp 159–176

Török TJ, Wang Q-Y, Gary GW, Yang C-F, Finch TM, Anderson LJ (1992) Prenatal diagnosis of intrauterine parvovirus B19 infection by the polymerase chain reaction technique. Clin Infect Dis 14:149–155

Turton J, Appleton H, Clewley JP (1990) Similarities in nucleotide sequence between serum and faecal human parvovirus DNA. Epidemiol Infect 105:197–201

Umene K, Nunoue T (1990) The genome type of human parvovirus B19 strains isolated in Japan during 1981 differs from types detected in 1986 to 1987: a correlation between genome type and prevalence. J Gen Virol 71:983–986 van Elsacker-Niele AMW, Salimans MMM, Weiland HT, Vermey-Keers Chr, Anderson MJ, Versteeg J (1989) Fetal pathology in human parvovirus B19 infection. Br J Obstet Gynecol 96:768–775

Weiland HT, Vermey-Keers Chr, Salimans MMM, Fleuren GJ, Verwey RA, Anderson MJ (1987) Parvovirus B19 associated with fetal abnormality. Lancet i:682–683

Weiland HT, Salimans MMM, Fibbe WE, Kluin PM, Cohen BJ (1989) Prolonged parvovirus B19 infection with severe anaemia in a bone marrow transplant recipient. Br J Haematol 71:300

White DG, Woolf AD, Mortimer PP, Cohen BJ, Blake DR, Bacon PA (1985) Human parvovirus arthropathy. Lancet i:419–421

Yaegashi N, Shiraishi H, Takeshita T, Nakamura M, Yajima A, Sugamura K (1989) Propagation of human parvovirus B19 in primary culture of erythroid lineage cells derived from fetal liver. J Virol 63:2422–2426

Young N (1988) Hematologic and hematopoietic consequences of B19 parvovirus infection. Semin Hematol 25:159–172

Zerbini M, Musiani M, Venturoli S, Gallinella G, Gibellini D, Gentilomi G, La Placa M (1990) Rapid screening for B19 parvovirus DNA in clinical specimens with a digoxigenin-labelled DNA hybridization probe. J Clin Microbiol 28:2496–2499

Chapter 23 Polymerase Chain Reaction for the Detection of Adenoviruses

Annika Allard and Göran Wadell

Summary

The use of the polymerase chain reaction (PCR) for the detection of human adenoviruses in diluted stool samples was investigated. Several primers, including ones specific for the hexon-coding region and for enteric adenovirus (EAd) types 40 and 41, were evaluated. The primers constitute three different PCR systems designed for the detection of all six adenovirus subgenera (A – F), the two EAds Ad40 and Ad41, and Ad40, respectively. The general system and the EAd-specific system also include nested primers for a two-step amplification. The limit of detection of the PCR assay in a two-step amplification was 1 single virus particle when the two sets of general hexon primers or EAd-specific primers were used. The sensitivity of the PCR, together with its simplicity and reduced time-scale compared with other detection methods, emphasize the potential of this technique as an additional method for routine diagnosis of human adenovirus infections.

Introduction

Adenovirus infections show high host species specificity and occur worldwide in human beings and in a variety of most studied animals. The Adenoviridae is a family of nonenveloped, icosahedral viruses classified into two main genera: Mastadenovirus (adenoviruses with a mammalian host) and Aviadenovirus (adenoviruses with an avian host). Several adenovirus serotypes have been detected in dogs, horses, pigs, sheep, cattle, and nonhuman primates. Human adenoviruses (hAds) were first detected in human adenoids by Rowe et al. (1953). They cause mainly respiratory, ocular, and gastrointestinal disease, but they have been associated with other syndromes as well. Adenovirus infections are also an increasing problem in immunocompromised hosts, contributing to their morbidity and mortality (Fiala et al. 1974; Keller et al. 1977; Stalder et al. 1977; Siegal et al. 1981; Harnett et al. 1982; Hammond et al. 1985; Shields et al. 1985; Ambinder et al. 1986; Flomenberg et al. 1987; Hierholzer et al. 1988; Ljungman et al. 1989).

Department of Virology, University of Umeå, 901 85 Umeå, Sweden

At present, there are 47 recognized hAd serotypes divided into six subgenera based on their DNA homology and their biological and biochemical properties. A great intraserotypic genomic variability was demonstrated for all serotypes analysed so far by the use of restriction enzyme analysis, and several distinct viral entities designated as genome types have been identified in each case (Wadell et al. 1980; Wigand et al. 1987; Hierholzer et al. 1988).

Subgenus A consists of Ad12, 18, and 31, which represent only 0.5% of the reported typed virus isolates. The majority of the isolates have been recovered from stools and most of them in cases of pediatric gastrointestinal disease. Ad31 has also been recovered from immunocompromised patients (Johansson et al. 1991).

Subgenus B consists of Ad3, 7, 11, 14, 16, 21, 34, and 35, which have been subclassified into two clusters of DNA homology. B1 comprises Ad3, 7, 16, and 21, which cause primarily outbreaks of respiratory disease. B2 comprises Ad11, 14, 34, and 35, which cause persistent infections of the urinary tract; they have also been reported to be serious pathogens in immunocompromised patients.

Subgenus C contains Ad1, 2, 5, and 6. The first three represent some of the most frequently isolated hAd types. They represent more than one-half of all isolates reported to WHO (Schmitz et al. 1983). The characteristic latent infection of lymphoid tissue described for adenoviruses is mostly associated with these types (Fox and Hall 1980). According to several published reports, this subgenus appears to express a greater genetic variability exhibiting a wide variety of genome types (Adrian et al. 1990).

Subgenus D consists of 23 well-established serotypes and 5 recently described serotypes: candidate Ad43, 44, 45, 46, and 47 (Hierholzer et al. 1988). They show a predilection for infecting the eye. The most important members of subgenus D are serotypes 8, 19, and 37, which are frequently associated with epidemic outbreaks of keratoconjunctivitis. Adenovirus strains corresponding to serotypes 43–47 have so far only been isolated from patients with AIDS (Hierholzer et al. 1988).

Subgenus E contains only one type based on sequence homology, restriction endonuclease cleavage patterns, and protein profiles: Ad4, which has been associated with both epidemic follicular conjunctivitis and respiratory disease. When the main recognized genome types Ad4p, 4a, and 4b are analyzed with restriction enzymes, only 25%–46% of the DNA restriction fragments comigrate. Such a pronounced genetic variability within one serotype is unique for Ad4 (Li and Wadell 1988).

Subgenus F contains the two so-called enteric or fastidious adenoviruses (EAds) serotypes 40 and 41. After rotavirus, the EAds are now recognized as the second most commonly identified agent in stools of infants and young children with gastroenteritis. The inability of EAd to grow in vitro initially hampered classification by the standard serological methods, such as neutralization and hemagglutination inhibition techniques. Alternative methods such as enzyme-linked immunosorbent assay, ELISA (Johansson et al. 1980, 1985), monoclonal antibodies (Herrmann et al. 1987; van der Avoort et al. 1989;

Wood et al. 1989), sodium dodecyl sulfate (SDS)-polyacrylamide gel electrophoresis (Wadell 1979), and DNA restriction enzyme analysis (Uhnoo et al. 1983; Wadell et al. 1980, 1986) were therefore employed for the identification and characterization of these agents.

By liquid hybridization, van Loon et al. (1985) determined the DNA homology between Ad40 and Ad41 to be 62%–69%. This degree of homology is not very high, but considering the other common characteristics such as their restricted host cell range, their association with gastroenteritis, and their cross-reactivity in immunological tests, Ad40 and Ad41 have been classified together in subgenus F. Restriction analysis has revealed that only 18 of 177 DNA restriction fragments of Ad40 and Ad41 comigrate (Wadell et al. 1986).

The polymerase chain reaction (PCR) is an in vitro method for primer-directed enzymatic amplification of specific target DNA sequences (Saiki et al. 1985). A few copies of target DNA can be amplified to a level detectable by gel electrophoresis or Southern blot hybridization. We have evaluated three different PCR primer systems for the detection of human adenoviruses including one-step and two-step amplifications in a nested fashion.

Material and Methods

The detection of adenovirus by PCR was developed with the aim of using the assay on stool specimens. Clinical samples were either assayed directly after a disruptive treatment or cultivated first in an adequate cell line.

Pretreatment of Stool Samples

The fecal specimens were collected in transport medium (phosphate buffered saline, PBS, 20 mM N-2-hydroxyethylpiperazine-N'-2-ethanesulfonic acid, HEPES, pH 7.3, 1% bovine serum albumin (BSA) 5% sucrose, 50 µg of gentamicin per ml) and clarified by centrifugation for 5 min at $8800 \times g$.

Three different treatments for disrupting the adenovirus capsid have been tested: phenol-cloroform, NaOH, and heating. This disruption is required to make the template viral DNA available for initiation of the PCR-mediated amplification in clinical specimens. The reactions were performed with clarified stool suspensions in transport medium.

Phenol-Chloroform Extraction. A 50-µl sample was incubated at 55 °C for 1 h in the presence of 60 µg of proteinase K per ml in 10 mM TRIS-hydrochloride (pH 7.5). The solution was extracted with phenol-chloroform-isoamyl alcohol (25:24:1), and the DNA was precipitated with ethanol in the presence of carrier tRNA (100 µg/ml).

NaOH Treatment. A 50-µl sample was treated with NaOH at a final concentration of 0.5 M to denature the capsids. After an incubation at 37 °C in a water bath for 15 min, the mixture was neutralized with HCl to a final concentration of 0.5 M.

Heating. A 10-µl sample, undiluted or diluted 10- or 100-fold, was mixed with all the reagents necessary for PCR amplification, including the Taq polymerase, and heated at 95 °C for 15 min directly in the programmable heat block (Hybaid, Teddington, UK, or Techne PHC-2, Cambridge, UK) used for the thermal cycling procedure. Evaluation of optimal methods for the pretreatment of the stool specimens will be discussed.

Preparation of Vira DNA from Infected Cells

For the preparation of viral DNA, about 25 µl of a clarified stool suspension was inoculated per 10 cm² area of monolayered 293 cells or HEp2 cells. These two cell types can be used for cultivation of samples to ensure the isolation of even fastidious adenoviruses. When cytopathic effect (CPE) was well developed, after 3–4 days, cells were harvested, and intracellular viral DNA was extracted by the method of Shinagawa et al. (1983).

Description of Primers

Locations and sequences of the adenovirus-specific primers used are shown in Fig. 1 and Table 1.

General Primers. The DNA sequences of the open reading frames of the hexon genes of Ad2 (Akusjärvi et al. 1984), Ad40 (Toogood et al. 1989), and Ad41 (Toogood and Hay 1988) were examined to locate suitable target sequences that could be used as general primers for the detection of adenoviruses. The specificity of the general primers, hexAA1885 and hexAA1913, in the hexon

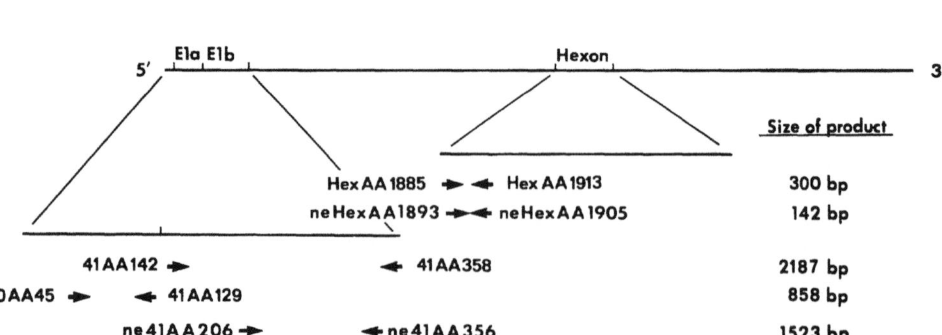

Fig. 1. Locations of the five sets of oligonucleotide primers in the 36-kb adenovirus genome

Table 1. Oligonucleotide primers for polymerase chain reaction (PCR) amplification of adenoviruses

Virus type (region)	Position	Primer[a]	Sequence	Amplimer length (bp)
Ad2 (hexon)				
Ad40 (hexon)	18858–18883[b]	hexAA1885	5'-GCCGCAGTGGTCTTACATGCACATC-3'[c]	300
Ad41 (hexon)	19136–19158[b]	hexAA1913	5'-CAGCACGCCGCGGATGTCAAAGT-3'[c]	
Ad2 (hexon)	18937–18960[b]	nehexAA1893	5'-GCCACCGAGACGTACTTCAGCCTG-3'	142
Ad2 (hexon)	19051–19079[b]	nehexAA1905	5'-TTGTACGAGTACGCGGTATCCTCGCGGTC-3'	
Ad41 (E1B)	1421–1446[d]	41AA142	5'-TCTGATGGAGTTTTGGAGTGAGCTA-3'	2187
Ad41 (E1B)	3585–3608[d]	41AA358	5'-AGAAGCATTAGCGGGAGGGTTAAG-3'	
Ad41 (E1B)	2061–2085[d]	ne41AA206	5'-GTCTGGTGGGCTGATTTGGAAGATG-3'	1523
Ad41 (E1B)	3561–3584[d]	ne41AA356	5'-CAGGGCCACTTTGGCAAACAAATC-3'	
AD40 (E1A)	453–477[e]	40AA45	5'-ATTGCTGTTGGCGCTTTTGACATAG-3'	858
Ad41 (E1A)	1297–1320[f]	41AA129	5'-TCAAGAGGACTTGGGGCGCTTTAA-3'	

[a] Primers were named in such a way as to describe the adenovirus type or adenovirus gene, the initials of the individual who discovered the primer, and the sequence position with the last figure deleted

[b] The sequence positions of the hexon primers are referred to the Ad2 hexon region (Akusjärvi et al. 1984)

[c] These primer sequences are shared among Ad2 (Akusjärvi et al. 1984), Ad40 (Toogood et al. 1989), and Ad41 (Toogood and Hay 1988)

[d] The sequence position is from Allard et al. (unpublished data)

[e] The sequence position is from van Loon et al. (1987)

[f] The sequence position is from Allard and Wadell (1988)

region of Ad2 (nucleotides 18858–19158) and the corresponding regions of the other human adenoviruses was tested on 18 different adenovirus types, representing all six subgenera. Satisfactory results were obtained both with a 1% agarose gel stained with ethidium bromide and with hybridization of a Southern blot of the gel with a mixture of six adenovirus probes representing all subgenera.

Specific Primers. For the specific detection of EAds Ad40 and Ad41, primers were chosen mainly from the genes encoding early regions E1A and E1B (van Loon et al. 1987; Allard and Wadell 1988). The specific primers for the two EAds, 41AA142 and 41AA358, which flank the region between nucleotides 1421–3608 in Ad41 and the corresponding region in Ad40, were tested against the same collection of 18 adenovirus types. Amplification was only detected when the two EAds Ad40 and Ad41 were used as templates. All PCRs for adenovirus types belonging to subgenera A – E were negative when assayed on ethidium bromide-stained agarose gels or on Southern blots.

To distinguish between the two enteric types, Ad40 and Ad41, we tested an Ad40-specific pair of primers, 40AA45 and 41AA129, agains the 18 adenovirus types. Ad40 was amplified, whereas the 17 other adenoviruses were negative on ethidium bromide-stained agarose gels or on hybridization of Southern blots. The reverse primer 41AA129 binds to both Ad41 and Ad40 E1A regions, but two mismatches to the Ad40 sequence can be found within the primer when the Ad41 sequence is used as a reference sequence. To match the Ad40 genome to the Ad41 genome, position 4 within the primer has to be changed from an A to a G, and position 8 has to be changed from a G to a C. However, the specificity results from the lack of homology of the upstream primer, 40AA45, to Ad41, due to a deletion in the Ad41 genome at that primer position (van Loon et al. 1987).

Nested Primers. In two-step amplification for detection of all adenovirus subgenera, two conserved nucleotide sequences localized within the first amplimer were identified. These two primers, nehexAA1893 and nehexAA1905, arranged in a nested fashion directed this second PCR step (Fig. 1). The specificity of the primers, which flank a region between nucleotides 18937–19079 in Ad2, was tested against 12 different adenovirus types representing all six subgenera. Positive results were obtained with all types when analyzed on an agarose gel. The Ad2 sequence was used as a reference sequence, and in matching this sequence to the Ad40 and Ad41 sequences, two mismatches and one mismatch, respectively, to the upstream primer nehexAA1893, and three mismatches and four mismatches, respectively, to the downstream primer nehexAA1905 were revealed. These mismatches do not seem to affect the amplification efficiency of the two EAd DNAs.

In the two-step amplification for the detection of enteric adenoviruses, a pair of primers internal of the two specific EAd primers was used (Fig. 1). The specificity of these internal primers, ne41AA206 and ne41AA356, which flank the region between nucleotides 2061 and 3584 in Ad41 and the corresponding region in Ad40, was tested on the same set of 12 adenovirus types. Amplification was only detected when the two EAds Ad40 and Ad41 were used as templates. A summary of the specificity of all the sets of primers analyzed on agarose gels is given in Fig. 2.

Polymerase Chain Reaction

Amplification of adenovirus DNA was carried out according to two different protocols with two different sets of primers.

1a. The general primer pair hexAA1885/hexAA1913 was used in a one-step amplification of the DNA prepared from infected cells or on pretreated stools (30 cycles).

1b. The general primer pair nehexAA1893/nehexAA1905 is positioned between the two hexAA1885/hexAA1913 primers. The former was used in combination with the latter in a nested primer two-step amplification of the adenovirus DNA (30 cycles hexAA1885/hexAA1913, 25 cycles nehexAA1893/nehexAA1905).

2a. The specific EAd primer pair 41AA142/41AA358 was used in a one-step amplification of the DNA prepared from infected cells or on pretreated stools (30–35 cycles).

2b. The other specific primer pair for EAd ne41AA206/ne41AA356 is positioned between the two 41AA142/41AA358 primers. The former was used in combination with the latter in a nested primer two-step amplification of the EAd DNAs (30–35 cycles 41AA142/41AA358, 25 cycles ne41AA206/ne41AA356).

For a typical one-step amplification reaction, 10 µl of a pretreated stool suspension diluted 10-fold in distilled water or 5–20 ng of extracted viral DNA were used. Amplification was carried out in a 50 µl reaction mixture containing 16.6 mM $(NH_4)SO_4$, 67 mM TRIS-HCl (pH 8.8 at 25 °C), 6.7 mM $MgCl_2$, 10 mM β-mercaptoethanol, 200 µM each dNTP, 100 µg BSA per ml, 0.08 µM of each primer needed for the specific reaction, and 2 U of thermostable Taq DNA polymerase (Perkin-Elmer Cetus). The samples were overlaid with 75 µl of mineral oil to prevent evaporation. Thermal cycling of the amplification mixture was performed in a programmable heat block (Hybaid or Techne PHC-2). In all PCR assays the first cycle of denaturation was carried out for 4 min at 94 °C.

With the use of two different sets of general hexon primers the denaturation was performed at 92 °C for 30 s, the annealing at 55 °C for 30 s, and the extension at 72 °C for 30 s. With the use of two different sets of specific EAd primers, which both yield amplified products of more than 1500 bp, a cycle comprises denaturation for 30 s at 92 °C, annealing for 30 s at 60 °C, and primer extension for 1 min at 72 °C. For a typical two-step amplification, 1 µl (1/50) was taken from a 30-cycled one-step amplification and added to a new batch of 50 µl PCR reaction mixture containing PCR buffer, the four nucleotides at a final concentration of 200 µM, 100 µg of BSA per ml, 2 U of Taq

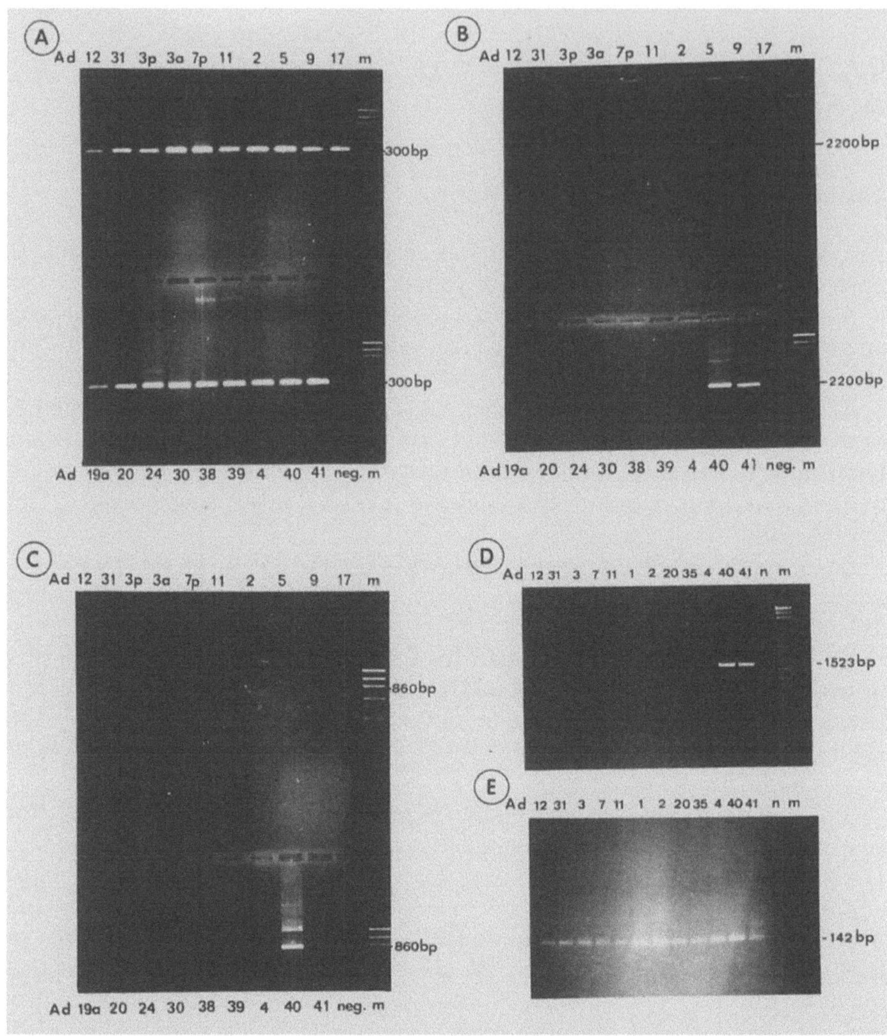

Fig. 2. A Polymerase chain reaction (PCR) amplification of DNA from 18 different adenovirus types, representing all six subgenera, with the general hexon region primers hexAA1885 and hexAA1913. Agarose gel electrophoresis of amplified products shows the characteristic 300-bp band. Two genome types of Ad3 were used. **B** Adenovirus DNA amplified with the EAd-specific primers 41AA142 and 41AA358. Amplified DNA was electrophoresed on a 1% agarose gel, resulting in a sequence of 2187 bp. **C** Adenovirus DNA amplified by PCR with the Ad40-specific primers 40AA45 and 41AA129. Amplified DNA resulted in a sequence of 858 bp. **D** PCR amplification of DNA from 12 different adenovirus types representing all six subgenera with the EAd-specific primers ne41AA206 and ne41AA356 used in the two-step amplification. Amplified DNA resulted in a sequence of 1523 bp. **E** adenovirus DNA amplified by PCR with the general hexon primers nehexAA1893 and nehexAA1905 used in the two-step amplification. Agarose gel electrophoresis of amplified products shows the characteristic 142-bp band. In all panels, *lane m* is molecular weight standards (lambda *Hind*III digest and/or ϕX174 *Hae*III digest)

polymerase, and 0.16 μM of each primer of nehexAA1893/nehexAA1905 or ne41AA206/ne41AA356, depending on which system was used. Some 25 additional cycles were then performed. As positive controls in all experiments, 10 pg of Ad2 or Ad41 DNA were used. PCR mixtures without DNA were used as negative controls.

Detection and Hybridization

Large fragments (1500–2200 bp) of amplified DNA were separated on a 1% SeaKem ME agarose gel, whereas small fragments of amplified DNA (142–300 bp) were separated on a 2% NuSieve GTG + 1% SeaKem ME agarose gel (FMC Bioproducts, Rockland, Mass.) by electrophoresis and stained with ethidium bromide (Sambrook et al. 1989). Gels were blotted onto nylon filters which were used for hybridization. The entire genomes of Ad31 (subgenus A), Ad3 (subgenus B), Ad2 (subgenus C), Ad19a (subgenus D), Ad4 (subgenus E) and Ad40 and Ad41 (subgenus F) were labeled with [α-^{32}P]dCTP by the multipriming DNA labeling technique to an activity of 10^8–10^9 dpm/μg and used as probes under stringent hybridization and washing conditions (T_m $-12\,°C$). Filters were exposed to an X-ray film for 16–40 h at $-70\,°C$.

Discussion

Apart from PCR, many other alternative methods for the detection of adenoviruses have been developed such as latex agglutination (Grandien et al. 1987), immunofluorescence, and dot-blot hybridization (Hyypiä 1984; Stålhandske et al. 1985). However, PCR provides a unique possibility to detect previously unrecognized adenoviruses or EAd that are fastidious, due to a lack of an optimal isolation system. The advantage with PCR is also its great sensitivity. With the use of a two-step amplification system on pretreated stool samples, the PCR method offers the possibility of detecting 1 copy of adenovirus DNA (Fig. 3).

Hexon Primers

A group-specific region for the detection of adenoviruses has been identified which partially codes for the basal part of the β-barrel-forming P1 domain within the hexon (Roberts et al. 1986). By the use of primers flanking this region of the hexon gene, 18 out of 18 adenovirus types representing all six subgenera could be detected by PCR. The hexon region of only four human adenovirus types has been sequenced so far, e.g., Ad2 (Akusjärvi et al. 1984), Ad5 (Kinloch et al. 1984), Ad40 (Toogood et al. 1989), and Ad41 (Toogood and Hay 1988). The general hexon primers hexAA1885 and hexAA1913 show

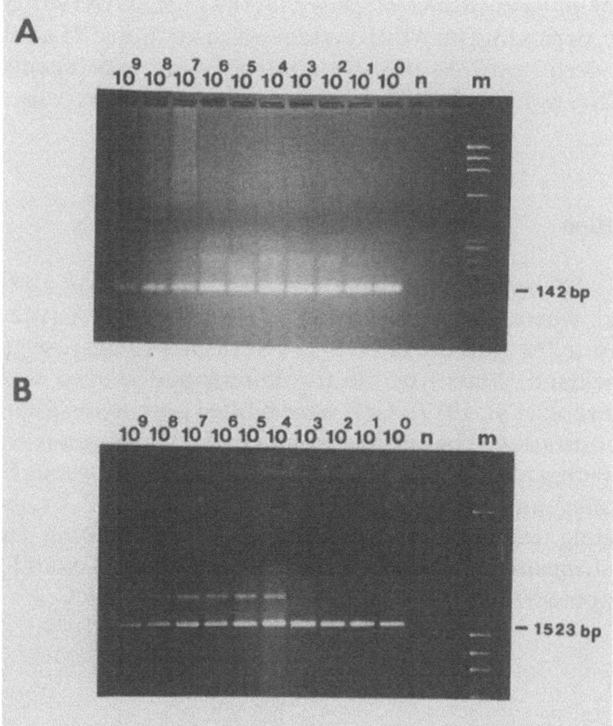

Fig. 3A, B. Sensitivity of polymerase chain reaction (PCR) for the detection of adenovirus DNA using plasmids containing the viral region to be amplified, diluted in supernatants of negative stool samples. The *numbers above the figures* represent the numbers of viral particles per genome copy contained in 10 µl (one-fifth) of the reaction mixture prior to amplification. **A** Ad41 DNA was amplified with the general hexon primers hexAA1885/hexAA1913. A 1/100 aliquot was transferred to a second amplification, primed by nehexAA1893/nehexAA1905 in a nested fashion. creating an amplimer of 142 bp. **B** Ad41 DNA was amplified with the enteric adenovirus-specific primers 41AA142/41AA358 and ne41AA206/ne41AA356 in a two-step amplification, as described in **A**, creating an amplimer of 1523 bp. *Lane n*, negative control: *lane m*, molecular weight standards (lambda *Hind*III digest and/or ϕX174 *Hae*III digest)

a 100% homology with the hexons of these four types. However, the upstream internal hexon primer used in the two-step amplification, nehexAA1893, has two mismatches in Ad40 and 1 mismatch in Ad41, and the downstream internal primer, nehexAA1905, has 3 mismatches in Ad40 and 4 mismatches in Ad41 compared with the sequences of Ad2 and Ad5, which are 100% homologous to the primers. These mismatches have not been shown to decrease the efficiency of annealing, at least not to detectable levels. Therefore, the annealing temperature was not altered.

The general hexon primers hexAA1885 and hexAA1913 could also be of great value in the veterinary diagnosis of bovine, ovine, and porcine aden-

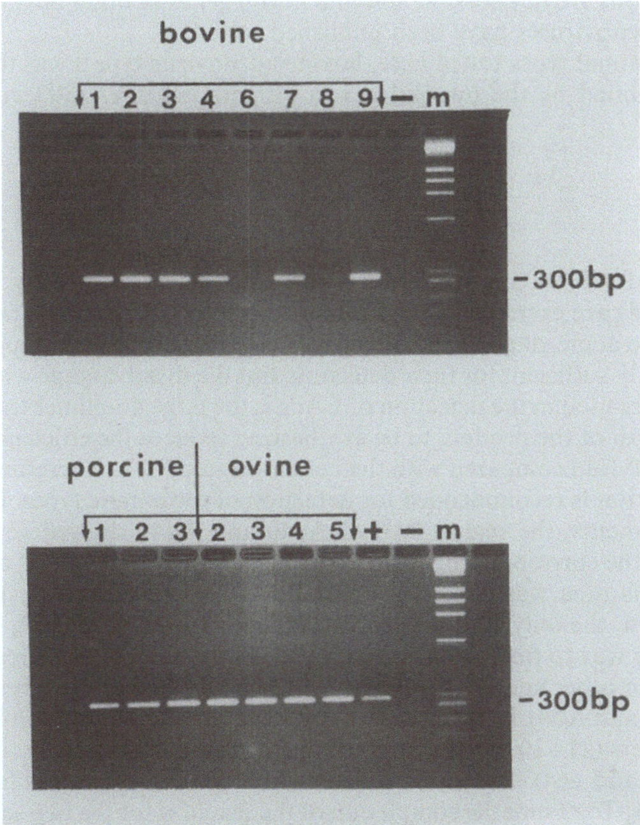

Fig. 4. Polymerase chain reaction (PCR) amplification of DNA from bovine adenovirus types 1–4, 6–9, porcine adenovirus types 1–3, and ovine adenoviruses types 2–5 with the general hexon primers hexAA1885 and hexAA1913. Agarose gel electrophoresis of amplified products shows the characteristic 300-bp band. *Lane m*, molecular weight standards (ϕX174 *Hae*III digest)

ovirus infections, since the region selected for the primers in the human hexon gene seems to share sequence homology with these species-specific animal adenoviruses. In PCR of viral DNA extracted from cells infected with bovine adenovirus types 1–4, 6–9, porcine adenovirus types 1–3, and ovine adenoviruses types 2–5, products were amplified to sizes of the amplimers corresponding to those synthesized with hAd DNA as a template (Fig. 4). In a comparison of the general hexon primers hexAA1885 and hexAA1913 with the corresponding sequence in the hexon gene of bovine adenovirus type 3 (Hu et al. 1984), 2 and 4 mismatches can be found, respectively, which do not seem to affect the efficiency of primer annealing.

Further comparisons are not possible since no hexon genes of other bovine, porcine, or ovine adenoviruses have been published so far.

Of the different animal types tested here, bovine adenovirus type 9 was the only one to be amplified by the internal hexon primers nehexAA1893 and nehexAA1905.

Specific Primers

The specific primers for EAd can be useful for their fast detection directly in stool specimens. EAds are excreted in large amounts, up to 10^{11} particles per gram of faeces, at the acute stage of the disease. Therefore, a one-step amplification is theoretically sufficient for their detection, but the disadvantage with the two sets of primers used in the detection of EAds is the large amplimer that is obtained. The length of the product to be synthesized reduces the efficiency of the PCR 10- to 100-fold compared with the hexon-primed PCR. Therefore, a two-step amplification is recommended for detection of the enteric types. In hybridization experiments, the region E1B of Ad40 and Ad41 showed very little homology with the corresponding region of other hAd types (Allard et al. 1985). Therefore, this area was chosen for selection of specific primers. In computer comparison, the only E1B sequences available are those of Ad2, 4, 5, 7, and 12. The aim was to find primers creating short amplimers, but areas with low homology to other adenovirus types seem to be located at the terminal ends of the Ad40 and Ad41 E1B genes. In a test with 12 different randomly chosen pairs of primers (21–30 mers) contained in the region amplified by the primers ne41AA206 and ne41AA356, we failed to find specific sequences for the detection of EAds. To ensure the completion of the reaction for the two sets of enteric-specific primers, 41AA142/41AA358 and ne41AA206/ne41AA356, we recommend an extension time of 1 min. To increase the specificity further, the annealing temperature can be increased from 55 °C to 60 °C, or the detection can be supplemented by hybridization under high stringency conditions.

Treatment of Samples

The proteinase K/phenol treatment was the most efficient one among the methods evaluated for pretreatment of stools. Both the NaOH method and heat treatment require a dilution of the samples to allow PCR amplification. A dilution reduces the effect of inhibiting factors present in the stools and of increased NaCl concentration caused by NaOH treatment that may affect the Taq polymerase. The disadvantage with the proteinase K/phenol treatment is that the method is laborious and time-consuming. Either one of the other two methods is recommended for preparation of stool samples on a large-scale basis, although they are not entirely efficient in breaking the capsid of the virions. In repeated experiments with the same samples, we sometimes had contradictory results. So the choice of treatment is a question of time versus accepting a loss of 1–2 positive per 50 positive analyzed samples. The stools have to be stored properly. To keep the virions as intact as possible, the stools should be kept at $-70\,°C$, preferably less than 1 year to avoid false-negative results. In long-term storage the viral capsid may be disrupted, and viral DNA become degraded, mainly due to lysis of bacteria in the stools resulting in release of proteolytic enzymes and DNase. Correct storage of stool samples is critical for the successful application of the EAd-specific primers, since they direct the synthesis of long amplimers of more than 2000 bp. If the viral DNA is degraded to some extent, the target sequence for the primer pair 41AA142/41AA358 may be damaged. When using this primer pair we have occasionally noted an extra amplified product of 530 bp. This product can be more or less dominant depending on the amount of DNA amplified and length of the storage time of the sample. Hybridization revealed that this fragment was located within the E1B region, but we cannot explain its origin since no further sequence homologies for the primers can be found in this region. Application of the two sets of general hexon primers requires only a 142-bp or a 300-bp template, which will probably not undergo the same degree of degradation as the longer templates and therefore still can produce a product.

Polymerase Chain Reaction Conditions

PCR buffer recommended by New England Biolabs was used in all reactions. A specific calibration for each pair of primers concerning different concentrations of the compounds that make up the buffer was not done in this study. We have compared the PCR buffers recommended by New England Biolabs, Promega (50 mM TRIS-HCl, pH 9.0 at 25 °C, 50 mM NaCl, 10 mM MgCl$_2$, 200 µM of each dNATP), and Perkin Elmer Cetus (50 mM KCl, 10 mM TRIS-HCl, pH 8.4 at room temperature, 1.5 mM MgCl$_2$, 100 µg/ml gelatin, 200 µM of each dNATP). In our hands, the buffer recommended by New England Biolabs was preferred for sensitivity. The primer concentration used of 0.08 µM corresponds to approximately 2×10^{12} molecules of each one. An increased primer concentration was not required since none of the primers are made by a compromise of alternative bases or inosine to create degenerated oligonucleotides. The hexon primers hexAA1885 and hexAA1913 were shown to be general, and the mismatches found in some of the other primers are not located too near the 3′-ends where the elongation starts. One of the most important steps in the PCR reaction itself is the dilution of stool samples, if the NaOH method or heat treatment is used.

When the two-step amplification is employed, one has to bear in mind the increase of the contamination risk caused by additional laborious steps compared with the one-step amplification reaction. Therefore, the use of aerosol-resistant tips is recommended, together with an additional number of negative controls placed in-between the samples to be analyzed. The two-step amplification can be performed in many different ways. We chose to transfer a small aliquot from the first amplification step to a new batch of PCR buffer contain-

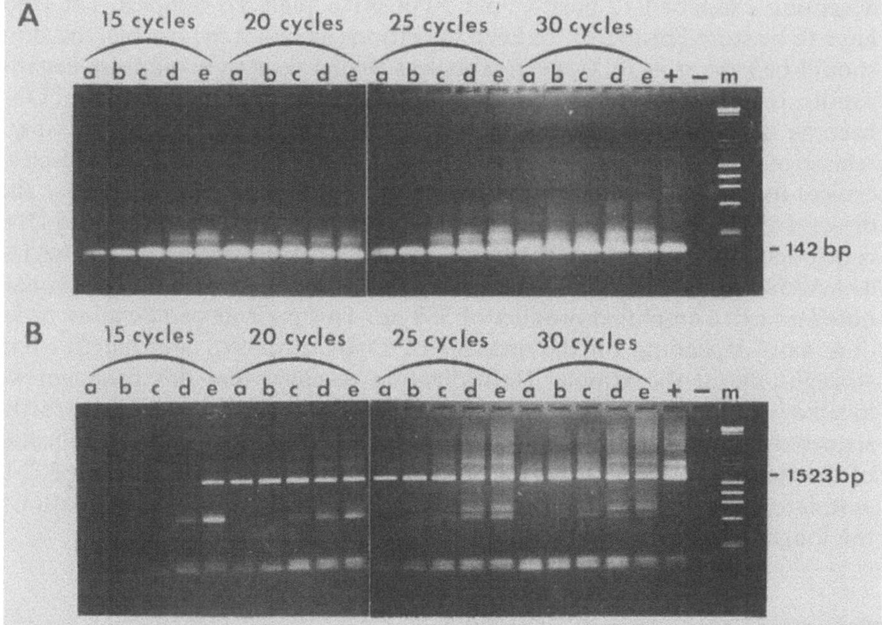

Fig. 5 A, B. Calibration of the two-step polymerase chain reaction (PCR) amplification. **A** Five different aliquots of amplified DNA from a NaOH-treated stool sample primed by the hexon primers hexAA1885/hexAA1913 (typed to be AD41 by DNA restriction enzyme analysis) were transferred to a second PCR amplification performed with a different number of cycles (15–30) using the nested hexon primers nehexAA1893/nehexAA1905. This was done to optimize the conditions for the two-step amplification system. *Lane a*, 0.1 µl; *lane b*, 0.5 µl; *lane c*, 1.0 µl; *lane d*, 2.5 µl; *lane e*, 5.0 µl; *lane m*, molecular weight standards (lambda *Hin*dIII digest and ϕX174 *Hae*III digest). **B** Calibration of the two-step amplification using the two sets of EAd-specific primers 41AA142/41AA358 and ne41AA206/ne41AA356, creating a final amplified DNA product of 1523 bp. Calibration was performed as described in **A**

ing the internal set of primers. We noted that the amount to be transferred from the first to the second step is more important than the number of cycles used in the second amplification. If a 1/100 aliquot of a given stool sample from the first amplification step of 30 cycles was transferred to the second amplification, another 15 cycles were sufficient for detection of hexon-primed DNA, whereas at least 20 cycles were required for detection of EAd-primed DNA (Fig. 5). If as much as 1/10 or 1/20 of an 30-cycled amplified stool sample was transferred to the second amplification step, a degradation of the final amplified product was seen. This phenomenon appeared in the two-step amplification primed by both the set of general hexon primers and the set of EAd-specific primers.

Sensitivity

The sensitivity of the different sets of primers differs with respect to the size of the amplimer produced. If pure Ad41 virions are treated with NaOH and amplified by use of EAd-specific primers, using both sets separately in one-step amplifications, it is possible to detect 10^3 virus particles by PCR. If the two different pairs of general hexon primers are used separately under the same conditions, the sensitivity is increased 10-fold to 10^2 virus particles.

The determination of sensitivity by using DNA from pure virions is not comparable with that obtained when virus DNA from stool samples is amplified consequently. It is necessary to simulate the characteristic conditions in the clinical sample. Therefore, plasmids containing the viral regions to be amplified were diluted in negative stool supernatants. In this PCR experiment, the sensitivity of all sets of primers was still the same as in the experiment with the virion DNA as template.

In order to determine the sensitivity of the two-step amplification, we continued with a second amplification of the products obtained from PCR of the plasmids diluted in negative stool supernatants. A two-step amplification using the internal set of EAd-specific primers lowered the detection limit to a single virus particle. By two-step amplification under the same conditions using the internal hexon primers, it was possible to detect one copy of the virus genome (Fig. 3).

Stool supernatants may inhibit the PCR. Eight different negative samples were mixed with a positive sample in a ratio of 1:1 in 8 independent reactions. One sample became negative in PCR amplification primed with the general hexon primers hexAA1885/hexAA1913. These results confirm assumptions of quantitative differences of inhibitor factors in different stool supernatants which may affect the activity of the Taq polymerase. All the tests of sensitivity were done after the routine performed NaOH treatment.

To eliminate the presence of inhibitor factors, the proteinase K/phenol treatment can be tried as an alternative method. A further dilution of the stool samples ($100 \times$) can also be recommended. Sometimes it may be necessary to use hybridization as a complementary method to increase the sensitivity and to confirm the specificity as well. In our hands, with the use of all the different sets of adenovirus primers, a hybridization with specific probes increased the sensitivity only slightly. What we have seen on Southern blots was also visualized on ethidium bromide-stained gels.

Extraction of viral DNA from cells infected with the patient specimen can furthermore enhance the detection sensitivity. In our hands, this additional step has been shown to give almost the same results as the two-step amplification system. It does not make sense to perform initial isolation attempts in cell cultures followed by detection of viral genomes by PCR, due to the enhanced risk of false-positive PCR amplifications.

                                           A. Allard and G. Wadell

Medical Importance

For routine PCR diagnosis of adenovirus particles in stool samples, a one-step amplification of 30–35 cycles primed by the general hexon primers hex-AA1885/hexAA1913 is recommended. The positive samples can further be confirmed to be enteric by use of the EAd-specific pair of primers, 41AA142/41AA358, or the Ad40-specific pair of primers, 40AA45/41AA129.

In an earlier study (Allard et al. 1990), the PCR method was compared with a commercial latex agglutination test (Adenolex). Sixteen stool samples were used that had previously been confirmed by DNA restriction enzyme analysis to contain adenovirus particles. Four samples were typed as Ad41, and 12 were found to be members of subgenera B and C. With the use of Adenolex directly on stools, only two samples turned out to be positive, both representing Ad41. By using a one-step amplification of 30 cycles primed by the general hexon primer pair hexAA1885/hexAA1913 on pretreated stools, 13 samples became positive. All four samples representing Ad41 were amplified with the use of the EAd-specific primers 41AA142/41AA358 in a one-step amplification of 30 cycles. At that time the primers for the two-step amplification were not available.

The two-step PCR amplification is very sensitive and can easily pick up persistent adenovirus particles. Adenoviruses should not be identified as the cause of illness just because they can be isolated from or detected in the stools of an individual with disease. Infection with members of subgenus C and Ad3 are characterized by a prolonged intermittent excretion (up to 906 days) and a frequent isolation rate (Brandt et al. 1985; Fox and Hall 1980; Kidd et al. 1982). If the aim is to look for persistent adenovirus particles or to detect adenoviruses in other sources than stool, such as liquor or biopsies, the two-step amplification procedure with general hexon primers could be useful. In a study including 50 healthy children and 50 healthy adults (unpublished data), pretreated stool samples were examined with a two-step amplification primed by the general hexon primers. Some 50% of the children and 18% of the adults were shown to be excreting adenovirus particles.

Information on possible persistence of EAds is scanty. We only have followed a 3-year-old boy infected with Ad41 for 4 months. In the acute stage of diarrheal disease and fever, 30 cycles of a one-step PCR amplification were sufficient for detection with both the general hexon primers hexAA1885/hex-AA1913 or the EAd-specific primers 41AA142/41AA358. After both 1 and 2 months, a two-step amplification primed by the general sets of hexon primers was required to detect virus particles in the stools. However, negative results were obtained already in the sample taken 1 month after the outbreak when the two sets of EAd specific primers were used in a two step amplification. After 3 months, no viral DNA was detected by PCR regardless of what primer system was used. Thus it seems that Ad41 particles are frequently shed in stools during the acute stage but that the excretion declines rapdily after recovery. The different sensitivity of the primer sets used in the two-step amplification may be due to the fact that the general hexon primers are more sensitive than the EAd specific primers or less likely that the general hexon primers detected persistent adenovirus particles from an earlier infection caused by a non-enteric type.

EAd particles are normally shed in high amounts during the acute phase of illness. Because of the rapid decrease of excretion, the PCR detection should be supplemented with a two-step amplification, DNA restriction enzyme analysis, or monoclonal antibodies. However, the use of EAd specific primers in a two-step amplified PCR represents the most sensitive method developed so far for the detection of enteric adenoviruses. Finally, the two-step PCR amplification system can also be used to assess the degree of fecal contamination in wells, sewage, or seawater.

References

Adrian T, Sassinek J, Wigand R (1990) Genome type analysis of 480 isolates of adenovirus types 1, 2, and 5. Arch Virol 112:235–248

Akusjärvi G, Aleström P, Pettersson M, Lager M, Jörnvall H, Pettersson U (1984) The genome for adenovirus 2 hexon polypeptide. J Biol Chem 259:13976–13979

Allard A, Wadell G (1988) Physical organization of the enteric adenovirus type 41 early region1A. Virology 164:220–229

Allard A, Wadell G, Evander M, Lindman K (1985) Specific properties of two enteric adenovirus 41 clones mapped within early region 1A. J Virol 54:145–150

Allard A, Girones R, Joto P, Wadell G (1990) Polymerase chain reaction for detection of adenoviruses in stool samples. J Clin Microbiol 28:2659–2667

Ambinder RF, Burns W, Forman M et al. (1986) Hemorrhagic cystitis associated with adenovirus infection in bone-marrow transplantation. Arch Intern Med 146:1400–1401

Brandt CD, Kim HW, Rodriguez WJ, Arrobio JO, Jeffries BC, Stallings EP, Lewis C, Miles AJ, Gardner MK, Parrott RH (1985) Adenovirus and pediatric gastroenteritis. J Infect Dis 151:437–443

Fiala M, Payne JE, Berne TV, Poolsawat S, Schieble J, Guze LB (1974) Role of adenovirus type 11 in hemorrhagic cystitis secondary to immunosuppression. J Urol 112:595–597

Flomenberg PR, Chen M, Munk G, Horwitz MS (1987) Molecular epidemiology of adenovirus type 35 infections in immunocompromised hosts. J Infect Dis 155:1127–1134

Fox JP, Hall CE (1980) Viruses in families. PSG, Littleton, pp 294–334

Grandien M, Pettersson CA, Svensson L, Uhnoo I (1987) Latex agglutination test for adenovirus diagnosis in diarrheal disease. J Med Virol 23:311–316

Hammond GW, Mauthe G, Joshua J, Hannon CK (1985) Examination of uncommon clinical isolates of human adenoviruses by restriction endonuclease analysis. J Clin Microbiol 21:611–616

Harnett GB, Bucens MR, Clay SJ, Saker BM (1982) Acute hemorrhagic cystitis caused by adenovirus type 11 in a recipient of a transplanted kidney. Med J Aust 1:565–567

Herrmann JE, Perron-Henry DM, Blacklow NR (1987) Antigen detection with monoclonal antibodies for the diagnosis of adenovirus gastroenteritis. J Infect Dis 155:1167–1171

Hierholzer JC, Wigand R, Anderson LJ, Adrian T, Gold JWM (1988) Adenoviruses from patients with AIDS: a plethora of serotypes and a description of five new serotypes of subgenus D (type 43–47). J Infect Dis 158:804–813

Hu SL, Hays WW, Potts DE (1984) Sequence homology between bovine and human adenoviruses. J Virol 49:604–608

Hyypiä T (1984) Detection of adenoviruses in nasopharyngeal specimens by radioactive and non-radioactive DNA probes. J Clin Microbiol 21:730–733

Johansson ME, Uhnoo I, Kidd AH, Madeley CR, Wadell G (1980) Direct identification of enteric adenovirus, a candidate new serotype, associated with infantile gastroenteritis. J Clin Microbiol 12:95–100

Johansson ME, Uhnoo I, Svensson L, Pettersson CA, Wadell G (1985) Enzyme-linked immunosorbent assay for detection of enteric adenovirus 41. J Med Virol 17:19–27

Johansson ME, Brown M, Hierholzer JC, Thörner Å, Ushijima H, Wadell G (1991) Genome analysis of adenovirus type 31 strains from immunocompromised and immunocompetent patients. J Infect Dis 163:292–299

Keller EW, Rubin RH, Black PH, Hirsch MS, Hierholzer JC (1977) Isolation of adenovirus type 34 from a renal transplant recipient with interstitial pneumonia. Transplantation 23:188–191

Kidd AH, Cosgrove BP, Brown RA, Madeley CR (1982) Faecal adenoviruses from Glasgow babies. Studies on culture and identity. J Hyg 88:463–474

Kinloch R, Mackay N, Mautner V (1984) Adenovirus hexon: sequence comparison of subgroup C serotypes 2 and 5. J Biol Chem 259:6431–6436

Li Q, Wadell G (1988) The degree of genetic variability among adenovirus type 4 strains isolated from man and chimpanzee. Arch Virol 101:65–77

Ljungman P, Gleaves CA, Meyers JD (1989) Respiratory virus infection in immunocompromised patients. Bone Marrow Transplant 4:35–40

Roberts MM, White JL, Grütter MG, Burnett RM (1986) Three-dimensional structure of the adenovirus major coat protein hexon. Science 232:1148–1151

Rowe WP, Huebner RJ, Gilmore LK, Parrott RH, Ward TG (1953) Isolation of a cytopathogenic agent from human adenoids undergoing spontaneous degeneration in tissue culture. Proc Soc Exp Biol Med 84:570–573

Saiki RK, Scharf S, Faloona F, Mullis KB, Horn GT, Erlich HA, Arnheim N (1985) Enzymatic amplification of β-globin genomic sequences and restriction site analysis for diagnosis of sickle-cell anemia. Science 230:1350–1354

Sambrook J, Fritsch EF, Maniatis T (1989) Molecular cloning: a laboratory manual, 2nd edn. Cold Spring Harbor Laboratory, Cold Spring Harbor

Schmitz H, Wigand R, Heinrich W (1983) Worldwide epidemiology of human adenovirus infections. Am J Epidemiol 117:455–466

Shields AF, Hackman RC, Fife KH, Corey L, Meyers JD (1985) Adenovirus infections in patients undergoing bone-marrow transplantation. N Engl J Med 312:529–533

Shinagawa M, Matsuda A, Ishiyama T, Goto H, Sato G (1983) A rapid and simple method for preparation of adenovirus DNA from infected cells. Microbiol Immunol 27:817–822

Siegal FP, Dikman SH, Arayata RB, Bottone EJ (1981) Fatal disseminated adenovirus type 11 pneumonia in an agammaglobulinemic patient. Am J Med 71:1062–1067

Stalder H, Hierholzer JC, Oxman MN (1977) New human adenovirus (candidate adenovirus type 35) causing fatal disseminated infection in a renal transplant recipient. J Clin Microbiol 6:257–265

Stålhandske P, Hyypiä T, Allard A, Halonen P, Pettersson U (1985) Detection of adenoviruses in stool specimens by nucleic acid spot hybridization. J Med Virol 16:213–218

Toogood CIA, Hay RT (1988) DNA sequence of the adenovirus type 41 hexon gene and predicted structure of the protein. J Gen Virol 69:2292–2301

Toogood CIA, Murali R, Burnett RM, Hay RT (1989) The adenovirus type 40 hexon: sequence, predicted structure and relationship to other adenovirus hexons. J Gen Virol 70:3203–3214

Uhnoo I, Wadell G, Svensson L, Johansson M (1983) Two new serotypes of enteric adenovirus causing infantile diarrhoea. Dev Biol Stand 53:311–318

van der Avoort HGAM, Wermenbol AG, Zomerdijk TPL, Kleijne JAFW, van Asten JAAM, Jensma P, Osterhaus ADME, Kidd AH, deJong JC (1989) Characterization of fastidious adenovirus types 40 and 41 by DNA restriction enzyme analysis and by neutralizing monoclonal antibodies. Virus Res 12:139–158

van Loon AE, Rozijn TH, deJong JC, Sussenbach JS (1985) Physiochemical properties of the DNAs of the fastidious adenovirus species 40 and 41. Virology 140:197–200 van Loon AE, Ligtenberg M, Reemst AMCB, Sussenbach JS, Rozijn TH (1987) Structure and organization of the left-terminal DNA regions of fastidious adenovirus types 40 and 41. Gene 58:109–126

Wadell G (1979) Classification of human adenovirus by SDS-polyacrylamide gel electrophoresis of structural polypeptides. Intervirology 11:47–57

Wadell G, Hammarskjöld ML, Winberg G, Varsanyi TM, Sundell G (1980) Genetic variability of adenoviruses. Ann NY Acad Sci 354:16–42

Wadell G, Allard A, Evander M, Li Q (1986) Genetic variability and evolution of adenoviruses. Chem Scr 26B:325–335

Wigand R, Adrian T, Bricout F (1987) A new adenovirus of subgenus D: candidate adenovirus type 42. Arch Virol 94:283–286

Wood DJ, Bijlsma K, deJong JC, Tonkin C (1989) Evaluation of a commercial monoclonal antibody-based enzyme immunoassay for detection of adenovirus types 40 and 41 in stool specimens. J Clin Microbiol 27:1155–1158

Chapter 24 Detection of Coronaviruses by the Polymerase Chain Reaction *

Janet N. Stewart, Samir Mounir and Pierre J. Talbot

Summary

A simple and reliable method for the amplification and specific detection of human coronavirus nucleotide sequences was developed, based on the synthesis of cDNA, the polymerase chain reaction, and the use of oligonucleotide probes in Southern blots. Regions from several genes of the two prototype strains of human coronavirus (229E and OC43) could be specifically amplified. This powerful technique was applied to clinical specimens, with appropriate controls, to study the tissue tropism of coronaviruses and their possible involvement in diseases other than the common cold. We have obtained preliminary evidence for the detection of the genome of a human coronavirus in central nervous system autopsy tissue from some multiple sclerosis patients.

Introduction

Human coronaviruses are among the causes of respiratory disease in man. They are responsible for 15%–35% of common colds (McIntosh 1990). This acute disease of the respiratory tract is highly prevalent; a large majority of the population is known to seroconvert at an early age. Moreover, the disease can be experimentally reproduced in volunteers (McIntosh 1990). Other disease associations have been suggested but are less well documented. A seroepidemiological study has linked coronavirus infection with some forms of pneumonia, perimyocarditis, meningitis, and radiculitis (Riski and Hovi 1980). A few

Virology Research Center, Institut Armand-Frappier, Université du Québec, Laval, Québec, Canada H7N 4Z3

* This work was supported by grant MT-9203 from the Medical Research Council of Canada to P.J.T. J.N.S. is grateful to the Institut Armand-Frappier for student support. P.J.T. gratefully acknowledges scholarship support from the National Sciences and Engineering Research Council of Canada (NSERC).

reports of their involvement in severe diarrhea have appeared (Resta et al. 1985; Battaglia et al. 1987). Moreover, coronaviruses belong to a long list of viruses suspected as etiological agents of multiple sclerosis (MS), the most widespread human demyelinating disease. This association stems from the observation of coronavirus-like particles in the brain of a MS patient (Tanaka et al. 1976), the isolation of coronaviruses from MS patients (Burks et al. 1980), the local synthesis of antibodies to human coronaviruses in the central nervous system (CNS) of MS patients (Salmi et al. 1982), and the preferential detection of coronavirus genomes in the brains of MS patients by *in situ* hybridization (Murray et al. 1992). Moreover, murine coronaviruses show diversified tropisms, and some strains have been used in rodents as a model system to study chronic and acute hepatic and neurologic diseases (ter Meulen et al. 1990), which further strengthens the possible involvement of coronaviruses in such diseases in humans. However, contradictory reports have been published on the association of coronaviruses with diseases other than the common cold, which emphasizes the importance of further research on their medical importance. The polymerase chain reaction (PCR) should provide an exquisite tool for that purpose.

Coronaviruses belong to the *Coronaviridae* family of enveloped RNA viruses (Cavanagh et al. 1990). They contain a single-stranded, positive-sense RNA molecule, with a molecular mass of approximately $6-8 \times 10^6$ daltons (Lai and Stohlman 1978; Lomniczi and Kennedy 1977) and an actual size, derived from nucleotide sequences, of $27-31$ kb (Boursnell et al. 1987; Lee et al. 1991). The virus-specific mRNAs in infected cells comprise a genomic-sized mRNA plus $5-8$ subgenomic mRNA species. These mRNAs are arranged in a 3'-coterminal nested-set structure, in which the sequence of every mRNA is contained within the sequence of the next larger mRNA (Lai 1990). The 5'-unique regions are translated into structural and nonstructural viral proteins. Three or four structural proteins have been identified: the nucleocapsid protein N, the membrane glycoprotein M, the surface peplomer glycoprotein S, and in some strains of coronaviruses such as the OC43 strain of human coronavirus, a surface hemagglutinin-esterase glycoprotein (Cavanagh et al. 1990; Holmes 1990).

The PCR technology was developed for the detection and amplification of individual nucleotide sequences from a single cell or small quantities of sample (Saiki et al. 1988). Convenient *in vitro* amplification of specific DNA sequences with remarkable efficiency is possible by using a thermostable DNA polymerase derived from the bacterium *Thermophilus aquaticus* (Taq) in an automated PCR. Reverse transcription coupled to the PCR (RT/PCR) can be used to amplify specific RNA sequences. The RNA is reverse-transcribed using a primer complementary to the target sequence (antisense primer) to create cDNA copies, which are amplified using sense and antisense primers.

In the present study, we report the use of RT/PCR methodology for the amplification of portions of four genes of two representative strains of human coronaviruses and the application of this technique to clinical specimens of CNS autopsy tissues.

Materials and Methods

Viruses and Cells. The 229E strain of HCV (HCV-229E) was propagated at 33 °C in L132 human fetal lung cells (American Type Culture Collection, ATCC, Rockville, MD), which were grown as monolayers at 37 °C in a humidified chamber with 5% (v/v) CO_2 in Earle's minimum essential medium: Hank's M199 (1:1 v/v) supplemented with 0.13% (w/v) sodium bicarbonate, 50 µg/ml gentamicin, and 5% (v/v) fetal bovine serum (reduced to 2% (v/v) for viral infections). The OC43 strain of HCV (HCV-OC43) was propagated at 37 °C in HRT-18 human rectal adenocarcinoma cells, a gracious gift from Drs. Jean-François Vautherot and Jacques Laporte (I.N.R.A., Jouy-en-Josos, France), grown in the same medium. Initial inocula of both human coronavirus strains were obtained from the ATCC.

Purification of RNA. Total cellular RNA from uninfected cells and human coronavirus-infected cells was prepared by disrupting the cells in 4 M guanidinium thiocyanate with a Polytron homogenizer and pelleting the RNA through a pad of 5.7 M cesium chloride (Chirgwin et al. 1979). The sample was then extracted with phenol/chloroform/isoamyl alcohol and the RNA precipitated by the addition of 0.3 M sodium acetate and 2.5 volumes of 100% (v/v) ethanol. After centrifugation, the pellet was washed with 70% (v/v) ethanol and dissolved in water.

For clinical specimens, total RNA was extracted by a more convenient method modified from Chomczynski and Sacchi (1987). Briefly, 50–300 mg of tissue were thawed and homogenized in 0.5 ml of 4 M guanidinium thiocyanate, extracted with phenol and chloroform and precipitated with ethanol twice. After washing with 70% (v/v) ethanol, the pellets were air-dried and resuspended in water.

Design of Primers and Probes. The oligonucleotides used to prime the extension of viral RNA were designed to amplify unique portions of the RNA coding for the following viral proteins: the integral membrane protein M, nonstructural (NS) proteins, or the spike glycoprotein S of HCV-OC43, NS proteins encoded by mRNA 4 of HCV-229E, or the nucleocapsid protein N of both HCV-229E and -OC43. These sequences were either based on published sequences of the corresponding viral genomes (HCV-229E N gene: Schreiber et al. 1989; HCV-229E mRNA 4 gene: Raabe and Siddell 1989; Jouvenne et al. 1992; HCV-OC43 N gene: Kamahora et al. 1989), or on unpublished sequences obtained in our laboratory (HCV-OC43 M, NS, and S genes). Oligonucleotides were synthesized by the phosphoramidite method on a Pharmacia-LKB "Gene Assembler Plus" DNA synthesizer (Pharmacia, Baie d'Urfé, Québec). Primers for the S gene of the HCV-OC43 strain were: 5'-GCGAT-TACCACTGGTTATCGG-3' (S2H, sense; unpublished data) and 5'-GGGC-GTGGCCTTAAGAAC-3' (S2l, antisense; unpublished data). Primers for the M and NS genes of HCV-OC43 were: 5'-CTGGACACCAGGAGTTAG-3'

(NS-M, sense; unpublished data) and 5'-TCGGCCCACTTGAGGATG3' (N, antisense, nucleotides 147–165; Kamahora et al. 1989). Primers for mRNA 4 of the HCV-229 strain were: 5'-CCACATACAGTAATGGCTCTAGGT-3' (229E-4, sense, nucleotides 37–60; Raabe and Siddell 1989; or 42–65; Jouvenne et al. 1992) and 5'-CACTATAAGCACCACACACCAGAG-3' (#5, antisense, nucleotides 756–780; Raabe and Siddell 1989; or 501–524; Jouvenne et al. 1992). Primers for the N gene of HCV-229E were: 5'-AGGCGCAAGAATTCAGAACCAGAG-3' (E1, sense, nucleotides 498–521; Schreiber et al. 1989) and 5'-AGCAGGACTCTGATTACGAGAAG-3' (E3, antisense, nucleotides 783–806; Schreiber et al. 1989). Primers for the N gene of the HCV-OC43 strain were: 5'-CCCAAGCAAACTGCTACCTCTCAG-3' (O1, sense, nucleotides 215–238; Kamahora et al. 1989) and 5'-GTAGACTCCGTCAATATCGGTGCC-3' (O3, antisense, nucleotides 497–520; Kamahora et al. 1989). Restriction enzyme cleavage sites were added to the 5'-end of some oligonucleotides for cloning purposes irrelevant to the present study.

The oligonucleotides used to prime the extension of control RNAs were located on different exons of the cellular genes, in order to distinguish between amplified RNA and DNA, Primers for myelin basic protein (MBP) RNA were: 5'-AGAACTGCTCACTACGGCTCCCTG-3' (M1, sense, nucleotides 274–307; Streicher and Stoffel 1989) and 5'-TCCAGAGCGACTATCTCTTCCTCC-3' (M3, antisense, nucleotides 550–573; Streicher and Stoffel 1989). Primers for γ-actin RNA were 5'-GACCTGACCGACTACCTCATGAAG-3' (A1, sense, nucleotides 1358–1381; Erba et al. 1988) and 5'-GGAGTTGAAGGTGGTCTCGTGGAT-3' (A3, antisense, nucleotides 1712–1735; Erba et al. 1988).

The PCR was carried out with a DNA Thermal Cycler (Perkin Elmer-Cetus, Montréal, Québec). The identity of the amplification products of the N genes of HCV-229E and HCV-OC43, as well as of the control RNAs, which were the target for amplification in clinical specimens, was confirmed in hybridization experiments with oligonucleotide probes located on the internal portion of the amplified sequences. The oligonucleotide probe sequences were as follows. For HCV-229E: 5'-ATGAAGGCAGTTGCTGCGGCTCTT-3' (E2, sense, nucleotides 693–716; Schreiber et al. 1989); for HCV-OC43: 5'-GATGGCAACCAGCGTCAACTGCTG-3' (O2, sense, nucleotides 419–442; Kamahora et al. 1989); for MBP: 5'-CTGTCCCTGAGCAGATTTAGCTGG-3' (M2, sense, nucleotides 406-429; Streicher and Stoffel 1989); for γ-actin: 5'-GAAATCGTGCGCGACATCAAGGAG-3' (A2, sense, nucleotides 1432–1455; Erba et al. 1988).

cDNA Synthesis. The RNA (10–100 ng) was reverse transcribed after initial heating at 65 °C for 7 min in a final volume of 20 µl, containing 1 × Taq DNA polymerase buffer (BIO/CAN Scientific, Mississauga, Ontario), 1 µl of 50 mM MgCl$_2$, 2 µl of dNTP mix (10 mM of each dNTP), 50 pmol of the antisense primer, 40 U of RNAguard (Pharmacia), and 20 U of Moloney murine leukemia virus reverse transcriptase (Pharmacia). The reaction mixture was incubated at 37 °C for 60 min.

Amplification. The cDNAs produced by RT were amplified using a modification of the original PCR method (Saiki et al. 1988). The 20-µl volume of cDNA was added to 80 µl of the PCR reaction mixture [1 × Taq DNA polymerase buffer, 60 pmol of each primer (antisense and sense), 200 µM of each dNTP, 1.3 mM MgCl$_2$, 2.5 U of Taq DNA polymerase (BIO/CAN Scientific)]. Thirty cycles of amplification were performed under the following conditions: denaturation at 94 °C for 1 min, annealing of the primers at 55 °C (cells) or 60 °C (clinical samples) for 2 min, chain extension at 72 °C for 2 min. After the last cycle, the samples were incubated at 72 °C for an additional – 10 min to complete all strands.

For clinical specimens, master mixes of the RT and PCR solutions were prepared containing the same final concentraitons of reagents as described above.

Gel Electrophoresis and Hybridization. Some 5 or 20 µl of the amplified cDNA from cellular or clinical sample RNA, respectively, were electrophoresed on agarose gels in TAE buffer (40 mM Tris-acetate, 1 mM ethylene diamine tetraacetic acid, pH 7.2), denatured, neutralized, and transferred to nitrocellulose sheets (Hybond-C Extra; Amersham, Oakville, Ontario) according to the method of Southern (1979). Blots were hybridized for 16 h with a purified, [γ-^{32}P]ATP end-labelled oligonucleotide probe (2 × 10^6 cpm/ml) at 50 °C, in a buffer containing 6 × SSC (1 × SSC = 0.15 M NaCl, 0.015 M sodium citrate, pH 7.0), 0.05% (w/v) pyrophosphate, 1 × Denhardt's solution, and 100 µg/ml denatured salmon sperm DNA. The blots were washed three times, for 20 min each time, at room temperature and once for 20 min at 60 °C in 6 × SSC, 0.05% (w/v) pyrophosphate, and exposed to X-ray film (Kodak) at − 70 °C.

Results and Discussion

Development of the Method

The RT-PCR was initially developed with continuous cell lines susceptible to HCV infection. The amplified DNA from infected cells was analyzed by agarose gel electrophoresis in parallel with RT-PCR products from uninfected cells. The specificity of the reaction was first based on the amplification of appropriately sized fragments. As shown in Fig. 1, RNA from cells infected with the HCV-229E and -OC43 strains generated cDNA bands of the predicted sizes for the target genes: 1870, 1501, and 729 bp (Fig. 1, lanes 3, 6, and 9, respectively). Uninfected cells (Fig. 1, lanes 2, 5, and 8) or water (Fig. 1, lanes 1, 4, and 7) was used as negative controls. The specificity of the PCR was further confirmed by Southern blot analysis of the 308- and 306-bp amplified products from the N genes of HCV-229E and HCV-OC43 with specific oligonucleotide primers homologous to an internal portion of each target sequence (Fig. 2).

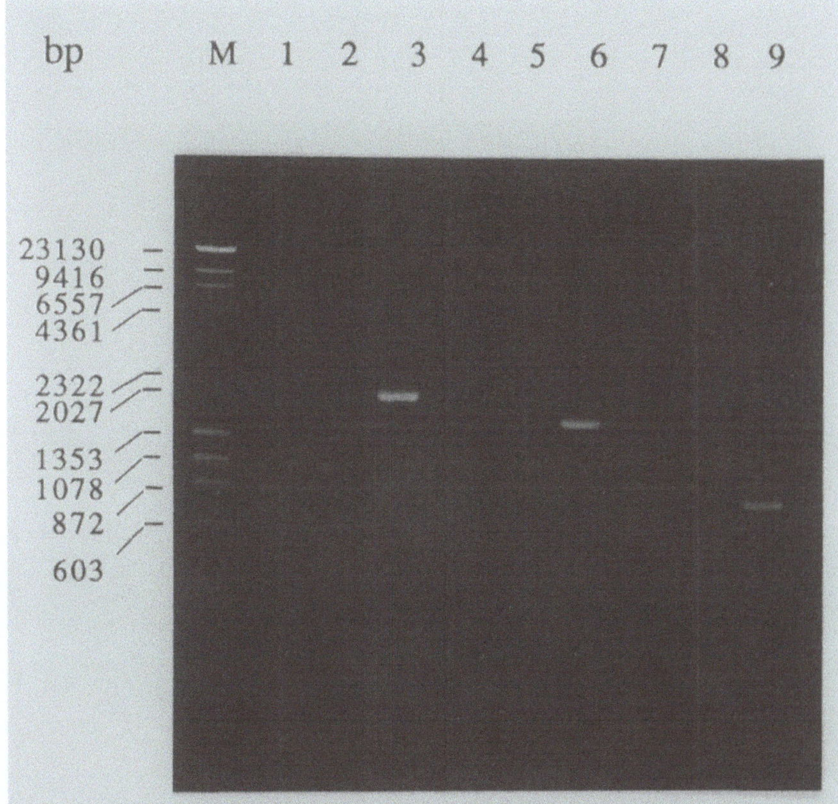

Fig. 1. Detection of various genes of the two prototype strains of human coronavirus (HCV) by *in vitro* amplification and agarose (1% w/v) gel analysis of the amplification products. *Lanes 1, 4 and 7*, H₂O control; *lanes 2 and 5*, HRT-18 cell line used to propagate HCV-OC43; *lane 3*, HCV-OC43 (1870-bp fragment from the S gene); *lane 6*, HCV-OC43 (1501-bp fragment from the M and NS genes); *lane 8*, L132 cell line used to propagate HCV-229E; *lane 9*, HCV-229E (729-bp gene from mRNA 4); *lane M*, *Hin*dIII-digested λ DNA mixed with *Hae*III-digested φX174 RF DNA used as molecular size markers. The numbers on the *left* indicate the lengths of fragments in bp

Application to Clinical Specimens

The PCR has made it possible to detect viral nucleic acids in archival tissues, facilitating the study of diseases of possible viral etiology. The distribution of HCVs in different tissues has been little characterized, and RT/PCR provides us with a rapid method for preliminary studies on this subject. Human CNS tissues are of particular interest due to the difficulty of obtaining adequate numbers of fresh specimens for virological studies.

The detection of RNA presents problems that do not occur in the detection of DNA. Formalin-fixed tissues can be studied for DNA content. However, we

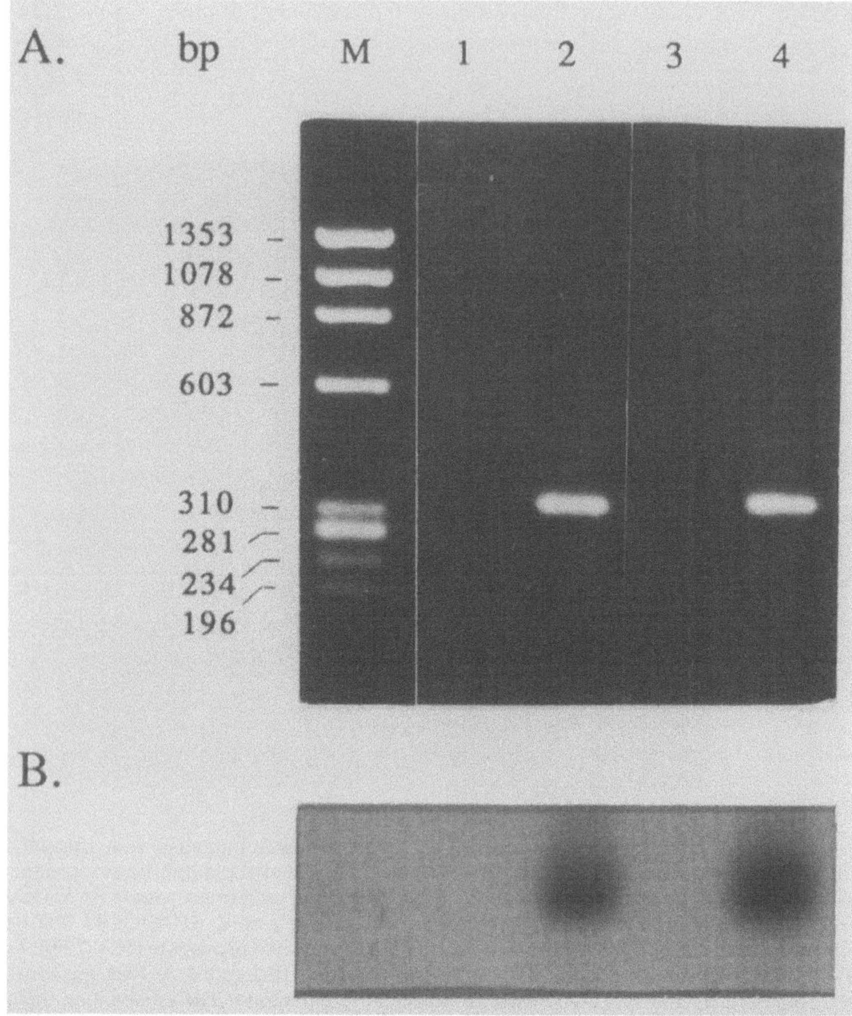

Fig. 2 A, B. Detection of the N genes of the two prototype strains of human coronaviruses (HCV) by *in vitro* amplification, followed by agarose (1% w/v) gel analysis of the amplification products (**A**) and identification of amplified sequences by hybridization with oligonucleotide probes (**B**). *Lane 1*, L132 cell line used to propagate HCV-229E; *lane 2*, HCV-229E; *lane 3*, HRT-18 cell line used to propagate HCV-OC43; *lane 4*, HCV-OC43; *lane M*, *Hae*III-digested ϕX174 RF DNA used as molecular size marker. The numbers on the *left* indicate the lengths of fragments in bp

have not been able to extract RNA suitable for amplification from tissues preserved in this manner. RNA is also highly susceptible to degradation by ribonucleases. In our laboratory we analyze RNA extracted from white matter obtained from autopsies which have been preserved at $-70\,^{\circ}$ to $-80\,^{\circ}$C for several years. Both the length of time between death of the individual and freezing of the tissue and repeated thawing of the tissues can contribute to RNA degradation. We perform an RT/PCR on each extract of RNA to detect cellular mRNAs to verify that significant degradation of RNA has not occurred. White matter specimens in which mRNA coding for MBP is not detectable are excluded from the study to avoid possible false-negative results due to RNA degradation. The mRNA coding for γ-actin is far more abundant than MBP mRNA in the specimens studied, and therefore we consider MBP mRNA detection a more sensitive indicator of partial RNA degradation in white matter (where MBP is expressed). As shown in Fig. 3, RNA prepared from clinical specimens yielded amplified cDNAs of the expected sizes (300 bp) for both MBP (lane 2) and γ-actin (lane 4). As expected, MBP RNA could not be amplified from lymphocytes (lane 1). Moreover, amplified fragments were shown to be specific by oligonucleotide hybridization. Since γ-actin is present in all cells, we use it for checking RNA extracted from other cells or tissues such as peripheral blood lymphocytes and gray matter. The primers used to amplify cellular control RNAs are located on different exons of their corresponding genes, and as such, RNA and DNA amplification can be distinguished on the basis of the size of the reaction product. An advantage in RNA detection is that viral mRNAs as well as genomes are detected, and since the former are present in much larger quantities, the test becomes more sensitive.

Several rapid methods of sample preparation for DNA amplification are available, but RNA extraction remains a time-consuming activity. Our extraction method is performed using Eppendorf tubes and a microfuge to extract up to 300 mg of tissue per tube. This produces $10-20$ µg of RNA per specimen, which is aliquoted for use in multiple tests. No DNA has yet been detected in the RNAs extracted. Several specimens can be prepared simultaneously in 1 day using this method (about 20 in our hands), rendering it applicable to research and clinical investigations but too cumbersome for use in diagnostics, at least until rapid and possibly automated RNA preparation techniques become available.

The contamination of PCR reaction mixtures with the products of previous amplifications is a well-known source of false-positive results (Persing 1991). An additional risk of contamination is the culture and cloning of the infectious agent in the same laboratory in which nucleic acids are prepared for PCR and the reactions are set up. We follow recommended guidelines (Kwok and Higuchi 1989) to avoid false-positive results arising from contamination. The extraction of RNA, reverse transcription, and the preparation of PCR reactions are performed using positive displacement pipettes in a separate laboratory in which HCVs are not studied. The analysis of reaction products by gel electrophoresis and hybridization is performed in the HCV laboratory.

The results of RT/PCR performed on specimens of white matter to detect the HCV-229E strain can be seen in Fig. 4. Positive controls of both low (lane 1) and high (lane 2) intensity and a negative control of brain RNA (lane 3), extracted at the same time as the tissue being tested, as well as water (not shown) are included in the RT/PCR reactions. A different aliquot of the same RNA extraction giving a positive hybridization signal (lane 4) is retested to

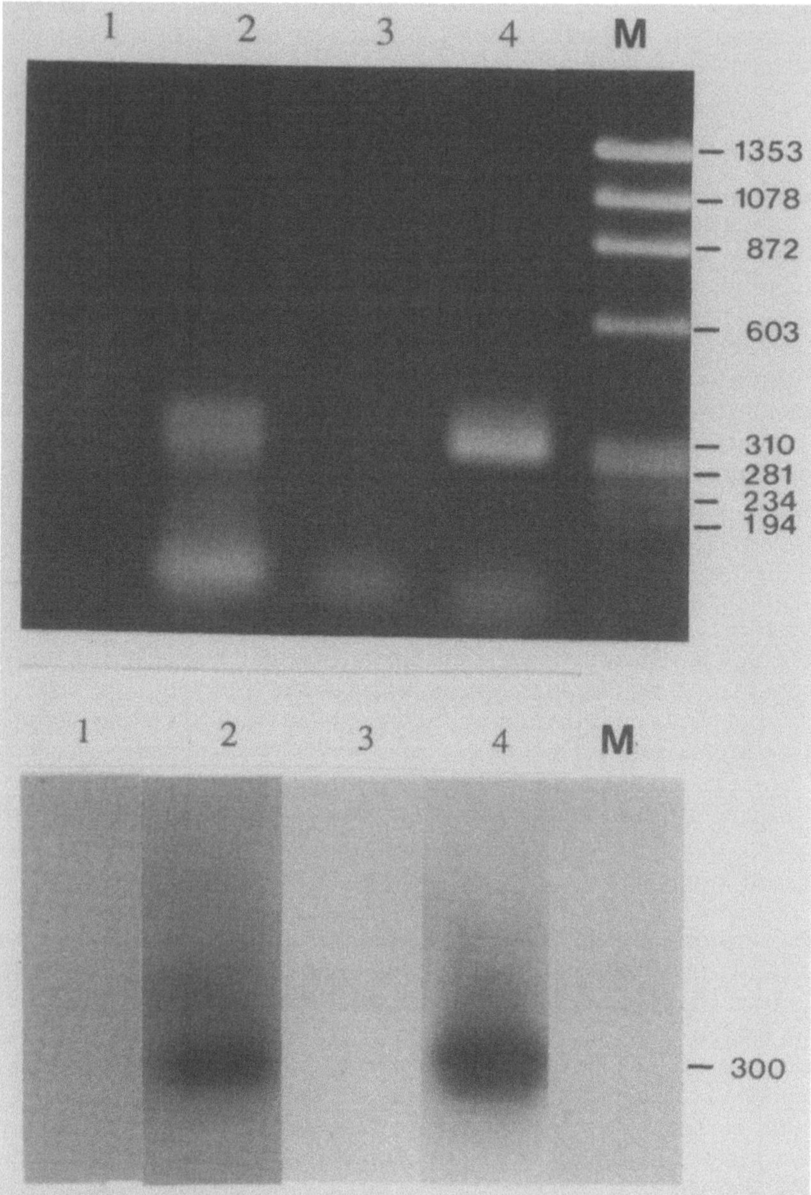

Fig. 3. Agarose (1.5% w/v) gel electrophoresis (*top*) and oligonucleotide hybridization (*bottom*) of amplification products of control mRNAs. *Lanes 1 and 2*, RT/PCR for myelin basic protein mRNA is performed on RNA extracted from human peripheral blood lymphocytes (*1*) and human central nervous system white matter (*2*) *lanes 3 and 4*, RT/PCR for human γ-actin mRNA is performed on H_2O (*3*) and on RNA extracted from human CNS white matter (*4*); *lane M*, HaeIII-digested φX174 RF DNA used as molecular size marker. The numbers on the right indicate the lengths of fragments in bp

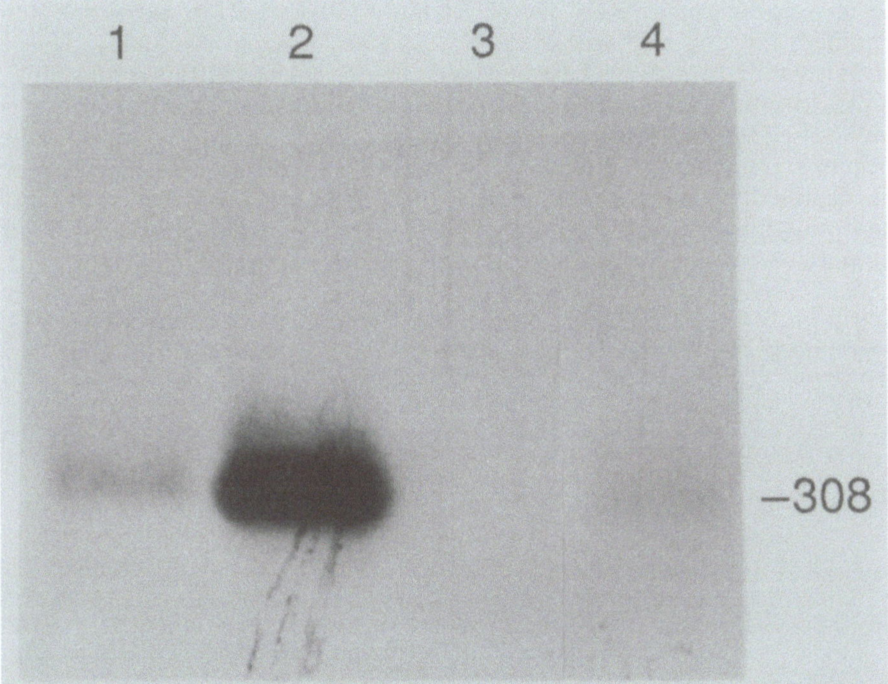

Fig. 4. Detection of the 229E strain of human coronavirus in clinical specimens by oligonucleotide hybridization of reverse transcription/polymerase chain reaction (RT/PCR) 308-bp amplification products from the N gene. *Lanes 1 and 2,* RNA isolated from mouse brain, to which 16 pg (1) or 160 pg (2) total RNA from virus-infected L132 cells was added; *lane 3,* RNA isolated from mouse brain tissue alone; *lane 4,* RNA extracted from human central nervous system white matter obtained from an MS patient

confirm a positive result. It is also possible to use a different set of primer pairs to confirm a positive result (not shown), but the sensitivity of the second set of primers must be similar to the first if reproducible results are to be obtained.

Similar tests using a transcription step prior to cDNA amplification have been developed for several RNA viruses, including rhinovirus (Gama et al. 1988), human immunodeficiency virus (Byrne et al. 1988, Murakawa et al. 1988), rubella virus (Carman et al. 1989), human picornavirus (Hyypia et al. 1989), equine arteritis virus (Chirnside and Spaan 1990), and avian infectious bronchitis virus (Lin et al. 1991). The procedure described in the present study has been repeated for several coronaviruses, and the results are reproducible.

Conclusion

RT/PCR is a relatively rapid method for locating viral nucleic acid in different tissues. However, the technique described in the present study remains restricted to research applications due to the labor-intensive RNA extraction and Southern blotting steps involved. Nevertheless, it could eventually replace virus isolation from clinical material and nucleic acid hybridization for simple, rapid, specific, and sensitive identification of coronaviruses. Moreover, it is well suited to the study of coronaviruses which exhibit multiple tissue tropisms. Using this technique, *in vivo* infections of different tissue types may be identified, opening up new avenues of research into the specific cell types harboring the virus and eventually into the pathogenic potential of HCVs as primary or secondary agents of diseases other than the common cold.

Acknowledgments. We thank Francine Lambert for excellent technical assistance and Lise Trempe and François Shareck for oligonucleotide synthesis.

References

Battaglia M, Passarani N, DiMatteo A, Gerna G (1987) Human enteric coronaviruses: further characterization and immunoblotting of viral proteins. J Infect Dis 155:140–143

Boursnell MEG, Brown TDK, Foulds IJ, Green PF, Tomley FM, Binns MM (1987) Completion of the sequence of the genome of the coronavirus avian infectious bronchitis virus. J Gen Virol 68:57–77

Burks JS, DeVald BL, Jankovsky LD, Gerdes JC (1980) Two coronaviruses isolated from central nervous system tissue of two multiple sclerosis patients. Science 209:933–934

Byrne BC, Li JJ, Sninsky J, Poiesz BJ (1988) Detection of HIV-1 RNA sequences by *in vitro* DNA amplification. Nucleic Acids Res 16:4165

Carman WF, Williamson C, Cunliffe BA, Kidd AH (1989) Reverse transcription and subsequent DNA amplification of rubella virus. J Virol Methods 25:21–30

Cavanagh D, Brian DA, Enjuanes L, Holmes KV, Lai MMC, Laude H, Siddell SG, Spaan W, Taguchi T, Talbot PJ (1990) Recommendations of the Coronavirus Study Group for the nomenclature of the structural proteins, mRNAs and genes of coronaviruses. Virology 176:306–307

Chirnside ED, Spaan WJM (1990) Reverse transcription and cDNA amplification by the polymerase reaction of equine arteritis virus (EAV). J Virol Methods 30:133–140

Chirgwin JM, Przybyla AE, MacDonald RJ, Rutter WJ (1979) Isolation of biologically active ribonucleic acid from sources enriched in ribonuclease. Biochemistry 18:5294–5299

Chomczynski P, Sacchi N (1987) Single step method of RNA isolation by acid guanidinium thiocyanate-phenol-chloroform extraction. Anal Biochem 162:156–159

Erba HP, Eddy R, Shows T, Kedes L, Gunning P (1988) Structure, chromosome location, and expression of the human γ-actin gene: differential evolution, location and expression of the cytoskeletal β- and γ-actin genes. Mol Cell Biol 8:1775–1789

Gama RE, Hughes PJ, Bruce CB, Stanway G (1988) Polymerase chain reaction amplification of rhinovirus nucleic acids from clinical material. Nucleic Acids Res 16:9346

Holmes KV (1990) Coronaviridae and their replication. In: Fields BN, Knipe DM et al. (eds) Raven, Virology, 2nd edn. New York, pp 841–856

Hyypia T, Auvinen P, Maaronen M (1989) Polymerase chain reaction for human picornaviruses. J Gen Virol 70:3261–3268

Jouvenne P, Mounir S, Stewart JN, Richardson CD, Talbot PJ (1992) Sequence analysis of human coronavirus 229E mRNAs 4 and 5: evidence for polymorphism and homology with myelin basic protein. Virus Res 22:125–141

Kamahora T, Soe LH, Lai MMC (1989) Sequence analysis of nucleocapsid gene and leader RNA of human coronavirus OC43. Virus Res 12:1–10

Kwok S, Higuchi R (1989) Avoiding false positives with PCR. Nature 339:237–238

Lai MMC (1990) Coronavirus: organization, replication, and expression of genome. Annu Rev Microbiol 44:303–333

Lai MMC, Stohlman SA (1978) The RNA of mouse hepatitis virus. J Virol 26:236–242

Lee HJ, Shieh CK, Gorbalenya AE, Koonin EV, La Monica N, Tuler J, Bagdzhadzhyan A, Lai MMC (1991) The complete sequence (22 kilobases) of murine coronavirus gene 1 encoding the putative proteases and RNA polymerases. Virology 180:567–582

Lin Z, Kato A, Kudou Y, Ueda S (1991) A new typing method for the avian infectious bronchitis virus using polymerase chain reaction and restriction enzyme fragment length polymorphism. Arch Virol 116:19–31

Lomniczi B, Kennedy I (1977) The genome of infectious bronchitis virus. J Virol 24:99–107

McIntosh K (1990) Coronaviruses. In: Fields BN, Knipe DM et al. (eds) Virology, 2nd edn. Raven, New York, pp 857–864

Murakawa GJ, Zaia JA, Spallone PA, Stephens DA, Kaplan BE, Wallace RB, Rossi JJ (1988) Direct detection of HIV-1 RNA from AIDS and ARC patient samples. DNA 7:287–295

Murray RS, Brown B, Brian D, Cabirac GF (1997) Detection of coronavirus RNA and antigen in multiple sclerosis brain. Ann Neurol 31:525–533

Persing DH (1991) Polymerase chain reaction. Trenches to benches. J Clin Microbiol 29:1281–1285

Raabe T, Siddell S (1989) Nucleotide sequence of the human coronavirus HCV-229E mRNA 4 and mRNA 5 unique regions. Nucleic Acids Res 17:6387

Resta S, Luby JP, Rosenfeld CR, Siegel JD (1985) Isolation and propagation of a human enteric coronavirus. Science 229:978–981

Riski N, Hovi T (1980) Coronavirus infections of man associated with diseases other than the common cold. J Med Virol 6:259–265

Saiki RK, Gelfand DH, Stoffel S, Scharf SJ, Higuchi R, Horn GT, Mullis KB, Erlich HA (1988) Primer-directed enzymatic amplification of DNA with a thermostable DNA polymerase. Science 239:487–491

Salmi A, Ziola B, Hovi T, Reunanen M (1982) Antibodies to coronaviruses OC43 and 229E in multiple sclerosis patients. Neurology 32:292–295

Schreiber SS, Kamahora T, Lai MMC (1989) Sequence analysis of the nucleocapsid protein gene of human coronavirus 229E. Virology 169:142–151

Southern E (1979) Gel electrophoresis of restriction fragments. Methods Enzymol 68:152–176

Streicher R, Stoffel W (1989) The organization of the human myelin basic protein gene. Biol Chem Hoppe Seyler 370:503–510

Tanaka R, Iwasaki Y, Koprowski HJ (1976) Intracisternal virus-like particles in the brain of a multiple sclerosis patient. J Neurol Sci 28:121–126

ter Meulen V, Massa PT, Dörries R (1990) Coronaviruses. In: McKendall (ed) Viral disease. Handbook of clinical neurology, vol 12. Elsevier, Amsterdam, pp 439–451

Section VI **Disease Causing RNA Viruses**

Chapter 25 **Detection of Human Enteroviruses Using the Polymerase Chain Reaction** *

Steven Tracy, Nora M. Chapman, and Jeanne M. Pistillo

Summary

Human enteroviruses are standardly detected and identified using cell culture and serology, techniques which are slow, expensive and may not always provide a conclusive answer. PCR primers have been designed to detect, identify, and clone enteroviral genomes, even though actual sequence data have not been derived in all cases. Enteroviral RNA can be detected using PCR primers directed against the relatively well-conserved 5′ nontranslated region in the enterovirus genome, while degenerate primers can be used to amplify most, if not all, of an enterovirus genome in overlapping fragments. These approaches should augment and in many cases replace the need for analytical or clinical isolation of infectious enteroviruses.

Introduction

A Brief Description of Enteroviruses

Enteroviruses (Picornaviridae) are small isosahedral viruses about 30 nm in diameter (Melnick 1985). The viral capsid is composed of 60 copies of four capsid proteins, and the structure of several enteroviruses has been solved at near atomic resolution (Rossmann et al. 1985; Hogle et al. 1985; Rossmann and Johnson 1989). The enteroviral genome is linear, single-stranded, of positive (message sense) polarity, and approximately 7400 nucleotides long. The 5′-terminus of the viral RNA is covalently linked to a small, virus-encoded protein, and the 3′-terminus is a poly(A) tract of about 80–120 residues in length. The genome is comprised of three primary regions: a 5′-nontranslated region (about 10% of the genome), the single open reading frame (in which all viral proteins are encoded), and the 3′-nontranslated region (about 1%–3% of the genome length) (Palmenberg 1989; Stanway 1990).

Department of Pathology and Microbiology, University of Nebraska, Medical Center, 600 S. 42nd Street, Omaha, NE 68198, USA

* This work was supported in part by funding from the Public Health Service (HL40403), the Nebraska Affiliate of the American Heart Association, and the University of Nebraska Medical Center.

There are about 70 serologically distinct enteroviruses, grouped as the polioviruses (PVs) coxsackieviruses of the A serogroup (CVAs) or B serogroup (CVBs), and echoviruses (EVs); newer isolates are not named but given numbers (e.g., enterovirus 70). Enteroviruses are etiologically suspected or proven to be causative agents in an extremely wide variety of human diseases (Grist et al. 1978; Melnick 1985; Modlin 1990). Perhaps most widely known of human enterovirus diseases, poliomyelitis has long been studied, and poliovirus clearly demonstrated to be its cause. Other etiologic connections, defined usually on the basis of serologic data, are less clear and in many instances implicate more than one enterovirus. An example is the role of enteroviruses in human myocarditis (inflammatory heart disease). Although enteroviral RNA has been demonstrated in a significant number of myocarditic heart samples and serologic evidence suggests an etiologic link between the coxsackie B viruses (CVBs) and the disease, relatively few enteroviruses have been actually isolated from the heart or specifically identified in the heart (reviewed in Tracy et al. 1990 b, 1991 a). Much of this confusion or inadequate identification of a given enterovirus with a specific disease stems from a lack of methods for the sensitive, rapid, and specific detection and identification of enteroviruses.

Difficulties in the Detection of Enteroviruses

Clinical detection of an enterovirus infection in a sample is standardly performed by propagation in sensitive cell cultures. Certain enteroviruses (e.g., the CVBs and PVs) replicate well in specific cells, providing discernible cytopathic effects within 6–48 h; on the other hand, many human enteroviruses (such as the CVAs and many EVs) do not replicate, or only poorly, in cell cultures. Although some information can be obtained on the identity of the enterovirus by observing in which cell culture it replicates, specific identification is not obtained in this fashion and clearly, a 1–2 day isolation time is not a rapid diagnosis. For identification of serotype, neutralization of the isolated virus is performed using specifically formulated pools of antibodies (Melnick 1985), but approximately 10% of the isolates cannot be typed. Titers of IgM and neutralizing antibodies are useful but at best inferential, as are isolations of an enterovirus from stool or throat, when the suspected infected tissue is internal (for example, muscle or the central nervous system).

Standard molecular approaches can be used to detect enteroviruses but suffer as well from the problem of being relatively slow (≥ 1 day) and highly labor-intensive. In situ hybridization has been used in experimental situations to detect enteroviral RNA in inflamed human heart muscle (Tracy et al. 1990 a; Easton and Eglin 1988); while precise and capable of providing data on tissue localization of the virus, it is complex and requires generally more than 1–2 days to complete (especially when highest sensitivity and low backgrounds are required). In situ transcription (Tecott et al. 1988) could be used clinically to detect, and thus verify the presence of, enteroviruses in infected cell cultures

(Carstens et al. 1991); while more rapid and simpler than in situ hybridization, it is used in conjunction with cell culture techniques for the propagation of virus and so does not represent a significant candidate for rapid enteroviral detection. Blot format nucleic acid hybridization has been employed for the detection of enteroviruses; this is generally more rapid and easier to accomplish than the in situ approach. However, sensitivity may also be a problem: approximately 10^4 or more copies of the viral genome in the sample are required for detection (Rotbart 1991), the equivalent of approximately 1–1000 infectious virus particles.

The Need for a Rapid Clinical Diagnostic Approach for Enteroviruses

The rapid detection of enteroviruses in the differential diagnosis of very serious acute diseases (such as meningitis) which require immediate clinical attention would be of value. As many other viruses and organisms can be agents of a disease such as meningitis (McGee and Baringer 1990), the inclusion of other likely etiologic agents should also be considered in such assays. A significant number of cases of myocardial inflammation are etiologically linked to enteroviruses (Bowles et al. 1986; Easton and Eglin 1988; Tracy et al. 1990 a, b), but it is not clear that rapid diagnosis in such diseases would represent a significant clinical value in terms of helping the physician to determine treatment. At present, there are no effective antienteroviral drugs. On the other hand, several pharmaceutical companies have an interest in compounds which can inhibit enteroviral replication in cell culture (Badger et al. 1989, Woods et al. 1989), and so it may be inferred that antienteroviral agents may become clinically available in the not-too-distant future. At such a time, a rapid, sensitive, and specific enteroviral diagnostic assay would have clear value. Long before industry perceives a profitable market for such a test, it is apparent that the ability to detect enteroviruses using enzymatic amplification techniques will be developed as a research tool in order to probe the relationship between virus and disease. Perhaps in this way the desired market will then be created.

The Polymerase Chain Reaction for the Detection of Human Enteroviruses

The technique of enzymatic amplification and the widespread availability of heat-stable DNA polymerases has significantly simplified the task of detecting enteroviruses. In general, enteroviruses are not found in clinical specimens in great quantity. Infectious titers in stool may range as high as one million per milliliter but are generally less. Environmental (e.g., river water) samples are even more dilute. Many samples (such as pediatric cerebrospinal fluid or endomyocardial biopsy samples) are small, difficult to obtain, and must be used to greatest efficiency in the clinical laboratory. Multiplex PCR (Chamberlain et al. 1990), in which two or more individual sequences are simultaneously detected in a single reaction, should be useful in such cases. The past decade has advanced our knowledge of the enteroviral genome in quantum fashion (Semler and Ehrenfeld 1989), and the large number of nucleotide sequences now available from human (and animal) enteroviral genomes has permitted the design and implementation of PCR for the detection of enteroviruses in general and, in some cases, the identification of specific enteroviral genotypes. It is clear that, with suitable primers, the approaches defined below are applicable as well to other picornaviruses.

Detection of Enteroviral Genotypes in General

The design of the enteroviral genome and studies designed to locate probe sequences useful for the detection of many or all enteroviruses (Tracy 1984; Rotbart et al. 1985) suggested that the relatively well conserved 5'-nontranslated region of the genome is the most logical target sequence for the detection of the greatest number of enteroviruses. Such "pan-enteroviral" primer sets (Torgersen et al. 1989; Hyypia et al. 1989; Chapman et al. 1990; Jin et al. 1990; Olive et al. 1990; Rotbart 1990) should be capable of detecting most, if not all, enteroviral genomes which conform to our ideas of the prototypic genomic sequence (Palmenberg 1989; Tracy et al. 1991 a). In one study (Chapman et al. 1990), such primers detected all enteroviruses tested with the exception of EV22; this virus has now been fully sequenced (Hyypia et al. 1991), and other primers for this viral genotype may be designed. In some cases (Hyypia et al. 1989; Torgersen et al. 1989), the sets were designed to take advantage of the close similarity between human enteroviral and human rhinoviral genomes for the detection of human rhinoviruses (closely related picornaviruses) in the same group or as a separate group. The sensitivity of such assays are on the order of 1 infectious unit of virus under conditions likely to be used in clinical settings (Olive et al. 1990). Several companies now market similar primers for use in the PCR.

Detection of Specific Enteroviral Genotypes

Specific genotypes may also be sought using the PCR. In this approach, one seeks a primer set which (ideally) is not present in all other enteroviral genomes or at least is significantly divergent so that the PCR criterion may effectively be adjusted to omit the detection of other enteroviral genomes. Such an approach is understandably limited to specific circumstances and in most cases would complement, not replace, the strategy of detecting enteroviruses in general. Nevertheless, one report (Weiss et al. 1991) specifically sought the presence of CVB3 RNA in human myocarditic heart biopsy samples rather than using this approach in conjunction with a screen for enteroviral genomic

RNA in general. Such an approach, for reasons discussed below, may be less than ideal.

Real and Potential Problems in the Detection of Specific Enteroviral Genotypes

The detection of specific enteroviral genotypes is still in its infancy, not so much due to the lack of basic enteroviral sequence information (which is vast and growing; Palmenberg 1989) but rather more due to the lack of sequence information on many different isolates of the *same* genotype. It is insufficient to design genotype-specific primers based upon the assumption that the target sequences are not found or are highly divergent in other ("prototypic") genomes, due to the variability of the enteroviral genome (indeed, likely due to the variability inherent in all RNA virus genomes; Steinhauer and Holland 1987; Tracy and Gauntt 1987). We reported on one set of primers which appeared to be CVB3 specific on the basis of the inability to detect non-CVB3 genomes and to detect five individual CVB3 genomes (Chapman et al. 1990). Subsequent studies on more than 15 CVB3 isolates, which had been derived over 40 years and minimally passaged in culture, demonstrated that these "CVB3-specific" primers detected a majority, but not every one, of these genomes (S. Tracy, unpublished data). One must wonder whether another CVB3-specific primer set (Weiss et al. 1991) would have shown a similar inability to detect all CVB3 genomes had it been similarly tested. Therefore, while such primers might be useful to seek out specific sequences within a genome, they should be used cautiously for detecting genomes of the same type unless the primers have been tested extensively against viral isolates from many years.

In addition to rapid genomic change due to RNA polymerase error during replication, enteroviral genomes can also recombine (King et al. 1987). A single specific primer set could not identify a recombinant genome (unless the site of recombination were between the primers and a specific probe or sequence analysis were used). Thus, in the design of specific enteroviral genotype identification strategies, one should also consider incorporating more than one primer set to avoid potentially false identification. This strategy could then also be used as an approach for screening naturally occurring isolates for recombinatory events. Because of our lack of knowledge about how various enteroviral genomes vary, specific identification of an enteroviral genotype remains an ill-defined science. Accumulation of sequence data for those viruses of greatest interest should one day permit accurate molecular identification.

Use of the PCR to Clone Enteroviral Genomes in Fragments

Because many enteroviruses grow poorly or not at all in cell culture, cell culture alone is an insufficient assay. The successful amplification of a DNA fragment using pan-enteroviral primers described above would indicate the presence of an enterovirus genome in a sample, even when viable virus cannot

Table 1. Primers for amplification of enteroviral cDNA

Primer	Sequence[a]	CVB3 genome position[b]	Same sense as virion RNA	No mismatches	1 Mismatch	2 Mismatches	3 Mismatches	Sites
V0[c]	TTAAAACAGC-CTGTG	N1–15, 5′NTR	Yes	CVA9, SVDV, CVB1, 3, 4			HRV1B 14	
V2	GCGTTGATAC-TTGAGCTCCC	N745–764	No	CVB3		CVA9, CVB1, PV3[d], SVDV	CVA21	Sac1
D	GAGCTCAAGT-NTCAWCRCA	N747–765, 1A aa2–7	Yes	CVB3, CVA9, SVDV	CVB1, PV3[d]	CVA21, CVB4 PV1, 2, 3[d]	HRV14	Sac1
E	CNATGCTNGG-YACACATGT	N2181–2199 1C aa148–154	Yes		CVB1, CVB3, HRV1B, 2, 14 PV2[d], PV3	BEV, CVA21 CVA9, PV1, PV2[d], SVDV		Afl3
G[5]	CCTGTTCCAT-GGCNTCNTCTTC	N3725–3746, 2A aa143–150	No	CVA21	BEV, CVB3, CVB4, PV1	CVB1, PV3	CVA9, PV2, SVDV	Nco1
H	AYTGYGGYGG-GATCCTMAG	N3621–3639, 2A aa109–15	Yes		CVB1, CVB3, PV2[d], PV3[d]	CVA9, CVA21 CVB4, SVDV, PV1, 3[d]	PV2[d], HRV14	BamH1
I[e]	GKCCYTGRAA-YAGCGCTTC	N5015–5033, 2C aa325-3A aa2	No		CVB1, 4 PV2[d], 3[d]	CVA21, CVB3 HRV14	BEV, PV1, 3[d] SVDV	Eco47 III
J	GYTGYCCACT-AGTRTGTG	N4884–4901 2C aa281–287	Yes	SVDV	CVA9, CVB4	CVB1, 3, HRV14, 1B, PV1, 2, 3	CVA21	Spe1
K[e]	ATATTGCCTC-YTCAA	N6071–6086, 3D aa54–59	No		CVB1, 3, 4, PV1, 2[d], SVD V	CVA9, PV2[d], PV3[d]	BEV, CVA21, PV2[d], 3[d]	Ssp1
L	ARGAACCAGC-TGTNCTCAG	N6024–6042, 3D aa8–14	Yes		CVA21, 9, CVB3, 4, PV2[d]	CVB1, PV1, 3 SVDV	BEV, HRV14, PV2[d]	Pvu2

| M | TCTAGAARGART-
CCAACCAYTTCCT | N7277–7301
3D aa456-termina-
tion codon | No | | CVA9, CVB3 | CVB1, 4
SVDV | *Xba*1 |
| PAN[f] | T$_{12}$ | 3'Poly(A) | No | All entero-
viruses | | | |

[a] R=A or G; Y=C or T; M=A or C; K=G or T; W=A or T; N=A, C, G, or T
[b] N, nucleotide number (the 5'-nucleotide is N1); aa, amino acid sequence number
[c] This primer actually has CATCGA preceding the sequence in order to create a *Cla*1 site for cloning purposes
[d] The number of mismatches varies with different strains of this serotype
[e] These primers actually have TCTAGA preceding the given sequence to create an additional *Xba*1 site for cloning
[f] This primer actually has GCGGCCGC preceding the sequence in order to create an additional *Not*1 site for cloning purposes
Enterovirus and rhinovirus serotypes are abbreviated as follows: BEV, bovine enterovirus; CVA coxsackievirus A; CVB, coxsackievirus B; HRV, human rhinovirus; PV, poliovirus; SVDV, swine vesicular disease virus. Sequences were obtained from Genbank Release 67.0

be propagated. Therefore, in order to preserve such genomes for later analysis, we have begun the investigation of PCR primers designed with a degree of degeneracy for the synthesis of overlapping enteroviral genomic fragments in the PCR. These primers were designed through comparison of known human enteroviral genomes and identification of regions in which sequence divergence was minimal. The primers' sequences and the extent of their identity to several known enteroviral genomes are shown in Table 1. By design, we included restriction endonuclease cleavage sites to facilitate subsequent cloning manipulations; these sites were engineered using codon wobble and did not incur changes in the coding sequence. Of the primers in Table 1, only V0 and V2 were made specifically for the CVB3 genome. Locations in the CVB3 genome (Tracy et al. 1991 b) are shown for reference; precise locations will likely vary slightly from genome to genome.

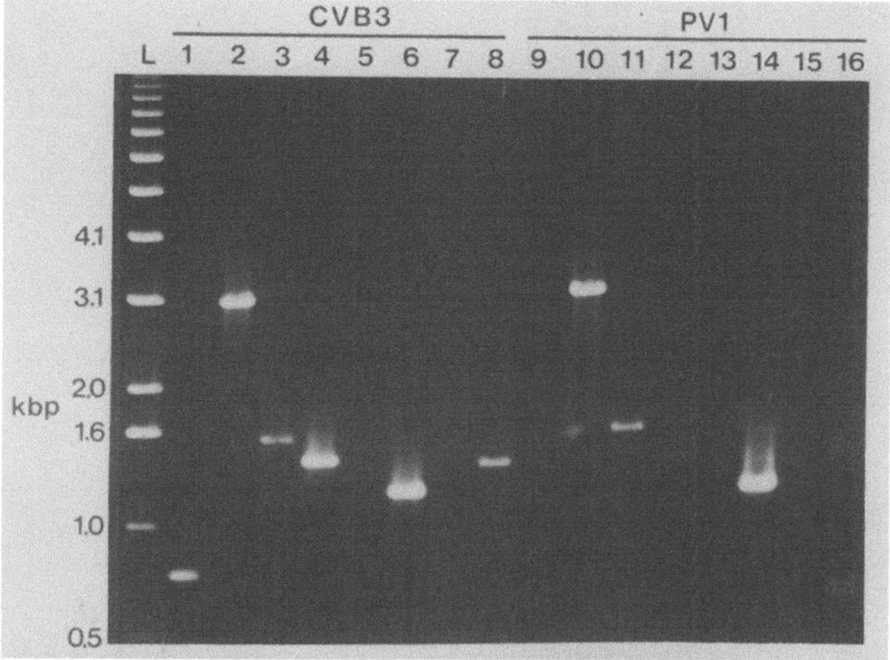

Fig. 1. Enzymatic amplification of subgenomic fragments from the genomes of coxsackievirus CVB3 and poliovirus PV1 with the aid of degenerate primers. Amplification reactions were assembled as described in the text. The template for the CVB3 reactions was linearized pCVB3–20 (Tracy et al. 1991 b) and for the PV1 reactions, linearized pESPV1M (this clone contains the cDNA copy of the PV1 Mahoney genome in the pES131 vector; Kuhn et al. 1987). The concentration of each DNA template was 5 ng/ml polymerase chain reaction (PCR). Primers for the PCRs shown are described in Table 1 and in the text. PCR products were separated on a 1% agarose gel, stained in ethidium bromide, and illuminated with UV light. *Lanes 1–8*, CVB3; *lanes 9–16*, PV1. Primers for lanes 1, 9 were V0 and V2; lanes 2, 10, D and G; lanes 3, 11, E and G; lanes 4, 12, H and I; lanes 5, 13, H and K; lanes 6, 14, J and K; lanes 7, 15, J and PAN; lanes 8, 16, L and PAN. *L*, 1 kbp ladder (BRL Inc.)

We amplified fragments from a myocarditic (in mice) CVB3 genome (Tracy et al. 1991 b) as well as from the cloned genome of PV type 1 (Semler et al. 1984) to examine the primers' efficacy. The PCRs were set up as described in the following section, except that cloned cDNA copies of the genomes of the viruses were used as substrates $(0.1 \times 10^9$ g/reaction) instead of reverse-transcribed cDNAs. In contrast to the rapid cycles used for detection purposes (see below) and with short fragments to be generated, longer cycles were employed. Typically following denaturation 10 cycles were performed ($37\,°C$, 1 s, increasing on a 2-min ramp to $72\,°C$, then $72\,°C$ for 2 min, followed by 10 s at $94\,°C$), followed by 40 cycles ($55\,°C$, 5 s; $72\,°C$, 2.5 min; $94\,°C$, 10 s). Results of a typical amplification are shown in Fig. 1. For both the CVB3 and the PV1 genomes (which diverge significantly from each other), these primers generally functioned well, generating fragments of the expected size suitable for molecular cloning; for each genome, 5 overlapping fragments were generated (D-G, E-G, H-I, J-K, and L-PAN). The primers V0 and V2 did not generate a fragment from PV1, likely due to the relatively divergent priming sequence. In some instances, other amplified fragments are evident; we have not determined their locations in the genome to date.

In other work (data not shown), we have used these and other primers to generate subgenomic fragments from cDNAs transcribed from the genomes of a nonmyocarditic strain of CVB3 ($CVB3_0$; Gauntt et al. 1979), more than 15 CVB3 isolates dating from 1950 through 1989, as well as PV2 (Lansing) (unpublished data). However, these primers have not been tested against enteroviral genomes for which a prototype genome has yet to be sequenced. Because they are generally located in the coding region of the genome, significant variation between enteroviral genomes may be expected and thus may obviate the utility of some of these primers. Nevertheless, these preliminary data suggest that these and similarly designed primers should prove useful in concert with the pan-enteroviral primers for the molecular cloning and identification of various human enteroviruses.

Techniques

What follows is a useful approach for the detection of enteroviruses in cell cultures and tissue samples. Genomic DNA may be removed from the preparation if desired for other reasons, but it does not have an apparent effect on the detection of the enteroviral RNA and is probably useful as a carrier molecule.

Nucleic Acid Isolation

If assay for infectious virus in cell culture is desired, tissues are homogenized in about 2–5 volumes (or a tractable volume) of complete tissue culture minimum essential medium (MEM, 10% FBS, 50 µg/ml gentamicin). A typical endomyocardial biopsy is about $1-2$ mm^3 and is homogenized in $0.05-0.1$ ml complete cell culture medium. Following homogenization, half is removed for PCR analysis (either directly into lysis buffer or frozen). The remainder is frozen and thawed three times to aid in the release of infectious virus, then assayed as usual on sensitive cell cultures.

Tissue homogenates are lysed at room temperature in minimal volumes by the addition of 2 volumes of lysis buffer [8 M NaSCN, 50 mM TRIS-HCl, pH 8, 5 mM EGTA, 114 mM β-mercaptoethanol, 2% w/v sarcosyl (N-lauryl sarcosinate)]. Three volumes of redistilled, buffer-equilibrated phenol are then added and the mixture rapidly mixed to form an emulsion. An equivalent volume of CHCl$_3$ is then added and mixed again. The concentration of NaSCN in the original volume is then reduced to about 1 M by the addition of an appropriate volume of 50 mM ethylene diamine tetra-acetic acid (EDTA), 50 mM TRIS-HCl, pH 8. Following mixing and centrifugation at room temperature to separate phases, the aqueous phase is reextracted twice with phenol-CHCl$_3$ and once with CHCl$_3$. The material is made up to 2.5 M in ammonium acetate and ethanol precipitated using purified glycogen as carrier. After warming to room temperature, the nucleic acids are collected by centrifugation and ethanol precipitated again. The nucleic acids are rinsed with 70% then 95% ethanol and dried, resuspended in sterile glass-distilled water, and stored at $-75\,°$C until used.

Reverse Transcription

We use murine leukemia virus (MLV) reverse transcriptase (RT) to generate the complementary enteroviral DNA prior to PCR. This has replaced use of the avian myeloblastosis virus (AMV) RT used previously (Chapman et al. 1990), as the cDNA product with the MLV RT is greater in length ($\geq 0.6-$ 1 kb) using random hexameric primers. Perkin-Elmer/Cetus markets a heat-stable DNA polymerase which, in the appropriate buffer, can function either as an RNA-dependent (for reverse transcription) or a DNA-dependent DNA polymerase (for the PCR). We have not yet characterized this enzyme for our use.

The MLV RT (SuperScript; BRL) can be used in the buffer supplied with the enzyme or in 1 × PCR buffer (30 mM KCl, 1.5 mM MgCl$_2$, 10 mM TRIS-HCl, pH 7.5, 0.01% gelatin). Concentrated (10 ×) PCR buffer may be purchased directly from Perkin-Elmer/Cetus and is inexpensive, and as the concentration of Mg^{2+} cation is critical, purchase of a quality controlled stock buffer can be advisable. The advantage of using the PCR buffer in the RT reaction is that it is not necessary to remove the RT buffer prior to PCR; for many applications, one may transfer $1-2$ µl of the RT reaction to an amplification reaction for the PCR. However, for greatest sensitivity, the cDNA is collected by ethanol precipitation prior to PCR. Deoxynucleotide triphosphates (dNTPs; 100 mM) and random hexamer oligonucleotides were purchased from Pharmacia. Random primers are suspended in water, adjusted to a concentration of 5 A_{260} units per ml; dNTPs are mixed at equimolar concentrations and diluted with water to stocks of 2 mM each. All RT and PCR reagents except viral nucleic acids are stored aliquoted at $-20\,^\circ$C.

The RT reaction is performed in 20 µl and contains 1 × PCR or RT buffer, 0.2 mM dNTPs, 0.25 A_{260} units/ml random hexamers, and 70 units MLV RT. Ten percent of the nucleic acids from a heart biopsy homogenate is routinely used in such a reaction, which leaves ample material for repeat experiments. Control reactions to ascertain sensitivity are performed with purified virion RNA diluted in the presence of 1 µg total uninfected HeLa cell nucleic acids per 20 µl reaction; in this manner, as few as 1000 viral genomes (ca. 0.01×10^{-12} g viral genomic RNA) can be detected by PCR. Reverse transcription is performed for 30–45 min. Nucleic acids are collected by ethanol precipitation in 2.5 M ammonium acetate with carrier glycogen and resuspended in 10 µl sterile water.

Enzymatic Amplification

The cDNA is amplified in 20 µl volumes containing 1 × PCR buffer, 0.125 A_{260} units/ml of each primer, 0.2 mM dNTPs, and 0.5 units *Thermus aquaticus* (Taq) DNA polymerase (Perkin-Elmer/Cetus). Small volumes permit the most rapid cycling times, and larger volumes are not necessary for the great majority of applications. The primers used for the detection of enteroviruses in general have been described in detail (Chapman et al. 1990) and yield a 195-bp fragment. Primer E1 is 5′-CACCGGATGGCCAATCCA, and primer E2 is 5′-TCCGGCCCCTGAATG. A third primer, E3 (5′-ACACGGA-CACCCAAAGTAGTCGGTTCC), is located between E1 and E2 and can be used either as a PCR primer with E2 (yielding a fragment length of 114 bp) or as a probe for the PCR product. For large-scale screening, mixtures can be assembled, aliquoted to tubes, then stored frozen until used. Reaction mixtures are thawed, 5 µl of nucleic acid sample and 1 µl of Taq polymerase (25 units/ml in the reaction) are added, the tube is then mixed, centrifuged, and a drop of mineral oil added.

Amplification is performed using a Perkin-Elmer/Cetus cycler. For the detection of fragments less than 1–1.5 kbp long, we normally cycle at two temperatures. For reactions containing E1 and E2, we denature the reactions initially at 94 °C for 20 s, then cycle 50 times with the cycler set at 55 °C for 1 s and 94 °C for 1 s. The final cycle incubates the tubes at 55 °C for 30 s, then at 72 °C for 30 s, and then holds the samples at 10 °C until ready for analysis. Setting the cycler at 1-s default times permits at least 8–10 s at each temperature, and the use of smaller volumes in the PCR (< 50 µl) permits rapid temperature equilibration in the reaction mixture.

Analysis of the PCR Product

Half of the PCR is analyzed by electrophoresis on a 1.5% agarose gel run in 0.2 × Tris-acetate-EDTA (TAE) buffer at 100 V. The gel is stained in ethidium bromide and examined under ultraviolet illumination. Alternatively, aliquots of the reactions can be applied to nylon membranes and probed using the E3 primer. The latter approach is more sensitive; 1–10 pg of DNA can easily be detected by hybridization versus approximately 10–50 ng needed to be visible as a band on an ethidium bromide-stained gel.

It is necessary to establish, then repeat during each run, the limit of sensitivity of the detection reaction. Unlike DNA sequences, which can be detected at the one or no copy level in a PCR, RNA sequences cannot be equivalently detected. Transcription of RNA into cDNA by RT is not completely efficient, requiring $\geqslant 1$ RNA sequence for detection by PCR. Negative results (the inability to detect virus) may be due to fewer copies in the sample than can be detected or to some other reason. Thus, it is also necessary to run a positive control to verify that there is RNA which can be transcribed and amplified in each sample. Any number of common gene transcripts can be used for this; α-tubulin is very useful, however, as the human and murine sequences are highly identical, permitting the use of such primers in human and in murine experiments.

References

Badger J, Krishnaswamy S, Kremer M, Oliveira M et al. (1989) Three-dimensional structures of drug-resistant mutants of human rhinovirus 14. J Mol Biol 207:163–174

Bowles N, Olsen E, Richardson P, Archard L (1986) Detection of coxsackie-B-virus specific RNA sequences in myocardial biopsy samples from patients with myocarditis and dilated cardiomyopathy. Lancet I:1120–1123

Carstens J, Tracy S, Chapman N, Gauntt C (1992) Detection of enteroviruses in cell culture using in situ transcription. J Clin Microbiol 30:25–35

Chamberlain J, Gibbs R, Ranier J, Caskey C (1990) Multiplex PCR for the diagnosis of Duchenne muscular dystrophy. In: Innis M, Gelfand D, Sninsky J, White T (eds) PCR protocols: a guide to methods and applications. Academic, New York, pp 272–281

Chapman N, Tracy S, Gauntt C, Fortmueller U (1990) Molecular detection and identification of enteroviruses using enzymatic amplification and nucleic acid hybridization. J Clin Microbiol 28:843–850

Easton A, Eglin R (1988) The detection of coxsackievirus RNA in cardiac tissue by in situ hybridization. J Gen Virol 69:285–291

Gauntt C, Trousdale M, LaBadie D et al. (1979) Properties of coxsackievirus B3 variants which are amyocarditic or myocarditic for mice. J Med Virol 3:207–220

Grist N, Bell E, Assaad F (1978) Enteroviruses in human disease. Prog Med Virol 24:114–157

Hogle J, Chow M, Filman D (1985) Three-dimensional structure of poliovirus at 2.9 Å resolution. Science 229:1358–1365

Hyypia T, Auvinen P, Maaronen M (1989) Polymerase chain reaction for human picornaviruses. J Gen Virol 70:3261–3268

Hyypia T, Horsnell C, Maaronen M, Khan M et al. (1991) A novel picornavirus group identified by sequence analysis. Proc Natl Acad Sci USA (submitted)

Jin O, Sole M, Butany J, Chia W et al. (1990) Detection of enterovirus RNA in myocardial biopsies from patients with myocarditis and cardiomyopathy using gene amplification by polymerase chain reaction. Circulation 82:8–16

King A, Ortlepp S, Newman J, McCahon D (1987) Genetic recombination in RNA viruses. In: Rowlands D, Mayo M, Mahy B (eds) The molecular biology of the positive strand RNA viruses. Academic, New York, pp 129–152

Kuhn R, Wimmer E, Semler B (1987) Expression of the poliovirus genome from an infectious cDNA is dependent upon arrangements of eukaryotic and prokaryotic sequences in recombinant plasmids. Virology 157:560–564

McGee Z, Baringer R (1990) Acute meningitis. In: Mandell G, Douglas Jr R, Bennett J (eds) Principles and practice of infectious diseases. Churchill Livingstone, New York, pp 741–755

Melnick J (1985) Enteroviruses: polioviruses, coxsackieviruses, echoviruses and newer enteroviruses. In: Fields B (ed) Virology. Raven, New York, pp 739–794

Modlin J (1990) Coxsackieviruses, echoviruses, and newer enteroviruses. In: Mandell G, Douglas R, Bennett J (eds) Principles and practice of infectious diseases, 3rd edn. Churchill Livingstone, New York, pp 1367–1382

Olive D, Al-mufti S, Al-mulla W, Khan M et al. (1990) Detection and differentiation of picornaviruses in clinical samples following genomic amplification. J Gen Virol 71:2141–2147

Palmenberg A (1989) Sequence alignments of picornaviral capsid proteins. In: Semler B, Ehrenfeld E (eds) Molecular aspects of picornavirus infection and detection. ASM, Washington, pp 211–241

Rossmann M, Johnson J (1989) Icosahedral RNA virus structure. Annu Rev Biochem 58:533–573

Rossmann M, Arnold E, Erickson J et al. (1985) Structure of a human common cold virus and functional relationship to other picornaviruses. Nature 317:145–153

Rotbart H (1990) Enzymatic RNA amplification of the enteroviruses. J Clin Microbiol 28:438–442

Rotbart H (1991) Nucleic acid detection systems. Clin Microbiol Rev 4:156–168

Rotbart H, Levin M, Villareal L, Tracy S et al. (1985) Factors affecting the detection of enteroviruses in cerebrospinal fluid with coxsackievirus B3 and poliovirus type 1 cDNA probes. J Clin Microbiol 22:222–226

Semler B, Ehrenfeld E (1989) Molecular aspects of picornavirus infection and detection. American Society for Microbiology, Washington

Semler B, Dorner A, Wimmer E (1984) Production of infectious poliovirus from cloned cDNA is dramatically increased by SV40 transcription and replication signals. Nucleic Acids Res 12:5123–5141

Stanway G (1990) Structure, function and evolution of picornaviruses. J Gen Virol 71:2483–2501

Steinhauer D, Holland J (1987) Rapid evolution of RNA viruses. Annu Rev Microbiol 41:409–433

Tecott L, Barachas J, Eberwine J (1988) In situ transcription: specific synthesis of complementary DNA in fixed tissue sections. Science 240:1661–1664

Torgersen H, Skern T, Blaas D (1989) Typing of human rhinoviruses based on sequence variations in the 5′ non-coding region. J Gen Virol 70:3111–3116

Tracy S (1984) A comparison of genomic homologies among the coxsackievirus B group: use of fragments of the cloned coxsackievirus B3 genome as probes. J Gen Virol 65:2167–2172

Tracy S (1988) The genome of group B coxsackieviruses. In: Bendinelli M, Friedman H (eds) Coxsackieviruses – a general update. Plenum, New York, pp 19–34

Tracy S, Gauntt C (1987) Phenotypic and genotypic differences among naturally occurring coxsackie virus B3 variants. Eur Heart J 8 (Suppl J):445–448

Tracy S, Chapman N, McManus B, Pallansch M, Beck M, Carstens J (1990a) A molecular and serologic evaluation of enteroviral involvement in human myocarditis. J Mol Cell Cardiol 22:403–414

Tracy S, Wiegand V, McManus B et al. (1990b) Molecular approaches to enteroviral diagnosis in idiopathic cardiomyopathy and myocarditis. J Am Coll Cardiol 15:1688–1694

Tracy S, Chapman N, Beck M (1991a) Molecular biology and pathogenesis of coxsackie B viruses. Rev Med Virol (in press)

Tracy S, Chapman N, Tu Z (1991b) Coxsackievirus B3 from an infectious cDNA copy of the genome is cardiovirulent in mice. Arch Virol (in press)

Weiss L, Movahed L, Billingham M, Cleary M (1991) Detection of coxsackievirus B3 RNA in myocardial tissues by the polymerase chain reaction. Am J Pathol 138:497–503

Woods M, Diana G, Rogge M, Otto M et al. (1989) In vitro and in vivo activities of WIN-54954, a new broad-spectrum antipicornavirus drug. Antimicrob Agents Chemother 33:2069–2074

Chapter 26 Polymerase Chain Reaction Detection and Typing of Rotaviruses in Fecal Specimens

Vera Gouvea[1] and Roger I. Glass[2]

Summary

A two-step polymerase chain reaction (PCR) procedure was developed to identify groups A, B, and C rotavirus and to type human group A strains in fecal extracts. Sample preparation consisted of deproteinization and extraction of viral dsRNA, followed by purification by adsorption to hydroxyapatite. In a first step, viral RNA was reverse transcribed into cDNA copies and amplified by Taq polymerase, using group-specific primers in an individual or in a combined detection assay. The amplified DNAs were then used in a second PCR procedure for confirmation of group B or C rotavirus or identification of human group A serotypes 1–4, 8, and 9. Results were interpreted by examining PCR products after ethidium bromide staining of agarose gels for the lengths characteristic of the expected products. Subpicograms of template RNA were diagnosed directly in human and animal fecal specimens. In addition, large quantities of cDNA suitable for sequencing, cloning, and other techniques of molecular analysis were rapidly obtained from field isolates.

This PCR assay, carried out with the appropriate primers, is applicable to the identification of other rotavirus traits, such as antigenic specificities defined by the vp4 protein (vp4 typing), subgrouping, host specificity, and virulence, as well as low-level identification of rotavirus in specimens of epidemiologic interest.

Introduction

Rotaviruses are widespread in nature and constitute important agents of gastroenteritis in humans and in several animal species (Kapikian and Chanock 1990). They are classified into seven groups on the basis of their antigenic

---

[1] Molecular Biology Branch, Division of Microbiology, Center for Food Safety and Applied Nutrition, Food and Drug Administration, 200 C St., SW, Washington, DC 20204, USA

[2] Viral Gastroenteritis Unit, Division of Viral and Rickettsial Diseases, Center for Infectious Diseases, Centers for Disease Control, 1600 Clifton Rd., NE, Atlanta, GA30333, USA

specificities and RNA electrophoretic profiles. Three groups (A, B, and C) infect humans. Group A rotavirus is the most common cause of diarrhea, which is often associated with vomiting and fever in infants and young children in both developing and developed countries. Severe rotavirus diarrhea leading to dehydration accounts for a large proportion of pediatric hospitalizations and diarrheal deaths among children 6 months to 2 years of age and has stimulated intense efforts toward the development of vaccines.

Groups B and C rotavirus appear to be more common in domestic animals than in humans (Saif 1989). Group B rotavirus infections of cattle, pigs, sheep, lambs, and rats are relatively common and widespread. In humans, group B virus (also referred to as adult diarrhea rotavirus or ADRV) has thus far been confined to China, where since 1982 it has caused yearly epidemics of severe, cholera-like disease affecting all age groups, including adults (Hung 1988). Group C rotavirus, a primary pathogen of swine, has been found sporadically in children with diarrheal illness in many parts of the world. Recent outbreaks among humans in the UK and Japan have suggested that group C rotavirus infection may be an emerging zoonosis (Saif 1989).

Rotaviruses belong to the Reoviridae family of segmented dsRNA viruses (Estes and Cohen 1989). They consist of 11 gene segments enclosed in a double-shelled capsid, forming a particle with distinctive morphologic features, as seen by negative-stain electron microscopy (EM). Three proteins are primarily responsible for the antigenic diversity of rotaviruses: a major inner capsid protein, vp6, specifies group and subgroup antigens; the two outer capsid proteins, vp7 and vp4, independently specify the serotypes, as both proteins are involved in virus neutralization. Six serotypes based on vp7 specificity have been identified among the human group A isolates by neutralization assays. Serotypes 1–4 are well established and widespread, whereas serotypes 8 and 9 have been isolated in rare instances, and their importance in disease remains unclear. Serotype determination has proved useful in understanding the epidemiology of strains circulating in a community and in considering the formulation and evaluation of rotavirus vaccines. Neutralizing monoclonal antibodies (MAb) to vp7 have been used in enzyme immunoassays (EIA) to serotype rotavirus. These assays constitute a practical serotyping technique that is well suited for clinical diagnosis and large epidemiologic studies. However, it leaves about 20% of strains untyped as a result of breakdown of complete virus particles or epitope variations (Ward et al. 1991).

Recently, Gorziglia et al. (1986, 1990) identified four other serotypes on the basis of vp4 specificity, but the frequency and distribution of vp4 types remain to be assessed. Because MAb specific for the vp4 types are not yet available, identification of these types requires tissue culture adaptation of viral isolates for neutralization assays.

We developed a PCR method to amplify the vp7 gene and to type human group A rotaviruses (Gouvea et al. 1990a). Later, we improved the sensitivity of the method by introducing a simple purification step and, as sequence data for group B and C became available, extended the method to detect these uncultivatable rotaviruses for which reagents are scarce or unavailable (Gouvea et al. 1991). In this chapter, we describe the PCR protocols, their application to clinical diagnosis and epidemiologic surveys, and potential future developments in the use of these methods.

Materials and Methods

Strategy and Primers

Our methods were designed on the basis of two strategies: (1) Primers were selected to generate DNA segments of distinct lengths so that they could be pooled in the same assay and the length would indicate the virus group or type; and (2) the first amplification was followed by a second confirmatory PCR assay using nested primers to avoid the use of radioactive material for hybridization with labeled internal probes.

Table 1 describes the oligonucleotide primers used for the first and second amplification assays. For group A rotavirus, primers were chosen from the vp7 gene (encoded by segment 7, 8, or 9, depending on the strain) of several prototypes of human rotavirus and the simian SA11 strain as indicated. Primers Beg9 and End9 were selected from the conserved 3'-ends to amplify the full-length segment of any group A rotavirus in the first amplification. We then designed six serotype-specific primers of the mRNA sense, using each of the six variable regions on the vp7 gene as the blueprint to detect each serotype (Fig. 1). These regions (A–F) are distinct among different serotypes but conserved within each given serotype (Glass et al. 1985; Green et al. 1987). Because they are at different distances from the 3'-end of the gene, the PCR segment generated by each serotype has a characteristic size that identifies the

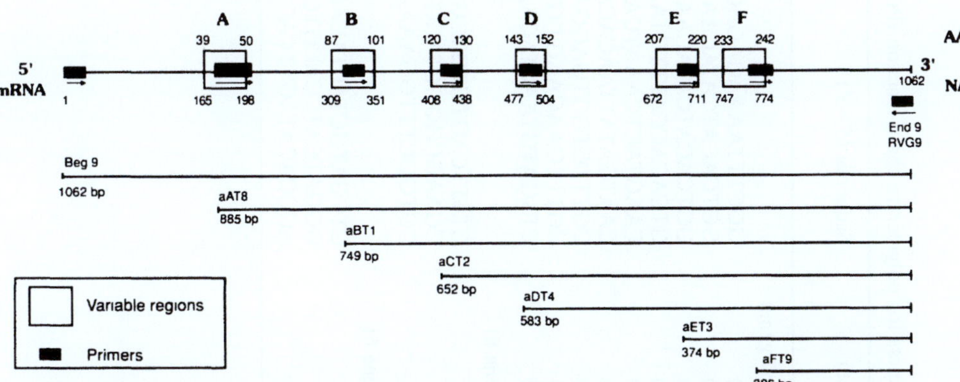

Fig. 1. The mRNA sense of the vp7 gene of group A rotavirus. Shown are locations of variable regions in nucleic acid (*NA*) and amino acid (*AA*) residues, position of primers, and expected lengths of amplified segments (reprinted from Gouvea et al. 1990a, with permission)

Table 1. Primers for polymerase chain reaction (PCR) detection and typing of rotavirus

Primer (sense)	Sequence (5′–3′)	Strain (type)	Position	Product size (bp)
Group A (vp7 gene)				
Beg9 (+)	GGCTTTAAAAGAGAGAATTTCCGTCTGG	WA (1)	1–28	1062 with End9
End9 (−)	GGTCACATCATACAATTCTAATCTAAG	SA11 (3)	1062–1036	
RVG9 (−)	GGTCACATCATACAATTCT	SA11 (3)	1062–1044	
aAT8 (+)	GTCACACCATTTGTAAATTCG	69M (8)	178–198	885 with RVG9
aBT1 (+)	CAAGTACTCAAATCAATGATGG	Wa (1)	314–335	749 with RVG9
aCT2 (+)	CAATGATATTAACACATTTTCTGTG	DS1 (2)	411–435	652 with RVG9
aDT4 (+)	CGTTTCTGGTGAGGAGTTG	ST3 (4)	480–498	583 with RVG9
aET3 (+)	CGTTTGAAGAAGTTGCAACAG	P (3)	689–709	374 with RVG9
aFT9 (+)	CTAGATGTAACTACAACTAC	WI61 (9)	757–776	306 with RVG9
Group B (gene 8)				
B1 (+)	CTATTCAGTGTGTCGTGAGAGG	ADRV	18–39	506 with B4
B3 (−)	CGAAGCGGGCTAGCTTGTCTGC	ADRV	430–451	451 with B1
B4 (−)	CGTGGCTTTGGAAAATTCTTG	ADRV	486–506	
Group C (gene 6)				
C1 (+)	CTCGATGCTACTACAGAATCAG	Cowden	994–1015	356 with C4
C3 (−)	GGGATCATCCACGTCATGCG	Cowden	1301–1320	327 with C1
C4 (−)	AGCCACATAGTTCACATTTCATCC	Cowden	1326–1349	

virus type. Thus, primer aBT1 maps to the variable region B of strain Wa (type 1), primer aCT2 to region C of strain DS1 (type 2), and so forth. The choice of the variable region to serve as type-specific primer was arbitrary; other combinations should work equally well. These primers were pooled with the common antisense primer RVG9, a short version of primer End9, for the second PCR typing (and confirmatory) assay.

Primers for groups C and B were selected from the available sequences of gene 6 of a porcine group C (Cowden) strain (Bremont et al. 1990) and from unpublished partial sequence data for gene 8 of a human group B strain of ADRV (kindly supplied by Dr. Malcolm McCrae, University of Warwick, Coventry, UK). Primer pairs B1–B4 and C1–C4 were chosen to generate segments of distinct lengths which, in combination with the primer pair Beg9-End9, would be indicative of the viral group. The nested primers B3 and C3 were then selected to be used with primers B1 and C1, respectively, in the second, confirmatory PCR assay.

Extraction of Viral RNA

Early in the development of the viral RNA extraction technique, some specimens seemed to contain substances that inhibited the reverse transcriptase reaction necessary to generate cDNA copies of the template and thus initiate the chain reaction catalyzed by Taq polymerase. These substances are common in fecal specimens, cannot be removed by repeated phenol-chloroform extractions and ethanol precipitations, and may be either polysaccharides or heme-containing substances. Dilution of samples results in removal of the inhibitory effect, with a proportional decrease in sensitivity. These inhibitors were effectively removed by selectively adsorbing dsRNA to, and eluting from, hydroxy-apatite (HA) crystals, as described here and elsewhere (Gouvea et al. 1991). Alternative methods using other solid matrices, i.e., glass powder and cellulose fiber (CF-11) (Xu et al. 1990; Wilde et al. 1990), and a commercial RNA purification kit (RNaid; Bio 101, La Jolla, Calif.) effectively remove these inhibitors, rendering the reverse transcription-amplification (RT-PCR) method extremely sensitive and well suited for viral detection.

Stool suspensions (10% in 10 mM phosphate-buffered saline, pH 7.4) or infected tissue culture fluids are extracted with freon and 400 µl of the clarified fluid incubated with 1% sodium dodecyl sulfate and 100 µg of proteinase K per ml for 30 min at 56 °C. The suspension is extracted with an equal volume of phenol-chloroform (1:1) and again with chloroform alone. The upper aqueous phase is transferred to microfuge tubes containing 50 µl of HA, prepared as described by Bernardi (1971). The suspension is mixed and left to sit for 30 min at room temperature, with occasional mixing to permit adsorption of the RNA to HA. After centrifugation for 5 min, the supernatant fluid is discarded and the crystals are washed three times with 10 mM potassium phosphate, pH 6.8. To the precipitated crystals, 40 µl of 200 mM potassium phosphate is added, mixed, and incubated at 42 °C for 5 min for complete elution of RNA. The suspension is centrifuged for 5 min; the clear RNA solution can be used immediately for the RT-PCR or stored at $-20\,^{\circ}$C.

PCR Detection Assay (First Amplification)

In the first amplification, viral dsRNA is used as template for reverse transcriptase to synthesize cDNA copies from both strands of the target viral gene, followed by PCR amplification. Both reactions are performed in a single, siliconized, 600-µl microfuge tube with a 50 µl total reaction mixture. A master mixture consisting of 10 mM TRIS, pH 8.3, 40 mM KCl, 1.5 mM MgCl, 0.2 mM each of dNTP, 7% dimethylsulfoxide (DMSO) (or 5% glycerol), and 500 nM of each primer of the desired primer pair is prepared in a volume sufficient for the number of samples to be tested and is divided into aliquots in reaction tubes. The pair Beg9-End9 is used for the detection of group A rotavirus, B1 – B4 for group B, and C1 – C4 for group C. Alternatively, all three pairs can be combined in the same reaction to detect simultaneously all three groups of human rotavirus. The eluted RNA solution (1 – 10 µl) is added to each reaction tube. The dsRNA is denatured by heating at 95° – 97 °C for 5 min and then cooled in ice for 5 min. Two microliters of a fresh solution containing 1 unit of avian myeloblastosis virus (Amv) reverse transcriptase (Molecular Genetics Resources, Tampa, Fla.) and 2.5 units of Taq polymerase (AmpliTaq; Perkin Elmer Cetus, Norwalk, Conn.) are prepared in a 10 mM TRIS (pH 8.3) and 2 mM dithiothreitol solution, added to the reaction, briefly centrifuged, and layered with 100 µl of sterile mineral oil. The tubes are placed in a thermo-

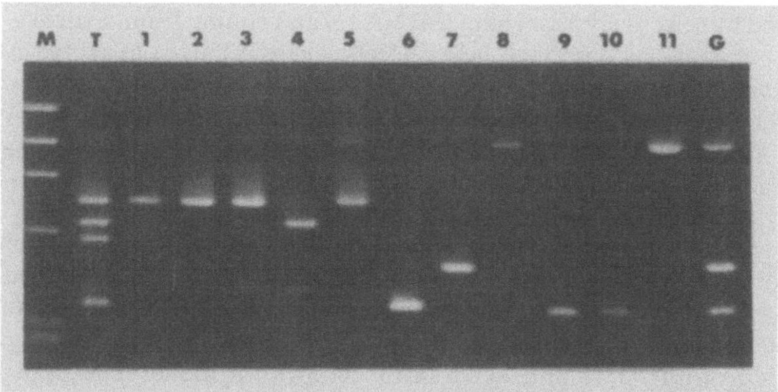

Fig. 2. Typing and grouping of rotavirus. The amplified products obtained from PCR assays of 11 fecal specimens were analyzed in a 1% agarose gel. Rotavirus strains were identified as type 1 (*lanes 1, 2, 3, and 5*), type 2 (*lane 4*), or type 3 (*lane 6*) in a PCR-typing assay (second amplification), or as group A (*lanes 8 and 11*), group B (*lane 7*), or group C (*lanes 9 and 10*) in a PCR detection assay (first amplification). A pool of DNA segments generated by standard types 1, 2, 3, and 4 rotavirus (*lane T*), groups A – C rotavirus (*lane G*), and ϕX174-*Hin*dIII-digested DNA fragments (*lane M*) served as molecular weight markers

cycler programmed for an initial incubation of 42 °C for 45 min, followed by 95 °C for 1 min, and 30 cycles of PCR at 94 °C for 1 min, 42 °C for 1 min, 72 °C for 1 min.

PCR Confirmation or Typing Assay (Second Amplification)

From the first amplification, 1–0.01 µl of the dsDNA product is subjected to a second confirmatory or typing PCR assay. The Taq polymerase is incorporated in the same master mixture as described above, which contains reduced concentrations (200 nM each) of the desired nested primers. The pair B1–B3 is used in the confirmatory assay for group B rotavirus, and C1–C3 for group C; a pool made of primers RVG9 and the six typing primers, aAT8, aBT1, aCT2, aET3, aDT4, and aFT9 are used for group A rotaviruses. Aliquots of the master mixture are placed in reaction tubes to receive the DNA template. Tubes are briefly centrifuged, layered with oil, and placed in the thermocycler for 25 cycles at 94 °C for 30 s, 42 °C for 30 s, 72 °C for 30 s, and a final 3-min incubation at 72 °C.

Analysis of Amplified Segments

PCR products (10 µl) are mixed with 3 µl of a gel-loading buffer (0.125% bromophenol blue, 20% sucrose), previously placed in the wells of a 96-well polyvinyl plate, and loaded in a 1% SeaKem agarose gel (FMC Bio Products, Rockland, Maine) in TBE buffer [0.089 M TRIS, 0.089 M boric acid, 0.002 M ethylene diamine tetra-acetic acid (EDTA), pH 8] containing 0.5 µl of ethidium bromide per ml. Molecular weight markers are used in all gels for proper assessment of the amplified segment size. Pools made with the amplified products generated by PCR typing of group A rotavirus prototype strains and PCR detection assay of known group A, B, and C strains are the most convenient markers for a precise identification of the segment sizes (Fig. 2). After electrophoresis at 140 V for 30 min, the gel is photographed under UV light on Polaroid 667 film.

Discussion

Sensitivity of the Methods

As little as 2 pg of total rotavirus RNA (or 10^5 genomic copies) can be detected in a combined assay for groups A and B rotavirus after adsorption of viral RNA onto hydroxyapatite (Gouvea et al. 1991). This method is about 1000-fold more sensitive than RNA electropherotyping (PAGE) and is comparable with results obtained by other investigators. Xu et al. (1990), using similar reaction conditions and RNA extracted by glass powder, identified 8×10^4 copies of a full-length vp7 gene. Wilde et al. (1990), and Eiden et al. (1991) identified $0.4\text{-}5 \times 10^4$ copies of group A rotavirus gene 6 or group B rotavirus genes 3 and 11 by an RT-PCR assay performed in the presence of excess reverse transcriptase but absence of DMSO or other nucleic acid destabilizer from viral RNA extracted with CF-11. Optimal reaction conditions vary with relative concentrations of template, primers, dNTPs, and enzymes. Because rotavirus is present in feces over a wide concentration range, we standardized the method to detect low levels of RNA and tested the viral RNA undiluted and at a 1:10 dilution.

Selection of primers, their position on the gene, and the size of the segment to be amplified also contribute to the efficiency of amplification. Additional nucleic acid sequences were recently reported for genes 9 and 11 of ADRV (Chen et al. 1990a, b) and gene 3 of a rat group B (IDIR) (Sato et al. 1989), and for gene 8 of a human and a porcine (Cowden) strain of group C rotavirus (Qian et al. 1991), providing alternative choices for new primers.

Application to Clinical Diagnosis

We applied the PCR method to a number of clinical and veterinary specimens obtained from laboratories in the USA and elsewhere (Gouvea et al. 1990a, b, 1991). Our PCR typing results were in complete agreement with the serotype assigned by EIA-MAb analysis in all specimens tested and proved to be particularly valuable for those that could not be serotyped by established EIA-MAb methods. An example of the results obtained with six such samples is shown in Fig. 2. In addition, large quantities of individual rotavirus genes suitable for sequencing and other types of genetic analyses were generated rapidly. PCR provides a simple and rapid means to identify the labile and uncultivatable group B and C rotaviruses; it is more sensitive than EM and requires no immune reagents that are limited or unavailable.

Although all fecal specimens in these studies were known to contain rotavirus by at least one of the current diagnostic tests (EM, PAGE, or EIA), the greater sensitivity of the PCR assay allows viral identification in specimens containing few copies of the viral genes. In a recent study, rotavirus shedding could be detected by RT-PCR a day earlier and 2–7 days later than by immunoassay in several hospitalized children (Wilde et al. 1991). The PCR method should, therefore, be particularly useful for studying virus transmission in special settings, such as hospital wards and nurseries, day-care centers, and schools. In addition, the greater sensitivity of the PCR method is well suited for identifying group C rotavirus, which is usually shed in small quantities in feces, for assessing the frequency of infection, and for defining its epidemiologic characteristics.

Future Developments

The approach described here is currently being developed to identify other antigenic characteristics of group A rotavirus. Primers specific for the subgroups I and II and for the serotypes defined by vp4 (types 1a, 1b, 2, and 3) have been designed, using sequence data from genes 6 and 4, respectively. Besides eliciting neutralizing antibodies, vp4 protein has been associated with virulence (Gorziglia et al. 1986, 1990). Strains obtained from asymptomatic neonates share a common vp4 allele (type 2 specificity), whereas strains obtained from ill children possess distinct vp4 specificities. Identification of vp4 alleles might further our understanding of the basis for rotavirus virulence.

References

Bernardi G (1971) Chromatography of nucleic acids on hydroxyapatite. Methods Enzymol 21:95–139

Bremont M, Chabanne-Vautherot D, Vannier P, McCrae MA, Cohen J (1990) Sequence analysis of the gene (6) encoding the major capsid protein (vp6) of group C rotavirus: higher than expected homology to the corresponding protein from group A virus. Virology 178:579–583

Chen G-M, Hung T, Mackow ER (1990a) cDNA cloning of each genomic segment of the group B rotavirus ADRV: molecular characterization of the 11th RNA segment. Virology 175:605–609

Chen G-M, Hung T, Mackow ER (1990b) Identification of the gene encoding the group B rotavirus vp7 equivalent: primary characterization of the ADRV segment 9 RNA. Virology 178:311–315

Eiden JJ, Wilde J, Firoozmand F, Yolken R (1991) Detection of animal and human group B rotaviruses in fecal specimens by polymerase chain reaction. J Clin Microbiol 29:539–543

Estes MK, Cohen J (1989) Rotavirus gene structure and function. Microbiol Rev 53:410–449

Glass RI, Keith J, Nakagomi O, Nakagomi T, Askaa J, Kapikian AZ, Chanock RM, Flores J (1985) Nucleotide sequence of the structural glycoprotein vp7 gene of Nebraska calf diarrhea virus rotavirus: comparison with homologous genes from four strains of human and animal rotaviruses. Virology 141:292–298

Gorziglia M, Hoshino Y, Buckler-White A, Blumentals I, Glass R, Flores J, Kapikian AZ, Chanock RM (1986) Conservation of amino acid sequence of vp8 and cleavage region of 84-kDa outer capsid protein among rotaviruses recovered from asymptomatic neonatal infection. Proc Natl Acad Sci USA 83:7039–7043

Gorziglia M, Larralde G, Kapikian AZ, Chanock RM (1990) Antigenic relationships among human rotaviruses as determined by outer capsid protein vp4. Proc Natl Acad Sci USA 87:7155–7159

Gouvea V, Glass RI, Woods P, Taniguchi K, Clark HF, Forrester B, Fang Z-Y (1990a) Polymerase chain reaction amplification and typing of rotavirus nucleic acid from stool specimens. J Clin Microbiol 28:276–282

Gouvea V, Ho M-S, Glass RI, Woods P, Forrester B, Robinson C, Ashley R, Riepenhoff-Talty M, Clark HF, Taniguchi K, Meddix E, McKellar B, Pickering L (1990b) Serotypes and electropherotypes of human rotavirus in the USA: 1987–1989. J Infect Dis 162:362–367

Gouvea V, Allen JR, Glass RI, Fang Z-Y, Bremont M, Cohen J, McCrae M, Saif LJ, Sinarachatanant P, Caul EO (1991) Detection of group B and C rotaviruses by polymerase chain reaction. J Clin Microbiol 29:519–523

Green KY, Midthun K, Gorziglia M, Hoshino Y, Kapikian AZ, Chanock RM, Flores J (1987) Comparison of the amino acid sequences of the major neutralization protein of four human rotavirus serotypes. Virology 161:153–159

Hung T (1988) Rotavirus and adult diarrhea. Adv Virus Res 35:193–218

Kapikian AZ, Chanock RM (1990) Rotaviruses. In: Fields BN, Knipe DM, Chanock RM, Hirsch MS, Melnick DL, Monath TP, Roizman B (eds) Virology. Raven, New York, pp 1353–1404

Qian Y, Jiang B, Saif LJ, Kang SY, Ishimaru Y, Yamashita Y, Oseto M, Green KY (1991) Sequence conservation of gene 8 between human and porcine group C rotaviruses and its relationship to the vp7 gene of group A rotaviruses. Virology 182:562–569

Saif LJ (1989) Non group A rotaviruses. In: Saif LJ, Theil KW (eds) Viral diarrheas of man and animals. CRC Press, Boca Raton, pp 73–95

Sato S, Yolken RH, Eiden JJ (1989) The complete nucleic acid sequence of gene segment 3 of the IDIR strain of group B rotavirus. Nucleic Acid Res 17:10113

Ward RL, McNeal MM, Clemens JD, Sack DA, Rao M, Huda N, Green KY, Kapikian AZ, Coulson BS, Bishop RF, Greenberg HB, Gerna G, Schiff GM (1991) Reactivities of serotyping monoclonal antibodies with culture-adapted human rotaviruses. J Clin Microbiol 29:449–456

Wilde J, Eiden J, Yolken R (1990) Removal of inhibitory substances from human fecal specimens for detection of group A rotaviruses by reverse transcriptase and polymerase chain reactions. J Clin Microbiol 28:1300–1307

Wilde J, Yolken R, Willoughby R, Eiden J (1991) Improved detection of rotavirus shedding by polymerase chain reaction. Lancet 337:323–326

Xu L, Harbour D, McCrae MA (1990) The application of polymerase chain reaction to the detection of rotaviruses in faeces. J Virol Methods 27:29–38

Chapter 27 Detection and Identification of Flaviviruses by Reverse Transcriptase Polymerase Chain Reaction

Dennis W. Trent and Gwong-Jen Chang

Summary

We developed a reverse transcriptase/polymerase chain reaction (RT/PCR) for the detection and identification of flavivirus RNA in a single-tube assay. Universal flavivirus RT/PCR amplimers amplify an 832-bp region in the carboxyl portion of NS5 (nt9166–9997) which contains nucleotide sequences conserved among all flaviviruses at the 3′- and 5′-extremities and internal virus-specific variable regions. Hybridization probes were designed which hybridize specifically with the genomic RNAs and amplimer DNAs of the four dengue serotypes, yellow fever, Japanese encephalitis, West Nile encephalitis, Murray Valley encephalitis, St. Louis encephalitis, and Powassan viruses. The virus-specific RT/PCR employs a flavivirus group down-amplimer and a virus-specific up-amplimer. Because the binding site for the virus-specific up-amplimer is unique, the size of the amplified DNA for each virus is different. Viruses are identified to serotype by agarose gel electrophoresis, ethidium bromide staining, and calculation of the fragment size. This RT/PCR test for medically important flaviviruses is specific, rapid, and simple.

Introduction

The flaviviridae family of viruses contains 68 antigenically distinct agents that have been classified into three groups based on their method of transmission (Calisher et al. 1989; Table 1). Viruses in this family include the etiologic agents of yellow fever (YF), Japanese encephalitis (JE), West Nile encephalitis (WNE), Murray Valley encephalitis (MVE), dengue, tick-borne encephalitis (TBE), and St. Louis encephalitis (SLE) (Monath 1990). All flaviviruses which cause human disease are transmitted to their enzootic, epidemic, and dead-end hosts by either ticks or mosquitoes (Turell 1988). Flaviviruses are widely distributed in nature where they are maintained in transmission cycles involving vertebrate reservoir hosts and arthropods which serve as biological vectors.

Molecular Biology Branch, Division of Vector-Borne Infectious Diseases, Centers for Disease Control, Fort Collins, CO 80522, USA

Table 1. Flavivirus antigenic complexes defined by close relationships in cross-neutralization tests with polyclonal antisera[a]

Principal vector	Antigenic complex	Viruses
Tick	Tick-borne encephalitis (TBE)	Russian spring-summer encephalitis, Central European encephalitis, Omsk hemorrhagic fever, louping-ill, Kyasanur forest disease
	Tyuleniy	Tyuleniy, Saumaurez Reef
Mosquito	Japanese encephalitis	Japanese encephalitis, St. Louis encephalitis. Murray Valley encephalitis, West Nile, Kunjin
	Ntaya	Ntaya encephalitis. Bagaza
	Uganda S	Uganda S, Banzi
	Dengue	Dengue 1, 2, 3, 4
None[b]	Rio Bravo	Rio Bravo, Entebbe bat, Dakar bat
	Modoc	Modoc, Cowbone Ridge

[a] Modified from Calisher et al. 1989
[b] Viruses transmitted directly between vertebrate hosts, principally bats and rodents. Two members of the TBE complex may have this mode of spread

Epizootics of flavivirus infection which can result in hundreds of human and animal infections occur sporadically throughout the temperate and tropical zones where they present serious public health problems (Bres 1988). Epidemics of YF in tropical areas may result in thousands of human cases with high rates of morbidity and mortality (Monath 1987, 1988). Dengue virus occurs as four distinct serotypes which cause an estimated 100 million cases per year worldwide (Gubler 1988; Halstead 1988). Both dengue and YF viruses are endemic in large regions of the tropics and subtropics where they are actively transmitted by *Aedes aegypti* mosquitoes that breed in fresh-water collections around human habitation. Human YF, JE, and TBE virus infections can be prevented by immunization; however, there are no vaccines for control of other flavivirus infections (Theiler and Smith 1937; Kunz et al. 1980; Hoke et al. 1988; Yu et al. 1988). Prevention of human infection, therefore, depends on the elimination of mosquitoes or the exercise of personal safety measures to prevent the vector from biting.

Flavivirus particles are enveloped, 37–50 nm in diameter, and have a chemical composition of 6% RNA, 66% protein, 17% lipid, and 9% carbohydrate (Trent and Naeve 1980). The nucleocapsid (C) protein forms a spherical icosahedral structure which encloses an infectious positive-strand RNA of approximately 11 kb (Rice et al. 1985). The lipid bilayer membrane surrounding the nucleocapsid contains the envelope (E) and membrane (M) proteins. The E protein of approximately 55 kDa is usually glycosylated and contains virus-specific antibody-neutralizing epitopes. It is also the virus hemagglutinin and contains the fusion domain involved in the initiation of infection (Henchal et al. 1985; Roehrig et al. 1983, 1989, 1990). Upon entry of virus into the cell and release of the viral RNA, the virus genome is translated into a polyprotein that is cleaved to form three structural proteins (C, M, and E) and five non-structural proteins designated NS1 through NS5 (Fig. 1). The latent phase of the virus growth cycle is approximately 12 h. Virus morphogenesis occurs in association with the rough endoplasmic reticulum and Golgi where viral protein cleavage and maturation of the virion are thought to occur (Brinton et al. 1986). Morphogenesis of the virion is not well understood, as virions are not formed by budding, and nucleocapsid structures do not accumulate within the infected cell. Virus particles are thought to mature by a process of condensation and fusion events associated with cytoplasmic membranes. Virions accumulate within intracytoplasmic vacuoles that fuse with the plasma membrane and are released into the extracellular space by reverse pinocytosis (Murphy 1980).

The entire genomes of YF (Rice et al. 1985), dengue 2 (Hahn et al. 1988; Deubel et al. 1988), dengue 3 (Osatomi and Sumiyoshi 1990), dengue 4 (Zaho et al. 1986), WNE (Castle et al. 1986; Castle and Wengler 1987; Wengler et al. 1985), JE (Hashimoto et al. 1988; Nitayaphan et al. 1990), TBE (Pletnev et al. 1990), Kunjin (Coia et al. 1988), and MVE (Dalgarno et al. 1986; Lee et al. 1990) viruses have been cloned and sequenced. Flavivirus genomic RNA is approximately 11 kb in length, contains a 5'-end type 1 cap structure and does not have a poly (A) tract at the 3'-end of the genome. Analysis of nucleotide and amino acid sequence data indicates that the genome contains a single uninterrupted reading frame which covers all but the noncoding regions at either end. At the 5'- and 3'-ends of the genome, there are nontranslated regions of approximately 100 and 500 nucleotides, respectively (Brinton et al. 1986; Brinton and Dispoto 1988). The amino acid sequence of regions within each protein is highly conserved among viruses in the family (Hahn et al. 1988;

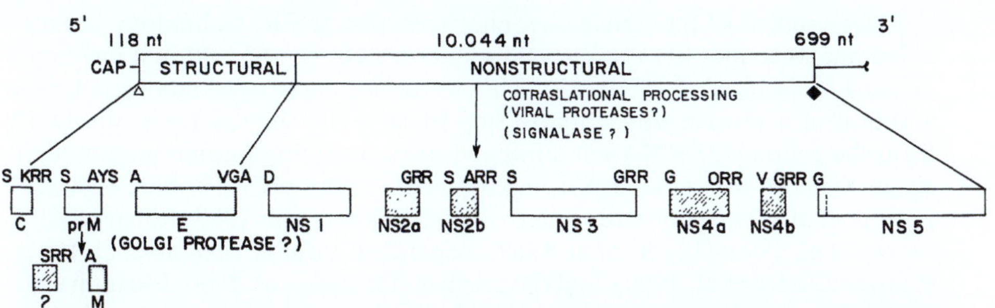

Fig. 1. Organization and processing of proteins encoded in the yellow fever virus genome. Untranslated regions are shown as *single lines* and the translated region as an *open box*. The protein nomenclature follows that of Rice et al. (1985), as modified by Speight et al. (1988). Single letter amino acid codes are used for sequences flanking assigned cleavage sites. (From Rice et al. 1985, reprinted with permission)

Pletnev et al. 1990). These include a 21 amino acid region in the R1 loop structure of the E protein thought to encode the fusion domain, a 35 amino acid sequence in the R3 region of the E protein near the C-terminus, a region of 56 amino acids in the C-terminal half of the NS1 protein, 7 regions containing short stretches of 7–15 amino acids in NS3, and 10 regions in NS5, including a region of over 100 amino acids near the C-terminal half of NS5. Hydrophobicity plots for each of the flavivirus proteins are similar. Structural protein prM/M and nonstructural proteins NS2A, NS2B, NS4A, and NS4B are the least conserved. Amino acid sequences of the nonstructural proteins NS5, NS3, and NS1 are the most conserved. In NS5, there are two highly conserved amino acid sequences, YADDTAGWDTRIT and SGDDCVV, that are present in other viral RNA-dependent RNA polymerase proteins (Pletnev et al. 1990). Cysteine residues are conserved throughout the E, NS1, and prM/M proteins, as are the two glycosylation sites in NS1. In addition to these coding regions, the nucleotide sequence of the 3'-noncoding region is highly conserved and can form a stable hairpin loop structure (Brinton et al. 1986).

Flaviviruses cause a wide variety of human diseases, including respiratory illness, hepatitis, febrile illness, rash, and aseptic meningitis and/or encephalitis (Bres 1988). Diagnosis of flavivirus encephalitis usually depends upon serology or isolation of virus from brain tissue of patients dying early in the infection (Calisher and Poland 1980). Attempts to isolate virus from serum or CSF of encephalitis patients are usually not productive. Yellow fever and dengue viruses, however, can often be isolated from serum during the febrile, early phase of illness. Serologic diagnosis of flavivirus infection depends upon the demonstration of a fourfold rise in antibodies by hemagglutination inhibition, complement fixation, or neutralization testing of acute and convalescent sera (Calisher and Poland 1980). The IgM-capture ELISA has replaced many of the classic tests; however, IgM antibodies indicate recent flavivirus infection but usually do not provide a type-specific diagnosis. Serologic identification of flavivirus isolates from human, animal, and mosquito tissues is usually accomplished by virus neutralization using either polyclonal or monoclonal antibodies (Calisher et al. 1989).

Development of the polymerase chain reaction (PCR) technology has provided means to amplify nucleic acid sequences 10^5 to 10^6 fold in a few hours in vitro (Saiki et al. 1985, 1988). This technology has been modified by the addition of a reverse transcriptase step to facilitate the synthesis of ssDNA from the genome of RNA-containing viruses, including human immunodeficiency virus (Kwok et al. 1987; Jackson et al. 1990), polio (Robart 1990a, b; Hyypia et al. 1989), nonpolio enteroviruses (Gama et al. 1989; Robart 1990a; Jansen et al. 1990; Hyypia et al. 1989), hepatitis C virus (Cristiano et al. 1991), measles (Godec et al. 1990), and flaviviruses (Deubel et al. 1990; Eldadah et al. 1991). The exquisite sensitivity of this procedure makes it an attractive approach for the rapid detection and identification of flaviviruses in mosquitoes and clinical specimens where cultivation of the virus is difficult and time consuming, and where diagnosis influences treatment and has epidemiological implications, including vaccination and mosquito control.

PCR techniques have been developed to detect and identify flavivirus RNA in infected brain, serum, and cell cultures (Deubel et al. 1990; Eldadah et al. 1991; Lanciotti et al. 1991). PCR-based assays for dengue, YF, and SLE viruses have been prompted by the recognized need for rapid and sensitive identification of virus and the availability of sequence information facilitating construction of PCR primers and hybridization probes. The techniques described have utilized amplimers which initiate synthesis in the structural gene region of the genome (Deubel et al. 1990; Eldadah et al. 1991). The RT/PCR technique described by Deubel et al. (1990) employs reverse type-specific hybridization to identify all four dengue virus serotypes. Southern blot analysis of the PCR products revealed specific reactivity with each of the dengue probes. The sensitivity of this assay was estimated to be 10^3 infecting particles/ml. Lanciotti et al. (1991) amplified a 511-bp segment of the dengue virus capsid and prM genes, which was then subjected to secondary amplification with nested primers to facilitate dengue virus serotype identification. Eldadah et al. (1991) amplified regions of both the envelope and nucleocapsid genes to obtain DNA which could be identified by agarose gel electrophoresis and confirmed by restriction endonuclease cleavage. This assay could differentiate SLE, JE, YF, dengue 2, and dengue 4 viruses. This test is specific and compares favorably with conventional assays for viral infectivity in suckling mice or cell culture.

Although several PCR assays for the detection of flaviviruses have been developed using different primers and methods for detection, a single-tube method for detection of all medically important flaviviruses has yet to be reported. We describe a PCR assay for the detection of 42 different flaviviruses and the use of virus-specific oligonucleotide hybridization probes to identify YF, dengue serotypes 1, 2, 3, 4, JE, MVE, WNE, SLE, and Powassan (POWA) viruses. The single-tube PCR uses flavivirus-universal down-amplimers and virus-specific up-amplimers to facilitate the synthesis of amplicons which have different sizes for each flavivirus (Table 2). With the availability of additional sequence information for the NS5 gene, this technique could be utilized for the rapid identification of any flavivirus for which this sequence information is available.

Materials and Methods

Virus Strains and RNA Extraction. Viruses used in this study are registered members of the family Flaviviridae obtained from the collection at the Division of Vector-Borne Infectious Diseases, Centers for Disease Control, Fort Collins, Colo. (Table 2). Viral RNA was extracted from seed virus stocks using a modification of the method of Chomczynski and Sacchi (1987), as described by Lanciotti et al. (1991).

Design and Synthesis of Oligonucleotide Probes and Amprimers. Flavivirus cross-reactive and viral-specific oligonucleotide primers used in RT/PCR am-

Table 2. Registered flaviviruses used to characterize flavivirus-reactive reverse transcriptase/ polymerase chain reaction (RT/PCR) amplification

Virus	Strain	Vectors[a]	Source[b]	Complex[c]
Alfuy	MRM 3929	M	SMB	JE
Apoi	Original	N	SMB	RB
Aroa	VenA 01809	M	SMB	N/A
Bagaza	DakArB 209	M	SMB	NTA
Banzi	SAH 336	M	SMB	UGS
Bouboui	DakArB 490	M	SMB	UGS
Bussuquara	BeAn 4073	M	SMB	N/A
Cowbone Ridge	W 10986	N	SMB	MOD
Dakar bat	IPDA 249	N	SMB	RB
Den1	Hawaii	M	TC	DEN
Den1	H 45487	M	SMB	DEN
Den2	NGC	M	TC	DEN
Den2	TR 1751	M	SMB	DEN
Den2	Jam	M	TC	DEN
Den3	H-87	M	SMB	DEN
Den3	RI-1	M	SMB	DEN
Den4	107-128-85	M	SMB	DEN
Den4	1351	M	TC	DEN
Den4	H-241	M	SMB	DEN
Edge Hill	AusC 281	M	SMB	UGS
Entebbe	UgIL 30	N	SMB	RB
Gadgets	CSIRO 122	T	SMB	N/A
I. trky men	Original	M	SMB	NTA
JEV	G 8924	M	SMB	JE+
JEV	NAKAYAMA	M	SMB	JE
JEV	HK 8256	M	SMB	JE
Kunjin	MRM 16	T	SMB	JE
Langat	TP 21	T	SMB	TBE
MVE[c]	11A	M	SMB	JE
MVE[c]	Original	M	SMB	JE
Powassan	1982–64	M, T	SMB	TBE
Powassan	CANADA	M, T	SMB	TBE
SLE[c]	MSI-7	M	TC	JE
SLE[c]	VP7	M	TC	JE
SLE[c]	TBH-28	M	DE	JE
SLE[c]	TNM11	M	TC	JE
WN[c]	Ar 248	M	SMB	JE
WN[c]	Eg 101	M	SMB	JE
YF[c]	17D	M	SMB	YF
YF[c]	788379 (Tr)	M	SMB	YF
YF[c]	Asibi	M	SMB	YF
YF[c]	DakAr 1279	M	SMB	YF

[a] M, mosquitoe-borne virus; T, Tick-borne virus; N, non-vector-associated virus

[b] SMB, suckling-mouse brain virus; TC, tissue-cultured virus; DE, duck embryo

[c] Antigenic complexes: JE, Japanese encephalitis; RB, Rio Bravo; NTA, Ntaya; UGS, Uganda S; MOD, Modoc; DEN, Dengue; TBE, tick-borne encephalitis; YF, yellow fever; MVE, Murray Valley encephalitis; SLE, St. Louis encephalitis; WN, West Nile; N/A, not assigned (Calisher et al. 1989)

plification (amplimers) of viral RNA were designed based on published flavivirus genome sequences (Rice et al. 1985; Deubel et al. 1988; Osatomi et al. 1988; Zaho et al. 1986; Castle et al. 1986; Wengler et al. 1985; Hashimoto et al. 1988; Pletnev et al. 1990; Coia et al. 1988; Dalgarno et al. 1986; Lee et al. 1990). SLE and POWA virus sequence information for the C-terminal one-fourth of the NS5 gene was obtained during this study. Computer programs DNASIS (Hitachi Software, Brisbane, Calif.) and OLIGO (National Biosciences, Hamel, Minn.) were used in the sequence analyses.

Criteria used in designing flavivirus cross-reactive up- and down-amplimers were: (a) sequences with maximum homology for all flaviviruses at the same genomic location, (b) low homology of this sequence with nucleotide sequence information for other genes in the GeneBank and EMBL databases, (c) a high melting temperature, (d) minimum primer dimer interaction, (e) amplification of a relatively short DNA fragment which contained variable regions suitable for development of individual virus-specific amplimers or probes.

Virus-specific amplimer/probes were designed to have: (a) minimum sequence homology with other flaviviruses and with regions of the same virus, (b) a high melting temperature, (c) minimum homodimer and heterodimer interaction, (d) virus-specific sequence regions located between nucleotides 9166 and 9977 in gene NS5. Oligonucleotide amplimers CF-DDJ9977 and FUDJ9166, with sequences that begin at nucleotides 9166 and 9977 in the dengue 2 sequence, were selected and designed to amplify all flavivirus RNAs in this region of the NS5 gene.

Oligonucleotides were synthesized on an Applied Biosystems DNA synthesizer model 380A using standard phosphoramidite chemistry (Applied Biosystems, Foster City, Calif.). Synthetic oligonucleotides were purified using OPC columns (Applied Biosystems) or by acrylamide gel electrophoresis.

Reverse Transcriptase and Taq Polymerase Chain Reactions. A single-vessel reverse transcriptase and Taq polymerase chain reaction, RT/PCR, was used to convert targeted viral RNA to complementary DNA (cDNA) with subsequent amplification of the cDNA fragment (amplicon). The RT/PCR reactions were done in 100- or 20-μl reaction volumes which contained the following components: 50 mM KCl, 10 mM TRIS-HCl, pH 8.5, 1.5 mM MgCl$_2$, 0.01% gelatin, 200 μM of each dNTP (Promega, Madison, Wis.), 2.5 mM dithiothreitol, 10 nM of up- and down-amplimer, 2.5 U/100 μl of RAV-2 reverse transcriptase (Amersham, Arlington Heights, Ill.), and 2.5 U/100 μl of Amplitaq polymerase (Perkin Elmer, Norwalk, Conn.). Reactions were incubated in a model 450 Thermalcycler (Perkin Elmer) programmed to incubate for 1 h at 50°C to facilitate the RT reaction and 4 min at 94°C for denaturation, followed by 35 cycles of denaturation (94°C for 1 min), amplimer annealing (55°C for 1 min), and amplimer extension (72°C for 3 min), and terminating with a 10-min incubation period at 72°C for long amplimer extension.

Characterization of Amplicons Produced with Flavivirus Universal Amplimers. The ability of the flavivirus cross-reactive up- and down-amplimers,

FUDJ9166 and CFDDJ9977, to direct the synthesis of viral cDNA and amplify the DNA was investigated using nucleic acid extracted from seed virus preparations (Table 2). Virus RNA in these extracts was amplified in a 100 µl RT/PCR reaction volume, as previously described (Chomczynski and Sacchi 1987). Following amplification, a 5-µl portion of the RT/PCR reaction mixture was electrophoresed on a 3% GTG agarose gel (FMC Bioproducts, Rockland, Me.) in TAE buffer (0.4 M TRIS-HCl, 0.05 M sodium acetate, 0.01 M ethylenediamine tetra-acetic acid (EDTA)-Na_2), and the DNA bands were othidium bromide stained with (0.1 µg/ml). The size of the amplicons for each virus was calculated using molecular weight standards and compared with the calculated values (Table 3).

Nucleotide Sequence Analysis of SLE and Powassan Amplified DNA Fragments with Universal Flavivirus Amplimers. The 832-bp DNA fragments obtained by amplification of SLE and POWA viral RNAs with amplimers CFDDJ9977 and FUDJ9166 were purified by agarose gel electrophoresis, tailed with G, and cloned into the *Pst*I site of M13/mp19 (Maniatis et al. 1982). Recombinant

Table 3. Reverse transcriptase/polymerase chain reaction (RT/PCR) amplification for the identification of flaviviruses

Designation[a]	Oligonucleotide sequence	Tm (°C)[b]	Range of homology (%)	Amplified DNA (bp)
Flavivirus cross-reactive down-amplimer:				
CFDDJ9977	5'-GCATGTCTTCCGTCGTCATCC	58.8	100.0–85.7	
Flavivirus cross-reactive up-amplimer:				
FUDJ9166	5'-GATGACACAGCAGGATGGGAC	56.3	100.0–87.0	838–832
Virus-specific up-amplimers:				
DEN1-J9243	5'-GCCTGAACATGCTCTATTGGCT	57.0	<55.0	761
DEN2-9452	5'-TCTTCAAAAGCATTCAGCACCT	55.4	<55.0	546
DEN3-9471	5'-CCCATCCGCTAGAGAAGAAAATTACAC	60.4	<55.0	522
DEN4-9586	5'-GCACTTCCCTCCTCTTCTTG	52.7	<55.0	405
JE-9574	5'-GACCACAACACTTGGAACAGCTAC	56.5	<55.0	546
MVE-9354	5'-TTGATGGAAGGTGAGCAGCGGAC	64.1	74.0 – <55.0	773
SLE-J9483	5'-ACGATTGGCCAAAGCGGTTGAG	63.8	<55.5	515
WN-9351	5'-TGGAGAACATCGACGTTTAGCG	58.7	<55.0	764
YF-9334	5'-ACAAGCAGTGATGGAAATGACA	54.3	<55.0	743
POWA-J9624	5'-CAAGGTTCGTAAGGATGTGGGA	57.5	<55.0	374
TBE-9569[c]	5'-CAAGACTACTTCGAGTGGAACGATGG	61.0	<55.0	516

[a] Amplimer designation is based on the particular viral sequence obtained. Number after virus designation indicates genomic position of the amplimer, except DEN 1, SLE and POWA which are based on DEN 2 Jamaica sequence

[b] T_m is calculated based on the method of Rychlik et al. (1990) at 50 mM NaCl concentration

[c] Proposed tick-borne encephalitis (TBE)-specific amplimer

phages were screened for size, orientation, and recombinant phage DNA from at least three independent clones having opposite orientations purified and sequenced in their entirety (Sanger et al. 1977). Sequence data were aligned and compared with those of other flaviviruses using the DNASIS (Hitachi Software, Brisbane, CA 94005) and KINSEQ (Kinney, personal communication) computer programs.

Virus-Specific Probe Hybridization Analysis of Amplicons Produced with Universal Flavivirus Amplimers. The RNA of 42 different flaviviruses was amplified using the flavivirus universal amplimers and 95 μl of each individual RT/PCR reaction mixture denatured at 100 °C for 20 min and immobilized on nitrocellulose membranes using a 96-well vacuum manifold (Schleicher & Schuell, Keene, N.H.). Viral cDNA was fixed to the membrane by drying at room temperature overnight, followed by baking at 85 °C for 30 min. Then 10 pmole of ^{32}P-5′-end-labeled virus-specific oligonucleotide or amplimer FUDJ9166 (Table 2) were added to a hybridization solution containing 5XSSPE, 5X Denhalt's solution, and 100 μg/ml of shared salmon sperm DNA. Hybridization reactions were performed at 65 °C for 1 h, then at 37 °C for 15 h. Specific hybridization was detected by autoradiography after washing the membranes with 1XSSC/0.1% SDS twice each at room temperature and at 37 °C and with 0.5XSSC/0.1% SDS twice each at 45 °C, 55 °C, and 65 °C.

Design of a Flavivirus Serotype-Specific RT/PCR Assay. A 25-μl sample of the RT/PCR reaction mixture containing viral RNA CFDDJ9977, the universal down-amplimer, and the other RT/PCR reaction components were dispensed into reaction vessels containing 10 nM of individual virus-specific amplimer of FUDJ9166, the universal up-amplimer. RT/PCR reactions were performed, and a 5-μl portion of the reaction products was electrophoresed on a 3% agarose gel and stained with ethidium bromide. Because of the specific priming site for each of the individual virus-specific amplimers (Table 2), only virus RNA homologous to the virus-specific up-amplimer was synthesized in each of the reactions. The RT/PCR synthesis-generated virus-specific DNA fragments have different molecular weights for each virus (Table 3).

Results

Characterization of Flavivirus NS5 Gene Amplicons Produced with the Universal Flavivirus Amplimers. A region of the flavivirus genome 832 nucleotides long, located between nucleotides 9166 and 9997 in the C-terminal portion of the NS5 gene, was identified as fulfilling the criteria of having both conserved and virus-specific regions. This portion of the NS5 gene is characterized as having two highly conserved stretches of flavivirus-specific sequence at both the 3′- and 5′-ends. The 832-bp DNA of 12 different flaviviruses has an average homology sequence of 66% among the flaviviruses for which sequence information is available (Table 4). This amplicon contains virus-specific variable regions as exemplified by nucleotide sequences located between nucleotides (nt) 9433 and 9516 in the dengue 2 Jamaica genome. The overall average sequence homology for this variable 83-nt region is 45% with other flaviviruses (Table 4). Flavivirus cross-reactive amplimers were engineered using sequence information at the 5'- and 3'-extremities of the regions nt 9166 and 9977, respectively. Virus-specific amplimer/probes were designed to react within this region of the NS5 gene (Table 3).

Virus RNA extracted from 42 different flavivirus seed stocks (Table 2) was subjected to RT/PCR amplification using the flavivirus universal amplimers FUDJ9166 and CFDDJ9977. These primers facilitated amplification of an 832-bp DNA fragment for all of the flaviviruses except Apoi, a nonvector-transmitted flavivirus from Japan (Table 2). Identity of the DNA fragments amplified in the RT/PCR reactions for the four serotypes of dengue, YF, MVE, WNE, SLE, and POWA viruses was confirmed by hybridization using virus-specific probes (Table 3, data not shown).

Nucleotide Sequence of the St. Louis and Powassan Virus NS5 Gene Amplicons. DNA synthesized in RT/PCR reactions using SLE and POWA virus RNAs and flavivirus universal amplimers CFDDJ9977 and FUDJ9166 were cloned into the *Pst*I site of M13/mp19 phage (Maniatis et al. 1982). The nucleotide sequence of these fragments was obtained by dideoxy chain termination sequencing (Sanger et al. 1977). Comparison of the nucleotide sequences of this region in the NS5 genes of SLE and POW viruses with those of other published flaviviruses revealed that the SLE virus has a high nucleotide sequence homology with viruses in the JE complex, and POWA virus is genetically related to viruses in the TBE complex (Table 4).

Characterization of the Flavivirus Serotype-Specific RT/PCR Assay. Single-tube RT/PCR amplifications of 10 different flavivirus RNAs were performed to determine the specificity of synthesis using the universal flavivirus down-amplimer CFDDJ997 and each of the virus-specific up-amplimers (Table 4). RT/PCR reactions in this experiment included each of the virus-specific amplimers and RNA from the four different serotypes of dengue, JE, SLE, MVE, WNE, YF, and POWA viruses. Following RT/PCR synthesis, reaction products were subjected to agarose gel electrophoresis, and the gels were examined after ethidium bromide staining to visualize the DNA bands. As seen in Fig. 2, RT/PCR reactions primed with the universal flavivirus down-primer and each of the virus-specific up-primers were specific, as evidenced by the DNA bands of appropriate size only in those lanes in which the specific primer and virus were homologous. In some cases, as exemplified by patterns seen for SLE and YF in Fig. 2, minor species of DNA were generated in the RT/PCR reactions which are not of the proper size and of low yield. This indicated that under the conditions employed JE, MVE, WN, and POWA amplimers may prime other flavivirus RNAs nonspecifically, resulting in generation of small amounts of nonspecific DNA. These nonspecific DNAs have sizes different from those

Table 4. Percentage homology between flavivirus NS5 genes in two regions: nucleotides 9166–9997 and 9433–9516

	DEN1	DEN2	DEN3	DEN4	JE	MVE	KUN	WN	SLE	POWA	TBE	YF
DEN1	–	68.8[a]	70.1	68.8	64.0	63.6	63.2	63.4	63.4	59.9	60.8	60.7
DEN2	42.5[b]	–	69.6	69.1	63.2	61.7	61.8	60.7	64.6	59.6	60.2	59.6
DEN3	44.4	43.3	–	73.2	64.4	64.0	64.5	63.1	63.5	61.0	59.2	61.3
DEN4	53.3	52.2	58.6	–	64.1	63.2	64.2	60.5	62.3	61.8	59.7	61.2
JE	48.9	46.7	41.1	52.2	–	74.5	71.4	72.7	66.0	63.9	61.6	63.0
MVE	43.3	42.2	40.0	55.6	65.6	–	74.3	71.5	67.2	61.0	60.0	62.3
KUN	46.7	43.3	43.3	53.3	57.8	55.6	–	78.4	69.0	62.1	62.2	61.4
WN	43.3	41.1	37.8	46.7	64.4	62.2	80.0	–	67.6	62.1	64.0	61.7
SLE	41.1	36.7	40.0	46.7	36.7	47.8	48.9	47.8	–	61.2	62.6	64.9
POWA	43.3	38.9	51.7	55.2	48.9	51.1	50.0	48.9	44.4	–	73.0	62.6
TBE	40.0	35.6	42.5	48.3	43.3	36.7	45.6	52.2	46.7	61.7	–	61.0
YF	36.7	34.4	32.2	40.0	46.7	42.2	46.7	47.8	41.1	37.8	36.7	–

[a] Homology in sequence between nt 9166 and 9997. Stippled area indicates homology between viruses within the same serocomplex
[b] Homology in sequence between nt 9433 and 9516. Stippled area indicates homology between viruses within the same serocomplex

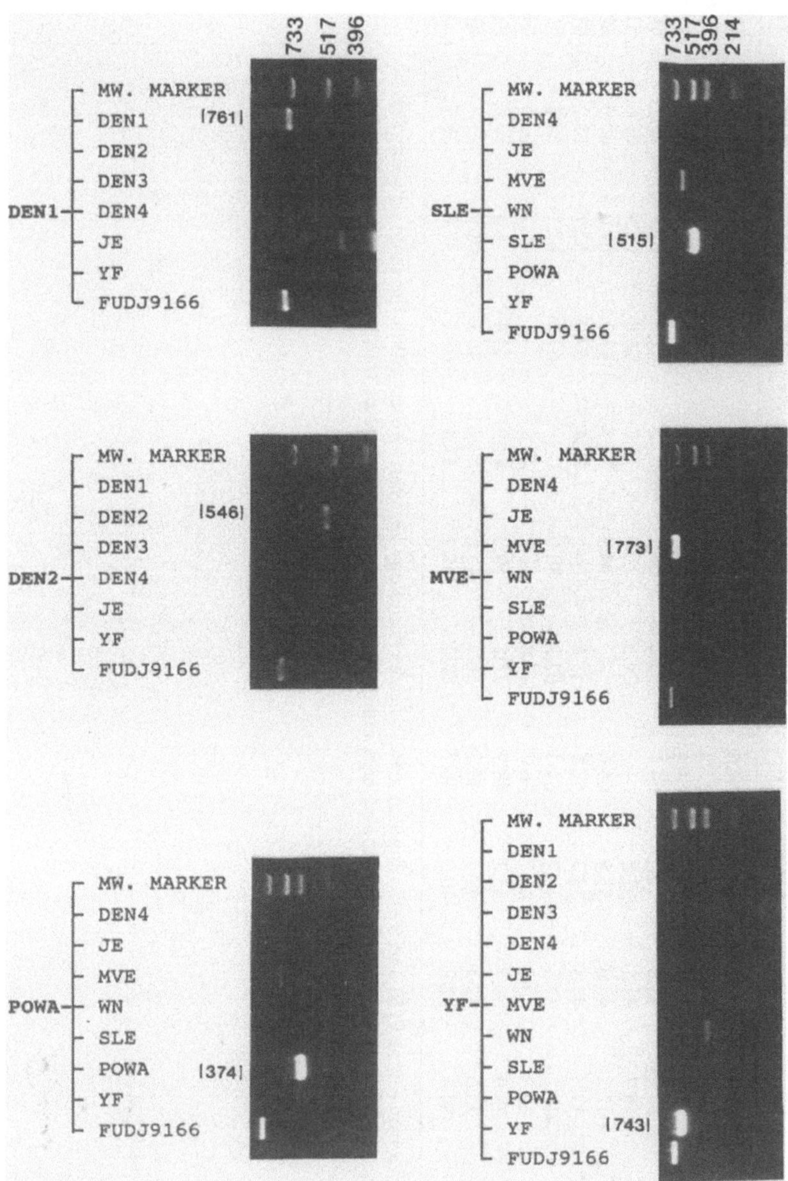

Fig. 2. Characterization of the flavivirus serotype-specific reverse transcriptase/polymerase chain reaction (RT/PCR) assay. Viral RNAs used (up-left to low-right) are dengue (*DEN1*, *DEN2*), Powassan (POWA), St. Louis encephalitis (SLE), Murray Valley encephalitis (*MVE*), and yellow fever (*YF*), respectively. Following the RT/PCR reaction using universal flavivirus down-amplimer and one of the virus-specific up-amplimers, reaction products were subjected to agarose gel electrophoresis, and the gels were examined after ethidium bromide staining to visualize the DNA bands. Size of the specific DNA fragment in which the amplimer and virus were homologous was indicated as base pair after that amplimer (Table 2)

produced by specific priming of viral RNA and amplification of the cDNA (Table 3). To improve the specificity of these RT/PCR reactions, additional studies need to be done to characterize the interaction of the virus-specific primers with the different flavivirus RNAs to realize optimal conditions for specific priming.

Discussion

Flaviviruses are important causes of human disease whose diagnosis and epidemiology are of major public health concern (Monath 1990). Surveillance and control of vector-borne flavivirus encephalitides such as SLE, MVE, and WN depend upon isolation of virus from mosquitoes and serology testing (Bowen and Francy 1980). Surveillance for dengue and YF virus during epidemic periods is investigated by virus isolation from vector mosquitoes or patient serum (Gubler 1988; Monath 1988). The cabability to detect and identify flaviviruses rapidly in mosquitoes and clinical materials would assist in determining the distribution and infection rate of vectors, establishing criteria for vector control, and providing information needed to determine whether or not to implement vaccination where possible. For example, the movement and introduction of new dengue virus serotypes must be detected as early as possible because of the potential threat of serious disease in populations who may have acquired immunity to another serotype (Halstead 1988). An accurate flavivirus PCR test for the detection and identification of viral RNA would significantly improve the current status of viral diagnosis and epidemiology.

PCR amplification of cDNA, coupled with virus-specific oligonucleotide hybridization, provides a powerful tool (Deubel et al. 1990; Eldadah et al. 1991; Lanciotti et al. 1991). This method is as sensitive as isolation of virus in cell cultures or mice but permits rapid identification of important human pathogens. Because several flavivirus pathogens may be transmitted in the same geographic area at the same time, more than one virus may be isolated from mosquitoes or clinical specimens (Table 5). It is therefore critical that flavivirus PCR assays be designed to detect any potential pathogen and that hybridization probes or other means be available to identify the amplified nucleic acid to species. From published nucleic acid sequence information, we have identified a region in the flavivirus NS5 gene which is sufficiently conserved at the 3'- and 5'-ends to permit one set of primers to be used to amplify cDNA of all known medically important flaviviruses. Virus-specific probes have been constructed which hybridize with specific sequences within this 832 base amplicon, facilitating virus identification. Because of this specificity, these probes can be used as up-amplimers, together with the universal down-amplimer, in a single-tube assay. In this format, amplification of flavivirus RNA is determined by the specificity of the up-amplimer engineered for each of the flaviviruses. Following amplification in the RT/PCR test, virus cDNA

Table 5. Identification of medically important flaviviruses based on geographic location of virus isolation and application of virus-specific RT/PCR amplification technique

Geographic location	Up-amplimer	
	Flavivirus	Virus-specific
North America	CFDDJ9977	SLE; POWA
Central and South America	CFDDJ9977	DEN1; DEN2; DEN3; DEN4; SLE; YF
Far East and SE Asia	CFDDJ9977	DEN1; DEN2; DEN3; DEN4; JE
Africa and Mid East Asia	CFDDJ9977	DEN1; DEN2; DEN3; DEN4; YF; WN
India and Sri Lanka	CFDDJ9977	DEN1; DEN2; DEN3; DEN4; JE; YF
Australia	CFDDJ9977	DEN1; DEN2; DEN3; DEN4; MVE; WN

RT/PCR, reverse transcriptase/polymerase chain reaction; SLE, St. Louis encephalitis; POWA, Powassan; DEN, dengue; YF, yellow fever; JE, Japanese encephalitis; WN, West Nile; MVE, Murray Valley encephalitis

can be analyzed by agarose gel electrophoresis and the virus identified by amplicon size. Although some nonspecific DNAs are occasionally observed in the gels, the size of these DNAs is other than expected, and they are always seen as minor components. Further analysis of the primer genome interaction to optimize hybridization temperatures and PCR conditions should eliminate this problem.

As with other RNA viruses, reverse transcription of genomic RNA to produce cDNA for amplification prior to the PCR determines the sensitivity of the test (Robart 1990b). Reverse transcription of viral RNA requires approximately $10^{2.5}-10^3$ infectious units of virus (Deubel et al. 1988; Eldadah et al. 1991; Gama et al. 1989; Robart 1990a). PCR amplification of the viral cDNA and identification of the DNA by oligonucleotide probe hybridization can detect as little as 10 ng of virus-specific cDNA. Therefore, the sensitivity of the technique is limited by the initial RT step, and for RNA-containing viruses it is approximately the same as that of the antigen capture assay (Tsai et al. 1988). A major advantage of PCR, however, is its capability to detect multiple flavivirus species, whereas antibody capture assays are virus-specific.

The technique we have developed has several advantages over other PCR techniques and conventional methods. The entire method can be completed within 24 h, in contrast to several days required for culture and immunologic identification. A single set of PCR primers is required for the detection of virus, which is identified to species with virus-specific amplimers. Nucleotide sequence analysis of the cDNA resulting from amplification of viral RNA using the flavivirus universal amplimers can be used to identify new viruses within a serocomplex. This sequence information can also be used to construct virus type-specific amplimers to facilitate the identification of the virus to species in the single-tube RT/PCR. Classic methods for virus identification rely on the growth of virus in mice or cell cultures, which may impose biological selection. The ability to directly examine the genetics of viruses within a population by sequence analysis, restriction mapping, and hybridization probing opens new avenues to investigate the molecular basis of pathogenesis.

References

Bowen GS, Francy BD (1980) Surveillance. In: Monath TP (ed) St. Louis encephalitis. American Public Health Association, Washington, pp 473–502

Bres P (1988) Impact of arboviruses on human and animal health. In: Monath TP (ed) The arboviruses: epidemiology and ecology, volume 1. CRC, Boca Raton, pp 2–17

Brinton MA, Dispoto JH (1988) Sequence and secondary structure analysis of the 5'-terminal region of flavivirus genome RNA. Virology 162:290–299

Brinton MA, Fernandez AA, Dispoto JH (1986) The 3'-nucleotides of flavivirus genomic RNA form a conserved secondary structure. Virology 153:113–121

Calisher CH, Poland JD (1980) Laboratory diagnosis. In: Monath TP (ed) St. Louis encephalitis. American Public Health Association, Washington, pp 571–601

Calisher CH, Karabatsos N, Dalrymple JM, Shope RE, Porterfield JS, Westaway EG, Brandt WE (1989) Antigenic relationships between flaviviruses as determined by cross-neutralization tests with polyclonal antisera. J Gen Virol 70:37–43

Castle E, Wengler G (1987) Nucleotide sequence of the 5'-terminal untranslated part of the genome of the flavivirus West Nile virus. Arch Virol 92:309–313

Castle E, Leidner U, Nowak T, Wengler G, Wengler G (1986) Primary structure of the West Nile flavivirus genome region coding for all nonstructural proteins. Virology 149:10–26

Chomczynski P, Sacchi N (1987) Single-step method of RNA isolation by acid guanidinium thiocyanate-phenol-chloroform extraction. Anal Biochem 162:156–159

Coia G, Parker MD, Speight G, Byrne ME, Westaway EG (1988) Nucleotide sequence and complete amino acid sequences of Kunjun virus: definitive gene order and characteristics of the virus-specified proteins. J Gen Virol 69:1–21

Cristiano K, DeBisceglie AM, Hoofnagle JH, Feinstone SM (1991) Hepatitis C viral RNA in serum of patients with chronic non-A, non-B hepatitis: detection by the polymerase chain reaction using multiple primer sets. Hepatology 14:51–55

Dalgarno L, Trent DW, Strauss JH, Rice CM (1986) Partial nucleotide sequence of the Murray Valley encephalitis virus genome: comparison of the coded polypeptides with yellow fever virus structural and non-structural proteins. J Mol Biol 187:309–377

Deubel V, Kinney RM, Trent DW (1988) Nucleotide sequence and deduced amino acid sequence of the nonstructural proteins of dengue type 2 virus, Jamaica genotype: comparative analysis of the full-length genome. Virology 165:234–244

Deubel V, Laille M, Hugnot J-P, Chungue E, Guesdon J-L, Drouet MT, Bassot S, Chevrier D (1990) Identification of dengue sequences by genomic amplification: rapid diagnosis of dengue virus serotypes in peripheral blood. J Virol Methods 30:41–54

Eldadah ZA, Asher DM, Godec MS, Pomeroy KL, Goldfarb LG, Feinstone SM, Levitan H, Gibbs CJ Jr, Gajdusek CD (1991) Detection of flaviviruses by reverse transcriptase polymerase chain reaction. J Med Virol 33:260–267

Gama RE, Horsnell PR, Hughes PJ, North C, Bruce CB, Al-Nakib W, Stanway G (1989) Amplification of rhinovirus specific nucleic acids from clinical specimens using the polymerase chain reaction. J Med Virol 28:73–77

Godec MS, Asher DM, Swoveland PT, Eldadah ZA, Feinstone SM, Goldrarb LG, Gibbs CJ Jr, Gajdusek DC (1990) Detection of measles virus genomic sequences in SSPE brain tissue by the polymerase chain reaction. J Med Virol 30:237–244

Gubler DJ (1988) Dengue. In: Monath TP (ed) The arboviruses: epidemiology and ecology, volume 2. CRC, Boca Raton, pp 223–260

Hahn YS, Galler R, Hunkapiller T, Dalrymple JM, Strauss JH, Strauss EG (1988) Nucleotide sequence of dengue 2 RNA and comparison of the encoded proteins with those of other flaviviruses. Virology 162:167–180

Halstead SB (1988) Pathogenesis of dengue: challenges to molecular biology. Science 239:476–481

Hashimoto H, Nomoto A, Watanabe K, Mori T, Takezawa T, Aizawa C, Takegami T, Hiramatsu K (1988) Molecular cloning and complete nucleotide sequence of the genome of Japanese encephalitis virus Bejing-1 strain. Virus Genes 1:305–317

Henchal EA, Putnak JR (1990) The dengue viruses. Clin Microbiol Rev 3:376–396

Henchal EA, McCown JM, Burke DS, Seguin MC, Brandt WE (1985) Epitopic analysis of antigenic determinants on the surface of dengue-2 virions using monoclonal antibodies. Am J Trop Med Hyg 34:162–169

Hoke CH, Nisalak A, Sangawhipa N, Jatanasen S, Laorakapongse T, Innis BL, Kotchasenee S, Gingrinch JB, Latendresse J, Fukai K, Burke DS (1988) Protection against Japanese encephalitis by inactivated vaccines. N Engl J Med 319:608–614

Hyypia T, Auvinen P, Maaronen M (1989) Polymerase chain reaction for human picornaviruses. J Gen Virol 70:3261–3268

Jackson JB, Kwok SY, Sninsky JJ, Hopsicker JS, Sannerud KJ, Rhame FS, Henry K, Simpson M, Balfour HH (1990) Human immunodeficiency virus type 1 detected in all seropositive symptomatic and asymptomatic individuals. J Clin Microbiol 28:16018

Jansen RW, Sigel G, Lemon SM (1990) Molecular epidemiology of human hepatitis A virus defined by an antigen-capture polymerase chain reaction method. Proc Natl Acad Sci USA 87:2867–2871

Kwok S, Mack DH, Mullis KB, Poiesz B, Ehrlich G, Blair D, Freidman-Kein A, Sninsky JJ (1987) Identification of human immunodeficiency virus sequences by using in vitro enzymatic amplification and oligomer cleavage detection. J Virol 61:1690–1694

Kunz C, Heinz FX, Hoffman H (1980) Immunogenicity and reactiongenicity of a highly purified vaccine against tick-borne encephalitis. J Med Virol 6:103–109

Lanciotti RS, Calisher CH, Gubler DJ, Chang G-J, Vorndam VA (1992) Detecting and typing dengue viruses by genomic amplification in a reverse transcriptase-polymerase chain reaction. J Clin Microbiol 30: 545–551

Lee E, Fernon C, Simpson R, Weir RC, Rice CM, Dalgarno L (1990) Sequence of the 3′ half of Murray Valley encephalitis virus genome and mapping of the nonstructural proteins NS1, NS3, and NS5. Virus Genes 4:197–213

Maniatis T, Fritsch EF, Sambrook J (1982) Molecular cloning: a laboratory manual. Cold Spring Harbor Laboratory, Cold Spring Harbor

Monath TP (1987) Yellow fever: a medically neglected disease. Rev Infect Dis 9:165–175

Monath TP (1988) Yellow fever. In: Monath TP (ed) The arboviruses: epidemiology and ecology, volume 5. CRC, Boca Raton, pp 139–232

Monath TP (1990) Flaviviruses. In: Fields BN, Knipe DN (eds) Fields virology, 2nd edn. Raven, New York, pp 763–814

Mullis KB, Faloona FA (1987) Specific synthesis of DNA in vitro via a polymerase-catalyzed chain reaction. Methods Enzymol 155:335–350

Murphy FA (1980) Togavirus morphology and morphogenesis. In: Schlesinger RW (ed) The togaviruses: biology, structure, replication. Academic, New York, pp 241–316

Nitayaphan S, Grant JA, Chang G-J, Trent DW (1990) Nucleotide sequence of the virulent SA-14 strain of Japanese encephalitis virus and its attenuated vaccine derivative, SA-14-14-2. Virology 177:514–522

Osatomi K, Sumiyoshi H (1990) Complete nucleotide sequence of dengue type 3 virus genomic RNA. Virology 176:643–647

Pletnev AG, Yamshchikov VF, Blinov VM (1990) Nucleotide sequence of the genome and complete amino acid sequence of the polyprotein of tick-borne encephalitis virus. Virology 174:250–263

Rice CM, Lenches EM, Eddy SR, Shin SJ, Sheets RL, Strauss JH (1985) Nucleotide sequence of yellow fever virus: implications for flavivirus gene expression and evolution. Science 229:726–733

Robart HA (1990a) Enzymatic RNA amplification of the enteroviruses. J Clin Microbiol 28:438–442

Robart HA (1990 b) Diagnosis of enteroviral meningitis with the polymerase chain reaction. J Pediatr 117:85–89

Roehrig JT, Mathews JH, Trent DW (1983) Identification of epitopes on the E glycoprotein of St. Louis encephalitis virus using monoclonal antibodies. Virology 128:118–126

Roehrig JT, Hunt AR, Johnson AJ, Hawkes RA (1989) Synthetic peptides derived from the deduced amino acid sequence of the E-glycoprotein of Murray Valley encephalitis virus elicit antiviral antibody. Virology 171:49–60

Roehrig JT, Johnson AJ, Hunt AR, Bolin RA, Chu MC (1990) Antibodies to dengue 2 virus E-glycoprotein synthetic peptides identify antigenic conformation. Virology 177:668–675

Rychlik W, Spencer WS, Rhoads RE (1990) Optimization of the annealing temperature for DNA amplification *in vitro*. Nucleic Acids Res 18:6409–6412

Saiki RK, Scharf S, Faloona F, Mullis KB, Horn GT, Erlich HA, Arnheim N (1985) Enzymatic amplification of B-globin genomic sequences and restriction site analysis for diagnosis of sickle cell anemia. Science 230:1350–1354

Saiki RK, Gelfand DH, Stoffel S, Scharf SJ, Higuchi R, Horn GT, Mullis KB, Erlich HA (1988) Primer-directed enzymatic amplification of DNA with a thermostable DNA polymerase. Science 239:487–491

Sanger F, Nicklen S, Coulson AR (1977) DNA sequencing with chain-terminating inhibitors. Proc Natl Acad Sci USA 74:5463–5467

Speight G, Cola G, Parker MD, Westaway EG (1988) Gene mapping and positive identification of the non-structural proteins NS2A, NS2B, NS3, NS4B, and NS5 of the flavivirus Kunjin and their cleavage sites. J Gen Virol 69:23–34

Theiler M, Smith HH (1937) Effect of prolonged cultivation in vitro upon pathogenicity of yellow fever virus. J Exp Med 65:767–786

Trent DW, Naeve C (1980) Biochemistry and replication. In: Monath TP (ed) St. Louis encephalitis. American Public Health Association, Washington, pp 159–199

Trent DW, Kinney RM, Johnson BJB, Vorndam AV, Grant JA, Deubel V, Rice CM, Hahn C (1987) Partial nucleotide sequence of St. Louis encephalitis virus RNA: structural proteins, NS1, ns2a and ns2b. Virology 156:293–304

Tsai TF, Happ CM, Bolin RA, Montoya M, Campos E, Francy DB, Hawkes RA, Roehrig JT (1988) Stability of St. Louis encephalitis virus antigen detected by enzyme immunoassay in infected mosquitoes. J Clin Microbiol 26:2620–2625

Turell MJ (1988) Horizontal and vertical transmission of viruses by insect and tick vectors. In: Monath TP (ed) The arboriviruses: epidemiology and ecology, vol 1. CRC Boca Raton, pp 128–152

Wengler G, Castle E, Leidner U, Nowak T, Wengler G (1985) Sequence analysis of the membrane protein V3 of the flavivirus West Nile virus and of its gene. Virology 147:264–274

Yu YX, Aang GM, Guo YP, Ao J, Li HM (1988) Safety of a live-attenuated Japanese encephalitis virus vaccine (SA-14-14-2) for children. Am J Trop Med Hyg 39:214–217

Zaho B, Mackow E, Buckler-White A, Markoff L, Channock RM, Lai C-J, Makino Y (1986) Cloning full-length dengue 4 viral DNA sequences: analysis of genes coding for structural proteins. Virology 155:77–88

Chapter 28 **Polymerase Chain Reaction for Detection of Hantaviruses** *

Ralf Stohwasser[1], Lutz B. Giebel[2], Karl Raab[3], E. K. F. Bautz[1],
and Gholamreza Darai[3, +]

This work is dedicated to the memory
of the great virologist
Professor Dr. Friedrich Deinhardt (1926–1992)

Summary

The diagnosis of hantaviruses as the etiologic agent of hemorrhagic fever with renal syndrome (HFRS) so far has relied on immunofluorescence assays which require cells infected with pathogenic viruses. In this chapter we describe the use of gene amplification by the polymerase chain reaction (PCR) to diagnose hantavirus infections rapidly. In combination with direct nucleotide sequence analyses or differential oligonucleotide hybridization, PCR can also be used to identify different hantavirus strains and even to detect previously unknown hantaviruses.

Introduction

Hantaviruses are a genus of human pathogenic viruses which cause a variety of clinically similar diseases collectively termed hemorrhagic fever with renal syndrome (HFRS). Hantavirus infections were first recognized during the Korean war between 1951 and 1954 when 3000 soldiers of the United Nations suffered from HFRS. Mortality was reported to be 5%–10% (Smadel 1953; Earle 1954). Hantaviruses are transmitted to humans by direct or indirect contact with subclinically infected rodents which serve as their natural reservoir. Today, many strains of hantaviruses have been isolated in virtually every part of the world. Five serologically distinct groups have been established: Hantaan (*Apodemus*) (Lee et al. 1978), Seoul (*Rattus*) (Sugiyama et al. 1984),

---

[1] Zentrum für Molekulare Biologie Heidelberg und Institut für Molekulare Genetik, Universität Heidelberg, Im Neuenheimer Feld 230, W-6900 Heidelberg, FRG

[2] ARRIS Pharmaceutical Corp., 385 Oyster Point Blvd., Suite 4, South San Francisco, CA 94080, USA

[3] Institut für Medizinische Virologie der Universität Heidelberg, Im Neuenheimer Feld 324, W-6900 Heidelberg, FRG

* This study was supported by the Bundesministerium für Forschung und Technologie (BMFT) Grant 0318973A/B

[+] To whom correspondence should be addressed.

Puumala (Hällnäs B1 [1]; CG18–20 [1]; *Clethrionomys*) (Yanagihara et al. 1984), Prospect Hill (*Microtus*) (Lee 1989), and Leaky (*Mus*) (Baek et al. 1989). These serotypes are more closely associated with the rodent genus (given in parenthesis) than with the geographical areas of their isolation. The Asian strains Hantaan and Seoul cause severe clinical symptoms in humans with mortality rates of up to 10%. European Puumala or Hällnäs B1 infections cause much milder symptoms with less than 1% mortality. In contrast, the American Prospect Hill and Leaky serotypes so far have not been associated with any known diseases in humans. Clinically, the infections are characterized by acute, sudden onset of fever with chills, conjunctival injection, prostration, anorexia, vomiting, abdominal pain, hemorrhagic manifestations, followed by proteinuria and hypotension. Renal disorders vary from mild symptoms to acute renal failure and may persist for several weeks.

In China, Hantaan virus infections account for 100000–150000 hospitalizations per year. Recent epidemiologic studies in several areas of Europe demonstrated antibody titers against the Puumula/Hällnäs B1 serotype in up to 5% of the local population (Lee 1989). The properties of genomic organization and coding capacities of hantaviruses are summarized in Table 1. The hantaviral large L RNA segment codes for the viral RNA-dependent RNA polymerase (250 kDa). The medium-size M RNA segments of hantaviruses encode a precursor polypeptide of approximately 126 kDa, which is cleaved into glycoproteins G1 (64 kDa) and G2 (54 kDa) of the viral envelope. Finally, unlike all the other genera of the Bunyaviridae family which utilize overlapping reading frames or ambisense coding strategies to encode a nucleocapsid protein N and nonstructural protein NS_s (Ihara et al. 1984; Elliott and McGregor 1989; Simons et al. 1990), the Hantavirus S RNA segment apparently only codes for the nucleocapsid protein. This nucleocapsid protein is the major antigenic determinant detected by sera of hantavirus-infected patients (Sheshberadaran et al. 1988; Zöller et al. 1989; Gött et al. 1991).

Detection of Hantavirus Infections by Immunological Assays

Detection of hantavirus infections in humans relies exclusively on immunological assays. Many different assays have been developed and successfully applied.

---

[1] Recent nucleotide sequence analyses of a variety of strains have shown that the isolate labeled nephropathia epidemica virus (NEV) Hällnäs B1 (provided by Dr. Pilaski, Düsseldorf, FRG, and sequenced in our laboratory; Giebel et al. 1989; Stohwasser et al. 1990, 1991) seems to be identical with the NEV strain CG18–20 (obtained from Dr. B. Niklasson, Stockholm, Sweden). The nucleotide sequence analysis of the cDNAs of the NEV Hällnäs B1 and CG18–20 strains from different laboratories is essential for final clarification of this discrepancy.

Table 1. Summary of genomic organization of hantaviruses

Virus strain	RNA segment	Size (bases)	Reference
Hantaan	S	1696	Schmaljohn et al. (1986)
	M	3616	Schmaljohn et al. (1987)
	L	6530	Schmaljohn (1990)
Hällnäs B1	S	1785	Stohwasser et al. (1990)
	M	3682	Giebel et al. (1989)
	L	6550	Stohwasser et al. (1991)
Seoul/DX	S	1765	Giebel et al. (1991)
	M	3651	Antic et al. (1991 b)
	L	6530	Antic et al. (1991 a)
Sapporo rat	S	1769	Arikawa et al. (1990)
	M	3651	Arikawa et al. (1990)
Prospect hill	S	1675	Parrington and Kang (1990)
	M	3707	Barrington et al. (1991)

The viral strain "Sapporo Rat" described by Arikawa et al. (1990) differs from strain Leakey. Leakey is a Mus musculus strain previously isolated and is so far not characterized on sequence level. Seoul virus (strain 80-39) was characterized by Antic et al. (1991 a, b) and is a rat strain different from the Sapporo rat strain (SR11). Virus isolate DX of unknown origin was shown to be identical to the strain 80-39 Seoul virus identified by sequence comparisons in our laboratory

Indirect Immunofluorescence. Different hantavirus isolates have been successfully adapted to the Vero E6 tissue culture cell line, and infected cells are used as a source of viral antigen in this procedure. The cells are incubated with patients' sera as a first antibody and subsequently stained with a human-specific IgG preparation conjugated with a fluorescent dye (Lee and Lee 1976; Zöller et al. 1989; Groen et al. 1989, 1991). This method is sensitive and reliable.

Indirect IgG ELISA. Hantavirus-infected Vero E6 cells are used as viral antigen to coat microtiter plates. Plates are then incubated with patients' sera as first antibody and subsequently with human-specific IgG conjugated to peroxidase or alkaline phosphatase. After addition of enzyme substrate the optical density is determined in a microtiter plate reader (Groen et al. 1989).

Indirect IgM (μ-Capture) ELISA. The μ-capture enzyme-linked immunosorbent assay (ELISA) detects IgM molecules during the acute phase of hantavirus infections. Microtiter plates are coated with human IgM-specific antibodies and incubated with patients' sera. Hantavirus-specific IgM molecules trapped to the plates are detected by the addition of hantavirus antigen (produced by recombinant DNA techniques; Gött et al. 1991) and incubation with a rabbit anti-hantavirus antibody. Wells are then stained by incubation with peroxidase-conjugated, rabbit-specific antiserum and enzyme substrate.

Immunoblot. Viral antigen is subjected to polyacrylamide gel electrophoresis, transferred to a membrane, and incubated with patients' sera. Hantavirus-specific antibodies are detected by incubation with a second human IgG-specific antibody conjugated to peroxidase or alkaline phosphatase and subsequent staining with enzyme substrate (Zöller et al. 1989).

Most of these immunological assays have several disadvantages. First, sensitivity is relatively low. This is a serious problem with the low antibody titer against European and American strains leading to high assay background and false-positive results. Second, all assays are indirect and rely on detection of antibodies in patients' sera, which could stem from previous viral infections since antibodies can persist for more than 30 years (Lee 1982). Therefore, the only immunological assay differentiating between a previous and a new infection is the μ-capture ELISA. In addition, these assays may allow us to differentiate between viruses of Asian and European origin; they do not, however, differentiate between strains of high sequence homology. Because of these disadvantages inherent to immunology-based assays, recent advances in DNA recombinant technology and in the application of the polymerase chain reaction (PCR) techniques to amplify genome segments of hantaviruses have provided interesting alternatives to identify hantaviruses.

Experimental Procedures

Extraction of Total RNA. Adjust sample of Vero E6 cells infected with hantaviruses with 4 M guanidinium isothiocyanate, 0.5% sodium lauryl sarcosinate, 5 mM sodium citrate, pH 7, 0.1% antifoam A (Sigma), and 0.1 M 2-mercaptoethanol. Bring mixture to 60 °C, add an equal volume of phenol preheated to 60 °C, and vortex. Add 0.5 volume of 0.1 M sodium acetate, pH 5.2, 10 mM TRIS-HCl, pH 7.4, 1 mM ethylene diamine tetra-acetic acid (EDTA), and an equal volume of chloroform and vortex. Cool on ice and centrifuge at 2000 g for 10 min at 4 °C. Recover aqueous phase and reextract with phenol/chloroform. Subsequently extract twice with chloroform. Add 2 volumes of ice-cold ethanol and precipitate at -20 °C for 1 h. Recover RNA by centrifugation at 12000 g for 20 min at 4 °C. Alternatively, layer the guanidinium isothiocyanate homogenate onto a 1.2-ml cushion of 5.7 M CsCl in 0.1 M EDTA, pH 7.5, in a Beckman SW50.1 polyallomer tube and centrifuge at 35000 rpm for 12 h at 20 °C. Discard supernatant, dry walls of centrifuge tubes thoroughly, and resuspend RNA pellet in 200 µl 0.3 M sodium acetate, ph 5.2. Precipitate twice with 2 volumes of ice-cold ethanol. RNA of infected tissue (like lung) and/or from pelletized urin can be extracted under the same conditions.

cDNA Synthesis. First strand cDNA synthesis was performed according to the procedure of Krug and Berger (1987). An aliquot of the purified RNA is denatured in 12 µl diethyl pyrocarbonate (DEPC)-treated H_2O at 60 °C for

10 min in the presence of 150 ng of the 14-mer oligonucleotide primer complementary to the 3′-termini of all three RNA segments [5′-TAGTAGTAGACN (C/T) C] of different hantavirus strains (Schmaljohn et al. 1985) and put on ice afterwards. Add 2 µl of 10 × buffer (10 × buffer is 500 mM TRIS-HCl, pH 8.3, 500 mM KCl, 100 mM MgCl$_2$, 10 mM dithiothreitol (DTT), 10 mM EDTA, 100 µg/ml bovine serum albumin, BSA), 1 µl 10 mM spermidine, 1 µl 80 mM pyrophosphate, 2 µl 10 mM of each dNTP, 40–80 units (1 µl) ribonuclease inhibitor, and 10–20 units (1 µl) of avian myeloblastosis virus (AMV) reverse transcriptase. Incubate at 42 °C for 40 min.

Polymerase Chain Reaction. The amplification of the single-stranded cDNA is performed by PCR (Saiki et al. 1988). Serogroup-specific primers (Table 2) and universal primers (Table 3) are synthesized. Nucleotide sequences of hantaviral primers are based on the most conserved hantaviral RNA sequences detected by cDNA sequence alignments. An aliquot of single-stranded cDNA (1–2 µl) is used in PCR. Single-stranded cDNA is incubated at 95 °C for 7 min to denature RNA-DNA hybrids. Denatured cDNA is cooled on ice. Hantavirus-specific cDNA segments are amplified in 50–100 µl volumes of 50 mM KCl, 10 mM TRIS-HCl, pH 8.3, 1.5 mM MgCl$_2$, 200 µM of each dNTP, 100 µg/ml gelatin, 100 pmoles of each primer, and 2.5 units Taq DNA polymerase (Perkin-Elmer Cetus) for 35 cycles in an automated temperature cycler (ERICOMP Inc., USA) which provides per cycle 2 min each of incubation at 94°, 46°–50°, and 72 °C. Aliquots of amplified cDNAs are size-fractionated on 1.0%–1.5% agarose gels in order to analyze product length and restriction pattern for hybridizing Southern blots using radioactively labeled oligonucleotides (Table 4).

Nucleotide Sequence Analysis of PCR Products. Amplification products are analyzed by direct nucleotide sequencing without subcloning. PCR reaction mixtures are extracted once with phenol/chloroform and ethanol precipitated. DNA is resuspended in 300 µl TE buffer (10 mM TRIS pH 8.0, 1 mM EDTA), and oligonucleotide primers and dNTPs are removed by spin column chromatography (Millipore Ultrafree-MC cellulose UF membrane) according to manufacturer's protocol. After 3 washes with TE buffer, the column is spun dry, and DNA is resuspended in 20 µl of TE. An aliquot is then sequenced using the Sequenase 2.0 dideoxy chain termination sequencing kit (United States Biochemicals, Cleveland, Ohio): The 5 µl of DNA is denatured in a 10 µl volume of 1 × reaction buffer and 10 pmoles of the appropriate oligonucleotide primer (Table 3) for 3 min at 96 °C and put on ice. Labeling and extension reactions are performed according to manufacturer's protocol using [α-^{35}S]dATP (specific activity > 1000 Ci/mol). Sequencing reactions are then separated on 6% denaturing polyacrylamide buffer gradient gels and autoradiographed.

Differential Oligonucleotide Hybridization. As an alternative to nucleotide sequence analysis, the genomes of hantaviruses can be identified by differential

Table 2. Physical properties of serogroup-specific primers for polymerase chain reaction (PCR) amplification of hantaviral genomic RNA

Name	Primer sequence	NEV seq-position	PCR product
L-NE1	TAGTAGTAGACTCCG	1–15	470 bp
L-NE2	GGGTTCTATCTCTGTAGTAATTGC	466–444	
L-NE3	GTCAAAGGATGGGGAGAGTGTG	6322–6339	220 bp
L-NE4	TAGTAGTAGACTCCGAGATA	6550–6531	
M-NE5	GCCAGAAACCTTAATGAGC	110–128	810 bp
M-NE6	TTAATCATGATCTCACCATGTGG	923–900	
M-NE7	AATGGGTTCAATGGTTTGTG	1687–1706	560 bp
M-NE8	CAGTGAATGCAGTTTTAAG	2256–2274	
S-NE9	GTAATGGGAGTGATTGGTTTT	697–718	550 bp
S-NE10	CCTTAGCTCAGGATCCATGTCATC	1252–1275	
L-NE1	TAGTAGTAGACTCCG	1–15	660 bp
L-SE2	TAATGCAGCTCTTTC	657–643	
L-SE3	GGGATGGGGGTGAGTCTG	6309–6326	220 bp
L-NE4	TAGTAGTAGACTCCGAGATA	6550–6531	

NE, *Nephropathia epidemica* virus (Giebel et al. 1989; Stohwasser et al. 1990, 1991); SE, Seoul virus (Antic et al. 1991 b). All primer positions according to the nucleotide sequence of the S, M, and L cDNAs of NEV

Table 3. Physical properties of universal oligonucleotide primers for polymerase chain reaction amplification of hantaviral genomic RNA segments

Primer pairs		Sequence amplified
P1:	5'-TAGTAGTAGACT-3'	*S segment*
P2:	5'-AGCTCNGGATCCAT(A/G)TCATC-3'	about 1270 bp
P3:	5'-GG(T/G)GATGA(C/T)ATGGATCC-3'	*S segment*
P4:	5'-TAGTAGTAN(A/G)CTCC-3'	about 520 bp
P5:	5'-TAGTAGTAGAC(T/A)CCGCAA-3'	*M segment*
P6:	5'-ATCATG(G/A)TCTTC(N)CCATG(C/T)GGTG-3	about 920 bp
P7:	5'-AAATCAGG(T/G)GAATGG-3'	*M segment*
P8:	5'-TAGTAGTAGACNCCGC-3'	about 340 bp
P9:	5'-TAGTAGTAGACTCC-3'	*L segment*
P10:	5'-ACT(A/G)GGCCATTGTG-3'	about 560 bp
P11:	5'-CT(T/G)GCTCAAAATAATAA-3'	*L segment*
P12:	5'-GCATC(T/A)GCACT(A/G)ACATACAT-3'	about 890 bp
P13:	5'-CAATATGATGCATATTGTGT-3'	*L segment*
P14:	5'-TC(A/T)CCCCATCC-3'	about 1180 bp

Primers P1, P5, and P9 are derived from the 5'-ends, and primers P4 and P8 are complementary to the 3'-ends of hantaviral S, M, and L RNA messenger-sense sequences, respectively. Sequence positions of cDNAs generated from S, M, and L RNA segments of nephropathia epidemica virus complementary to primer sequences: P1, 1–12; P2, 1271–1251; P3, 1251–1267; P4, 1785–1772; P5, 1–18; P6, 919–897; P7, 3544–3358; P8, 3682–3665; P9, 1–14; P10, 558–545; P11, 2047–2063; P12, 2954–2935; P13, 5161–5180; P14, 6338–6328
N indicates nucleotide position containing each of the four dNTPs

Table 4. Physical properties of oligonucleotides used for differential hybridization

Name	Oligonucleotide	RNA	Strain	Reference
HY-1	5′-GGCAGAGTCTCCTATTGCT	S	DX[a]	Giebel et al. 1991
HY-2	5′-GGCGGAGTCTCTTATTGCC	S	SR	Arikawa et al. 1990
HY-3	5′-TCCAGATACAGCAGCAGTT	S	HT	Schmaljohn et al. 1986
HY-4	5′-AAAGCCAGAAGTTAAACCT	S	B1	Stohwasser et al. 1990
HY-5	5′-AAATGAGCCACGTCCTGGA	S	PH	Parrington et Kang 1990
HY-6	5′-AAAGCCTGAAGTAAAACCA	S	PU	Giebel et al. 1991
HY-7	5′-TTTGATCCCTACTTTAGTG	M	B1	Giebel et al. 1989
HY-8	5′-TGTTCTAAAACACAAAATG	M	HT	Schmaljohn et al. 1987
HY-9	5′-TATGTTAAAGCATAGAATG	M	SE	Antic et al. 1991 a
HY-10	5′-TCAAAGAAAGGGGTTTTTG	L	B1	Stohwasser et al. 1991
HY-11	5′-AAAAGTGGTGAAAAGCATG	L	HT	Schmaljohn et al. 1990
HY-12	5′-TGAACCTCGAGAAAATCTG	L	SE	Antic et al. 1991 b

Oligonucleotide sequences are derived from genomic regions that differ the most between various hantavirus strains (B1, nephropathia epidemica virus Hällnäs B1; HT, Hantaan; SE, Seoul; SR, Sapporo rat; PU, Puumala; PH, Prospect Hill). The S segment oligonucleotides correspond to strain DX position 784–802 in the sequence alignment published by Giebel et al. (1991). The M segment oligonucleotides correspond to Hällnäs B1 position 391–409 in the sequence alignment published by Giebel et al. (1989). The L segment oligonucleotides correspond to Hällnäs B1 position 1210–1228 (Stohwasser et al. 1991). All oligonucleotides are messenger-sense

[a] The comparison of the nucleotide sequences of the M and L cDNAs of hantavirus DX strain and Seoul virus strain 80–39 revealed that the compared parts of sequences of both hantavirus strains were completely identical. Therefore, it is obvious that the unknown DX strain is identical with Seoul strain 80–39

oligonucleotide hybridization using strain-specific oligonucleotide probes (Kogan and Gitschier 1990; Wood et al. 1985). PCR amplification products (10 µl) are heat-denatured in 200 µl volumes of 0.4 M NaOH and 25 mM EDTA and transferred to nylon membranes by a sample filtration manifold device. Slots are rinsed with 400 µl 20 × SSPE (standard saline phosphate-EDTA buffer). Membranes are prehybridized for 2 h at 37 °C in 6 × SSC (standard saline citrate buffer) 50 mM sodium phosphate, pH 6.8, 5 × Denhardt's, and 0.1 mg/mol salmon sperm DNA. The radioactive endlabeled oligonucleotide is added at 1–3 ng/ml and hybridized for 6 h at 37 °C in prehybridization buffer containing 100 ng/ml dextran sulfate. After hybridization membranes are rinsed three times at 4 °C with precooled 6 × SSC for 10 min. The filters are then rinsed with TMACl solution to exchange salts and incubated twice for 20 min at 57 °C to wash off all oligonucleotides that are not perfectly matched to their target sequence. Membranes are then autoradiographed. TMACl solution is 3 M tetramethyl-ammoniumchloride (TMACl, Aldrich, Milwaukee, USA), 50 mM TRIS-HCl, pH 8, 2 mM EDTA, 0.1% SDS. To prepare a roughly 5 M stock of TMACl, add 400 ml H_2O to 500 g TMACl and dissolve at 68 °C. Determine exact molarity with refractometer by the formula: molarity = (reading of solution − reading of H_2O)/0.018. The oligonucleotides are hantavirus strain-specific 19-mers, designed such that the number of mismatches between different strains is as large as possible, and mismatches are located towards the center of the sequence (Table 4). The 19-mers will only hybridize to the PCR amplification products if there are perfect matches. A single nucleotide difference will be detected. Thus, even closely related hantavirus strains can be distinguished.

Results

Strategy for Amplification of Hantaviral cDNAs

Two methods of PCR amplification of S, M, and L cDNAs were used in this study. One approach is based on the complementarity of hantaviral RNA segment termini (Fig. 1 A). On oligonucleotide complementary to both the viral genomic RNA and the hantaviral mRNA of S, M, and L RNA segments was synthesized to prime cDNA synthesis. For PCR amplification, a terminal primer (tp) and a second internal primer (ip) were chosen. Using different internal primers but only one terminal primer, both termini of each RNA segment can be amplified (Fig. 1 B). Alternatively, randomly primed cDNA was used for PCR amplification based on two internal primers. The latter approach circumvents termination of cDNA synthesis caused by the terminal secondary

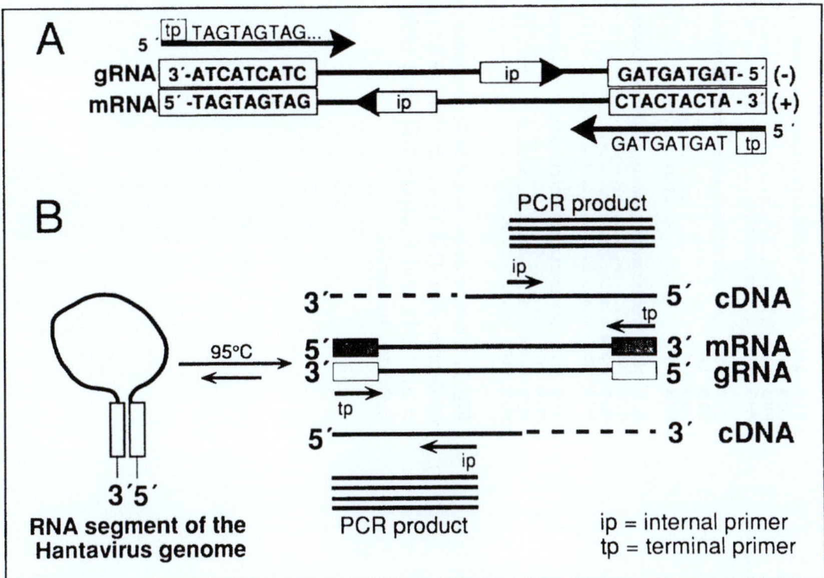

Fig. 1. Schematic diagram of genomic arrangement of Hantavirus RNA segments (*A*) and the strategy for cDNA synthesis and polymerase chain reaction (PCR) amplification (*B*)

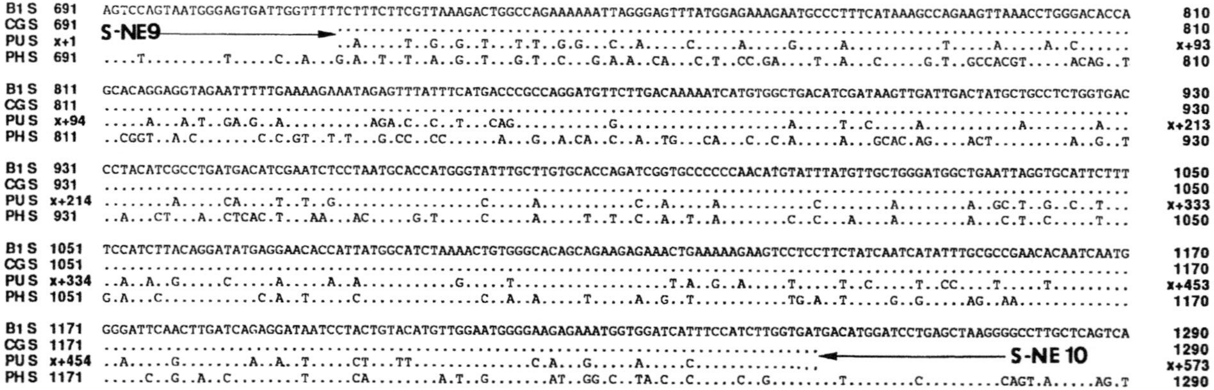

```
B1 S  691   AGTCCAGTAATGGGAGTGATTGGTTTTTCTTTCTTCGTTAAAGACTGGCCAGAAAAAATTAGGGAGTTTATGGAGAAAGAATGCCCTTTCATAAAGCCAGAAGTTAAACCTGGGACACCA   810
CG S  691   S-NE9 ------------------------>  .............................................................................................................   810
PU S  x+1                             ..A.....T..G..G..T...T.T..G.G...C..A.....C.....A.....G.....A...........T.....A.....A..C......                              x+93
PH S  691   ....T...........T.....C..A...G.A..T..T..A..G..T...G.T..C...G.A.A..CA...C.T..CC.GA....T..A...C.....G.T..GCCACGT.....ACAG..T   810

B1 S  811   GCACAGGAGGTAGAATTTTTGAAAAGAAATAGAGTTTATTTCATGACCCGCCAGGATGTTCTTGACAAAAATCATGTGGCTGACATCGATAAGTTGATTGACTATGCTGCCTCTGGTGAC   930
CG S  811   .............................................................................................................................   930
PU S  x+94  .....A...A.T..GA.G..A..........AGA.C..C..T...CAG...........G...........A.....T..C....A.........A........A...                 x+213
PH S  811   ..CGGT..A.C.......C.C.GT..T.T...G.CC..CC......A...G..A.CA..C..A..TG...CA...C..C.A.....A...GCAC.AG....ACT.........A..G..T      930

B1 S  931   CCTACATCGCCTGATGACATCGAATCTCCTAATGCACCATGGGTATTTGCTTGTGCACCAGATCGGTGCCCCCCAACATGTATTTATGTTGCTGGGATGGCTGAATTAGGTGCATTCTTT   1050
CG S  931   .............................................................................................................................   1050
PU S  x+214  ........A.....CA....T..T..G.................C.....A..........C..A....A...........C.........A...........A..GC.T..G..C..T...   x+333
PH S  931   ..A...CT...A..CTCAC.T...AA...AC.....G.T.....C.....A.....T..T..C..A..T..A.........C..C...A...A...........A...C.T..C.....T...   1050

B1 S  1051  TCCATCTTACAGGATATGAGGAACACCATTATGGCATCTAAAACTGTGGGCACAGCAGAAGAGAAACTGAAAAAGAAGTCCTCCTTCTATCAATCATATTTGCGCCGAACACAATCAATG   1170
CG S  1051  .............................................................................................................................   1170
PU S  x+334  ..A..A..G.....C.....A.....A..A.........G.....T.................T.A..G..A.....T.....T..C....T..CC....T.....T.........          x+453
PH S  1051  G.A...C..........C.A..T.....C..........C..A..A.....T.....A..G..T.........TG.A..T.....G..G.....AG..AA............              1170

B1 S  1171  GGGATTCAACTTGATCAGAGGATAATCCTACTGTACATGTTGGAATGGGGAAGAGAAATGGTGGATCATTTCCATCTTGGTGATGACATGGATCCTGAGCTAAGGGGCCTTGCTCAGTCA   1290
CG S  1171  .............................................................................................................................   1290
PU S  x+454  ..A.....G.......A..A..T....CT...TT...........C.A..G.....A.....C.............,  <----------------- S-NE 10                    x+573
PH S  1171  .....C..G..A..C........T.....CA.......A.T..G.......AT..GG.C..TA.C..C.....C..G.........T........C.........CAGT.A.....AG.T      1290
```

Fig. 2. Comparative nucleotide sequence analysis of cDNA sequences generated from S RNA segments of nephropathia epidemica virus (NEV) strain Hällnäs B1 (*B1*; Stohwasser et al. 1990), NEV strain CG18–20 (*CG*), Puumala virus (*Pu*; Giebel et al. 1991), and Prospect Hill virus (*PH*; Parrington and Kang 1991). The position of oligonucleotide primers used are indicated by *arrows*

structure of RNA segments of all hantaviruses (Giebel et al. 1989; Stohwasser et al. 1990). Those RNA stem-loop structures compete with oligonucleotides during annealing of the terminal primer. The first approach was used in this study since the 3'-termini of RNA segments belong to the most conserved sequence regions in the hantaviral genome. Additionally, cDNA generated with one primer can be used for amplification of subfragments of all three RNA segments.

Genomic Variation of Hantaviruses

First experiments in our laboratory revealed that PCR primers shown in Table 2 and derived from sequences of the NEV amplify S, M, and L cDNAs of closely related viruses. The corresponding sequences of the prototype member of the serogroup Puumala virus and a second serogroup (Prospect Hill virus; sequenced by Parrington and Kang 1990) revealed 63% identity at the nucleotide level (Fig. 2). Additionally, it was found that oligonucleotides would amplify closely related hantaviral sequences even if they contained several mismatches.

As shown in Fig. 3 for the amplification of L RNA segment-specific cDNAs, the recognition of target sequences of Puumala virus by primers L-NE1 (4)/2 (3) is limited. Whereas the target sequence of hantavirus strain CG18–20 was recognized, resulting in PCR products of 466 and 220 base pairs in length, L RNA segment termini of Puumala virus were amplified only partially (lane 6, Fig. 3), and L RNA segment termini of DX virus (Giebel et al. 1991) were not amplified at all (lanes 9–12, Fig. 3). Therefore, the inter-

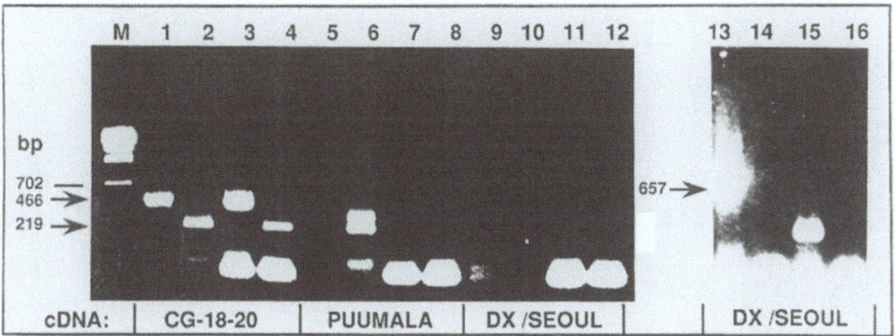

Fig. 3. Characterization and identification of polymerase chain reaction (PCR) products by gel electrophoresis. The DNA fragments were separated on agarose slab gels (1.2%), stained with ethidium bromide, and photographed under UV light. Physical properties of individual oligonucleotide primers are described in Table 2. The following primers were used: L-NE1/L-NE2 (*lanes 1, 5, and 9*), L-NE1/L-NE3 (*lanes 2, 6, and 10*), L-NE4/L-NE2 (*lanes 3, 7, and 11*), L-NE4/L-NE3 (*lanes 4, 8, and 12*), L-NE1/L-DX2 (*lane 13*), L-NE4/L-DX2 (*lane 14*), L-NE1/L-DX3 (*lane 15*), and L-NE4/L-DX3 (*lane 16*). The positions of specifically amplified DNA fragments are indicated by *arrows*

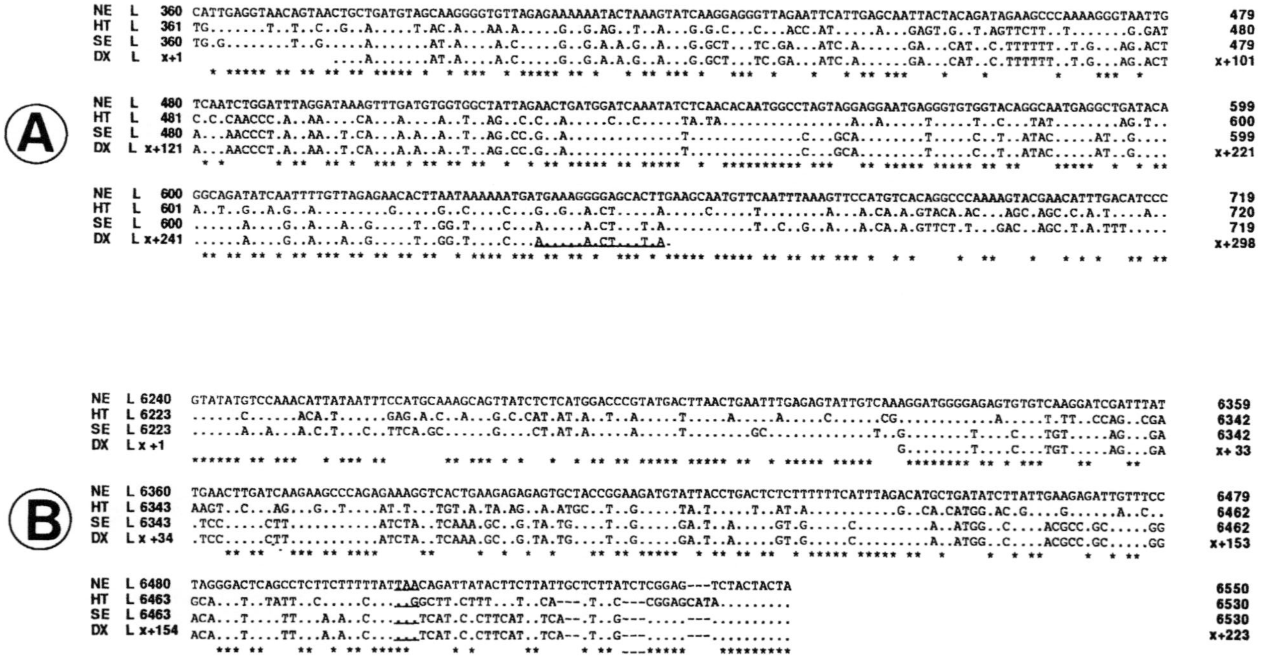

A

```
NE  L  360   CATTGAGGTAACAGTAACTGCTGATGTAGCAAGGGGTGTTAGAGAAAAAATACTAAAGTATCAAGGAGGGTTAGAATTCATTGAGCAATTACTACAGATAGAAGCCCAAAAGGGTAATTG   479
HT  L  361   TG.......T..T..C..G..A.....T.AC.A...AA.A.....G..G.AG..T..A...G.G.C...C...ACC.AT....A...GAGT.G..T.AGTTCTT..T.......G.GAT    480
SE  L  360   TG.G.......T..G.....A.......AT.A....A.C.....G..G.A.A.G..A...G.GCT...TC.GA...ATC.A.....GA...CAT..C.TTTTTT..T.G...AG.ACT    479
DX  L  x+1   ....A.......AT.A....A.C.....G..G.A.A.G..A...G.GCT...TC.GA...ATC.A.....GA...CAT..C.TTTTTT..T.G...AG.ACT                  x+101
             * ***** ** ** ** ** ***** * * *** * ***** ** *  * ** *** *   *** *   *  * ** ***   *   *  *      *    *    * *** *

NE  L  480   TCAATCTGGATTTAGGATAAAGTTTGATGTGGTGGCTATTAGAACTGATGGATCAAATATCTCAACACAATGGCCTAGTAGGAGGAATGAGGGTGTGGTACAGGCAATGAGGCTGATACA   599
HT  L  481   C.C.CAACCC.A..AA....CA...A....A..T..AG..C.C.A.....C..C....TA.TA.................A..A.....T.....T..C...TAT......AG.T..       600
SE  L  480   A...AACCCT.A..AA...T.CA...A.A..A..T..AG.CC.G..A.................T................C...GCA.........T.....C..T..ATAC.....AT..G....   599
DX  L  x+121 A...AACCCT.A..AA...T.CA...A.A..A..T..AG.CC.G..A.................T................C...GCA.........T.....C..T..ATAC.....AT..G....   x+221
             * *  * ** ** *** ** ** ** *** ****** ** ***** *  ********** *** ** ***** ***** ** **  ***** * * **

NE  L  600   GGCAGATATCAATTTTGTTAGAGAACACTTAATAAAAAATGATGAAAGGGGAGCACTTGAAGCAATGTTCAATTTAAAGTTCCATGTCACAGGCCCAAAAGTACGAACATTTGACATCCC   719
HT  L  601   A..T..G..A.G..A.....G.....G..C....C...G..G..A.CT.....A.....C.....T........A...A.CA.A.GTACA.AC...AGC.AGC.C.A.T....A..        720
SE  L  600   ......A....G.A....A.G.....T..GG.T....C....A.CT...T.A...........T..C..G..A...A.CA.A.GTTCT.T...GAC..AGC.T.A.TTT.....           719
DX  L  x+241 ......A....G.A....A.G.....T..GG.T....C...A.....A.CT...T.A                                                                x+298
             ** ** ** * ** *** ** ** ** ** *  **** *** **  * *** * ***** ***** ** ** ** *** * * *       * **     *  * * *    ** **
```

B

```
NE  L 6240   GTATATGTCCAAACATTATAATTTCCATGCAAAGCAGTTATCTCTCATGGACCCGTATGACTTAACTGAATTTGAGAGTATTGTCAAAGGATGGGGAGAGTGTGTCAAGGATCGATTTAT   6359
HT  L 6223   ......C......ACA.T......GAG.A.C..A...G.C.CAT.AT.A..T..A.....T.....A.....A.....C......CG...........A.....T.TT..CCAG..CGA   6342
SE  L 6223   ......A..A...A.C.T...C..TTCA.GC......G....CT.AT.A.....A....T........GC............T..G.........T....C...TGT.....AG...GA   6342
DX  L x +1                                                                    G.........T....C...TGT.....AG...GA   x+ 33
             ****** ** ***  * *** **       ** *** * *  * * ** ** ***** ***** **  * ***** *****   ******** ** * ***   **   **

NE  L 6360   TGAACTTGATCAAGAAGCCCAGAGAAAGGTCACTGAAGAGAGAGTGCTACCGGAAGATGTATTACCTGACTCTCTTTTTTCATTTAGACATGCTGATATCTTATTGAAGAGATTGTTTCC   6479
HT  L 6343   AAGT..C...AG...G..T....AT.T...TGT.A.TA.AG..A.ATGC..T..G.....TA.T.....T..AT.A...........G..CA.CATGG.AC.G....G......A..C..   6462
SE  L 6343   .TCC.....CTT..........ATCTA..TCAAA.GC..G.TA.TG....T..G.....GA.T..A.....GT.G.....C.........A..ATGG..C....ACGCC.GC.....GG   6462
DX  L x +34  .TCC.....CTT..........ATCTA..TCAAA.GC..G.TA.TG....T..G.....GA.T..A.....GT.G.....C.........A..ATGG..C....ACGCC.GC.....GG   x+153
             ** ** ` *** ** ****    **     * * * *   ** ** *****  * ** ** ** * ***** ***** **  *     * * **     *  * **

NE  L 6480   TAGGGACTCAGCCTCTTCTTTTTATTAACAGATTATACTTCTTATTGCTCTTATCTCGGAG---TCTACTACTA   6550
HT  L 6463   GCA...T..TATT..C.....C.....GGCTT.CTTT...T..CA---.T..C---CGGAGCATA.........   6530
SE  L 6463   ACA...T....TT...A.A..C.....TCAT.C.CTTCAT..TCA--.T..G---.....---.........     6530
DX  L x+154  ACA...T....TT...A.A..C.....TCAT.C.CTTCAT..TCA--.T..G---.....---.........     x+223
             *** **    ** * ** *****    * *      **     * ** ---***** ********
```

Fig. 4 A, B. Comparative analysis of cDNA nucleotide sequences from 5′- (**A**) and 3′- (**B**) parts of the L RNA segments of nephropathia epidemica virus (*NE*), Hantaan virus (*HT*), Seoul virus (*SE*), and DX hantavirus (*DX*; Giebel et al. 1991). The cDNA sequence of DX strain (identical to Seoul strain) was obtained by polymerase chain reaction amplification (products shown in Fig. 3, lanes 13 and 15)

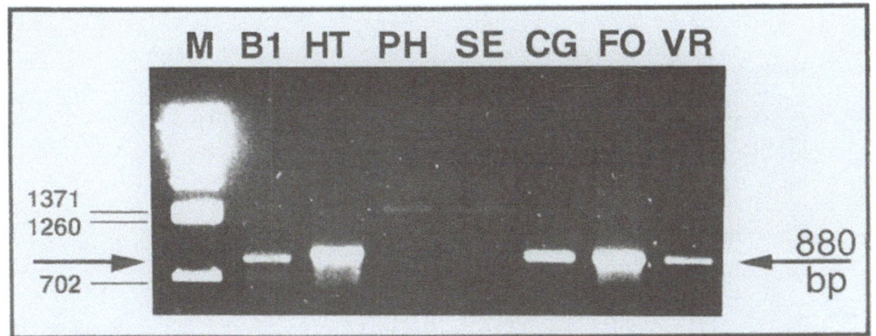

Fig. 5. Characterization and identification of polymerase chain reaction (PCR) products by gel electrophoresis. The DNA fragments were separated on agarose slab gels (1.2%), stained with ethidium bromide, and photographed under UV light. cDNA synthesis was performed by specific oligonucleotide priming (5′-TAGTAGTAGACN(C/T)C-3′) using total RNA of Vero E6 cells infected with nephropathia epidemica virus Hällnäs B1 (*B1*; 6 µg), Hantaan virus (*HT*; 2 µg), Prospect Hill virus (*PH*; 25 µg), Seoul virus (*SE*; 36 µg), nephropathia epidemica virus CG18–20 (*CG*; 4 µg), Yugoslavian hantavirus strains Foijnica (*FO*; 15 µg) and Vranica (*VR*; 22 µg). The amplification of hantaviral cDNAs was performed using universal oligonucleotides P11/P12 (annealing temperature 50 °C). The amplified cDNAs of hantaviral L RNA segment according to B1 sequence position 2024–2954 (Stohwasser et al. 1991) are indicated by the arrows

nal primers L-NE2 and L-NE3 were replaced by oligonucleotides L-SE2 and L-SE3, whose sequences were derived from DX L cDNA clones (Stohwasser, data not shown). Using such internal DX L segment-specific primers and terminal L-NE1 15-mer primer, the 3′- and 5′-termini of the DX L RNA segment were amplified (lanes 13 and 15, Fig. 3).

Sequence analysis of parts of those DX-specific amplification products (Fig. 4A, B) revealed a perfect identity of the nucleotide sequence from genome parts of the DX and the Seoul 80-32 virus (Antic et al. 1991 b). Furthermore, the alignment of the 5′- and 3′-terminal regions of L RNA segments from NEV (Stohwasser et al. 1991), Hantaan virus (Schmaljohn 1990), and Seoul virus (Antic et al. 1991 b) revealed 53% nucleotide sequence identity for the 5′-sequences (Fig. 4A) and 56% for the 3′-sequences (Fig. 4B).

Application of Universal Hantaviral Primers. To test the potency of universal primers to amplify cDNAs of a variety of hantaviral isolates single-stranded cDNA was synthesized from total cellular RNA of Vero E6 cells infected by different hantavirus strains. The oligonucleotide used for starting cDNA synthesis was degenerated in positions where aberrations from the hantaviral consensus 3′-terminus were reported (TAGTAGTAGACN(C/T)C). cDNAs

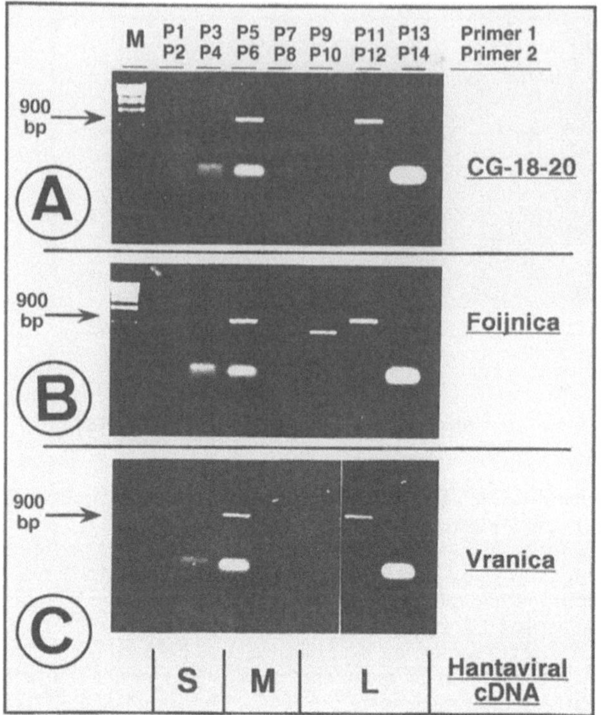

Fig. 6 A–C. Characterization and identification of polymerase chain reaction (PCR) products by gel electrophoresis. The DNA fragments were separated on agarose slab gels (1.2%), stained with ethidium bromide, and photographed under UV light. cDNAs derived from S, M, and L RNA segments of hantavirus strains CG18–20 (**A**), Foijnica (**B**), and Vranica (**C**) were amplified using oligonucleotide primers P1–P14 (annealing temperature 46°C). Physical properties are described in Table 3. PCR products approximately 900 bp in length derived from both M cDNAs (using P5/P6) and L cDNAs (using P11/P12) are indicated by *arrows*. Additionally, a PCR product of 560 bp, amplified by use of primers P9/P10, was observed in **B**

derived from Vero E6 RNA infected with seven different hantaviruses were subject to amplification with primer set P11/P12 representing both internal primers (Fig. 5; Table 3). The observed PCR products of about 900 bp in length represent positions 2047–2954 of the hantaviral L RNA segment (Stohwasser et al. 1991). This indicates that the specific cDNA synthesis of the L RNA segments of Hällnäs B1, Hantaan, CG18–20, Foijnica and Vranica (Foijnica and Vranica were isolated originally by A. Gligic in Yugoslavia from *Apodemus flavicollis* and *Clethrionomys glareolus*, respectively) hantavirus strains was successful in the generation of single-stranded PCR templates about 3 kb in length. Therefore, these templates were used for further screening of universal primer sets. As shown in Fig. 6 all primers of Table 3 were used in PCR to amplify cDNAs of CG18–20, Foijnica, and Vranica. Specific PCR amplification was observed using longer degenerated primers (P5/P6, P11/P12;

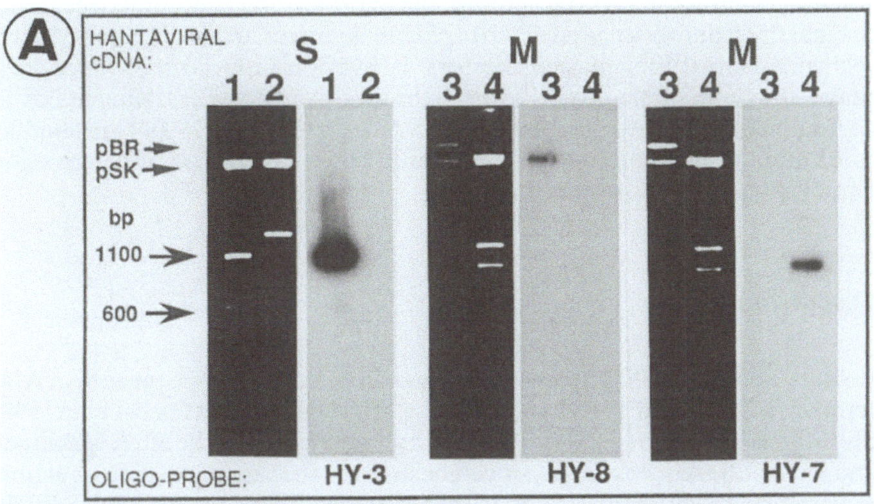

lane	plasmid	cloning site	insert	restriction site		restriction fragments
1	pSK-HT-S	Pst I	1–1696	Pst I	1100	1100 + 600
2	pSK-NE-S54	EcoRI	29–1553	BamHI	1260	1260 + 293
3	pBR-HT-M	PstI	1–3682	PstI	---	3682
4	pSK-NE-M111	EcoR I	50–2200	Hind III	1239	890 + 1310

Fig. 7 A, B. Differential oligonucleotide hybridization for the identification of hantaviral cDNA sequences. Plasmid DNA of hantaviral cDNA clones (physical properties shown in **B**) was digested using restriction enzymes which cut at assigned cloning sites and at internal restriction sites. DNA fragments were separated by gel electrophoresis on agarose slab gels (1.2%) and transferred to nitrocellulose filters. Hybridization was performed at 37°C using 5′-endlabeled 19-mer oligonucleotides HY-3, HY-8, or HY-7 (specific activity 3.5×10^6 cpm/µg; see experimental procedures and hybridization oligonucleotides summarized in Table 4). *pBR*, DNA of plasmid vector pBR322; *pSK*, DNA of plasmid vector pSK-Bluescript (Stratagene)

Table 3), whereas shorter oligonucleotides (P1/P2; P7/P8; P13/P14; Table 3) were not successful when tested under the same conditions.

Identification of PCR Products by Oligonucleotide Hybridization

Identification of PCR products is important for the efficient application of PCR technology. In order to verify the specificity of PCR products, we used 19-mer oligonucleotides as radioactively labeled hybridization probes. The oligonucleotides were derived from the most variable regions of hantaviral

RNA segments (Table 4). To optimize the hybridization conditions (see Experimental Procedures), we used cDNAs of Hantaan virus and NEV S and M RNA segments cloned into plasmid vectors as hybridization targets. Oligonucleotides shown in Table 4 specifically hybridize to restriction fragments of the corresponding cDNAs of hantaviral RNA segments (Fig. 7). Background hybridization of oligonucleotides to vector DNA or to cDNAs of heterologous hantaviruses was not observed.

Discussion

In our laboratory PCR technology was used to clone the genomic RNA termini of S, M, and L RNAs of NEV strain Hällnäs B1 (Giebel et al. 1989; Stohwasser et al. 1990, 1991). Recently, we reported that the PCR techniques and nucleotide sequence analysis can be used for the rapid detection of genomic variation in different strains of hantaviruses (Giebel et al. 1990). In this report, we describe the general use of the PCR for identification of hantaviral genomic RNA segments. Universal oligonucleotide primers were constructed derived from the most conserved sequences between known hantaviral genome RNA segments. This approach guarantees amplification of new strains even if there are several mismatches between primer and target sequence (Giebel et al. 1990). Universal hantavirus primers were successfully tested in PCR experiments. The amplification of hantaviral RNA segments using PCR is suitable for the identification and detection of unknown hantaviruses. Application of this method in routine diagnosis has several significant advantages compared with the established immunological detection systems. PCR amplification is able to detect directly and reliably the presence of hantaviral particles in serum, urine or tissue samples (Xiao et al. 1991) with an unprecedented sensitivity. The direct sequence analysis of amplification products can exactly determine the source of a strain. Direct nucleotide sequencing of PCR products is convenient if the number of samples to be analysed is relatively small. Otherwise, differential oligonucleotide hybridization can be used to analyse very large numbers of PCR samples. Nonradioactive alternatives may be more suitable for routine diagnosis. The results reported in this chapter demonstrate that the PCR technique is a powerful tool to detect rapidly genomic variations in hantaviruses. Accordingly, the development of specific diagnostic and prevention systems based on PCR technology is of particular interest. The probability of double infections of hantaviruses and generation of RNA segment reassortment had been pointed out (Bishop 1985). The application of PCR technology is unambiguously the method of choice for the precise characterization of reassorted hantaviruses.

References

Antic D, Lim BU, Kang CY (1991 a) Molecular characterization of the M genomic segment of the Seoul 80-39 virus; nucleotide and amino acid sequence comparisons with other hantaviruses reveal the evolutionary pathway. Virus Res 19:51–58

Antic D, Lim B-U, Kang CY (1991 b) Nucleotide sequence and coding capacity of the large (L) genomic RNA segment of Seoul 80-39 virus, a member of the hantavirus genus. Virus Res 19:59–66

Arikawa J, Lapenotire HF, Iacono-Connors L, Wang M, Schmaljohn CS (1990) Coding properties of the S and M genome segments of Sapporo rat virus: comparison to other causative agents of hemorrhagic fever with renal syndrome. Virology 176:114–125

Baek LJ, Yanagihara R, Gibbs CJ Jr, Miyazaki M, Gajdusek DC (1989) Leaky virus: a new Hantavirus isolated from *Mus musculus* in the United States. J Gen Virol 69:3129–3132

Bishop DHL (1985) Bunyaviridae. In: Fields B, Knipe DM, Chanock RM, Melnik JL, Roizman B, Shope RE (eds) Virology. Raven, New York, pp 1083–1118

Earle DP (1954) Analysis of sequential physiologic derangements in epidemic hemorrhagic fever. Am J Med 16:690–709

Elliott RM, McGregor A (1989) Nucleotide sequence and expression of the small (S) RNA segment of Maguari Bunyavirus. Virology 171:516–524

Giebel LB, Stohwasser R, Zöller L, Bautz EKF, Darai G (1989) Determination of the coding capacity of the M genome segment of nephropathia epidemica virus strain Hällnäs B1 by molecular cloning and nucleotide sequence analysis. Virology 172:498–505

Giebel LB, Zöller L, Bautz EKF, Darai G (1990) Rapid detection of genomic variations in different strains of Hantaviruses by polymerase chain reaction techniques and nucleotide sequence analysis. Virus Res 16:127–136

Giebel LB, Raab K, Zöller L, Bautz EKF, Darai G (1991) Identification and characterization of a Hantavirus strain of unknown origin by nucleotide sequence analysis of the cDNA derived from the viral S RNA segment. Virus Genes 5:111–120

Gött P, Zöller L, Yang S, Stohwasser R, Bautz EKF, Darai G (1991) Antigenicity of Hantavirus nucleocapsid proteins expressed in *E. coli*. Virus Res 19:1–16

Groen J, Van der Groen G, Hood G, Osterhaus ADME (1989) Comparison of immunofluorescence and enzyme-linked immunosorbent assays for the serology of Hantaan virus infections. J Virol Methods 23:195–203

Groen J, Jordans HGM, Clement JPG, Rooijakkers EJM, UytdeHaag J, Dalrymple J, Van der Groen G, Osterhaus ADME (1991) Identification of Hantavirus serotypes by testing of post-infection sera in immunofluorescence and enzyme-linked immunosorbent assays. J Med Virol 33:26–32

Ihara T, Akashi H, Bishop DHL (1984) Novel coding strategy (ambisense genomic RNA) revealed by sequence analysis of Punta Toro phlebovirus S RNA. Virology 136:293–306

Kogan SC, Gitschier J (1990) Genetic prediction of haemophilia A. In: Innis MA, Gelfand DH, Sninsky JJ, White TJ (eds) PCR protocols. A guide to methods and applications. Academic, New York

Krug MS, Berger SL (1987) Preparation of cDNA and the generation of cDNA libraries: first strand cDNA synthesis. In: Berger S, Kimmel A (eds) Methods in enzymology, vol 152. Academic, New York, pp 316–324

Lee HW (1982) Korean hemorrhagic fever. Progr Med Virol 28:96–113

Lee HW (1989) Hemorrhagic fever with renal syndrome. In: Lee HW, Dalrymple JM (eds) Manual of hemorrhagic fever with renal syndrome. WHO Collaborating Center for Virus Reference and Research, Institute for Viral Diseases, Korea University, pp 13–18

Lee HW, Lee PW (1976) Korean hemorrhagic fever. I. Demonstration of causative antigen and antibody. Korean J Intern Med 19:371–383

Lee HW, Lee PW, Johnson KM (1978) Isolation of the etiologic agent of Korean hemorrhagic fever. J Infect Dis 137:298–308

Parrington MA, Kang CY (1990) Nucleotide sequence analysis of the S genomic segment of Prospect Hill virus: comparison with the prototype Hantavirus. Virology 175:168–175

Parrington MA, Lee PW, Kang CY (1991) Molecular characterization of the Prospect Hill virus M RNA segment: a comparison with M RNA segments of other hantaviruses. J Gen Virol 72:1845–1854

Saiki RK, Gelfand DH, Stoffel S, Scharf SJ, Higuchi R, Horn GT, Mullis KB, Erlih HA (1988) Primer directed enzymatic amplification of DNA with a thermostable DNA polymerase. Science 239:487–491

Schmaljohn CS (1990) Nucleotide sequence of the Hantaan virus L RNA segment. Nucl Acids Res 18:6728

Schmaljohn CS, Jennings GB, Hay J, Dalrymple JM (1986) Coding strategy of the S genome segment of Hantaan virus. Virology 155:633–643

Schmaljohn CS, Schmaljohn AL, Dalrymple JL (1987) Hantaan virus M RNA: coding strategy, nucleotide sequence, and gene order. Virology 151:31–39

Sheshberadaran H, Niklasson B, Tkachenko EA (1988) Antigenic relationship between Hantaviruses analysed by immunoprecipitation. J Gen Virol 69:2645–2651

Simons JF, Hellman U, Petersson RF (1990) Uukuniemi virus S RNA segment: ambisense coding strategy, packaging of complementary strands into virions, and homology to members of the genus Phlebovirus. J Virol 64:247–255

Smadel JE (1953) Epidemic hemorrhagic fever. Am J Public Health 43:1327–1338

Stohwasser R, Giebel LB, Zöller L, Bautz EKF, Darai G (1990) Molecular characterization of the S RNA segment of nephropathia epidemica virus strain Hällnäs B1. Virology 174:79–86

Stohwasser R, Raab K, Darai G, Bautz EKF (1991) Primary structure of the large (L) RNA segment of nephropathia epidemica virus strain Hällnäs B1 coding for the viral RNA polymerase. Virology 183:386–391

Sugiyama K, Matsuura Y, Morita C, Morikawa S, Komatsu T, Shiga S, Akao Y, Kitamura T (1984) Isolation of a virus related to hemorrhagic fever with renal syndrome from urban rats in a nonendemic area. J Infect Dis 149:472–473

Wood WI, Gitschier J, Lasky LA, Lawn RM (1985) Base composition-independent hybridization in tetramethylammonium chloride: a method for oligonucleotide screening of highly complex gene libraries. Proc Natl Acad Sci USA 82:1585–1588

Xiao S-Y, Yanagihara R, Godec MS, Eldadah ZH, Johnson BK, Gajdusek DC, Asher DM (1991) Detection of hantavirus RNA in tissues of experimentally infected mice using reverse transcriptase-directed polymerase chain reaction. J Med Virol 33:277–282

Yanagihara R, Goldgaber D, Lee PW, Amyx HL, Gajdusek DC, Gibbs CJ Jr, Svedomir A (1984) Propagation of nephropathia epidemica virus in cell culture. Lancet 1:1013

Zöller L, Scholz J, Stohwasser R, Giebel LB, Sethi KK, Bautz EKF, Darai G (1989) Immunoblot analysis of the serologic response in Hantavirus infections. J Med Virol 27:231–237

Chapter 29 Polymerase Chain Reaction Technology for Rabies Virus

N. Tordo, H. Bourhy, and D. Sacramento

Summary

The polymerase chain reaction (PCR) amplification technique of viral ribonucleic acids has been evaluated as an alternative protocol for the diagnosis, typing, and molecular epidemiology of the rabies virus. Convenient regions of the viral genome have been delineated for each purpose. In its current form, PCR presents the same sensitivity but is slower than the routine diagnosis techniques by immunodetection of viral nucleocapsid. It can be already considered as a suitable confirmatory method for diagnosis and promises to be more competitive after further improvement. On the other hand, it advantageously compares with monoclonal antibody analysis for typing by restriction fragment length polymorphism. Limited panels of restriction enzyme can be defined to distinguish the vaccinal strains, as well as wild viral isolates of different geographic provenance. Finally, PCR suffers no concurrent technique in the ultimate identification of the viral isolate by direct sequence determination of the amplified segment. This allows us to undertake a worldwide molecular epidemiological study of rabies, and to reinvestigate the taxonomy of the *Lyssavirus* genus.

Introduction

Rabies is one of the most feared diseases, and its history intermingles with that of the human race (Théodoridès 1986). Its dark fame is mainly because bite is the favored mode of transmission. Despite its immemorial presence, Pierre-Victor Galtier (1846–1908) and subsequently Louis Pasteur (1822–1895) were the first to undertake a serious scientific approach to the disease in an attempt to find an effective vaccine. The illuminating successes that were achieved in this domain (Pasteur 1885) should not mask the fact that the deep-seated mechanisms of the disease have remained largely obscure. Rabies virus is elective for the nervous system of mammals, but the reasons for this neurotropy are still not understood. Effective infection invariably leads to a fatal

Unité de la Rage, Institut Pasteur, 25, rue du Docteur Roux, 75724 Paris Cedex 15, France encephalitis, although the altered neuronal functions have not been characterized. These unresolved questions make it impossible to propose an efficient therapy. One century after Louis Pasteur, prevention by vaccination remains the only defense against the threat.

Preventive vaccination against rabies is recommended only for particularly exposed persons, such as researchers, veterinarians, and wildlife agents. In the great majority of cases, the vaccination is performed after exposure. The physician is generally faced with two basic situations. If the exposure is severe, a vaccination combined with immunotherapy is initiated immediately. If the exposure is less severe and the biting animal available, a clinical survey or a laboratory diagnosis are performed, depending on whether the animal is alive or dead. The physician may wait for the diagnostic results before either undertaking a vaccination or, more frequently, stoping unnecessary treatment. Clearly, this means that rapidity is the major requirement of rabies diagnosis.

During the last 30 years, the techniques used for rabies diagnosis have advanced significantly (Bourhy et al. 1989; Sureau et al. 1991; Trimarchi and Debbie 1991). The histological detection of Negri bodies in the brain has been progressively replaced by immunological methods for detecting evidence of viral nucleocapsid antigens. For virus isolation, the tissue culture inoculation test (RTCIT) was preferred to the newborn mouse intracerebral inoculation test (MIT) because it is both faster (1 day instead of 10) and more ethical. Today, three very efficient techniques are available for rabies diagnosis: (a) the direct detection of viral nucleocapsid inclusions on brain smears with a fluorescent polyclonal antibody (FAT), (b) the immunocapture of viral nucleocapsids in brain extracts by an ELISA technique (rapid rabies enzyme immunodiagnosis, RREID), (c) the viral isolation in very sensitive neuroblastoma cell cultures (RTCIT). These three techniques are used routinely in the National and WHO Collaborative Reference Center for Rabies at the Pasteur Institute, Paris, France (Bourhy et al. 1989; Bourhy and Sureau 1990). This laboratory is also in charge of the development of new techniques. In particular, it is important to find methods for typing the isolates and appreciating their variability throughout the world.

A progressive polymerase chain reaction (PCR) technique that takes these diverse questions into account has recently been developed (Sacramento et al. 1991). It already appears to be a promising alternative tool for diagnosis and a very efficient means of typing and tool in molecular epidemiology studies.

Material and Methods

The principal steps of the progressive PCR method (Sacramento et al. 1991), illustrated in Fig. 1 are described below in detail.

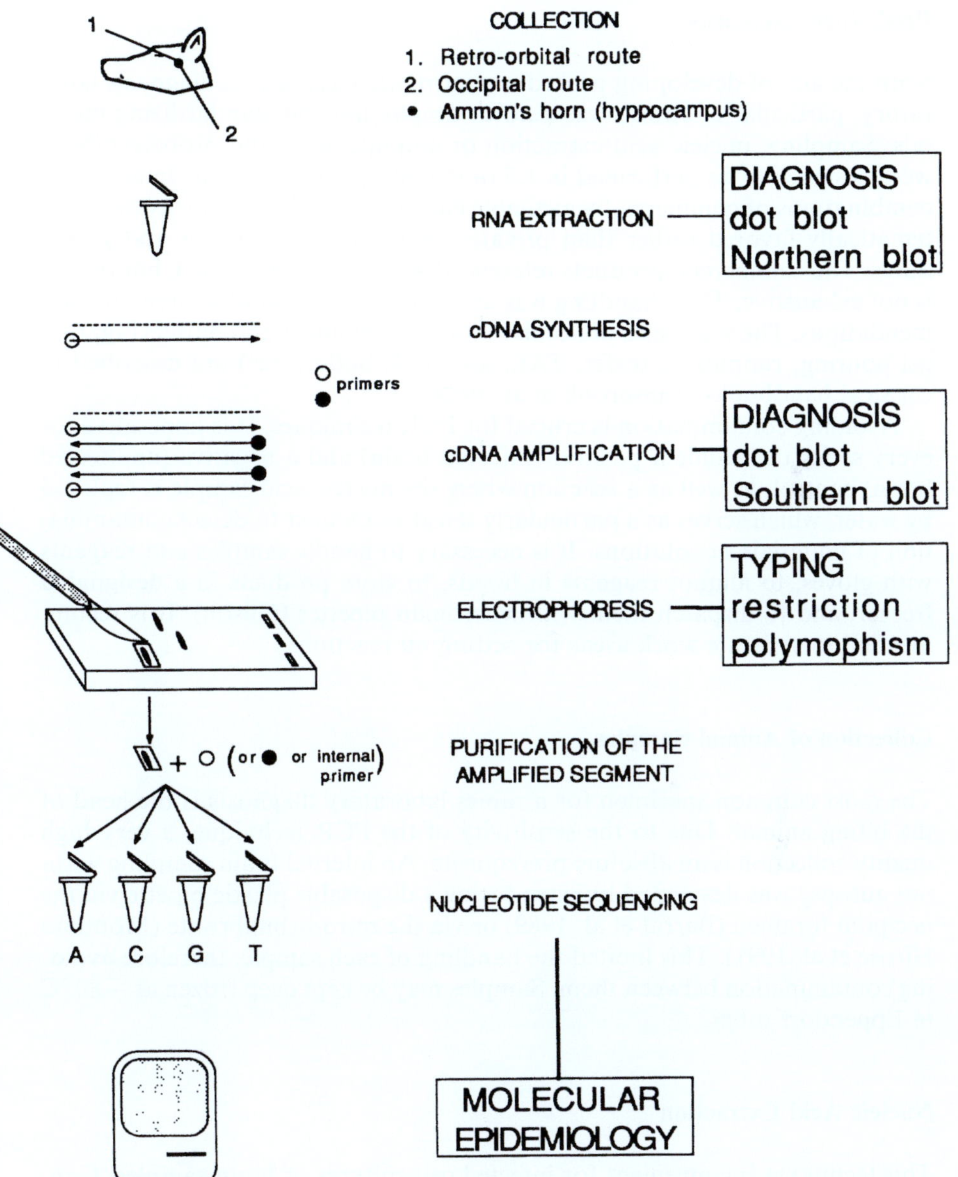

Fig. 1. Schematic view of the progressive PCR method for the characterization of the rabies transcripts possibly present in a suspect animal sample: from diagnosis to typing and molecular epidemiology

Preliminary Remarks

With the aim of developing procedures currently feasible in a diagnosis laboratory, particular attention was paid to simplifying and standardizing methods. Sampling, nucleic acid extraction or amplification, and probe synthesis were designed to be performed in 1.5 or 0.5 ml Eppendorf tubes. In addition, combinations of commercially available enzymes, kits, or apparatus were systematically favored rather than private laboratory preparations and procedures. The commercial products referenced were successfully used, but the list is not exhaustive. Their handling was according to the manufacturer's recommendations. The very basic molecular biology techniques and buffers (agarose gel pouring, running, transfer, TAE, and TBE buffer, etc.) are described in classical handbooks (Sambrook et al. 1989).

Avoiding contamination is crucial for PCR techniques. It is imperative for every series to include a positive (infected brain) and a negative (uninfected brain) control, as well as a reaction where the nucleic acid sample is replaced by water, which serves as a particularly sensitive manner to detect contamination of materials or solutions. It is necessary to handle samples and reagents with gloves, to aliquot reagents in hoods, to store products in a designated freezer, and to dispatch them with Microman pipettes (Gilson). It is recommended to separate work areas for setting up reactions.

Collection of Animal Samples

The most common specimen for a rabies laboratory diagnosis is the head of the biting animal. Due to the sensitivity of the PCR technique, a very high quality collection is an absolute prerequisite. An internal brain sampling without autopsy was developed by introducing a disposable plastic pipette via the occipital foramen (Barrat et al. 1989) or via the retro-orbital route (Montano Hirose et al. 1991). This limited the handling of each sample, therefore avoiding contamination between them. Samples may be kept deep frozen at $-80\,^{\circ}\mathrm{C}$ in Eppendorf tubes.

Nucleic Acid Extraction

This technique is convenient for infected cell cultures or brain samples. Confluent monolayer cells are washed with PBS and covered with extraction buffer (1% SDS, 1% Nonidet P40, 1 mM EDTA pH$=$8.0, 50 mg/ml dextran sulphate) (5 ml for 75 cm^3 flask). Brain samples (approximately 2 g) are crushed to homogeneity in 0.5 ml of extraction buffer using a plastic pestle adapted to the 1.5 ml Eppendorf tube. Up to four successive protein extractions of the lysate (once with phenol, twice with phenol-chloroform (50/50), once with chloroform) are performed by the addition of an equal volume of solvent, mixing vigorously, and separating aqueous and organic phases by centrifugation at

13 000 g for 30 min at room temperature. Finally, sodium acetate (pH = 5.2) is added to give a final concentration of 0.3 M, and the total RNA is precipitated by the addition of two volumes of ethanol. After centrifugation at 13 000 g for 10–30 min, the pellets are washed twice in 70% ethanol, dried, and resuspended in pyrolysed water at 1 µg/µl (absorbance at 260 nm).

cDNA Synthesis and Amplification of the Viral Transcripts by PCR

Total RNA (1 µg) is annealed with the primer (100 ng) in a total volume of 3 µl at 65 °C for 3 min, and chilled on ice. The reaction mixture is made up to 10 µl with 50 mM tris-HCl pH = 8.3, 75 mM KCl, 3 mM MgCl$_2$, 10 mM DTT, 1 mM each dNTP, 0.4 U/µl RNAsin (Promega Biotec), 20 U/µl Moloney murine leukemia virus reverse transcriptase (Bethesda Research Laboratories) and incubated for 90 min at 42 °C. The resulting RNA/cDNA hybrid is diluted 10-fold in 10 mM tris-HCl pH = 8.3, 1 mM EDTA.

Diluted RNA/cDNA hybrid (10 µl) is made up to 100 µl with 10 mM tris-HCl pH = 8.3, 50 mM KCl, 1.5 mM MgCl$_2$, 10% DMSO, 0.2 mM each dNTP, 0.01% gelatin, 100 ng of each primer and 2 U of AmpliTaq DNA polymerase (Perkin-Elmer Cetus). The mixture is covered with 100 µl of paraffin oil and subjected to consecutive cycles of denaturation (D), annealing (A) and elongation (E) in an automatic thermal reactor (Hybaid).

The oligodeoxynucleotide priming the cDNA synthesis must never be located downstream from any of the two primers used for amplification. However, it can be identical to the upstream PCR primer. The setting of the PCR apparatus should be modified according to the primer length, composition and complementarity to the target region. An empirical adjustment is frequently required. For a 20–25 residue long primer set, the following values could be considered as basic: five initial cycles (D = 1 min at 94 °C, A = 1.5 min at 45 °C then 0.3 min at 50 °C; E = 1.5 min at 72 °C); 30 additional cycles (but D = 0.5 min; E = 1 min); a last cycle (but E = 10 min).

Northern, Southern and Dot Blotting

For northern blots, total RNA (5–10 µg) is denatured in 10 mM Na$_2$HPO$_4$-NaH$_2$PO$_4$ pH = 7.5, 0.5 mM EDTA, 50% formamide for 5 min at 65 °C, electrophoresed through a denaturing 0.9% agarose gel buffered in 10 mM Na$_2$HPO$_4$-NaH$_2$PO$_4$ pH = 7.5, 1.1 M formaldehyde, then transferred onto nylon membranes using 15 mM Na-citrate, 150 mM NaCl, pH = 7.0. For Southern blots, the PCR products (10 µl) are separated on 1%–3% agarose gel in TEA buffer containing ethidium bromide (10 mg/ml), photographed and transferred onto nylon membranes using 0.4 N NaOH. For dot blots, total RNA (5–10 µg) or PCR products (5 µl) are diluted to 100 µl in 10 mM tris-HCl pH = 8.3, 1 mM EDTA, heat-denaturated at 95 °C for 10 min, chilled on ice and filtered onto a nylon membrane using a multiwell vacuum filtration unit (Biodot Apparatus Biorad). The nylon membranes (Hybond-N Amersham) used in the blotting techniques are air dried before covalent binding of the nucleic acids by UV illumination at 312 nm for 3 min.

Probe Labeling, Membrane Hybridization and Revelation

A purified DNA fragment (100 ng) internal to the amplified region is prepared by restriction hydrolysis of cDNA clones of various lyssavirus (Bourhy et al. 1989; Tordo et al. 1986a; Tordo et al. 1986b; Tordo et al. 1988). The non-radioactive DNA labeling by random priming with digoxigenin-dUTP, membrane hybridization, immunological detection of digoxigenin, and enzymatic revelation with alkaline phophatase are performed as described in the DNA labeling and detection kit nonradioactive (Boehringer).

The radioactive DNA labeling is done by random priming using α-^{32}P-dNTP and the Multiprime labeling kit (Amersham). Membranes are prehybridized (1 h), then hybridized for 16 h at 65 °C in 0.5 M NaH$_2$PO$_4$-Na$_2$HPO$_4$ pH = 7.5, 1 mM EDTA, 7% SDS, 1% bovine serum albumin, in a sealed plastic bag. During hybridization, 10^6 cpm of denatured ^{32}P-labeled probe are added per 5 ml buffer. After three washings for 15 min at 65 °C in 40 mM NaH$_2$PO$_4$-Na$_2$HPO$_4$ pH = 7.5, 1 mM EDTA, 1% SDS, the membranes are dried and exposed to Kodak XAR-5 film with intensifying screens at -70 °C from between 1 h to overnight.

Analysis of the PCR Products:
Restriction Polymorphism and Nucleotide Sequencing

For typing, 2–10 µl of the PCR products are digested by a panel of selected restriction enzymes and separated by electrophoresis on a convenient agarose gel in TAE buffer containing ethidium bromide (10 mg/ml). The pattern is analyzed on a print of the gel viewed under UV illumination (312 nm).

Before nucleotide sequencing, the PCR products (10–30 µl) are preferentially purified on a 0.7–2% NuSieve GTG (FMC) agarose gel with ethidium bromide (10 mg/ml), run at 4 °C to avoid melting. When visibly separated from the excess of primer and the (possible) nonspecific products, the amplified fragment is cut out from the gel, melted at 95 °C, aliquoted into 10 µl fractions in Eppendorf tubes and can be kept for months at -20 °C. One tube is convenient for one four-track sequence reaction by the dideoxy chain termination technique (Sanger et al. 1977) using a classical kit (T7 Sequencing Kit, Pharmacia). The differences with the manufacturer's procedure are: (a) the primer (50 ng in 2 µl of any oligodeoxynucleotide complementary to one strand of the fragment are used), and (b) the reaction temperature (maintained at 37 °C instead of room temperature in order to avoid agarose polymerization). To gain sensitivity, the sequencing reactions are lyophilized in a Speed Vac Concentrator (Savant) to 1–3 µl, before loading onto a denaturing sequencing gel (60 cm × 0.1 mm). A gradient gel from 0.5 × TBE – 5% acrylamide urea to 5 × TBE – 7% acrylamide urea is used (Sambrook et al. 1989). The bottom and top buffer are 1 × and 0.5 × TBE, respectively. The sequence reactions are separated for approximately 5 h at 50 W (around 2000 V). The gel is fixed in aqueous 10% ethanol, 10% acetic acid, dried on Whatman 3MM filter paper and exposed to Kodak XAR-5 film overnight at room temperature.

Discussion

Comparative Analysis of Lyssavirus Genomes: Purpose of Each Genomic Area

Rabies disease is caused by a group of neurotropic viruses describing the *Lyssavirus* genus of the Rhabdoviridae family (Francki et al. 1991). Most of them were ranked on the basis of serological and antigenic relationships into four serotypes (Dietzschold et al. 1988; King and Crick 1988; WHO 1991). The current vaccines are exclusively produced from rabies viruses belonging to serotype 1. Members of the other serotypes as well as unclassified isolates recently performed on European bats constitute the so-called rabies-related viruses (Bourhy et al. 1990). The more divergent is the Mokola virus (serotype 3) against which the rabies vaccines provide absolutely no protection. Consequently, a comparative study of the rabies and Mokola viruses is illustrative of the maximum variability within the *Lyssavirus* genus (Sacramento et al. 1991).

We have cloned and analyzed their respective genome (Bourhy et al. 1989; Tordo et al. 1986a; Tordo et al. 1986b; Tordo et al. 1988). From 3′ to 5′, six nonoverlapping genes, corresponding to the N nucleoprotein, M1 phosphoprotein, M2 matrix protein, G glycoprotein, Ψ pseudogene and L polymerase, are encoded (Fig. 2). The Ψ pseudogene is a substantial remnant protein gene (450 nucleotides) typical of the *Lyssavirus* genus among unsegmented negative stranded RNA viruses (Mononegavirales order). The expression rate decreases from the 3′ encoded to the 5′ encoded cistrons. Sequence comparison between rabies and Mokola genomes shows that the nonprotein coding regions are logically less conserved than the coding ones (Fig. 2). In the latter, the divergence increases in the order N → M2 → G → M1 proteins. The sequence of the L protein of Mokola virus is not complete, but assuming the general conservation of polymerases among unsegmented negative strand genomes (Poch et al. 1990), it probably contains domains surpassing the conservation rate of the nucleoprotein.

An ideal diagnosis tool should detect minute traces of infection by a maximal spectrum of lyssaviruses (Sacramento et al. 1991). As far as PCR is concerned, the most suitable targets are the highly conserved and expressed genes, which are two qualities shown by the nucleoprotein gene. On the other hand, the principal quality of a typing or molecular epidemiology tool is efficiency in differentiating and classifying isolates. The very divergent Ψ pseu-

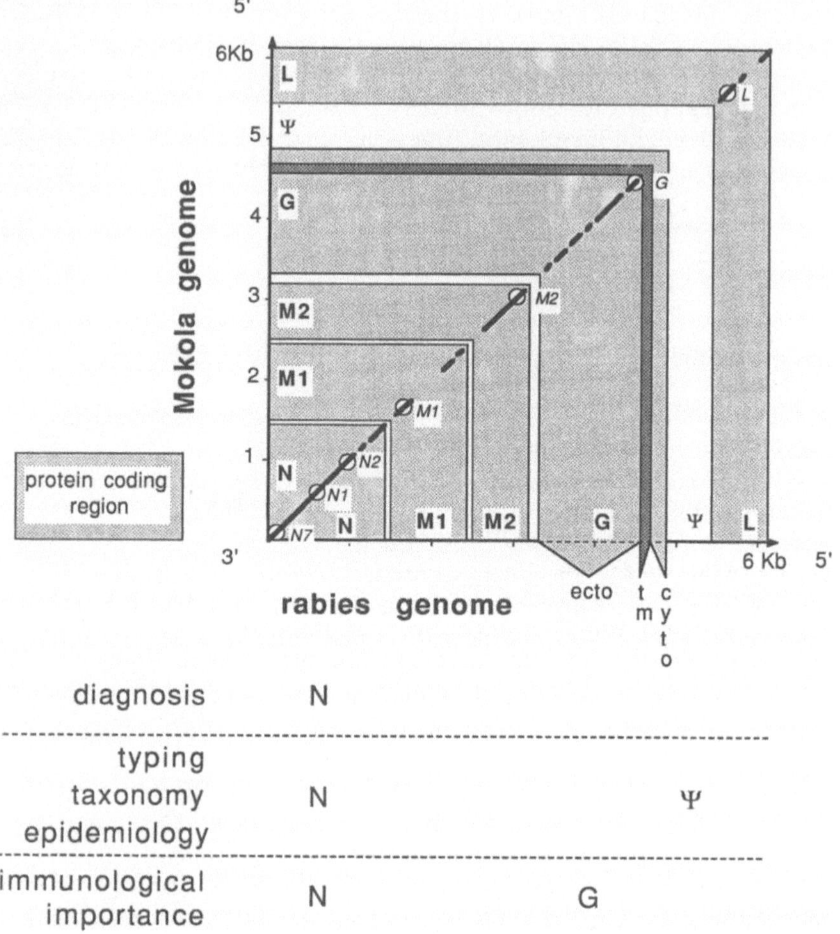

Fig. 2. Comparative analysis of the 6000 first 3′ nucleotides of the rabies virus (PV strain) and the Mokola virus (MOK5 isolate) genome. Diagonal lines indicates the homologous regions. The oligodeoxynucleotide primers cited in the text and shown in Fig. 3 are located. The different domains of the G glycoprotein (ectodomain of immunological importance, transmembrane domain, cytoplasmic domain protruding at the iner side of the viron membrane) are shown. The lower part of the figure suggests convenient genes to use for defined purposes

dogene seems the more sensitive target, because its great susceptibility to mutations is representative of the viral evolution outside any external selective pressure. It is the best candidate to differentiate isolates of the same serotype (Sacramento et al. 1992). However, its variability is too important to permit a significant alignment between serotypes. The N gene is alternatively selected for two reasons: (a) for comparative purposes, because the nucleoprotein was the main target antigen in taxonomic studies (Dietzschold et al. 1988), and (b) to approach the molecular basis of the antigenicity and cross-protection between serotypes, since the nucleoprotein is the second viral antigen involved in the protective immunity (Celis et al. 1990; Tollis et al. 1991). Finally, primer sets were also found to analyze the entire glycoprotein, the only antigen able to elicit neutralizing antibodies (Wiktor et al. 1973).

Choice of the Oligodeoxynucleotide Primers

Primer sets were chosen in particularly stable sequences flanking the targeted regions, in order to obtain suitable amplification of most lyssaviruses (Sacramento et al. 1991). Until now, the criteria for stability were mostly based on the comparison of the rabies (serotype 1) and Mokola (serotype 3) genomes, since they constitute the only available data (Fig. 2). Unpublished results obtained from other lyssaviruses confirm the global validility of this approach, but nonetheless local mismatches could exist in certain groups of isolates, limiting the usage of the corresponding primer sets. The sequence of the proposed primers are shown in Fig. 3.

For diagnosis, two primer sets ($N_1 - N_2$ and N_7-M1) were selected. $N_1 - N_2$ was synthesized according to the Mokola sequence, although this virus is absent from European countries. This choice of template-primer heterogeneity was deliberate in order to provide for genuine variability of isolates. The specific advantage of the $N_1 - N_2$ set is the short distance separating primers (450 residues), promising a high amplification rate, even on damaged samples. The N_7 and M1 primers, although more distant (1500 residues), allow the amplification of the entire nucleoprotein coding region for subsequent study. As previously noted, the $N_1 - N_2$ primer set is very efficient for European isolates of serotype 1 or for isolates of serotype 3, but hardly amplifies either the European bat isolates or several non-European isolates of serotype 1. On the other hand, the N_7-M1 primer set is convenient for all tested isolates of serotype 1 and for the European bat isolates but is nonefficient for isolates of serotype 3 (unpublished results).

The amplification of the Ψ region in most lyssaviruses is more complex due to its strong variability. At the G gene side, a suitable G primer could not be found until the conserved ectodomain of the glycoprotein. Despite its relative divergence between rabies and Mokola genomes (five mismatches for 23 residues), the G primer terminates with five very stable nucleotides. They correspond to the unique tryptophane codon and the two first residues (avoiding the "wobble" position) of a very conserved glycine codon. At the L gene side, a L primer was defined in a motif conserved in the polymerase gene of all unsegmented negative stranded RNA viruses (Poch et al. 1990). To amplify the entire glycoprotein gene, a M2 primer was selected following the same criteria. The M2-L set produces a huge 2500 residue fragment. An easier alternative way is to divide in two small overlapping amplifications with intermediate primers (not shown). The M2, G and L primers are synthesized according to the PV strain rabies sequence (Tordo et al. 1986 b; Tordo et al. 1988).

```
                                      M
PV            55    ATGTAACACCTCTACA ATG    73
(ERA)-SADB19        ATGTAACACCcCTACA ATG                          N7
CVS-AvO1            ATGTAACACCcCTACA ATG                          (+ sense)
MOK           55    ATGTAACACtcCTACA ATG    73                    5'-->3'


                    F   E   T   A   P   F
PV            587   TTT GAG ACa GCc CCT TTT G   605               N1
(ERA)-SADB19        TTT GAG ACa GCc CCT TTT G                     (+ sense)
CVS-AvO1            TTc GAG ACa GCa CCT TTT G                     5'-->3'
MOK           587   TTT GAG ACT GCT CCT TTT G   605


                    V   G   C   Y   M   G
PV            1013  CAT CCT ACG ATA TAC CC   1029                 N2
(ERA)-SADB19        CAT CCT ACG ATA TAC CC                        (- sense)
CVS-AvO1            CAa CCT ACG ATg TAC CC                         3'<--5'
MOK           1013  CAT CCT ACG ATA TAC CC   1029


                    E   M   A   E   E   T
PV            1568  CTC TAC CGA CTT CTT TGA   1585                 M1
ERA-SADB19          CTC TAC CGA CTT CTT TGA                        (- sense)
CVS-AvO1            CTC TAC CGg CTT CTc TGA                         3'<--5'
MOK           1578  CTt TAC CGt CTT CTc TGA   1595


                    W   C   I   N   M   N   S
PV            3000  TGG TGT ATC AAC ATG AAC TC   3019             M2
ERA-SADB19          TGG TGT ATC AAC ATG AAC cC                    (+ sense)
(CVS)-AvO1          TGG TGT ATC AAC tcG AAC TC                    5'-->3'
MOK           3020  TGG TGc ATt AAC ATG AAC TC   3039


                    D   L   G   L   P   N   W   G
PV            4665  GAC TTG GGT CTC CCG AAC TGG GG   4687
ERA-SADB19          GAC TTG GGT CTC CCG AAC TGG GG                G
CVS-AvO1            GAC cTG GGT CTC CCG AAC TGG GG                (+ sense)
HEP                 GAC cTG GGT CTC CCG AAa TGG GG                5'-->3'
MOK           4675  GAC cTG GGT CTg CCt cAt TGG GG   4697


                    D   Y   N   L   N   S   P   L
PV            5520  CTG ATG TTA GAG TTG AGA GGA AAC   5543         L
(ERA)-SADB19        CTG ATG TTA GAG TTG AGA GGA AAC                (- sense)
MOK           5545  CTG ATG TTg GAc TTG AGA GGA AAC   5568          3'<--5'
```

Fig. 3. Comparison of the primer structure in all the available Lyssaviruses, i.e., rabies vaccinal strains (serotype 1) and Mokola virus (serotype 3). The primers used for cDNA synthesis or amplification are *underlined*. Data were obtained from Tordo et al. (1986a), Tordo et al. (1986b), and Tordo et al. (1988) for PV, from Anilionis et al. (1981), Larson and Wunner (1990), and Rayssiguier et al. (1986) for ERA, from Conzelmann et al. (1990) for SADB19, from Larson and Wunner (1990), Mannen et al. (1991), Sumner et al. (1991) for CVS, from Poch et al. (1988) for AvO1, from Morimoto et al. (1989) for HEP, and from unpublished results for MOK. Strains for which the sequence of the corresponding primer is not available are in parenthesis. The primer position is given respectively to the position in the PV rabies virus and Mokola virus genome

Diagnosis Methods with Nucleic Acid Probes

The rabies genome transcription and replication produce exclusively RNA molecules (Tordo and Poch 1988). Their presence in the sample can be evidenced either directly, by hybridization with DNA probes complementary to the targeted region, or indirectly, after a preliminary amplification step. The latter alternative should comprise, in succession, the reverse transcription of the target viral transcript into cDNA and then the cDNA amplification by PCR (Figs. 1 and 4).

These different methods were checked over one hundred suspect brain samples received in the National Reference Center for Rabies in Paris (Sacramento et al. 1991). The direct detection of viral RNA either on dot or northern blot is not, in its current form (Ermine et al. 1988), sensitive enough, whatever the radio- or non-radiolabeled probe used. In contrast, the dot or Southern blot analysis of N_1-N_2 or N_7-M1 amplified products is an efficient diagnosis method, 100% correlative with the routine (FAT; RTCIT; RREID) techniques in systematic blind trials. The dot blotting method should be preferred for several reasons: (a) because of its simplicity and rapidity, and (b) because it can be hybridized with the two types of labeled probes (^{32}P or digoxigenin), while the probing of Southern blots is uncertain with a nonradiolabeled probe. Southern blot is, however, an obligatory step since the direct observation of the amplified segment on agarose gel with ethidium bromide is insufficient because several nonspecific bands can comigrate in highly damaged samples.

In summary, dot blot analysis of the N_1-N_2 or N_7-M1 amplified products is an alternative protocol for rabies diagnosis. A statistical improvement of its sensitivity is currently in progress using a larger number of samples. The digoxigenin-labeled probes are favored because they are more stable than the ^{32}P-labeled (several months rather than 2 weeks), while they allow detection of as little as 1 pg, i.e., 10^5 molecules of amplified products.

Comparative Evaluation with Routine Diagnosis Techniques, Development

Rapidity is the major quality of a rabies diagnosis method. The delay of availability of results for the different methods are compared in Fig. 4. In the National Reference Center for Rabies (Bourhy et al. 1989; Bourhy and Sureau 1990), the suspected samples are received and autopsied every morning before 10 a.m. The FAT is completed within 2–3 h, and its results are the first available, by 1 p.m, and therefore remain the main reference of the diagnosis. The RREID (4 h) delivers its results by 3 p.m., while the RTCIT requires an overnight incubation on neuronal cell cultures whose positivity is revealed on day 2, at 11 a.m., by FAT. By comparison, the PCR technique is clearly slower and may only compete with RTCIT when dot blots and digoxigenin-labeled probes are used. Indeed, if the internal brain collection is rapid, the total RNA preparation is hardly completed before 1 p.m. because of the successive extractions and precipitation. Since the cDNA synthesis takes 1 h 30 min, the PCR

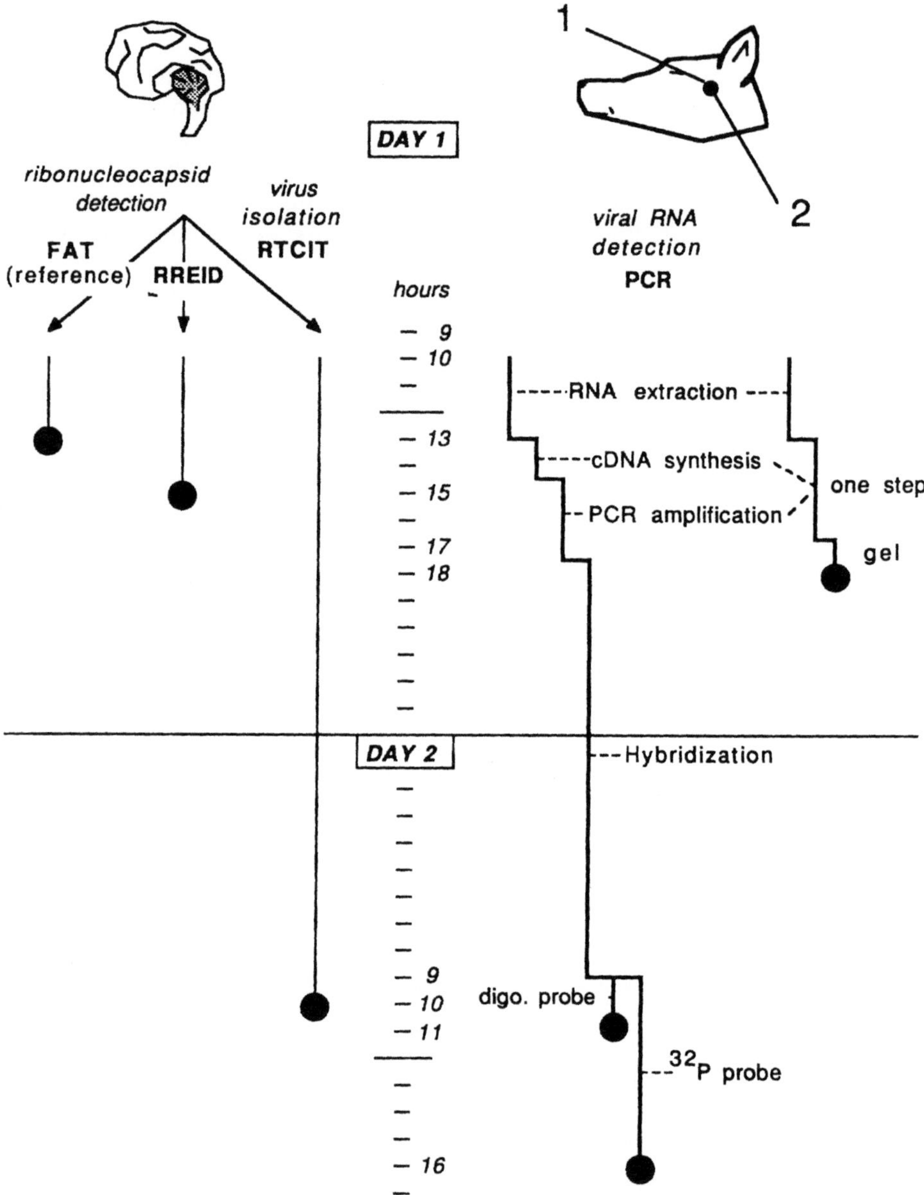

Fig. 4. Comparative evaluation of the delay for completing the three routine diagnosis techniques (*FAT*, fluorescent antibody test; *RREID*, rapid rabies enzyme immunodiagnosis; *RTCIT*, rapid tissue culture inoculation test), and the new PCR technique

3 h, and the dot blotting 30 min, the blots are not ready for hybridization before 6 p.m., when the technicians are leaving the laboratory. Therefore, the hybridization has to be performed overnight, and the immunoenzymatic determination is completed at 11–12 a.m. on day 2. The slowness is still greater using a radiolabeled probe.

In order to reduce this time handicap several alternatives are under investigation: (a) simplification of the nucleic acid extraction procedure, (b) combination of cDNA synthesis and PCR amplification in a single step, which also results in an increased sensitivity since the N gene and the N messenger RNA are simultaneously amplified, and (c) direct characterization of the amplified products by restriction fragment length polymorphism (RFLP), in order to shunt the hybridization step. Whatever this time refinement, it is unlikely that it will ever become faster than the FAT technique. Therefore, the real future of PCR for diagnosis purposes would reside in its ability to surpass the sensitivity of routine techniques, particularly for intra-vitam diagnosis of human and animal secretions during clinical survey or quarantine, when laboratory diagnosis is not simple.

The second requirement of a diagnostic technique is its competence whatever the infecting lyssavirus. The parallel use of N_1-N_2 and N_7-M1 primer sets was shown to be efficient for rabies, Mokola and European bat lyssaviruses isolates (Sacramento et al. 1991). Among the classical techniques, only the immunocapture of viral nucleocapsids by ELISA has been recently extended to viruses divergent from serotype 1 (RREID-Lyssa) (Perrin et al. 1992).

Comparison of Typing by Restriction Fragment Length Polymorphism and Monoclonal Antibodies

The very divergent Ψ pseudogene is the most suitable to differentiate isolates of the same serotype. The G-L primer set was found efficient for the fixed or wild rabies viruses (serotype 1), but also for the Mokola viruses (serotype 3), which are so divergent that internal Ψ probes are incapable of reciprocal hybridization (Sacramento et al. 1991). Such differential hybridization could be regarded as a preliminary approach in typing. However, RFLP with typical enzymes is a more simple and precise way that was recently checked both on the Ψ (Sacramento et al. 1991) and on the N (Smith et al. 1991) genes. It offers numerous advantages with classical antigenic analysis with monoclonal antibodies (Mabs): it is easier and quicker (completed the day of sample reception); it does not require virulent sample; it preserves the natural situation, avoiding the possible mutations during cell adaptation that could be required before Mabs analysis; it appreciates the variation in the total sequence length rather than being limited to immunologic epitopes.

We compared the performance of the two methods in differentiating the principal fixed rabies virus strains used as vaccine seed (serotype 1) (Sacramento et al. 1991) (Fig. 5): Pasteur virus (PV), Evelyn-Rokitnicki-Abelseth (ERA), south Alabama Dufferin (SADB19), Flury high egg passage (HEP), challenge

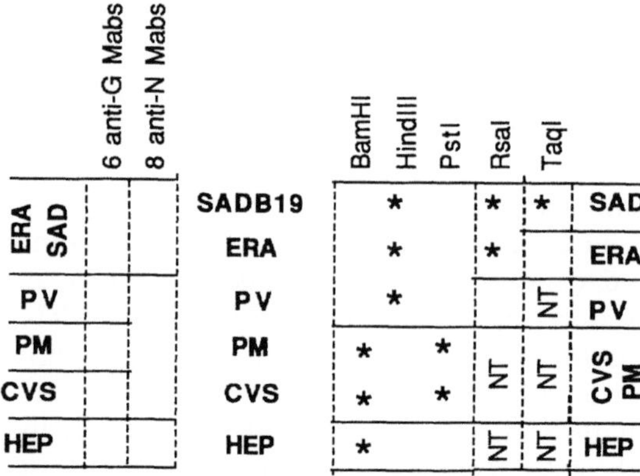

Fig. 5. Comparative efficiency of typing by Mab (*left part*) and RFLP (*right part*) evaluated on the main fixed strains used in vaccines (serotype 1): Pasteur virus (PV), Evelyn-Rokitnicki-Abelseth (ERA), south Alabama Dufferin (SADB19 or SAD), Flury high egg passage (HEP), challenge virus strain (CVS), and Pitman-Moore (PM). Two successive panels of 8 anti-N and 6 anti-G Mabs allow distinguishing between all strains but ERA-SAD. Full horizontal lines outline the distinction. The RFLP is analyzed in three steps: (a) *Bam*HI, *Hin*dIII and *Pst*I patterns distinguish HEP, CVS-PM and PV-ERA-SAD; (b) *Rsa*I pattern separqates PV from ERA-SAD; and (c) *Taq*I pattern distinguishes ERA from SAD. Finally, only CVS-PM cannot be distinguished by this method. A star symbolizes the sensitivity to an enzyme. The enzymes were selected to provide a distinction by "cleavage" or "no cleavage," except *Taq*I which is distinguished by the number of sites. NT significates that the analysis of the cleavage pattern is not necessary following the results of the (*Bam*HI-*Hin*dIII-*Pst*I) pattern

virus strain (CVS), and Pitman-Moore (PM). A panel of five basic restriction enzymes (*Bam*HI, *Hin*dIII, *Pst*I, *Rsa*I, *Taq*I) is sufficient to distinguish between all strains, except CVS and PM that show only three mismatches in the amplified segment. A similar distinction is obtained using two successive panels of 8 anti-N and 6 anti-G Mabs (P. Sureau, personal communication), but two different techniques are required: the fluorescent antibody test and neutralizing test. In addition, the latter test obligates to previous cell culture adaptation.

Using the N_7-M1 primer set, it was similarly shown that RFLP with *Dde*I or *Rsa*I is a convenient tool for differentiating isolates from the four known serotypes, but also the still unclassified EBL1 and EBL2 isolates (Bourhy et al. 1992).

Other panels of restriction enzymes can be defined for particular discriminations (geographic or host specificity). Modification of cleavage sites characterizes the emergence of mutant strains and designate sample of interest for further sequence studies.

Molecular Epidemiology Studies

The protocol described allows direct sequencing of any double-stranded DNA purified only by excision from NuSieve agarose gel. Specific primers complementary to any region of the fragment can efficiently anneal to the targeted region after a single heat denaturation. The fidelity of the technique was verified by repeatedly sequencing the same region amplified in different PCR reactions. Its competitiveness is evident both in term of rapidity (results ready for computer assisted analysis only 3 days after reception) and quality, because of the definitive characterization of the isolate. It allows the determination of the consensus sequence of the viral population but is not convenient for quasispecies studies that require cloning of individual amplified molecules.

In preliminary worldwide studies of serotype 1 isolates by the Ψ gene variability, a parallelism was observed between homology and geographic proximity. For example, the rabies viruses currently circulating in wildlife in France form a very homogeneous group (3.5% of intrinsic divergence) but significantly diverge from the vaccinal strains (15%) (Sacramento et al. 1992). Such divergence could be expected since the classical vaccine seeds derive from isolates performed 40 to 100 years ago, even though they still provide full protection in France. Up to now, the more divergent isolates of serotype 1 were found in Thailand, Brazil, and Guyana (35%) (unpublished results).

The taxonomy was also reinvestigated using nucleoprotein gene variability as a criteria. Four genotypes, correlating with the four predefined serotypes, were found. In addition, EBL1 and EBL2 isolates are sufficiently typical to justify the delineation of two new independent genotypes. PCR technique reveals its great sensitivity in appreciating intergenotype relationships (Bourhy et al. 1992). This work leads to the proposal of more consistent taxonomic criteria for the future.

Conclusion

The PCR technique appears suitable for rabies diagnosis, but still needs to be improved for routine use. On the other hand, it can already be considered as the most convenient and powerful tool available for typing and studying the molecular epidemiology of lyssaviruses, even on degraded samples. This invites worldwide molecular epidemiological studies, notably in countries infected with divergent lyssaviruses against which current vaccines are suspected or known not to provide protection. Such studies could help, if necessary, to select new vaccinal strains offering a larger spectrum of protection or more adapted to local problems.

References

Anilionis A, Wunner WH, Curtis P (1981) Structure of the glycoprotein gene in rabies virus. Nature 294:275–278
Barrat J, Artois M, Bourhy H (1989) Two bats diagnosed rabid in France. Rabies Bulletin Europe. WHO Collaborative Center for Rabies Surveillance and Research, Tübingen, FRG 13:10–11

Bourhy H, Rollin PE, Vincent J, Sureau P (1989) Comparative field evaluation of the fluorescent-antibody test, virus isolation from tissue culture, and enzyme immunodiagnosis for rapid laboratory diagnosis of rabies. J Clin Microbiol 27:519–523

Bourhy H, Sureau P (1990) Laboratory methods for rabies diagnosis. Institute Pasteur, Paris

Bourhy H, Sureau P, Tordo N (1990) From rabies to rabies-related viruses. Vet Microbiol 23:115–128

Bourhy H, Tordo N, Lafon M, Sureau P (1989) Complete cloning and molecular organization of a rabies-related virus: Mokola virus. J Gen Virol 70:2063–2074

Bourhy H, Kissi B, Lafon M, Sacramento D, Tordo N, (1992) Antigenic and molecular characterization of bat rabies virus in Europe. J Clin Microbiol (in press)

Celis E, Rupprecht CE, Plotkin SA (1990) New and improved vaccines against rabies. In: Woodrow GC, Levine MM (ed) New generation vaccines. Dekker, New York, pp 419–439

Conzelmann KK, Cox JH, Schneider LG, Thiel HJ (1990) Molecular cloning and complete nucleotide sequence of the attenuated rabies virus SAD B19. Virology 175:485–489

Dietzschold B, Rupprecht CE, Tollis M, Lafon M, Mattei J, Wiktor TJ, Koprowski H (1988) Antigenic diversity of the glycoprotein and nucleocapsid proteins of rabies and rabies-related viruses: implications for epidemiology and control of rabies. Rev Infect Dis 10:S785–S798

Ermine A, Tordo N, Tsiang H (1988) A rapid diagnosis of the rabies infection through a dot hybridization assay. Mol Cell Probes 2:75–82

Francki RIB, Fauquet CM, Knudson DL, Brown F (1991) Classification and Nomenclature of Viruses. Fifth Report of the International Committee on Taxonomy of Viruses. Springer, Vienna

King A, Crick J (1988) Rabies-related viruses. In: Campbell JB, Charlton KM (ed) Rabies. Kluwer, Boston, pp 177–200

Larson JK, Wunner WH (1990) Nucleotide and deduced amino acid sequence of the nominal nonstructural phosphoprotein of the ERA, PM and CVS-11 strains of rabies virus. Nucleic Acids Res 18:7172

Mannen K, Hiramatsu K, Mifune K, Sakamoto S (1991) Conserved nucleotide sequence of rabies virus cDNA encoding the nucleoprotein. Virus Genes 5:69–73

Montano Hirose JA, Bourhy H, Sureau P (1991) Retro-orbital route for the collection of brain specimens for rabies diagnosis Vet Record 129:291–292

Morimoto K, Ohkubo A, Kawai A (1989) Structure and transcription of the glycoprotein gene of attenuated HEP-Flury strain of rabies virus. Virology 173:465–477

Pasteur L (1885) Méthode pour prévenir la rage après morsure. CR Acad Sci Paris 101:765–774

Perrin P, Gontier C, Lecocq E, Bourhy H (1992) A modified rapid enzyme immunoassay for the detection of rabies and rabies-related viruses: RREID-lyssa. Biologicals 20:51–58

Poch O, Blumberg BM, Bougueleret L, Tordo N (1990) Sequence comparison of five polymerases (L proteins) of unsegmented negative-strand RNA viruses: theoretical assignments of functional domains. J Gen Virol 71:1153–1162

Poch O, Tordo N, Keith G (1988) Sequence of the 3386 3' nucleotides of the genome of the AvO1 strain rabies virus: structural similarities of the protein regions involved in transcription. Biochimie 70:1019–1029

Rayssiguier C, Cioe L, Withers E, Wunner WM, Curtis P (1986) Cloning of rabies virus matrix protein mRNA and determination of its amino acid sequence. Virus Res 5:275–278

Sacramento D, Badrane H, Bourhy H, Tordo N (1992) Molecular epidemiology of rabies in France: comparison with vaccinal strains. J Gen Virol 73:1149–1158

Sacramento D, Bourhy H, Tordo N (1991) PCR technique as an alternative method for diagnosis and molecular epidemiology of rabies virus. Mol Cel Probes 6:229–240

Sambrook J, Fritsch EF, Maniatis T (1989) Molecular cloning: a laboratory manual. Cold Spring Harbor, New York

Sanger F, Nicklen S, Coulson AR (1977) DNA sequencing with chain-terminating inhibitors. Proc Natl Acad Sci USA 74:5463–5467

Smith JS, Fishbein DB, Rupprecht CE, Clark K (1991) Unexplained rabies in three immigrants in the United States. A Virologic Investigation. N Engl J Med 324:205–211

Sumner JW, Fekadu M, Shaddock JH, Esposito JJ, Bellini WJ (1991) Protection of mice with vaccinia virus recombinants that express the rabies nucleoprotein. Virology 183:703–710

Sureau P, Ravisse P, Rollin P (1991) Rabies diagnosis by animal inoculation, identification of Negri bodies, or ELISA. In: Baer GM (ed) The natural history of rabies. CRC Press, Boca Raton, pp 203–217

Théodoridès J (1986) Histoire de la rage. Masson, Paris

Tollis M, Dietzschold B, Buona Volia C, Koprowski H (1991) Immunization of monkeys with rabies ribonucleoprotein (RNP) confers protective immunity against rabies. Vaccine 9:134–136

Tordo N, Poch O (1988) Structure of rabies virus. In: Campbell JB, Charlton KM (eds) Rabies. Kluwer, Boston, pp 25–45

Tordo N, Poch O, Ermine A, Keith G (1986a) Primary structure of leader RNA and nucleoprotein genes of the rabies genome: segmented homology with VSV. Nucleic Acids Res 14:2671–2683

Tordo N, Poch O, Ermine A, Keith G, Rougeon F (1986b) Walking along the rabies genome: is the large G-L intergenic region a remnant gene? Proc Natl Acad Sci USA 83:3914–3918

Tordo N, Poch O, Ermine A, Keith G, Rougeon F (1988) Completion of the rabies virus genome sequence determination: highly conserved domains along the L (polymerase) proteins of unsegmented negative-strand RNA viruses. Virology 165:565–576

Trimarchi CV, Debbie JG (1991) The fluorescent antibody in rabies. In: Baer GM (ed) The natural history of rabies. CRC Press, Boca Raton, pp 219–233

W.H.O. (1991) Eighth Report of the WHO Expert Committee on Rabies. WHO, Geneva

Wiktor TJ, Gyorgy E, Schlumberger HD, Sokol F, Koprowski H (1973) Antigenic properties of rabies virus components. J Immunol 110:269–276

Section VII Summary

Chapter 30 Detection of Viruses in Clinical Materials Is Enhanced by the Polymerase Chain Reaction: Current State of Knowledge

Yechiel Becker

Summary

The rapid advances in virus gene cloning and sequencing coupled with the rapid detection of viruses by polymerase chain reaction (PCR) have revolutionized clinical diagnostic virology. The rapid detection of viruses and viral genomes and messenger RNA may make it possible to detect viruses in clinical specimens and transplanted organs. PCR technology provides a new tool for basic clinical and molecular virological research and offers the potential of identifying the role played by viruses in human diseases.

The studies on the utilization of polymerase chain reaction (PCR) technology for the detection of viruses in clinical materials presented in this book indicate that diagnostic virology has entered into a new era. The rapidly executed PCR diagnostic technique makes it possible to detect and quickly identify disease-causing viruses in clinical materials. This method allows for rapid decisions on the strategy for the treatment of infectious diseases. Unfortunately, only a few effective antiviral drugs are available for the curative treatment of patients. In addition, the PCR can be of great value in testing for the presence of viruses in organs used for human transplantation. This is already done in the testing of blood and blood products for contamination by viruses. The PCR analysis of viruses involved in human cancers (as well as the expression of oncogenes and protooncogenes) might lead to new treatment modalities in cancer. The chapters of this book open for the reader a new frontier of virology and new horizons for the rapid detection of viruses with the hope that research on the prevention of human virus diseases will be enhanced.

The development of the PCR had an immediate impact on molecular virology. During the last decade the accumulation of the nucleotide sequences of viral nucleic acids made the utilization of the PCR technology very practical for the study of virus genes as well as for the detection of viruses in clinical materials. The possibility of amplifying viral-specific sequences from viral nucleic acids in impurified materials by PCR (Eisenstein 1990; Erlich et al.

Department of Molecular Virology, Faculty of Medicine, Hebrew University of Jerusalem, Jerusalem, Israel

1988; Mullis and Faloona 1987; Suiki et al. 1985) led to a marked improvement in viral diagnosis, as can be seen from the studies collected in this book. Weiner et al. (this volume) state in their chapter on hepatitis C virus (HCV), "The advent of polymerase chain reaction (PCR) technology ... could not have come at a more opportune time for HCV research and diagnostics since detecting and cloning low levels of cDNA synthesized from HCV RNA present in small volume samples by conventional techniques is difficult, if not impossible." In a way, this is generally true for the diagnosis of viruses in clinical material. The need to isolate infectious virus from clinical material by infecting cell cultures and identifying the virus with specific antibodies or by nucleic acid hybridization, a technique still valid for a more complete analysis of the biological properties of the virus, is time-consuming. The development of a PCR diagnosis of pathogenic viruses (Table 1) will completely change the diagnosis of viruses in clinical materials from patients. The diagnosis of viruses in clinical materials from patients. The diagnostic virus laboratory might be able to provide the physician with information on the infecting virus the same day the patient is admitted to the hospital. This information is highly valuable for decision-making about the patient's care and to protect the attending medical staff. Moreover, this information is critical for the decision on the use of antiviral agents for treatment. Due to the lack of antiviral agents for the treatment of many viral diseases, it is possible that the advancement in virus diagnosis will stimulate antiviral research and lead to the testing of new antiviral drugs.

Table 1. Medical diagnosis of human virus diseases requires rapid and accurate diagnosis of the infecting virus to allow effective curative treatment

Procedure step	Delay
1. Onset of disease in a patient (before medical attention)	days
2. Admission of a patient to the emergency room in hospital and physical examination of the patient. Clinical samples are sent to the diagnostic laboratory for virus diagnosis	hours
3. Diagnosis	
a) Direct detection of virions by EM (if possible)	hours
b) Virus isolation in cell cultures	1–7 days
c) Detection of viral nucleic acids	1–2 days
i) Extraction of nucleid acids	
ii) Hybridization with a viral nucleic acid probe (biotinylated/radioactive)	
iii) Detection of virus by PCR, nested PCR/radioactive PCR; electrophoresis of amplified DNA. Decision on antiviral treatment	hours
d) Serological tests (Western blot, Elisa, complement fixation)	hours
e) Paired sera for the testing of the rise in antibody titer to a specific virus	2 weeks
4. Clinical decision	immediate
5. Treatment of the patient with antiviral agents upon receiving lab results (chemotherapy, immunotherapy)	immediate

The studies presented in this book provide a comprehensive collection of research on virus diagnosis by PCR. Some of the technologies used by the authors are summarized in Table 2. It can be seen that viruses can be captured by specific antibodies, the nucleic acid extracted from the isolated virions or from infected cells in the clinical specimen, and PCR technology carried out to identify the virus by the use of specific oligonucleotide primers. Thus, the progress in virus diagnosis by PCR still depends on the development of molecular virological studies and the identification of the virus-specific genes in all known virus isolates. Such studies will allow the detection of mutations in virus genes, especially during epidemics (e.g., nucleotide changes in the HIV-1 genome during the current AIDS epidemic or in the influenza virus). Yet, the lack of an editing enzyme coupled to the reverse transcriptase or the Taq polymerase leaves open the possibility that the insertion of incorrect nucleotides into the cDNA or the amplified DNA molecules may occur, a problem that needs further study and improvement.

Another field in which PCR diagnosis is of critical value is organ transplantation. The PCR diagnosis of viruses present in the organism in a latent form might improve the chances of a successful organ transplantation. Several of the possibilites of detecting viruses are summarized in Table 3. The role of latent virus in posttransplant lymphoproliferative disorder (PTLD) is discussed by Cen et al. (1991). From the organs for transplantation listed in Table 3, blood and blood products are studied by PCR analysis for several viruses

Table 2. PCR diagnosis of viruses in clinical materials

---

1. Capturing of extracellular virus particles
   a) Virions captured by antibodies specific for a structural viral antigen present on the surface of the virion
   b) Binding of enveloped virions to immobilized lectins
   c) Magnetic beads with virion-specific antibodies to capture virions

2. Extraction/release of viral nucleic acid from virions/infected cells/infected tissue biopsy
   a) Phenol extraction
   b) NaCl extraction
   c) Isolation of viral polyA RNA with magnetic boads linked to oligo dT
   d) Heat denaturation

3. Extraction of viral nucleic acids from cells is possible from:
   a) Blood lymphocytes
   b) Tissue biopsies
   c) Pathological preparations

4. Polymerase chain reaction
   a) Elimination of secondary structure of viral RNA
   b) Elimination of double strandedness in the viral nucleic acid
   c) Nested PCR
   d) PCR with a nucleoside [^{32}P] triphosphate

5. Analysis of the PCR product
   a) Agarase gel/acrylamide gel electrophoresis
   b) Hybridization with a DNA/RNA probe

Table 3. Detection of viruses in organs from donors prior to transplantation to patients

1. Detection of latent viruses in blood/plasma/T and B cells/precursors of dendritic cells/macrophages/viruses: hepatitis A, B, C; flaviviruses, HIV-1, HIV-2, HTLV-I, HTLV-II, EBV, herpes 6, herpes 7

2. Detection of viruses in organs prior to transplantation
   Cornea: HSV-1
   Kidney: BK-JC virus, HSV-1 (Dummer et al. 1987), EBV (Cen et al. 1991),
        HCMV (Ho et al. 1975)
   Liver: hepatitis B
   Lung: HCMV
   Heart: HSV-1, HCMV
   Bone marrow: B19 parvovirus, HIV-1, EBV (Gratama et al. 1990; Zutler et al. 1988)
   Skin: HIV-1 (in Langerhans cells), papillomavirus

3. Viruses in human semen for in vitro fertilization: HIV-1, HIV-2, HTLV-I

as discussed in relevant chapters in this book (hepatitis viruses, HTLV, and HIV-1). The testing of blood donations to blood banks could use the PCR detection of HIV-1 to identify seronegative donors who have latent HIV-1 genomes in their lymphocytes to eliminate seronegative blood donations in which the lymphocytes are HIV-1 genome positive. Another aspect is the presence of latent viruses in organs donated for organ transplantation. Organs which harbor latent viruses (Table 3; e.g., kidney with BK-JC virus, HCMV, or EBV infected cells, liver with a hepatitis virus, skin with HIV-1 in Langerhans cells, bone marrow with B19 parvovirus) can be reactivated in the organ in the recipient and lead to a virus disease which could be life-threatening. The presence of HIV-1 in lymphocytes in human semen requires PCR diagnosis of HIV-1 DNA in semen donations for in vitro fertilization. Thus, the knowledge of latent or dormant virus infections in humans will lead to the development of PCR diagnostic tests for viruses in transplantable organs, a technique which will require the appropriate legislation to assure its use for the benefit of the transplanted patient and the medical staff who handle the allografts.

Viruses are implicated to be associated with tumors in humans (Table 4). Specific chapters in this book deal with the PCR diagnosis of viruses in tumor biopsies. Although this identification does not lead to better treatment, it is possible that studies on the specific genes related to the virus pathogenicity will provide information which might be useful for anticancer treatment.

Virus-specific nucleic acids which are integrated in the human chromosomal DNA are detectable by PCR. This report opens a new horizon for the study of the connection between such viral nucleic acids and medical conditions in humans (like cancer). Muranyi and Flugel (this book) describe the spumaviruses, which are complex retroviruses capable of transcribing single-stranded RNA genomes into double-stranded DNA. PCR studies on spumaviruses will provide an approach to the study of latent retroviruses in the chromosomal DNA of individuals to determine the role, if any, of these nucleotide sequences in medical states.

Table 4. PCR diagnosis of viruses in human tumors

Tumors and diseases	Virus	Reference
1. Laryngeal squamous papillomas, squamous carcinomas of the pharynx and larynx, verrucous carcinomas of the pharynx, nasal inverted papillomas	HPV 6, HPVM	Bryan and Crocker (Chapt. 16)
2. Hodgkins' disease (HD), angioimmunoblastic lymph-adenopathy (AILD), lymphocytic/mixed thymoma, reactive lymph node hyperplasia (HP)	EBV	Knecht et al. (Chapt. 13)
3. Human T-cell leukemia (ATL), lymphoma, tropical spastic paraparesis (TSP), myelopathy	HTLV-1	Matumoto and Nishioka (Chapt. 3)
Sjörgen's syndrome, polymyositis, chronic inflammatory arthropathy	HTLV-1	Hjelle (Chapt. 4)
4. Hairy cell leukemia, chronic fatigue syndrome	HILV II	
5. Hepatoma	Hepatitis B virus	

The present book marks the beginning of a new era in molecular virology. As in the past, when a new technology was adopted for molecular virus research, new knowledge on viruses from the basic to the applied will be generated. *Frontiers of Virology* will continue to follow the developments in this field in the forthcoming volumes.

References

Cen H, Breinig MC, Atchison RW, Ho M, McKnight JLC (1991) EBV transmission via the donor organs in solid organ transplantation: PCR and restriction length polymorphism analysis of IR2, IR3 and IR4. J Virol 65:976–980

Drummer JS, Armstrong J, Somers S, Kusne S, Carpenter BJ, Rosenthal JT, Ho M (1987) Transmission of infection with herpes simplex virus by renal transplantation. J Infect Dis 155:202–206

Eisenstein BI (1990) The polymerase chain reaction. A new method of using molecular genetics for medical diagnosis. N Engl J Med 322:178–183

Erlich HA, Gelfand DH, Saiki RK (1988) Specific DNA amplification. Nature 331:461–462

Gratama JW, Oosterveer MAP, Lepoutre JMM, Van Rood JJ, Zwaan FE et al. (1990) Serological and molecular studies of Epstein-Barr virus infection in allogenic marrow graft recipients. Transplantation 49:725–730

Ho M, Suwansirikul S, Dowling JN, Youngblood LA, Amstrong J (1975) The transplanted kidney as a source of cytomegalovirus infections. N Engl J Med 293:1109–1112

Mullis KB, Faloona FA (1987) Specific synthesis of DNA in vitro via a polymerase-catalyzed chain reaction. Methods Enzymol 155:335–350

Saiki R, Scharf S, Faloona F et al. (1985) Enzymatic amplification of b-globin genomic sequences and restriction site analysis for diagnosis of sickle cell anemia. Science 130:1350–1354

Zutter MM, Martin PJ, Sale GE, Shulman HM, Fisher L et al. (1988) Blood 72:520–529

Subject Index

MIX
Papier aus verantwortungsvollen Quellen
Paper from responsible sources
FSC® C105338

If you have any concerns about our products,
you can contact us on
ProductSafety@springernature.com

In case Publisher is established outside the EU,
the EU authorized representative is:
Springer Nature Customer Service Center GmbH
Europaplatz 3, 69115 Heidelberg, Germany

Printed by Libri Plureos GmbH
in Hamburg, Germany